T0140099

Lecture Notes in Networks and Systems

Volume 451

The series "Lecture Notes in Networks and Systems" publishes the latest developments in Networks and Systems—quickly, informally and with high quality. Original research reported in proceedings and post-proceedings represents the core of LNNS.

Volumes published in LNNS embrace all aspects and subfields of, as well as new challenges in, Networks and Systems.

The series contains proceedings and edited volumes in systems and networks, spanning the areas of Cyber-Physical Systems, Autonomous Systems, Sensor Networks, Control Systems, Energy Systems, Automotive Systems, Biological Systems, Vehicular Networking and Connected Vehicles, Aerospace Systems, Automation, Manufacturing, Smart Grids, Nonlinear Systems, Power Systems, Robotics, Social Systems, Economic Systems and other. Of particular value to both the contributors and the readership are the short publication timeframe and the world-wide distribution and exposure which enable both a wide and rapid dissemination of research output.

The series covers the theory, applications, and perspectives on the state of the art and future developments relevant to systems and networks, decision making, control, complex processes and related areas, as embedded in the fields of interdisciplinary and applied sciences, engineering, computer science, physics, economics, social, and life sciences, as well as the paradigms and methodologies behind them.

Indexed by SCOPUS, INSPEC, WTI Frankfurt eG, zbMATH, SCImago.

All books published in the series are submitted for consideration in Web of Science.

For proposals from Asia please contact Aninda Bose (aninda.bose@springer.com).

More information about this series at https://link.springer.com/bookseries/15179

Leonard Barolli · Farookh Hussain ·
Tomoya Enokido
Editors

Advanced Information Networking and Applications

Proceedings of the 36th International
Conference on Advanced Information
Networking and Applications (AINA-2022),
Volume 3

 Springer

Editors
Leonard Barolli
Department of Information
and Communication Engineering
Fukuoka Institute of Technology
Fukuoka, Japan

Farookh Hussain
University of Technology Sydney
Sydney, NSW, Australia

Tomoya Enokido
Faculty of Bussiness Administration
Rissho University
Tokyo, Japan

ISSN 2367-3370   ISSN 2367-3389 (electronic)
Lecture Notes in Networks and Systems
ISBN 978-3-030-99618-5   ISBN 978-3-030-99619-2   (eBook)
https://doi.org/10.1007/978-3-030-99619-2

This Springer imprint is published by the registered company Springer Nature Switzerland AG
The registered company address is: Gewerbestrasse 11, 6330 Cham, Switzerland

Welcome Message from AINA-2022 Organizers

Welcome to the 36th International Conference on Advanced Information Networking and Applications (AINA-2022). On behalf of AINA-2022 Organizing Committee, we would like to express to all participants our cordial welcome and high respect.

AINA is an international forum, where scientists and researchers from academia and industry working in various scientific and technical areas of networking and distributed computing systems can demonstrate new ideas and solutions in distributed computing systems. AINA was born in Asia, but it is now an international conference with high quality thanks to the great help and cooperation of many international friendly volunteers. AINA is a very open society and is always welcoming international volunteers from any country and any area in the world.

AINA International Conference is a forum for sharing ideas and research work in the emerging areas of information networking and their applications. The area of advanced networking has grown very rapidly, and the applications have experienced an explosive growth especially in the area of pervasive and mobile applications, wireless sensor networks, wireless ad-hoc networks, vehicular networks, multimedia computing and social networking, semantic collaborative systems, as well as grid, P2P, IoT, big data, and cloud computing. This advanced networking revolution is transforming the way people live, work, and interact with each other and is impacting the way business, education, entertainment, and health care are operating. The papers included in the proceedings cover theory, design, and application of computer networks, distributed computing, and information systems.

Each year AINA receives a lot of paper submissions from all around the world. It has maintained high-quality accepted papers and is aspiring to be one of the main international conferences on the information networking in the world.

We are very proud and honored to have two distinguished keynote talks by Prof. Mario A. R. Dantas, University of Juiz de Fora, Minas Gerais, Brazil, and Prof. Isaac Woungang, Ryerson University, Toronto, Ontario, Canada, who will present their recent work and will give new insights and ideas to the conference participants.

An international conference of this size requires the support and help of many people. A lot of people have helped and worked hard to produce a successful AINA-2022 technical program and conference proceedings. First, we would like to thank all authors for submitting their papers, the session chairs, and distinguished keynote speakers. We are indebted to program track co-chairs, program committee members and reviewers, who carried out the most difficult work of carefully evaluating the submitted papers.

We would like to thank AINA-2022 General Co-chairs, PC Co-chairs, and Workshops Co-chairs for their great efforts to make AINA-2022 a very successful event. We have special thanks to Finance Chair and Web Administrator Co-chairs.

We do hope that you will enjoy the conference proceedings and readings.

Organization

AINA-2022 Organizing Committee

Honorary Chair

Makoto Takizawa   Hosei University, Japan

General Co-chairs

Farookh Hussain   University of Technology Sydney, Australia
Tomoya Enokido   Rissho University, Japan
Isaac Woungang   Ryerson University, Canada

Program Committee Co-chairs

Omar Hussain   University of New South Wales, Australia
Flora Amato   University of Naples "Federico II," Italy
Marek Ogiela   AGH University of Science and Technology,
                                  Poland

Workshops Co-chairs

Beniamino Di Martino   University of Campania "Luigi Vanvitelli," Italy
Omid Ameri Sianaki   Victoria University, Australia
Kin Fun Li   University of Victoria, Canada

International Journals Special Issues Co-chairs

Fatos Xhafa   Technical University of Catalonia, Spain
David Taniar   Monash University, Australia

Award Co-chairs

Arjan Durresi   Indiana University Purdue University in
                                               Indianapolis (IUPUI), USA
Fang-Yie Leu   Tunghai University, Taiwan

Publicity Co-chairs

Markus Aleksy   ABB AG, Germany
Lidia Ogiela   AGH University of Science and Technology,
                                               Poland
Hsing-Chung Chen   Asia University, Taiwan

International Liaison Co-chairs

Nadeem Javaid   COMSATS University Islamabad, Pakistan
Wenny Rahayu   La Trobe University, Australia

Local Arrangement Co-chairs

Rania Alhazmi   University of Technology Sydney, Australia
Huda Alsobhi   University of Technology Sydney, Australia
Ebtesam Almansour   University of Technology Sydney, Australia

Finance Chair

Makoto Ikeda   Fukuoka Institute of Technology, Japan

Web Co-chairs

Phudit Ampririt   Fukuoka Institute of Technology, Japan
Kevin Bylykbashi   Fukuoka Institute of Technology, Japan
Ermioni Qafzezi   Fukuoka Institute of Technology, Japan

Steering Committee Chair

Leonard Barolli   Fukuoka Institute of Technology, Japan

Tracks and Program Committee Members

1. Network Protocols and Applications

Track Co-chairs

Makoto Ikeda   Fukuoka Institute of Technology, Japan
Sanjay Kumar Dhurandher   Netaji Subhas University of Technology,
                                               New Delhi, India
Bhed Bahadur Bista   Iwate Prefectural University, Japan

TPC Members

Admir Barolli	Aleksander Moisiu University of Durres, Albania
Elis Kulla	Okayama University of Science, Japan
Keita Matsuo	Fukuoka Institute of Technology, Japan
Shinji Sakamoto	Kanazawa Institute of Technology, Japan
Akio Koyama	Yamagata University, Japan
Evjola Spaho	Polytechnic University of Tirana, Albania
Jiahong Wang	Iwate Prefectural University, Japan
Shigetomo Kimura	University of Tsukuba, Japan
Chotipat Pornavalai	King Mongkut's Institute of Technology Ladkrabang, Thailand
Danda B. Rawat	Howard University, USA
Amita Malik	Deenbandhu Chhotu Ram University of Science and Technology, India
R. K. Pateriya	Maulana Azad National Institute of Technology, India
Vinesh Kumar	University of Delhi, India
Petros Nicopolitidis	Aristotle University of Thessaloniki, Greece
Satya Jyoti Borah	North Eastern Regional Institute of Science and Technology, India

2. Next-Generation Wireless Networks

Track Co-chairs

Christos J. Bouras	University of Patras, Greece
Tales Heimfarth	Universidade Federal de Lavras, Brazil
Leonardo Mostarda	University of Camerino, Italy

TPC Members

Fadi Al-Turjman	Near East University, Nicosia, Cyprus
Alfredo Navarra	University of Perugia, Italy
Purav Shah	Middlesex University London, UK
Enver Ever	Middle East Technical University, Northern Cyprus Campus, Cyprus
Rosario Culmone	University of Camerino, Camerino, Italy
Antonio Alfredo F. Loureiro	Federal University of Minas Gerais, Brazil
Holger Karl	University of Paderborn, Germany
Daniel Ludovico Guidoni	Federal University of São João Del-Rei, Brazil
João Paulo Carvalho Lustosa da Costa	Hamm-Lippstadt University of Applied Sciences, Germany
Jorge Sá Silva	University of Coimbra, Portugal

Apostolos Gkamas   University Ecclesiastical Academy of Vella,
                                       Ioannina, Greece
Zoubir Mammeri   University Paul Sabatier, France
Eirini Eleni Tsiropoulou   University of New Mexico, USA
Raouf Hamzaoui   De Montfort University, UK
Miroslav Voznak   University of Ostrava, Czech Republic
Kevin Bylykbashi   Fukuoka Institute of Technology, Japan

3. Multimedia Systems and Applications

Track Co-chairs

Markus Aleksy   ABB Corporate Research Center, Germany
Francesco Orciuoli   University of Salerno, Italy
Tomoyuki Ishida   Fukuoka Institute of Technology, Japan

TPC Members

Tetsuro Ogi   Keio University, Japan
Yasuo Ebara   Osaka Electro-Communication University, Japan
Hideo Miyachi   Tokyo City University, Japan
Kaoru Sugita   Fukuoka Institute of Technology, Japan
Akio Doi   Iwate Prefectural University, Japan
Hadil Abukwaik   ABB Corporate Research Center, Germany
Monique Duengen   Robert Bosch GmbH, Germany
Thomas Preuss   Brandenburg University of Applied Sciences,
                                       Germany
Peter M. Rost   NOKIA Bell Labs, Germany
Lukasz Wisniewski   inIT, Germany
Angelo Gaeta   University of Salerno, Italy
Graziano Fuccio   University of Salerno, Italy
Giuseppe Fenza   University of Salerno, Italy
Maria Cristina   University of Salerno, Italy
Alberto Volpe   University of Salerno, Italy

4. Pervasive and Ubiquitous Computing

Track Co-chairs

Chih-Lin Hu	National Central University, Taiwan
Vamsi Paruchuri	University of Central Arkansas, USA
Winston Seah	Victoria University of Wellington, New Zealand

TPC Members

Hong Va Leong	Hong Kong Polytechnic University, Hong Kong
Ling-Jyh Chen	Academia Sinica, Taiwan
Jiun-Yu Tu	Southern Taiwan University of Science and Technology, Taiwan
Jiun-Long Huang	National Chiao Tung University, Taiwan
Thitinan Tantidham	Mahidol University, Thailand
Tanapat Anusas-amornkul	King Mongkut's University of Technology North Bangkok, Thailand
Xin-Mao Huang	Aletheia University, Taiwan
Hui Lin	Tamkang University, Taiwan
Eugen Dedu	Universite de Franche-Comte, France
Peng Huang	Sichuan Agricultural University, China
Wuyungerile Li	Inner Mongolia University, China
Adrian Pekar	Budapest University of Technology and Economics, Hungary
Jyoti Sahni	Victoria University of Technology, New Zealand
Normalia Samian	Universiti Putra Malaysia, Malaysia
Sriram Chellappan	University of South Florida, USA
Yu Sun	University of Central Arkansas, USA
Qiang Duan	Penn State University, USA
Han-Chieh Wei	Dallas Baptist University, USA

5. Web-Based and E-Learning Systems

Track Co-chairs

Santi Caballe	Open University of Catalonia, Spain
Kin Fun Li	University of Victoria, Canada
Nobuo Funabiki	Okayama University, Japan

TPC Members

Jordi Conesa	Open University of Catalonia, Spain
Joan Casas	Open University of Catalonia, Spain
David Gañán	Open University of Catalonia, Spain
Nicola Capuano	University of Basilicata, Italy
Antonio Sarasa	Complutense University of Madrid, Spain
Chih-Peng Fan	National Chung Hsing University, Taiwan
Nobuya Ishihara	Okayama University, Japan
Sho Yamamoto	Kindai University, Japan
Khin Khin Zaw	Yangon Technical University, Myanmar
Kaoru Fujioka	Fukuoka Women's University, Japan
Kosuke Takano	Kanagawa Institute of Technology, Japan
Shengrui Wang	University of Sherbrooke, Canada
Darshika Perera	University of Colorado at Colorado Spring, USA
Carson Leung	University of Manitoba, Canada

6. Distributed and Parallel Computing

Track Co-chairs

Naohiro Hayashibara	Kyoto Sangyo University, Japan
Minoru Uehara	Toyo University, Japan
Tomoya Enokido	Rissho University, Japan

TPC Members

Eric Pardede	La Trobe University, Australia
Lidia Ogiela	AGH University of Science and Technology, Poland
Evjola Spaho	Polytechnic University of Tirana, Albania
Akio Koyama	Yamagata University, Japan
Omar Hussain	University of New South Wales, Australia
Hideharu Amano	Keio University, Japan
Ryuji Shioya	Toyo University, Japan
Ji Zhang	The University of Southern Queensland
Lucian Prodan	Universitatea Politehnica Timisoara, Romania
Ragib Hasan	The University of Alabama at Birmingham, USA
Young-Hoon Park	Sookmyung Women's University, Korea
Dilawaer Duolikun	Cognizant Technology Solutions, Hungary
Shigenari Nakamura	Tokyo Metropolitan Industrial Technology Research Institute, Japan

7. Data Mining, Big Data Analytics and Social Networks

Track Co-chairs

Omid Ameri Sianaki	Victoria University, Australia
Alex Thomo	University of Victoria, Canada
Flora Amato	University of Naples "Frederico II," Italy

TPC Members

Eric Pardede	La Trobe University, Australia
Alireza Amrollahi	Macquarie University, Australia
Javad Rezazadeh	University Technology Sydney, Australia
Farshid Hajati	Victoria University, Australia
Mehregan Mahdavi	Sydney International School of Technology and Commerce, Australia
Ji Zhang	University of Southern Queensland, Australia
Salimur Choudhury	Lakehead University, Canada
Xiaofeng Ding	Huazhong University of Science and Technology, China
Ronaldo dos Santos Mello	Universidade Federal de Santa Catarina, Brazil
Irena Holubova	Charles University, Czech Republic
Lucian Prodan	Universitatea Politehnica Timisoara, Romania
Alex Tomy	La Trobe University, Australia
Dhomas Hatta Fudholi	Universitas Islam Indonesia, Indonesia
Saqib Ali	Sultan Qaboos University, Oman
Ahmad Alqarni	Al Baha University, Saudi Arabia
Alessandra Amato	University of Naples "Frederico II," Italy
Luigi Coppolino	Parthenope University, Italy
Giovanni Cozzolino	University of Naples "Frederico II," Italy
Giovanni Mazzeo	Parthenope University, Italy
Francesco Mercaldo	Italian National Research Council, Italy
Francesco Moscato	University of Salerno, Italy
Vincenzo Moscato	University of Naples "Frederico II," Italy
Francesco Piccialli	University of Naples "Frederico II," Italy

8. Internet of Things and Cyber-Physical Systems

Track Co-chairs

Euripides G. M. Petrakis	Technical University of Crete (TUC), Greece
Tomoki Yoshihisa	Osaka University, Japan
Mario Dantas	Federal University of Juiz de Fora (UFJF), Brazil

TPC Members

Akihiro Fujimoto	Wakayama University, Japan
Akimitsu Kanzaki	Shimane University, Japan
Kawakami Tomoya	University of Fukui, Japan
Lei Shu	University of Lincoln, UK
Naoyuki Morimoto	Mie University, Japan
Yusuke Gotoh	Okayama University, Japan
Vasilis Samolada	Technical University of Crete (TUC), Greece
Konstantinos Tsakos	Technical University of Crete (TUC), Greece
Aimilios Tzavaras	Technical University of Crete (TUC), Greece
Spanakis Manolis	Foundation for Research and Technology Hellas (FORTH), Greece
Katerina Doka	National Technical University of Athens (NTUA), Greece
Giorgos Vasiliadis	Foundation for Research and Technology Hellas (FORTH), Greece
Stefan Covaci	Technische Universität Berlin, Berlin (TUB), Germany
Stelios Sotiriadis	University of London, UK
Stefano Chessa	University of Pisa, Italy
Jean-Francois Méhaut	Université Grenoble Alpes, France
Michael Bauer	University of Western Ontario, Canada

9. Intelligent Computing and Machine Learning

Track Co-chairs

Takahiro Uchiya	Nagoya Institute of Technology, Japan
Omar Hussain	UNSW, Australia
Nadeem Javaid	COMSATS University Islamabad, Pakistan

TPC Members

Morteza Saberi	University of Technology Sydney, Australia
Abderrahmane Leshob	University of Quebec in Montreal, Canada
Adil Hammadi	Curtin University, Australia
Naeem Janjua	Edith Cowan University, Australia
Sazia Parvin	Melbourne Polytechnic, Australia
Kazuto Sasai	Ibaraki University, Japan
Shigeru Fujita	Chiba Institute of Technology, Japan
Yuki Kaeri	Mejiro University, Japan
Zahoor Ali Khan	HCT, UAE
Muhammad Imran	King Saud University, Saudi Arabia

Ashfaq Ahmad	The University of Newcastle, Australia
Syed Hassan Ahmad	JMA Wireless, USA
Safdar Hussain Bouk	Daegu Gyeongbuk Institute of Science and Technology, Korea
Jolanta Mizera-Pietraszko	Military University of Land Forces, Poland

10. Cloud and Services Computing

Track Co-chairs

Asm Kayes	La Trobe University, Australia
Salvatore Venticinque	University of Campania "Luigi Vanvitelli," Italy
Baojiang Cui	Beijing University of Posts and Telecommunications, China

TPC Members

Shahriar Badsha	University of Nevada, USA
Abdur Rahman Bin Shahid	Concord University, USA
Iqbal H. Sarker	Chittagong University of Engineering and Technology, Bangladesh
Jabed Morshed Chowdhury	La Trobe University, Australia
Alex Ng	La Trobe University, Australia
Indika Kumara	Jheronimus Academy of Data Science, Netherlands
Tarique Anwar	Macquarie University and CSIRO's Data61, Australia
Giancarlo Fortino	University of Calabria, Italy
Massimiliano Rak	University of Campania "Luigi Vanvitelli," Italy
Jason J. Jung	Chung-Ang University, Korea
Dimosthenis Kyriazis	University of Piraeus, Greece
Geir Horn	University of Oslo, Norway
Gang Wang	Nankai University, China
Shaozhang Niu	Beijing University of Posts and Telecommunications, China
Jianxin Wang	Beijing Forestry University, China
Jie Cheng	Shandong University, China
Shaoyin Cheng	University of Science And Technology of China, China

11. Security, Privacy and Trust Computing

Track Co-chairs

Hiroaki Kikuchi	Meiji University, Japan
Xu An Wang	Engineering University of PAP, China
Lidia Ogiela	AGH University of Science and Technology, Poland

TPC Members

Takamichi Saito	Meiji University, Japan
Kouichi Sakurai	Kyushu University, Japan
Kazumasa Omote	Univesity of Tsukuba, Japan
Shou-Hsuan Stephen Huang	University of Houston, USA
Masakatsu Nishigaki	Shizuoka University, Japan
Mingwu Zhang	Hubei University of Technology, China
Caiquan Xiong	Hubei University of Technology, China
Wei Ren	China University of Geosciences, China
Peng Li	Nanjing University of Posts and Telecommunications, China
Guangquan Xu	Tianjing University, China
Urszula Ogiela	AGH University of Science and Technology, Poland
Hoon Ko	Chosun University, Korea
Goreti Marreiros	Institute of Engineering of Polytechnic of Porto, Portugal
Chang Choi	Gachon University, Korea
Libor Měsíček	J.E. Purkyně University, Czech Republic

12. Software-Defined Networking and Network Virtualization

Track Co-chairs

Flavio de Oliveira Silva	Federal University of Uberlândia, Brazil
Ashutosh Bhatia	Birla Institute of Technology and Science, Pilani, India
Alaa Allakany	Kyushu University, Japan

TPC Members

Rui Luís Andrade Aguiar	Universidade de Aveiro (UA), Portugal
Ivan Vidal	Universidad Carlos III de Madrid, Spain
Eduardo Coelho Cerqueira	Federal University of Pará (UFPA), Brazil

Christos Tranoris	University of Patras (UoP), Greece
Juliano Araújo Wickboldt	Federal University of Rio Grande do Sul (UFRGS), Brazil
Yaokai Feng	Kyushu University, Japan
Chengming Li	Chinese Academy of Science (CAS), China
Othman Othman	An-Najah National University (ANNU), Palestine
Nor-masri Bin-sahri	University Technology of MARA, Malaysia
Sanouphab Phomkeona	National University of Laos, Laos
Haribabu K.	BITS Pilani, India
Shekhavat, Virendra	BITS Pilani, India
Makoto Ikeda	Fukuoka Institute of Technology, Japan
Farookh Hussain	University of Technology Sydney, Australia
Keita Matsuo	Fukuoka Institute of Technology, Japan

AINA-2022 Reviewers

Abderrahmane Leshob
Abdullah Al-khatib
Adil Hammadi
Admir Barolli
Adrian Pekar
Ahmad Alqarni
Aimilios Tzavaras
Akihiro Fujihara
Akihiro Fujimoto
Akimitsu Kanzaki
Akio Doi
Akira Sakuraba
Alaa Allakany
Alex Ng
Alex Thomo
Alfredo Cuzzocrea
Alfredo Navarra
Amita Malik
Angelo Gaeta
Anne Kayem
Antonio Esposito
Antonio Loureiro
Apostolos Gkamas
Arcangelo Castiglione
Arjan Durresi
Ashutosh Bhatia
Asm Kayes

Baojiang Cui
Beniamino Di Martino
Bhed Bista
Caiquan Xiong
Carson Leung
Chang Choi
Christos Bouras
Christos Tranoris
Danda Rawat
David Taniar
Dimitris Apostolou
Dimosthenis Kyriazis
Eirini Eleni Tsiropoulou
Elis Kulla
Enver Ever
Eric Pardede
Ernst Gran
Eugen Dedu
Evjola Spaho
Farookh Hussain
Fatos Xhafa
Feilong Tang
Feroz Zahid
Flavio Silva
Flora Amato
Francesco Orciuoli
Francesco Palmieri

Funabiki Nobuo
Gang Wang
Goreti Marreiros
Guangquan Xu
Hideharu Amano
Hiroaki Kikuchi
Hiroshi Maeda
Hsing-Chung Chen
Indika Kumara
Irena Holubova
Isaac Woungang
Jana Nowaková
Javad Rezazadeh
Ji Zhang
Jianxin Wang
Jolanta Mizera-Pietraszko
Jordi Conesa
Jorge Sá Silva
Kazunori Uchida
Kazuto Sasai
Keita Matsuo
Kevin Bylykbashi
Kin Fun Li
Kiyotaka Fujisaki
Koki Watanabe
Konstantinos Tsakos
Kosuke Takano
Kouichi Sakurai
Leonard Barolli
Leonardo Mostarda
Libor Mesicek
Lidia Ogiela
Lucian Prodan
Luigi Coppolino
Makoto Ikeda
Makoto Takizawa
Marek Ogiela
Mario Dantas
Markus Aleksy
Masakatsu Nishigaki
Masaki Kohana
Mingwu Zhang
Minoru Uehara
Miralda Cuka

Mirang Park
Miroslav Voznak
Nadeem Javaid
Naeem Janjua
Naohiro Hayashibara
Nobuo Funabiki
Norimasa Nakashima
Omar Hussain
Omid Ameri Sianaki
Othman Othman
Øyvind Ytrehus
Paresh Saxena
Pavel Kromer
Philip Moore
Pornavalai Chotipat
Purav Shah
Quentin Jacquemart
Ragib Hasan
Ricardo Rodríguez Jorge
Rosario Culmone
Rui Aguiar
Ryuji Shioya
Safdar Hussain Bouk
Salimur Choudhury
Salvatore Venticinque
Sanjay Dhurandher
Santi Caballé
Satya Borah
Sazia Parvin
Shahriar Badsha
Shigenari Nakamura
Shigeru Fujita
Shigetomo Kimura
Shinji Sakamoto
Somnath Mazumdar
Sriram Chellappan
Stefan Covaci
Stefano Chessa
Takahiro Uchiya
Takamichi Saito
Tarique Anwar
Tetsuro Ogi
Tetsuya Oda
Tetsuya Shigeyasu

Thomas Dreibholz
Tomoki Yoshihisa
Tomoya Enokido
Tomoya Kawakami
Tomoyuki Ishida
Urszula Ogiela
Vamsi Paruchuri
Vinesh Kumar
Wang Xu An

Wei Ren
Wenny Rahayu
Winston Seah Isaac Woungang
Xiaofeng Ding
Yaokai Feng
Yoshitaka Shibata
Yuki Kaeri
Yusuke Gotoh
Zahoor Khan

AINA-2022 Keynote Talks

Data Intensive Scalable Computing in Edge/Fog/Cloud Environments

Mario A. R. Dantas

University of Juiz de Fora, Minas Gerais, Brazil

Abstract. In this talk are presented and discussed some aspects related to the adoption of data intensive scalable computing (DISC) paradigm considering the new adoption trend of edge/fog/cloud environments. These contemporaneous scenarios are very relevant for all organizations in a world where billion of IoT and IIoT devices are being connected, and an unprecedent amount of digital data is generated. Therefore, they require special processing and storage.

Resource Management in 5G Cloudified Infrastructure: Design Issues and Challenges

Isaac Woungang

Ryerson University, Toronto, Canada

Abstract. 5G and Beyond (B5G) networks will be featured by a closer collaboration between mobile network operators (MNOs) and cloud service providers (CSPs) to meet the communication and computational requirements of modern mobile applications and services in a mobile cloud computing (MCC) environment. In this talk, we enlighten the marriage between the heterogeneous wireless networks (HetNets) and the multiple clouds (termed as InterCloud) for a better resource management in B5G networks. First, we start with an overview of the building blocks of HetNet and InterCloud, and then we describe the resource managers in both domains. Second, the key design criteria and challenges related to interoperation between the InterCloud and HetNet are described. Third, the state-of the-art security-aware resource allocation mechanisms for a multi-cloud orchestration over a B5G networks are enlighten.

Contents

LSTM-Based Reinforcement Q Learning Model for Non Intrusive Load Monitoring

Kalthoum Zaouali[(✉)], Mohamed Lassaad Ammari, and Ridha Bouallegue

Higher School of Communication of Tunis-Sup'Com, Innov'Com Laboratory,
Carthage University, Carthage, Tunisia
kalthoum.zaouali@isitc.u-sousse.tn, mlammari@gel.ulaval.ca,
ridha.bouallegue@supcom.rnu.tn

Abstract. Smart meters have been widely used in smart homes to provide efficient monitoring and billing to consumers. While providing customers with usage information at the device level can lead to energy savings, modern smart meters can only provide useful data for the whole house with low accuracy. Therefore, machine learning applied to the problem of energy disaggregation has gained wide attention. In this paper, an intelligent and optimized recurrent Long Short-Term Memory (LSTM) reinforcement Q-learning technique was evaluated on a large-scale household energy use dataset for Non-Intrusive Load Monitoring (NILM). Our proposed model can maximize energy disaggregation performance and is able to predict new observations from previous ones. The design of such a deep learning model for energy disaggregation is examined in the universal REDD smart meter dataset and compared to reference model. The experimental results demonstrate that the accuracy of the energy prediction in terms of accuracy was significantly improved in 99% of cases after using LSTM-based reinforcement Q learning, compared to the deep learning approach TFIDF-DAE [1] with an accuracy of 85%.

1 Introduction

Technological advances and the growing interest of new generations for investment in household appliances are leading to an increase in sales of modern appliances and their integration into our daily life. Home Energy Management manages the use of electricity in smart homes [2]. The variation in electricity production costs depends on changing the timing of consumption rather than the amount of energy consumed. In this case, savings on electricity consumption bills can be achieved by scheduling the load on a timescale to reduce energy demand

L. Barolli et al. (Eds.): AINA 2022, LNNS 451, pp. 1–13, 2022.
https://doi.org/10.1007/978-3-030-99619-2_1

during peak hours [3]. This type of aggregate load pricing causes consumers to adjust their device usage to take advantage of lower prices during specific periods [4]. Reinforcement learning is an area of machine learning where a learning agent interacts with an environment and builds states with the best action that would result in expected future rewards. These agents have decision-making skills in complex environments without any prior knowledge of the domain [5].

In this paper, we will merge deep learning, such as the Long Short-Term Memory (LSTM) with reinforcement learning, such as Q-learning, in order to prove the performance of this new deep learning by Q reinforcement in the disaggregation problem of the general load curve. The proposed work aims to apply Deep Reinforcement Learning (DRL) techniques to the non-intrusive load scenario from a general smart meter. The results obtained are to be compared with the deep learning approach TFIDF-DAE proposed in [1]. The reinforcement deep learning agent transfers the load in question in order to optimize its energy consumption by distributing the predicted loads of each device in specific periods in order to monitor its energy behavior. This technique reduces the total load on the electricity grid as well as the cost of energy consumption in smart homes.

Our main contribution presented in this paper is the application of LSTM deep learning by Q reinforcement in the problem of energy disaggregation. The proposed learning model outperforms the traditional methods presented in the literature. Figure 1 describe the principal synoptic of smart meter energy disaggregation approach. The proposed intelligent algorithm is assessed using several metrics and tested on a public accessible REDD dataset. The results demonstrate that our method outperforms existing machine learning in all of the tested circumstances. Providing the following contributions, we will study the effectiveness of the deep learning technique in the framework of reinforcement learning for the Non-Intrusive Load Monitoring (NILM) problem. We will use the LSTM recurrent network in an DRL algorithm to approximate the Q function, extending the state with historical partial observations. We demonstrate their performance for the disaggregation of the general load course. We compare the performance of the proposed model with other models in the literature. The paper is structured as follows. Section 2 presents some related works. Section 3 describes the proposed deep learning technique used to extract relevant features based on observation sequences and used in DLR. The experimental evaluation and the obtained results are given in Sect. 4. Finally, Sect. 5 draws conclusions and discusses some perspectives.

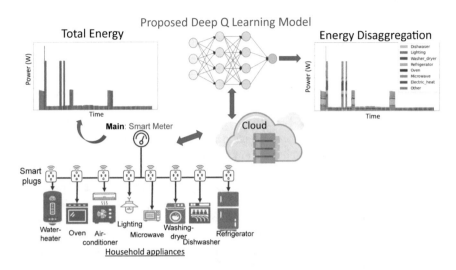

Fig. 1. Principal synoptic of smart meter energy disaggregation

2 Related Work

The developed algorithm can infer a variety of appliances using a transferred model with low-frequency data. Specifically, the developed algorithm integrates the reinforcement learning technique along with the LSTM deep neural networks to extract the lower level spatial and temporal features from the measurements and to identify and classify the appliance type [6]. The authors in [7] proposed a Deep Reinforcement Learning (DRL) model for demand response and for human activity recognition. A virtual agent gets feedback for every taken action to learn about the environment and to make smarter decisions later. Based on scattered observations of its environment, a demand response agent must extract a relevant set of characteristics and find an optimal sequence of decisions by applying deep learning techniques on sequences of observations and of past actions stored in the state vector. To find almost optimal characteristics for a residential heating system and an electric heater, the authors in [8] proved that feeding time-series sequences to recurrent LSTM model is more efficient than CNN or DNN models.

The challenges that hinder the application of the residential demand response are the partial observability of the environment [9], where part of the state remains hidden from the agent because of the limited number of sensors. Reinforcement learning and predictive control have developed a set of different techniques to solve the optimal control problem. Reinforcement learning approaches store sequences of the past in interactions with their environment in memory and extracts relevant characteristics based on this memory [10]. The ability to combine reinforcement learning and predictive control can result in complex control characteristics with supervised learning.

Prior knowledge of the environment and disturbance is not readily available or maybe too costly to obtain to manage the response to residential demand. DRL techniques are applied extra in the field of residential demand response, as they do not require any prior knowledge or any step of system identification and can be applied blind [11]. More recently, the authors have used an automatic feature extraction method based on convolutional neural networks which use as input the complete state of the cluster mapped by a clustering algorithm [12]. The application of DRL to solve decision and control problems in power supply systems is extensively studied and analyzed in the reviews [13]. Motivated by the promising Deep Q method, we will demonstrate how an LSTM network can be combined with a deep Q network to manage observability in the problem of load curve disaggregation.

The NILM provides insight into the electricity consumption of household appliances using only a building smart meter. Most of the learning techniques for NILM require a significant amount of power data supplied by the various connected devices. However, it is very difficult to collect such data which forms a major challenge in applying a generalized approach of NILM. The NILM approach is often achieved using supervised learning methods based on learning device-specific and global power measurements taken simultaneously over specific time periods [14]. The NILM problem remains a challenge considering its complexity due to the limitation of data available for training algorithms, the unknown number of devices in each household, the specific characteristics of each device, as well as their different modes of consumption. Many methods have been proposed to solve this problem, such as classical machine learning algorithms like SVM [16], and advanced statistical learning methods like Hidden Markov Models (HMM) and Bayesian Statistical Models [17]. Recently, deep learning algorithms are established in NILM [18]. These models are applied for general load curve disaggregation prove that their performance depends on how the training data represents the true distribution of energy consumption characteristics of a household [19]. Long-term data training and testing sets should be used to ensure a good approximation of the actual distributions, for a better evaluation of performance. However, each data set can only provide a small portion of the full distribution of power consumption. In this context of energy disaggregation using deep learning, Eduardo Gomes et al. [20] have proposed an architecture based on convolutional neural networks and deep neural networks DNN models have been used very well in the NILM approach. These deep learning methods require an abundance of training data, however, high-quality labeling datasets are still not available [21]. Unlike conventional model-based approaches, our proposed DRL techniques do not require knowledge of system dynamics or solutions, nor a system identification step. Expert knowledge can also be incorporated using the Q policy adjustment method.

3   Design of the Proposed Intelligent Algorithm

In our case study, we implemented our recurrent LSTM model combined with deep reinforcement Q-learning dedicated to time series forecasting and classification.

3.1   Recurrent Long Short-Term Memory (LSTM)

We are particularly interested in RNNs with LSTM which make it possible to overcome a certain number of difficulties encountered with the standard RNN model [22]. The input sequence, $\mathbf{x} = x_1, x_2, ..., x_T$, is a particular time series vector of dimension E involving the power of the counter smart price at every timescale. We aim to predict smart meter output power in time scale test and identify power consumption of electrical outlets. At each time t, the LSTM model observes a sample x_t and updates its time memory C_t which describes the state of the time series at time t, and produces a temporal feature vector h_t which summarizes the temporal information of LSTM after the observation x_t. The LSTM functionalities are:

$$i_t = \sigma(\mathbf{W}_i[\mathbf{h}_{t-1}, \mathbf{x}_t + \mathbf{b}_i)$$
$$f_t = \sigma(\mathbf{W}_f[\mathbf{h}_{t-1}, \mathbf{x}_t + \mathbf{b}_f)$$
$$o_t = \sigma(\mathbf{W}_o[\mathbf{h}_{t-1}, \mathbf{x}_t + \mathbf{b}_o)$$
$$\tilde{\mathbf{C}}_t = tanh(\mathbf{W}_C[\mathbf{h}_{t-1}, \mathbf{x}_t + \mathbf{b}_C)$$
$$\mathbf{C}_t = f_t * \mathbf{C}_{t-1} + i_t * \tilde{\mathbf{C}}_t$$
$$\mathbf{h}_t = o_t * tanh(\mathbf{C}_t) \tag{1}$$

or: i_t is the input ('input gate') which decides the amplitude of the flow of information in the memory which depends on time t, noted $\mathbf{C}_t$, using the sigmoid activation function with a weight $\mathbf{W}_i$ and a bias $\mathbf{b}_i$. f_t is the 'Forget gate' which determines the amount of information to be removed from $\mathbf{C}_t$ using the weight $\mathbf{W}_f$ and the bias $\mathbf{b}_f$. o_t is the output of the LSTM at time t using the weight $\mathbf{W}_o$ and the bias $\mathbf{b}_o$. $\mathbf{h}_t$ is the temporal characteristic extracted at time t. At each instant t, the memory is updated by $\tilde{\mathbf{C}}_t$ as a nonlinear function parameterized by $\mathbf{W}_C$ and $\mathbf{b}_C$.

3.2   Reinforcement Learning: Q-Reinforcement Learning

Reinforcement learning obtains improved signals in interaction with the external environment, then dynamically adjusts the parameters in order to improve the action strategy and adapt to the new environment [23]. The decision-making process of reinforcement learning can be seen as a Markov decision-making process:

$$\phi = (\mathbf{S}, \mathbf{A}, \mathbf{P}, \mathbf{R}, \mathbf{S}) \tag{2}$$

with: $\mathbf{S}$ is a set of finite states, $\mathbf{A}$ is a finite set of actions, $\mathbf{P}$ is the transition probability matrix and $\mathbf{R}$ is the return of the completed action. Reinforcement learning mainly evaluates the value of each state (*state*) or *action* through the state value function (*state_value*) and the action value function (*action_value*), then select the strategy. The *state_value* function represents the expected return of the current state (*current_state*) according to the current strategy:

$$V(s) = E(\mathbf{R}_t | \mathbf{S}_t = s) \tag{3}$$

The state behavior value function (*state_behavior_value*) represents the expected return of the current policy after performing an action from the current state, that is:

$$Q(s, a) = E(\mathbf{R}_t | \mathbf{S}_t = s, \mathbf{A}_t = a) \tag{4}$$

In the function $\mathbf{R}_t = r_{t+1} + \gamma r_{t+2} + \gamma^2 r_{t+3} + ... + \gamma^{Tt-1} r_T$, it indicates the cumulative return since time t. The optimal state value function (*optimal_state_value*) and the optimal action value function (*optimal_action_value*) are calculated by:

$$V(s) = max_{\mathbf{a}} \sum_{s} \mathbf{P}_{ss}^a [\mathbf{R}_{ss}^a + \gamma \mathbf{V}^*(s')] \tag{5}$$

$$Q(s, a) = \sum_{s} \mathbf{P}_{ss}^a [\mathbf{R}_{ss}^a + \gamma max_a \mathbf{Q}^*(s', a')] \tag{6}$$

In the function, $\mathbf{P}$ indicates the probability that the execution action a changes from state s to s'. $\mathbf{R}$ indicates the return of the execution action a from state s to s'. In the process of the Agent's interaction with the environment, the *state_value* function is updated. The Sarsa primitive value function update formula is:

$$Q(s, a)_{t+1} = Q(s, a)_t + \alpha(\mathbf{R}_{ss}^a + \gamma Q(s', a')_t - Q(s, a)_t) \tag{7}$$

Our proposed model allows first state of charge analysis via the LSTM algorithm and establishes a domestic HMM model with M loads:

$$\lambda = (\mathbf{X}, \mathbf{Y}, \mathbf{T}, \mathbf{O}, \pi) \tag{8}$$

with: $\mathbf{X}$ is a finite set of states, Y is a finite set of observations, $\mathbf{T}$ is a state transition probability matrix, $\mathbf{O}$ is the observation matrix and π is the initial probability vector. Based on the HMM model established in our work, the optimal state value function and the optimal action value function can be rewritten as follows:

$$\mathbf{V}^*(\mathbf{x}_i) = max_{\mathbf{y}_i} \sum_{\mathbf{x}_j} T_{ij} [O_j + \gamma \mathbf{V}^*(\mathbf{x}_j)] \tag{9}$$

$$\mathbf{Q}^*(\mathbf{x}_i, \mathbf{y}_i) = \sum_{\mathbf{x}_j} T_{ij} [O_j + \gamma max_{\mathbf{y}_i} \mathbf{Q}^*(\mathbf{x}_j, \mathbf{y}_j)] \tag{10}$$

$$\mathbf{Q}(\mathbf{x}_i, \mathbf{y}_i)_{t+1} = \mathbf{Q}(\mathbf{x}_i, \mathbf{y}_i)_t + \alpha [O_j + \gamma \sum_{\mathbf{x}_j} T_{ij} \mathbf{Q}(\mathbf{x}_j, \mathbf{y}_j)_t - \mathbf{Q}(\mathbf{x}_i, \mathbf{y}_i)_t] \tag{11}$$

3.3   Proposed Model: Reinforcement Learning Based on Recurrent LSTM

In this paper, we will summarize the choice of the algorithm adopted for the disaggregation of the load curve and the prediction of electrical uses. Our LSTM recurrent neural network-based reinforcement learning model is able to predict new observations from previous ones by performing a deep learning process of the collected data. Thus, this proposed method can be used for the modeling and forecasting of time series adequate to the problem of disaggregation of the energy curve from a smart meter in order to optimize the accuracy of time series forecasts. Figure 2 shows the proposed model.

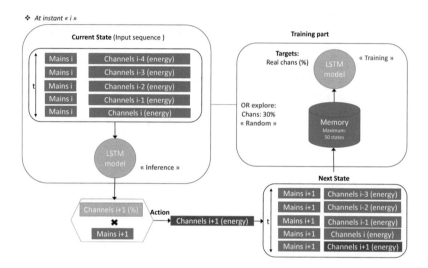

Fig. 2. Reinforcement learning solution based on LSTM

We design our recurrent LSTM model for this problem as follows: Our model has a visible layer with an input, a hidden layer with 5 LSTM blocks ('LSTM blocks'), and an output layer that gives the predicted input. The sigmoid activation function is used for LSTM blocks. The LSTM model is trained for 15 periods ('epochs') with a 'batch size' of 128. The LSTM model tends to estimate a function f_θ parameterized by $\theta \in \mathbf{R}^p$ which directly maps an input to the actual output of the problem. Consider a training dataset $\mathbf{D}_{train} = (\mathbf{x}_1, \mathbf{y}_1), (\mathbf{x}_2, \mathbf{y}_2), \ldots, (\mathbf{x}_n, \mathbf{y}_n)$ which contains n training samples $(\mathbf{x}_i, \mathbf{y}_i)$ with the input x_i corresponding to the real output/label y_i, and a set of test data $\mathbf{D}_{test} = (\mathbf{x}_{n+1}, \mathbf{y}_{n+1}), (\mathbf{x}_{n+2}, \mathbf{y}_{n+2}), \ldots, (\mathbf{x}_{n+m}, \mathbf{y}_{n+m})$ with m unobserved test samples. The goal is to learn the optimal parameter θ where the average distance between $f_\theta * (x)$ and $\mathbf{y}$ is the lowest for all samples $(\mathbf{x}, \mathbf{y}) \in \mathbf{D}_{train}$. The test error is the average error between the trained $f_\theta * (x)$ and $\mathbf{y}$ for all $(\mathbf{x}, \mathbf{y}) \in \mathbf{D}_{test}$. The deep learning algorithm proposed in our project combines the recurrent LSTM algorithm with reinforcement learning. It classifies the state of the load through LSTM and it establishes the state space, then updates it through reinforcement learning. Reinforcement learning is learned through the enhanced signal from the environment. When new data is entered, the parameter improvement action plan can be dynamically adjusted to suit the environment, so that prior knowledge is less dependent. The recurrent LSTM-based reinforcement learning algorithm is divided into two parts: *state_space* training and *state_value* function training. The learning of the *state_space* determines whether to add the new state to the new representative point of the state space by judging whether the minimum distance between the new appearing state and the existing discrete state is greater than the contribution threshold g. Learning the *state_value* function computes the corresponding *action_value* function based on the known degree of the current state. Then the system selects the action based on this value and goes to the next state. At the same time, it calculates the corresponding update amount according to the known degree of the next state. The proposed deep learning model estimates the state behavior values of recurrent LSTM in the current state $\mathbf{x}_i$ to estimate $\mathbf{Q}(\mathbf{x}_i, \mathbf{y}_i)$. The corresponding value function update formula can be rewritten as:

$$\mathbf{Q}(\mathbf{x}_i, \mathbf{y}_i)_{t+1} = \mathbf{Q}(\mathbf{x}_i, \mathbf{y}_i)_t + \alpha[O_j + \gamma \sum_{\mathbf{x}_j} \mathbf{g}_j \mathbf{Q}\ (\mathbf{x}_j, \mathbf{y}_j)_t - \mathbf{Q}(\mathbf{x}_i, \mathbf{y}_i)_t] \qquad (12)$$

The algorithm first enters historical data for each electrical load and analyzes the state of the charge through the LSTM algorithm, then generates the initial HMM model. It solves the optimal state transition strategy through reinforcement learning algorithm based on the recurrent LSTM model. The following activity diagram (Fig. 3) shows the applied deep learning method which combines the recurrent LSTM model and Q-reinforcement learning.

3.4   Description of the Training Database for NILM

REDD (Reference Energy Disaggregation Dataset) is a database built for the purpose of energy disaggregation, using readings from 6 different houses. The frequency of the readings is approximately 1 Hz, but in some cases the readings are taken every 3 s.

The data contains the actual energy consumption for all homes as well as for each individual circuit in the house. REDD includes power readings of low frequency data, as well as, high frequency voltage data, which can also be used for disaggregation. However, since our algorithm only deals with household power consumption, we will ignore high frequency data. The only house that completely includes the data of all electrical appliances we need is Smart Home 4. For this reason, we used Smart Homes 1, 2, 3, 5, and 6 to train and Smart Home 4 to test in the REDD dataset.

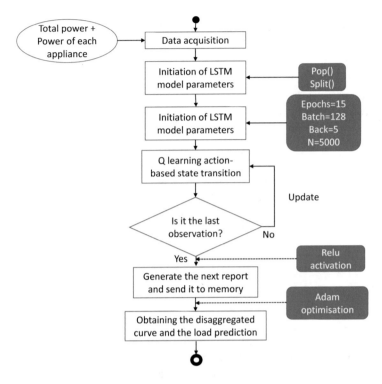

Fig. 3. Algorithm activity diagram

3.5 Evaluation Metrics

Many performance metrics have been proposed to evaluate the performance of NILM algorithms. The higher values indicate a better match of assigned power. In this paper, we propose the following metrics to quantify model performance: Accuracy (Acc): It measures the ratio of successfully recognized devices and to evaluate the performance of an appliance identification system.

$$Acc = \frac{true\ positives + true\ negatives}{true\ positives + true\ negatives + false\ positives + false\ negatives} \tag{13}$$

Cosine Similarity (Cosine sim): It uses the cosine angle between two vectors to quantify the differences. This metric can highlight the electric feature relevance of appliances in NILM.

$$Cos(\phi) = \frac{\sum_i x_i y_i}{\sqrt{\sum_i x_i} \sum_i \sqrt{\sum_i y_i}} \tag{14}$$

4 Simulation Results

Our proposed NILM algorithm is executed in Python. The optimization algorithm was adaptive moment estimation (Adam), and the performance of the learning algorithm is validated by MSE. Using the optimization function, this loss function learns to reduce the prediction error. The MSE is the sum of the squared distances between the target variable and the predicted values. More details of our proposed model parameters are exposed in Table 1.

Table 1. LSTM-based Reinforcement Q Learning model parameters

Item	Value
Number of blocks	5
Activation function	relu
Number of epochs	15
Batch size	128
Loss	'mse'
Optimizer	Adam
Learning rate	0.0001
Data preprocessing	min-max scaler

Table 2 shows accuracy and cosinus similarity of NILM. The learning result of our system by measuring accuracy over epochs for different rewards functions. The accuracy trend over epochs on training set becomes stable after 15 epochs. According to this Table 2, accuracy of our recurrent LSTM reinforcement Q-learning-based NILM approach is up to 99% for most Smart Homes for an exploration vs. exploitation ratio of 30% that infer the optimal Reinforcement Learning policy. These results show a significant improvement in learning

Table 2. Evaluation metrics for the proposed LSTM-based Reinforcement Q Learning method

	Inference			
	10%	20%	30%	40%
Smart Home 1	Acc = 0.76 Cosine sim = 0.96 Reward = −13.2	Acc = 0.78 Cosine sim = 0.96 Reward = −13.28	Acc = 0.82 Cosine sim = 0.96 Reward = −13.5	Acc = 0.78 Cosine sim = 0.96 Reward = −13.4
Smart Home 2	Acc = 0.93 Cosine sim = 0.92 Reward = −5.4	Acc = 0.995 Cosine sim = 0.983 Reward = −4.84	Acc = 0.99 Cosine sim = 0.99 Reward = −4.9	Acc = 0.93 Cosine sim = 0.98 Reward = −6.2
Smart Home 3	Acc = 0.95 Cosine sim = 0.993 Reward = −13.72	Acc = 0.95 Cosine sim = 0.993 Reward = −13.72	Acc = 0.99 Cosine sim = 0.99 Reward = −11.27	Acc = 0.94 Cosine sim = 0.98 Reward = −13.74
Smart Home 5	Acc = 0.998 Cosine sim = 0.988 Reward = −16.6	Acc = 0.998 Cosine sim = 0.98 Reward = −16.65	Acc = 0.99 Cosine sim = 0.98 Reward = −16.6	Acc = 0.998 Cosine sim = 0.97 Reward = −16.65
Smart Home 6	Acc = 0.715 Cosine sim = 0.84 Reward = −9.94	Acc = 0.86 Cosine sim = 0.80 Reward = −9.92	Acc = 0.90 Cosine sim = 0.58 Reward = −11.1	Acc = 0.84 Cosine sim = 0.60 Reward = −9.2

compared to the literature review (*Accuracy* = 85%). Figure 4 shows the performance of our proposed disaggregation algorithm applied to estimate the power consumption of different households. For Smart Home 2, we can distinguish the consumption of the 5 households (dishwasher, lighting, washing machine-dryer, refrigerator and microwave. The Main curve seems very close to the sum of these consumptions. Applying our proposed model, we got about 99% accuracy for this Smart Home 2, which is a very good performance.

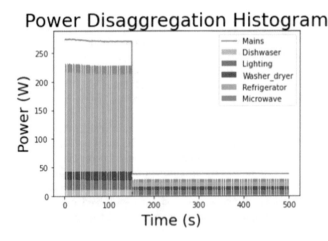

Fig. 4. Power disaggregation for Smart Home 2.

5   Conclusion

In recent years, the development of new technology to better regulate energy use inside buildings becomes a necessity. In this context, several IoT solutions exist that enable resource monitoring in smart homes by connecting an increasing number of equipment and sensors. Hence, it is crucial to develop methods that promote rational and intelligent energy use through the integration of new technologies to discover efficient solutions. In this paper, we proposed an intelligent solution, using a recurrent LSTM reinforcement Q learning technique, for the disaggregation of electrical power using the non-intrusive load monitoring (NILM) technique. This technique is used to identify smart home loads in a non-intrusive way. The evaluation results, show the performance of the proposed method (*Accuracy* = 99%) against an existing deep learning technique [1] (*Accuracy* = 85%).

References

1. Siddiqui, A., Sibal, A.: Energy disaggregation in smart home appliances: a deep learning approach. Energy (2020, in press). Elsevier. hal-02954362

2. Jaradat, A., Lutfiyya, H., Haque, A.: Smart home energy visualizer: a fusion of data analytics and information visualization. IEEE Can. J. Electr. Comput. Eng. **45**(1), 77–87 (2022)
3. Siddiqui, S.A., Ahmad, M.O., Ahmed, J.: Smart home for efficient energy management. In: Agarwal, P., Mittal, M., Ahmed, J., Idrees, S.M. (eds.) Smart Technologies for Energy and Environmental Sustainability. GET, pp. 97–103. Springer, Cham (2022). https://doi.org/10.1007/978-3-030-80702-3_6
4. Kim, H., Choi, H., et al.: A systematic review of the smart energy conservation system: from smart homes to sustainable smart cities. Renew. Sustain. Energy Rev. **140**, 110755 (2021)
5. Pinto, G., Wang, Z., Roy, A., Hong, T., Capozzoli, A.: Transfer learning for smart buildings: a critical review of algorithms, applications, and future perspectives. Adv. Appl. Energy **5**, 100084 (2022)
6. Popa, D., Pop, F., Serbanescu, C., Castiglione, A.: Deep learning model for home automation and energy reduction in a smart home environment platform. Neural Comput. Appl. **31**(5), 1317–1337 (2018). https://doi.org/10.1007/s00521-018-3724-6
7. Mathew, A., Roy, A., Mathew, J., et al.: Intelligent residential energy management system using deep reinforcement learning. IEEE Syst. J. **14**(4), 5362–5372 (2020)
8. Ruelens, F., Claessens, B.J., Vrancx, P., et al.: Direct load control of thermostatically controlled loads based on sparse observations using deep reinforcement learning. CSEE J. Power Energy Syst. **5**(4), 423–432 (2019)
9. Arroyo, J., Manna, C., Spiessens, F., et al.: Reinforced model predictive control (RL-MPC) for building energy management. Appl. Energy J. **309**, 118346 (2022)
10. Dargazany, A.: DRL: deep reinforcement learning for intelligent robot control-concept. literature and future. arXiv preprint arXiv:2105.13806 (2021)
11. Ren, M., Liu, X., Yang, Z., et al.: A novel forecasting based scheduling method for household energy management system based on deep reinforcement learning. Sustain. Urban Areas **76**, 103207 (2022)
12. Claessens, B.J., Vrancx, P., Ruelens, F.: Convolutional neural networks for automatic state-time feature extraction in reinforcement learning applied to residential load control. IEEE Trans. Smart Grid **9**(4), 3259–3269 (2016)
13. Huang, Q., Huang, R., Hao, W., et al.: Adaptive power system emergency control using deep reinforcement learning. IEEE Trans. Smart Grid **11**(2), 1171–1182 (2019)
14. Bucci, G., Ciancetta, F., Fiorucci, E., et al.: State of art overview of non-intrusive load monitoring applications in smart grids. Sensors **18**, 100145 (2021)
15. Deshpande, R., Hire, S., Mohammed, Z.A.: Smart energy management system using non-intrusive load monitoring. SN Comput. Sci. **3**(2), 1–11 (2022)
16. Figueiredo, M., De Almeida, A., Ribeiro, B.: Home electrical signal disaggregation for non-intrusive load monitoring (NILM) systems. Neurocomputing **96**, 66–73 (2012)
17. Bonfigli, R., Principi, E., et al.: Non-intrusive load monitoring by using active and reactive power in additive Factorial Hidden Markov Models. Appl. Energy **208**, 1590–1607 (2017)
18. Guo, L., Wang, S., Chen, H., et al.: A load identification method based on active deep learning and discrete wavelet transform. IEEE Access **8**, 113932–113942 (2020)
19. Nalmpantis, C., Gkalinikis, V., et al.: Neural Fourier energy disaggregation. Sensors **22**(2), 473 (2022)

20. Gomes, E., Pereira, L.: PB-NILM: pinball guided deep non-intrusive load monitoring. IEEE Access **8**, 48386–48398 (2020)
21. Pereira, L., Nunes, N.: Performance evaluation in non-intrusive load monitoring: datasets, metrics, and tools - a review. Wiley Interdisc. Rev. Data Min. Knowl. Discov. **8**(6), 1265 (2018)
22. Khodayar, M.: Learning deep architectures for power systems operation and analysis. Electrical Engineering Theses and Dissertations. 41 (2020)
23. Li, H.: A Non-intrusive home load identification method based on adaptive reinforcement learning algorithm. IOP Conf. Ser. Mater. Sci. Eng. **853**(1), 012030 (2020)

Machine Learning for Student QoE Prediction in Mobile Learning During COVID-19

Besma Korchani and Kaouthar Sethom[✉]

INNOVCOM LAB, SUPCOM, University of Carthage, Tunis, Tunisia
kaouthar.sethom@enicar.u-carthage.tn

Abstract. The emergence of COVID-19 has shaped a new type of learning that is called "digital learning", which can be approached through e-learning and m-learning technologies. An important element in reducing student dropout rates in a virtual learning is to understand the degree of engagement and satisfaction experienced by students in meaningful activities. The major novel contribution of this study is the combination of network Quality of Service (QoS) parameters and student engagement behavior (through eye-tracking) to automatically estimate the Mean Opinion Score (MOS) for mobile learning synchronous activities. Several machine learning algorithms were compared for best QoE prediction.

1 Introduction

Within and beyond the era of COVID-19, educational institutions might consider the m-learning as an effective approach for digital learning during the current and future crises. The COVID-19 pandemic might not be the last crisis to challenge educational institutions stakeholders. Hence, the preparation for m-learning infrastructure would be the key to handle future global risks. Mobile learning not only offer students online learning space, but also enable them to quick access to learning activities and materials anytime, anywhere, and anyhow and create pioneering opportunities for innovative learning. Hence, mobile learning has attracted many researchers' attention and have been introduced into many fields. It is therefore imperative to discover what among e-learning service quality attributes are the most important ones that have impacts on overall e-learning service quality, and to evaluate the relationship between overall e-learning service quality, e-learning student satisfaction, and e-learning student loyalty [1].

In this study, we propose to develop a framework to estimate the quality of experience (QoE) felled by student during mobile visio learning. We thus conduct a series of subjective measurements, to collect QoE data and evaluate the impact of different network parameters (delay, bandwidth, video types, ...), student profile and engagement degree on mobile learning experience. Based on this, the system will be able to predict student QoE in future scenario and help teachers to have feedbacks from students and point-out whose have difficulties and may be adapt some course's activities. To evaluate student engagement during the online synchronous class, we detect eye gaze movements during the synchronous course through the mobile camera (Fig. 1). The system will be able to automatically inform the teacher about student experience with a specific course.

L. Barolli et al. (Eds.): AINA 2022, LNNS 451, pp. 14–22, 2022.
https://doi.org/10.1007/978-3-030-99619-2_2

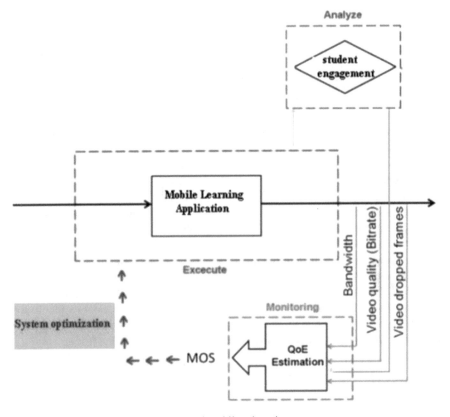

Fig. 1. Proposed mobile e-learning system

2   QoE Definition and Metric

In the past, QoS based measurements that consider network parameters (e.g., packet loss, delay and throughput) were used to define the level of satisfaction/performance of a service. As QoS metrics are not explicitly and directly linked with a customer's satisfaction of a service, user-centric Quality of Experience (QoE) metrics (e.g. Mean Opinion Scores) have been used in recent years to assess the quality of multimedia services, in addition to QoS metrics. QoE considers the user's subjectivity towards a specific service which can be defined as "the degree of delight or annoyance of the user of an application or service. It results from the fulfillment of his or her expectations with respect to the utility and/or enjoyment of the application or service in the light of the users' personality and current state" [2]. The understanding of the users' expectations and experiences from a service is vital for the success of a service.

QoE metrics are subjective and thus more difficult to define than Quality of Service (QoS) metrics which are objective and measurable. QoE management remains a challenging task due to many issues that influence it. We can categorize them in our case into three different aspects: The first aspect is the consideration of a variety of networks parmaters (delay,bandwidth,...). The second is the multiple type of mobile devices that can be used by students to access the plateform. And finally, the student behaviors and degree of investiments. There is a long literature covering QoE-based monitoring approaches, a good summary is provided in [3]. When it comes to the specific case of mobile learning, we can find multiple studies mapping either network or device/application measurements to QoE metrics [4–7]. They use different Machine Learning (ML) algorithms to investiagte the influence of radio network characteristics on user experience. However, no research has been conducted to explore the relationship between QoS metrics, student engagement and QoE metrics at the same time. We think that student engagement is a very important metric that influence the QoE in the case of mobile learning and we decide to integrate it in our framework. A sensor camera (present in the mobile device) will be used to acquire and recognize the degree of student concentration through eye-tracking. Generally, user's QoE estimation is done in terms of MOS (Mean Opinion Score) score (see ITU-T P.910) [8]. However, it isn't always easy to estimate the MOS due to the complexity of performing subjective test campaign in addition to customer involvement issue. We use ML technique to predict the MOS in advance and train the IA algorithm based on previous collected dataset information.

3   Proposed Framework

This framework is composed by four modules (Fig. 1):

- QoS monitoring module: For monitoring the used network, video, and devices QoS parameters., We emulate via Mininet [9] the desired network and upload 4 short course videos from YouTube with several resolutions.
- Student engagement module (described in Sect. 3.1): it is used for estimating the student concentration during the online synchronous course through eye-tracking technique.
- Dataset: The dataset is collected using YouTube JavaScript player and Video camera as a sensor. Mininet is used to emulate different network scenarios. 50 participants (man and women) aged from 18 to 40 years are invited to attend a set of tests and give an evaluation on the quality of the watched videos (MOS). The user's MOS score can be:(5 -> Excellent, 4 -> Good, 3 -> Fair, 2 -> Poor, 1 -> Bad).
- QoE estimation module: Based on the collected information on the dataset, QoE estimation is done using a machine learning algorithm. ML methods learn automatically from the past observations in the dataset to make accurate predictions in the future. In our case, the model output is instantaneous user's perceived quality in terms of predicted MOS score (Fig. 2).

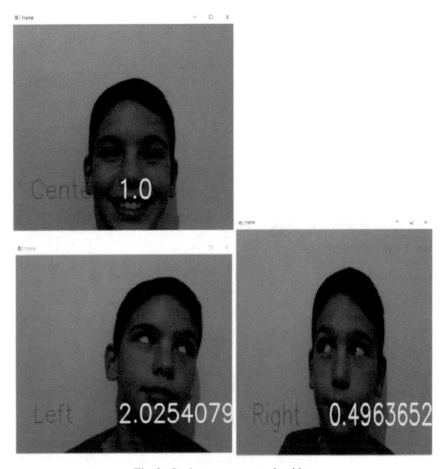

Fig. 2. Student engagement algorithm

The originality of our approach stems from the fact that our system can consider the actual dynamics of the network and the detected eye position by a camera to estimate the instantaneous student's viewpoint. Moreover, our work differs from [10] in the sense that we don't care about all students 'faces emotions, only the eye characteristics are used for student engagement calculation. We think that happy or sad emotions are not enough correlated to the student concentration level wheras the eye position can indicate if he or she is following the course (Fig. 3).

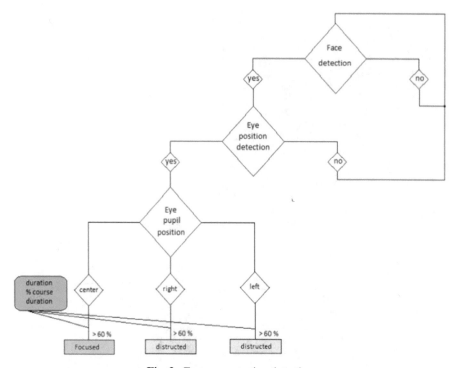

Fig. 3. Eye concentration detection

3.1 Student Engagement Detection Module

In facts, professors cannot recognize students' engagement perfectly in mobile learning compared with teaching in class. The challenge for the teacher in virtual classrooms is the adjustment between the presentation and the explanation of the lesson and the student engagement control.

A three-level model for student engagement classify them as: High, Nominal, and Low. We decide to simplify the classification on a two-level model: engaged and low-engaged. In a typical m-learning environment, students have a smartphone with a built-in camera. Here, we propose the use of the built-in webcam to grab real time information about the eye's movements (eye tracking) of the students. This information will be used to determine a concentration level, hence helping the instructor to see how engaged (or not) the student is.

To produce the concentration index presented here, in real-time the system operates according to the following main steps:

- The student logs into the m-learning plateform and the camera starts image acquisition.
- The face is detected, the eyes region is detected and cropped.
- The student's attention state is classified in "Distracted" or "Focused".
- If the student is focused, the resultant concentration index is calculated as the total Focused time par the total video watching time.

- Finally, the student's engagement level is determined depending on the value of concentration index (Fig. 4).

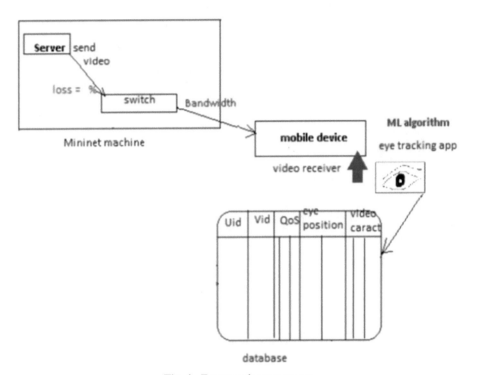

Fig. 4. Framework components

The system is based on three main steps, the face detection sub system, the eye gaze movements detection, and the decision-making step.

1. Face detection procedure: We used the Dlib library to for the face and the eyes detection. For frontal face, which is exactly what we need because the student must be in front of his mobile device cam, Dlib give very accurate detection. The Dlib detector is build using the histogram of oriented gradient (HOG) and the linear Support Vector Machine (SVM).
2. Eye gaze tracking procedure: The eye gaze tracking is based on distinguish the black region of the eye from the white region after detecting the eyes from the face. The Dlib face detection concept is pointing the face depending on edges in the gray image. This method can also detect all face components like the eyes, the nose and the mouth… as a result, every face image processed by this algorithm will reformatted to a set of points that shape the origin face and each point have its own number.
3. Decision making: If the eye gaze is in the center of the eye, the student is focusing and if it is in the right or in the left, the student is distracted.

According to the eye pupil position in the whole eye, we get three main cases:

- eye pupil in the right side: if the gaze-ratio < 0.9
- eye pupil in the center of the eye: if 0.9 <= the gaze ratio < 1.7
- eye pupil in the left side: if the gaze-ratio >= 1.7

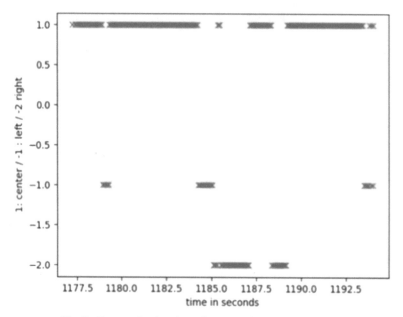

Fig. 5. Eye monitoring through an experience of visio-class.

4 QoE Estimation for MOS Prediction

We conduct different tests with 4 different supervised ML algorithms and compare the result of each method to deduce the best algorithm. We construct decision tree (DT), SVM, k-nearest neighbor (KNN) and Linear regression models in Python from SKlearn Library. Before the data is processed, the data set is split into two parts by a ratio of 75:25, which 75% to training and 25% to testing. Training data used to construct the model. Training data used were 1148 samples, 6 predictor, and 5 classes (MOS), with cross-validation 10-fold and repeated 3 times.

- Decision tree: This method is considered non-parametric, making no assumption on the distribution of data and the structure of the true model. It requires less data cleaning and is not influenced by outliers and multicollinearity to some fair extent. It can be used for both regression and classification problems and provide a useful and simple tool for interpretation.

- Support Vector Machines (SVM) is a supervised machine learning algorithm which can be used for classification or regression problems. The goal of SVM is to identify an optimal separating hyperplane which maximizes the margin between different classes of the training data.
- KNN: Predictions are made for a new instance (x) by searching through the entire training set for the K most similar instances (the neighbors) and summarizing the output variable for those K instances.
- LR: It predicts the probability of the occurrence of an event by fitting data to the algorithm. Since, it predicts the probability, its output values lies between 0 and 1.

The prediction efficiency of the model was analyzed using statistical error measures such as the Root Mean Square Error (RMSE) and the Outlier Ratio (OR) (Fig. 5). The RMSE is used to test the accuracy of the models; smaller RMSE means greater accuracy. The OR tests consistency: a smaller value of OR means the predictions of the model are more consistent. To provide more insight into the subjective test, we evaluate the model performance using the epsilon insensitive RMSE (RMSE *), which considers the uncertainty of the subjective scores [8]. This is a RMSE that considers the confidence interval of the individual MOS scores. It is calculated like the traditional RMSE, but small differences to the target value are not counted. As shown in Fig. 6, the best ML algorithm in our case is decision tree with RMSE at 0.408, RMSE * is 0.242 and OR is 0.167.

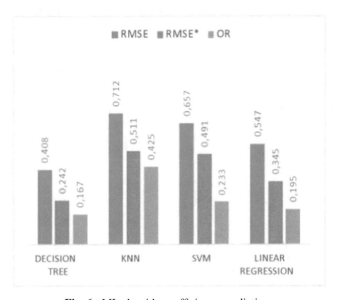

Fig. 6. ML algorithms efficiency predictions

5   Conclusion

Mobile learning offers to students online learning space and enable pioneering opportunities for innovative learning. The major novel contribution of this study is the combination of Quality of Service (QoS) parameters and student engagement (eye-tracking) to estimate the QoE through Mean Opinion Score (MOS) for mobile visio-conferencing. Future works will focus on adding students' comments and likes in the overall estimation.

References

1. Alsabawy, Y.A., Cater-Steel, A., Soar, J.: Identifying the determinants of e-learning service delivery quality. In: 23rd Australian Conference on Information Systems, 3–5 Dec 2012, Greelong (2012)
2. Qualinet White Paper on Definitions of Quality of Experience (2012). European Network on Quality of Experience in Multimedia Systems and Services (COST Action IC 1003), Patrick Le Callet, Sebastian Möller and Andrew Perkis, eds., Lausanne, Switzerland, Version 1.2 (2013)
3. Ahmad, W.R.A., Kara, P.A., Atzori, L., Martini, M.G., Raake, A, Sun, L.: Challenges of future multimedia qoe monitoring for internet service providers. Multimedia Tools Appl. **76**(21), 22 243–22 266 (2017)
4. Taha, M., Abu Ali, N., Ran Chi, H., Radwan, A.: MEC resource offloading for QoE-Aware has video streaming. In: Communications ICC 2021, pp. 1–5 (2021)
5. Anchuen, P., Uthansakul, P., Uthansakul, M.: QoE analysis for improving multimedia services based on different operating situations on cellular networks. Access IEEE **9**, 116215–116230 (2021)
6. Seufert, M., Wehner, N., Wamser, F., Casas, P., D'Alconzo, A., TranGia, P.: Unsupervised QoE field study for mobile youtube video streaming with yomoapp. In: QoMEX, pp. 1–6 (2017)
7. Katsarakis, M., Teixeira, R.C., Papadopouli, M., Christophides, V.: Towards a causal analysis of video QoE from network and application QoS. In: Proceedings of the Internet QoE'16. ACM (2016)
8. ITU-T Rec. P.10/G.100: Vocabulary for performance and quality of service. Amendment 5: New definitions for inclusion in Recommendation ITU-T P.10/G.100 (2016)
9. www.mininet.org

XceptionUnetV1: A Lightweight DCNN for Biomedical Image Segmentation

Mohammad Faiz Iqbal Faiz[1][(✉)] and Mohammad Zafar Iqbal[2]

[1] Tezpur University, Tezpur, Assam, India
faiziqbal.faiz066@gmail.com
[2] W-Pratiksha Hospital, Gurgaon, India

Abstract. This paper proposes a Deep Convolutional Neural Network (DCNN) architecture for biomedical image segmentation. The proposed work explores the role of separable convolution within the context of biomedical image segmentation. The proposed architecture, XceptionUnetV1, is an amalgam of the Xception and U-Net models. Additionally, the model uses dilated convolution to increase the field view of the filters. The proposed model achieves better results than various versions of U-Net. The proposed DCNN architecture requires much lesser parameters, approximately $1/4^{th}$ of the U-Net model. The proposed model has been tested on chest X-rays for lung segmentation.

1 Introduction

Computer vision plays a vital role in the domain of image recognition, classification, and segmentation. Its application ranges from object detection in traffic control scenarios to image segmentation in biomedical applications. In this work, we discuss the latter. In the past few years, we have seen significant growth in computer vision tools in biomedical applications. Especially with the introduction of DCNN, image segmentation has become easier within the context of biomedical applications. The advantage of DCNN was two-fold. Firstly, we can do detailed segmentation on biomedical images such as chest X-rays. Secondly, the process has become faster and more efficient.

The first DCNN architecture proposed for biomedical image segmentation is the U-Net model, discussed in the seminal paper [1]. The basic architecture of U-Net is an encoder-decoder model. The U-Net also uses data augmentation to alleviate the problem of data shortage in biomedical applications. The encoder module extracts the spatial information from a given image. The decoder module then uses that spatial information to do image localization by applying pixel-wise prediction in the image. The U-Net model also uses skip connections from the encoder stage to the decoder stage. These skip connections concatenate the corresponding feature map from the encoder stage.

L. Barolli et al. (Eds.): AINA 2022, LNNS 451, pp. 23–32, 2022.
https://doi.org/10.1007/978-3-030-99619-2_3

Over the years, many modifications have been proposed to the U-Net architecture. The authors in [2], introduced residual units to the U-Net architecture to get the benefit of residual learning. In [3], authors used dilated convolution in place of standard convolution to increase the field-of-view of the kernels. In [4], authors used Atrous Spatial Pyramid Pooling (ASPP) in the bridge section of the U-Net to get the benefits of Atrous convolution and Spatial Pyramid Pooling (SPP). We use SPP to classify regions of an arbitrary scale by resampling convolutional features extracted at a single scale.

2 Motivation

The primary motivation of this work is the success of depthwise separable convolutions introduced in the seminal paper [5]. The author converted the residual blocks into Xception units where the standard convolution is replaced with separable convolution. The blocks in the Xception architecture did have residual connections. The application of Xception units in the broader context of image segmentation was first discussed in [6]. Nevertheless, its application in the context of biomedical image segmentation is still unexplored.

As we know, that X-ray is the first step in many forms of diagnosis in developing countries as its cheap and fast. However, during the recent academic COVID 19, the radiologists were burdened with manually analyzing hundreds and thousands of chest X-ray images, which slowed the diagnosis. Various Deep Learning (DL) tools are developed to aid radiologists. However, there is one inherent disadvantage in using DL algorithms. These algorithms take time to complete the task and consume a heavy amount of computing resources. Despite their high efficiency, this hurdle impedes using DL algorithms at full scale, especially in developing countries.

So, a DL algorithm that can finish the task quickly and with fewer resources is necessary. With this idea in mind, we develop a modified U-Net architecture that takes a significantly lesser number of parameters than the standard U-Net model and gives better performance than U-Net.

The contributions of this paper are as follows:

1. A modified version of U-Net, XceptionUnetV1, uses separable convolution instead of standard convolution, is proposed.
2. We also added dilation in the separable convolution to get the benefit of dilated convolution.
3. The blocks in the XceptionUnet have residual connections, as in Xception and ResNet [7].

3 Proposed Model

The proposed architecture in this paper for image segmentation is an amalgam of the Xception architecture and the U-Net architecture, hence the name XceptionUnetV1. The architecture is divided into several blocks. We can broadly divide the blocks into (i) encoder and (ii) decoder. Both the encoder and decoder have similar architecture, except that we use transposed convolution and no max-pooling operation in the decoder. The encoder and decoder parts are connected by a bridge block that is also an Xception block. The proposed architecture is shown in Fig. 1. We briefly discuss the critical aspects of the proposed architecture, XceptionUnet, in the following subsections.

3.1   Xception Units

The Xception units/blocks are one of the significant components of this architecture. The typical Xception unit consists of two separable convolution layers, three batch normalization layers, and a max-pooling layer. The distinct aspect of the Xception unit is that the separable convolution layers use depthwise separable convolution in place of standard convolution. The separable convolution layer separates the learning of spatial features and channel-wise features. In this case, we assume that the spatial locations in the input are highly correlated, leading to significantly fewer parameters and faster computation, resulting in faster models. Another advantage of using separable convolution is learning better representations using fewer data. Since, in biomedical application, we are often struck with the shortage of training data. So, using separable convolution is a suitable option. An Xception block is shown in Fig. 2.

3.2   Residual Learning

We have added residual connections in the proposed model between the Xception units. The concept of residual connection is first proposed in the seminal paper [7]. Residual connections are used to tackle two common problems found in large-scale deep learning models: vanishing gradients and representational bottlenecks.

3.3   Atrous Convolution

Atrous convolution is also known as dilated convolution. Atrous convolution is used to negate the effect of repeated use of max-pooling and striding at consecutive layers of a DCNN [8,9]. Atrous convolutions enable us to enlarge the field-of-view of filters/kernels at any layer of a DCNN without increasing the number of parameters or the amount of computation. Thus, it gives us the best trade-off between accurate localization and context assimilation. We have added dilation with a dilation_rate set to 2 to the separable convolution in the proposed model. Thus, increasing the kernel size of a $k \times k$ from k to $k_e = 2k - 1$.

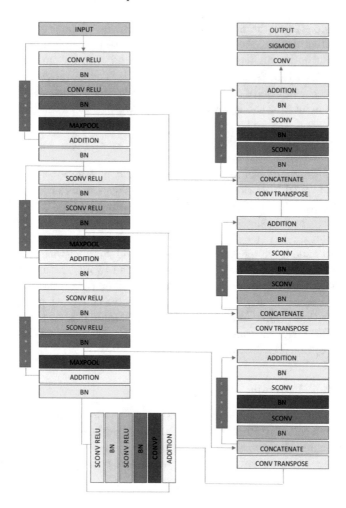

Fig. 1. Architecture of proposed model, XceptionUnetV1.

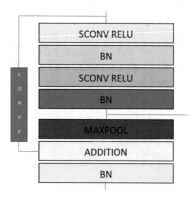

Fig. 2. An Xception block in XceptionUnetV1.

Table 1. Network architecture of XceptionUnetV1.

	Unit level	Conv layer	Filter	Stride	Dilation	Output size
Input						$256 \times 256 \times 1$
	Level 1	Conv 1	$3 \times 3/64$	1	2	$256 \times 256 \times 64$
		Conv 2	$3 \times 3/64$	1	2	$256 \times 256 \times 64$
Encoding	Level 2	SConv 3	$3 \times 3/128$	1	1	$128 \times 128 \times 128$
		SConv 4	$3 \times 3/128$	1	1	$128 \times 128 \times 128$
	Level 3	SConv 5	$3 \times 3/256$	1	1	$64 \times 64 \times 256$
		SConv 6	$3 \times 3/256$	1	1	$64 \times 64 \times 256$
Bridge	Level 4	SConv 7	$3 \times 3/512$	1	1	$32 \times 32 \times 512$
		SConv 8	$3 \times 3/512$	1	1	$32 \times 32 \times 512$
	Level 5	SConv 9	$3 \times 3/256$	1	1	$64 \times 64 \times 256$
		SConv 10	$3 \times 3/256$	1	1	$64 \times 64 \times 256$
Decoding	Level 6	SConv 11	$3 \times 3/128$	1	2	$128 \times 128 \times 128$
		SConv 12	$3 \times 3/128$	1	2	$128 \times 128 \times 128$
	Level 7	SConv 13	$3 \times 3/64$	1	2	$256 \times 256 \times 64$
		SConv 14	$3 \times 3/64$	1	2	$256 \times 256 \times 64$
Output		Conv 15	1×1	1		$256 \times 256 \times 1$

4 Experimental Results

4.1 Dataset

We evaluated our proposed DCNN, XceptionUnetV1, on a chest X-ray dataset. The dataset is a combination of two datasets. The first dataset is obtained from [10, 12, 13]. The second dataset is obtained from [11–13]. Since our task is lung segmentation, the dataset contains chest X-rays and masks. In the dataset, chest X-rays are either normal or abnormal with manifestations of tuberculosis. The dataset consists of 800 chest X-ray images and masks, but some chest X-rays do not have lung masks, and some are duplicated. After careful inspection, the dataset is reduced to 704 images - all in JPEG format. In our training and testing, we keep the size of the images as 256×256. The training set, validation set, and test set are kept at 8:1:1 proportion. We use sklearn's train_test_split function to generate random train and test subsets.

4.2 Network Architecture

This section briefly discusses the architecture details of the proposed model, XceptionUnetV1. The main components of this architecture are already discussed in Sect. 3. Table 1 shows the network structure of the XceptionUnetV1. The max-pooling layers and pointwise convolution layers are not shown. However, these can be derived easily from Table 1. We use the "same" padding option for all of our convolution and separable convolution layers. The depth multiplier for separable convolution layers is one. In the upsampling layer, we use transposed convolution. We use transposed convolution when the transformation is going in the opposite direction of a standard convolution. Since we use dilated convolution in all of our convolution and separable convolution layers, we have a dilation rate parameter that controls a filter's field of view.

Table 1 shows the main layers of our network. XceptionUnetV1 has 15 convolution and separable convolution layers. Conv denotes the convolution layer in the given table, and SConv stands for separable convolution layer. The network structure is divided into three stages. The first stage is called the encoding stage. The encoding stage consists of three levels. The first level has two Conv layers. The second and the third level consist of two SConv layers. The encoding stage uses 64,128,256 filters of size 3 × 3. The bridge stage consists of two SConv layers that use 512 filters of size 3 × 3. The decoding stage also consists of three levels. The first level has two SConv layers that use 256 filters of size 3 × 3. The second level has two SConv layers that use 128 filters of size 3 × 3, and the third level also has two SConv layers that use 64 filters of size 3 × 3. Finally, the output of the last Conv layer is a 256 × 256 × 1 image. Thus, input and output images have the same size.

4.3 Loss Function

We use two metrics for our image segmentation: (i) Dice coefficient (ii) Jaccard's index. Dice coefficient is defined as twice the area of overlap divided by the total number of pixels in both images. It is calculated as:

$$\text{Dice coefficient} = \frac{2 \times Intersection}{Union + Intersection} = \frac{2 \times TP}{2TP + FN + FP} \quad (1)$$

where TP is the true-positives implying the area of overlap, FN is the area of ground truth mask. FP is the area of the predicted mask. The Dice loss will be $1 - $ Dice coefficient. Jaccard's index or Intersection over Union (IoU) is defined as the area of overlap between the predicted segmentation and the ground truth divided by the area of union between the predicted segmentation and the ground truth. It is calculated as:

$$\text{Jaccard's index} = \frac{Intersection}{Union} = \frac{TP}{TP + FN + FP} \quad (2)$$

The loss function using Jaccard's index will be $1 - $ Jaccard's index, also known as the Jaccard's distance. Both Dice Coefficient and Jaccard's index vary from 0 to 1.

4.4 Training and Testing

The proposed model is trained on a dataset of 566 chest X-ray images and their masks. The input image size is 256 × 256 × 1. We use Adam [14] as the optimizer in the training algorithm. The initial learning rate was kept at $2e - 4$. During the training process, the learning rate is reduced by a factor of 0.5 when the metric stops improving. The network is trained independently using Dice loss and Jaccard's distance. Figure 3 shows the training and validation accuracy, and the training and validation loss for the network when we train the network using the Dice loss. Figure 4 shows the training and validation accuracy, and the training and validation loss for the network when we train the network using the Jaccard's distance. We train the network for 50 epochs with a batch size of 8. The whole implementation was done in Keras [15], on the Google Colab Pro+ platform, with 52 GB RAM and Google Compute Engine Backend (GPU). The training process took an hour to complete.

Figure 5 shows the predicted masks showing the segmented lung when the network is trained using the Dice loss. Figure 6 shows the predicted masks when the network is trained using the Jaccard's distance. The leftmost column shows the chest X-ray images, the middle column shows the ground truth masks, and the rightmost column shows the predicted masks generated by the XceptionUnetV1.

4.5 Comparative Results

Table 2 shows the metric values of the Dice coefficient, Jaccard's index for different versions of U-Net, and the XceptionUnetV1. XceptionUnetV1 outperforms other models on both metrics. Additionally, XceptionUnetV1 requires approximately 2 million parameters for training, whereas U-Net requires approximately 7.8 million parameters for training. The number of trainable parameters is reduced by $1/4^{th}$ in our model, a significant reduction.

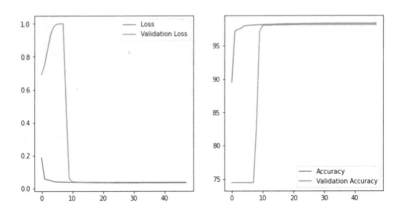

Fig. 3. Training and validation loss (left), training and validation accuracy (right) (using Dice loss).

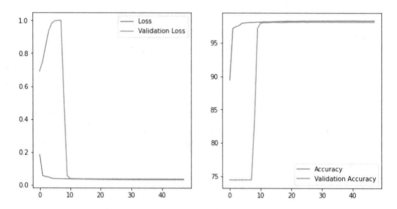

Fig. 4. Training and validation loss (left), training and validation accuracy (right) (using Jaccard's distance).

Table 2. Evaluation metrics of models.

	Dice coefficient	Jaccard's Index
U-Net	0.95	0.91
U-Net (Pretrained)	0.96	0.92
XceptionUnetV1	0.97	0.99

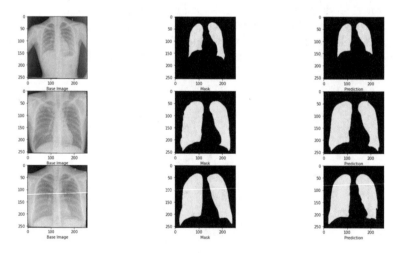

Fig. 5. Predicted masks (using Dice loss).

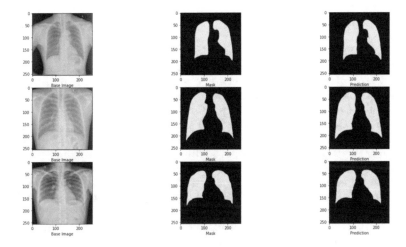

Fig. 6. Predicted masks (using Jaccard's distance).

5   Conclusion

In this work, we proposed a DCNN architecture, XceptionUnetV1, an amalgam of the Xception and U-Net models. The proposed model requires significantly fewer parameters to train and can work well with less data, making it a suitable model for biomedical image segmentation. We tested the model on a chest X-ray dataset and achieved better performance. We made the comparison only on two evaluation metrics, i.e., dice coefficient and Jaccard's index. In future works, we plan to expand the evaluation to other metrics. We also plan to explore the role of separable convolution in other scenarios of biomedical image segmentation and how ASPP will sit with separable convolutions.

References

1. Ronneberger, O., Fischer, P., Brox, T.: Convolutional Networks for Biomedical Image Segmentation, U-Net (2015)
2. Zhang, Z., Liu, Q., Wang, Y.: Road extraction by deep residual U-Net. IEEE Geosci. Remote Sens. Lett. **15**(5), 749–753 (2018)
3. Wang, S., et al.: U-Net using stacked dilated convolutions for medical image segmentation (2020)
4. Wang, J., Lv, P., Wang, H., Shi, C.: SAR-U-Net: squeeze-and-excitation block and atrous spatial pyramid pooling based residual U-Net for automatic liver segmentation in Computed Tomography (2021)
5. Chollet, F. Xception: deep learning with depthwise separable convolutions (2017)
6. Chen, L.-C., Zhu, Y., Papandreou, G., Schroff, F., Adam, H.: Encoder-decoder with atrous separable convolution for semantic image segmentation. In: Ferrari, V., Hebert, M., Sminchisescu, C., Weiss, Y. (eds.) ECCV 2018. LNCS, vol. 11211, pp. 833–851. Springer, Cham (2018). https://doi.org/10.1007/978-3-030-01234-2_49

7. He, K., Zhang, X., Ren, S., Sun, J.: Deep residual learning for image recognition (2015)
8. Chen, L., Papandreou, G., Kokkinos, I., Murphy, K., Yuille, A.: Semantic image segmentation with deep convolutional nets and fully connected CRFs (2016)
9. Chen, L., Papandreou, G., Kokkinos, I., Murphy, K., Yuille, A.: DeepLab: semantic image segmentation with deep convolutional nets, atrous convolution, and fully connected CRFs (2017)
10. Jaeger, S., Candemir, S., Antani, S., Wáng, Y.X.J., Lu, P.X., Thoma, G.: Two public chest X-ray datasets for computer-aided screening of pulmonary diseases. Quant. Imaging Med. Surg. 4(6), 475 (2014)
11. Higgins, G., et al.: Final report of the meeting "modeling & simulation in medicine: towards an integrated framework": July 20-21, 2000, national library of medicine, National Institutes of Health, Bethesda, Maryland, USA. Comput. Aided Surg. 6(1), 32–39 (2001)
12. Jaeger, S., et al.: Automatic tuberculosis screening using chest radiographs. IEEE Trans. Med. Imaging 33, 233–245 (2014)
13. Candemir, S., et al.: Lung segmentation in chest radiographs using anatomical atlases with nonrigid registration. IEEE Trans. Med. Imaging 33, 577–590 (2014)
14. Kingma, D., Ba, J.: A Method for Stochastic Optimization. Adam (2017)
15. Chollet, F., et al.: Keras (2015). https://keras.io

A Proposed Intrusion Detection Method Based on Machine Learning Used for Internet of Things Systems

Neder Karmous[✉], Mohamed Ould-Elhassen Aoueileyine, Manel Abdelkader, and Neji Youssef

Innov'COM Laboratory, SUPCOM, Carthage University, Tunis, Tunisia
{nader.karmous,mohamed.ouldelhassen,neji.youssef}@supcom.tn

Abstract. This paper presents an improved method using supervised machine-learning techniques of the Internet of things (IoT) systems to ensure security in deployments devices. The method increases accuracy and efficiency, identifies patterns, and makes decisions with significantly reduced error. In this work, we compare previous works by our improved ML method for both binary and multi-class classification on some IoT datasets. Based on metric parameters such as accuracy, precision, recall, F1 score, and ROC-AUC, the simulation results reveal that Classification and Regression Trees (CART) outperforms on all types of attacks in binary classification with an accuracy of 99% and with an accuracy between 21% and 37% higher than the original one. However, in multi-class classification, Naive Bayes (NB) outperforms other ML algorithms with an accuracy of 99% and an accuracy between 1% and 4% higher than the others works.

Keywords: Internet of things · Machine learning · Cybersecurity · Intrusion detection system · Artificial Intelligence

1 Introduction

With the IoT revolution [1], we have exponential growth in the deployment of devices. Billions of physical devices are connected to the Internet to collect and send information. This interconnection may allow hackers to gain control, usurp, damage devices functionality or steal their data. Minimizing the risk and attacks of a compromised IoT device is the biggest challenge in the future [2, 3].

Today, Artificial Intelligence (AI) and IoT have been widely dominated by Machine Learning (ML) [4], which is used for intrusion detection systems (IDS) for IoT devices to facilitate the detection of cyber-attacks [5]. In this topic, we build an IDS based on an improved ML method, which creates a model simulating regular activity and then compares new activity with the existing model to detect abnormal patterns. In the First work, we select seven datasets for IoT sensors devices. Furthermore, we propose a method without combining datasets, and at the same time, it ameliorates the results of intrusion detection. The improvement is based on classification methods, the most popular methods used for IDS development [6, 7]. These methods are k-nearest neighbour (KNN)

© The Author(s), under exclusive license to Springer Nature Switzerland AG 2022
L. Barolli et al. (Eds.): AINA 2022, LNNS 451, pp. 33–45, 2022.
https://doi.org/10.1007/978-3-030-99619-2_4

[8], support vector machine (SVM) [9], Classification and Regression Trees (CART) [10], naive Bayes (NB) [11], random forest (RF) [12], Logistic regression (LR) [13] and Linear discriminant analysis (LDA) [14]. We use the cross-validation K-Fold technique [15] to verify that our model is not overfitting also to estimate the performance of our machine-learning model. Then, we use the classification metrics: accuracy, precision, F1-Score, recall, Area under the Curve (AUC), and the Prediction Time in second to evaluate the performance of our model. Finally, we compare our improved method with other related binary and multi-class classification works.

2 Methodology

In our work, we use The TON_IoT datasets [16], which are datasets for multiple IoT sensors devices. They are the new generation of IoT and Industrial IoT (IIoT) datasets for evaluating the fidelity and efficiency of different cybersecurity applications based on AI. Hence, the datasets include heterogeneous data sources collected from Telemetry datasets of IoT and IIoT sensors, Operating systems datasets of Windows 7 and 10, and Ubuntu 14 and 18 TLS and Network traffic datasets.

The folder selected from the folders of TON_IoT is Train Test_IoT_datasets. This folder contains samples datasets of seven IoT sensors in CSV format; these datasets were selected to evaluate the fidelity and efficiency of new cyber security application-based AI and ML algorithms. The number of records, including normal (non-attack) and attack types. The attack types of these seven datasets are backdoor, DDoS, injection, normal, password, ransomware and XSS.

The selected datasets are characterized by seven IoT sensors: IoT Fridge, garage door, GPS tracker, MODBUS, motion light, thermostat, and weather. These IoT sensors datasets are composed of columns or features representing pieces of data that can be used for analysis. The IoT Features-Descriptions of these sensors exist in a folder named Description_stats_IoT_dataset with their Statistics records.

3 Related Works

Numerous studies have been conducted to develop efficient intrusion detection systems for IoT devices. Abdullah Alsaedi and Nour Moustafa [17] discuss some of their prior work on these seven datasets. The authors developed a measure for evaluating TON-IoT datasets. They proposed combining IoT datasets using machine learning algorithms for binary and multi-class classification such as Logistic Regression (LR) [13], Linear Discriminant Analysis (LDA) [14], K-Nearest Neighbors (KNN) [8], Random Forest (RF) [12], Classification and Regression Tree (CART) [10], Naive Bayes (NB) [11], Support Vector Machine (SVM) [9], and Long Short-Term Memory (LSTM) [18]. Mohamed Amine Ferrag, Lei Shu, Hamouda Djallel, and Kim Kwang Raymond Choo conducted another investigation [19]. The authors proposed three deep learning-based IDS models for multi-class categorization in smart agricultural IoT sensors: CNN [20], RNN [21], and DNN [22]. These efforts are related, and we may compare their methodology and results to our suggested binary and multi-class classification approaches.

3.1 Binary Classification

The author of [17] discovers that both RF and CART get fairly comparable results for binary classification due to their designs' tree-based structure. Likewise, Logistics Regression (LR) and Linear Discriminant Analysis (LDA) provide findings that are quite identical for the majority of datasets. This might be because both algorithms are linear-based and produce similar results. Naive Bayes (NB) obtains a decent result with an accuracy of 0.86 for the GPS dataset, but only an average performance for the other datasets. Support Vector Machines (SVM) values typically range between 0.63 and 0.67 for most of the datasets. In terms of training and testing timeframes, LSTM takes the longest compared to other models, followed by SVM, which takes the second longest. Both LR and LDA require the least amount of training and testing times.

3.2 Multi-class Classification

In [19], the result shows that deep learning techniques give better results than other ML strategies regarding important performance indicators, including detection rate, false alarm rate, precision, F1-score, recall, true negative rate, false accept rate, ROC curve and accuracy. This paper can conclude the performance of deep learning approaches relative to normal and various types of attacks in the TON-IoT dataset for multi-class classification. We can observe that all three deep learning techniques provide a higher true negative rate. RNN gives a higher detection ratio for three attack types: injection, password, and scanning. The CNN gives the higher detection ratio for five attacks: DDoS, Backdoor, ransomware, XSS and scanning. DNN have lower detection than RNN and CNN.

4 Evaluation Metrics for Classification Models

This section will mention the algorithms that we adopted in our improved method and the Evaluation Metrics, which is used to evaluate it.

4.1 ML Algorithms

The description of these algorithms is as shown below.

Logistic Regression. LR is a supervised learning technique. It is used for classification problems; their prediction function uses a sigmoid function to predict the output class. LR estimates the probabili based on the equation below:

$$Y = \frac{e^{(\beta_0 + \beta_1 X)}}{1 + e^{(\beta_0 + \beta_1 X)}} \tag{1}$$

Where X is the input value, Y is the predicted output, β_0 is the bias or intercept term, and β_1 is the coefficient for the single input value (X).

Random Forest. RF is a classification model that consists of individual and uncorrelated decision trees, where each tree produces a prediction. The class with the most 'votes' becomes the model's prediction. To get accurate predictions, the RF must have features with predictive input and the trees in the random forest must be uncorrelated. Our work set the number of trees in the forest to 10.

Classification and Regression Trees. Classification and Regression Trees or CART is a supervised learning technique and refers to the decision tree algorithm. It was introduced by Leo Breiman and used for classification problems. CART output is a decision tree where each fork is split in a predictor variable, and each end node contains a prediction for the outcome variable. The CART algorithm splits node into two child nodes and searches for the best homogeneity for the child nodes using the help of the Gini Index criterion, which is written by:

$$GI = 1 - \sum_{i=0}^{c} P_i^2 \tag{2}$$

Where c is the total number of classes, P_i is the probability of class I and GI is the Gini Index criterion.

K-Nearest Neighbour. The KNN classifier is a supervised learning algorithm used to solve classification problems. It works as follows: Load the dataset, then choose the K value. In our work, K value was set to 5, and the Euclidean distance between the two instances x and y was computed as follows:

$$D(x, y) = \sum_{i=1}^{n} \sqrt{(x_i - y_i)^2} \tag{3}$$

Where, x_i is the ith featured element of the instance x, y_i is the ith featured element of the instance y, n is the total number of features. D(x,y) is The Euclidean distance between the two instances x and y. The next step is adding the distance and index of the example into a collection of points and sorting the points in ascending order by the distances. After that, select the first K entries from the collection (now sorted) and retrieve the labels from the K entries. Finally, Return the mode of the K labels.

Support Vector Machines. SVM is a supervised machine-learning model that is often used for binary classification; however, it does not support multi-class classification. SVM is based on the concept of separating hyperplanes SVM transforms data using a kernel trick approach and then determines the optimal boundary between potential classes based on these transformations. Hyperplanes are decision boundaries that aid in the classification of data points. The equation of the hyperplane is defined as:

$$F(x) = \sum_{i=1}^{n} W_i * x_i + b_0 \tag{4}$$

Where F(x) is the hyperplane function, W_i a weight vector, x_i n is the number of features and b_0 a bias.

Naive Bayes. NB is a supervised machine learning model. It uses for classification problems. Furthermore, it uses the Bayes Theorem. It works by predicting the probabilities for each class. The class with the highest probability is named by maximum a posterior's probability and considered the target class. the Naive Bayes' posterior probability is defined as:

$$P(c|x) = \frac{P(x|c)P(c)}{P(x)}$$ (5)

Where P(c | x) is the posterior probability, P(x | c) is the likelihood, P(c) is the class prior probability, and P(x) is the predictor prior probability.

Linear Discriminant Analysis. LDA is a supervised classification that uses Bayes' Theorem to estimate probabilities. They make predictions based on the probability that a new input dataset belongs to each class. The class which has the highest probability is considered the output class. The probability of the data belonging to each class is defined as:

$$Pr(Y = k|X = x) = \frac{Pi_k * f_k(x)}{\sum_{i=1}^{n} Pi_k * f_k(x)}$$ (6)

Where x is the input, k is the output class. Pi_k So is the prior probability in Bayes 'Theorem. Y is the response variable, $Pr(X = x \mid Y = k)$ is the posterior probability, and $f_k(x)$ the estimated probability of x belonging to class k.

4.2 Evaluation Fundamental

Confusion Matrix. As seen in Table 1, the confusion matrix is a N × N matrix that is used to evaluate the performance of a classification model, where:

N: denotes the number of target classes, TP the rule matched and attacks present, FP denotes the rule matched and no attack present, TN is no rule matched and no attack present, and FN denotes no rule has been matched and attacks have been detected.

Table 1. Confusion matrix.

	Predicted+	Predicted-
Actual+	True positive (TP)	False-negative (FN)
Actual-	False-positive (FP)	True Negative (TN)

Accuracy, Recall, Precision, and F1 Scores, AUC-ROC are metrics used to evaluate our proposed method's performance with defined as shown in Table 2.

Table 2. Metrics equations.

Metrics names	Accuracy	Precision	Recall	F1-Score
Equation	$\frac{TP+TN}{TP+FN+FP+TN}$	$\frac{TP}{TP+FP}$	$\frac{TP}{TP+FN}$	$\frac{2*precision*recall}{precision+recall}$

The Area Under the Receiver Operating Characteristics. The AUC-ROC is one of the most critical evaluation metrics for checking classification model's performance.

5 Proposed Method and Simulation

5.1 Proposed Method Motivation

A proposed ML method for IoT devices used for preventive measuring and avoiding attacks. This enhanced method increases the accuracy of these datasets for binary and multi-class classification and reduces the processing times. Compared to other related works, we use the non-combine datasets to distinguish the type of attack in multi-class classification. We try the popular machine learning algorithms used for intrusion detection [6, 7] such as LR, LDA, KNN, CART, NB, SVM, and RF. Finally, we combine the features and use cross-validation to give a better approximation accuracy and less overfitting.

We use the default parameters for majorities of ML algorithms, which gives our proposed model the best results. For example, we select ten trees for the random forest, and for CART, the threshold is 0.5 for logistic regression. The value of k was set to 5 for the KNN algorithm, and we used 5 k fold for cross-validation, which has more efficiency and less execution time taken by the model, especially during the testing.

6 Implementation

Our proposed method is as shown in the below steps:

Step 1: Dropping duplicate columns: This operation will keep the first row and delete all other duplicates. It will increase time processing.

Step 2: Combine date and time columns into a single date-time column: This operation will reduce the number of features.

Step 3: Encoding the levels of categorical features into numeric values. This operation is essential because our models can only accept numerical variables.

Step 4: Balanced data: Whilst data is already balanced for binary classification, it is approximately 50% skewed for multi-class classification. The imbalanced data will provide biased outcomes, with the majority of non-attacks predicted. Therefore, we need to balance our data. To balance the classes, our approach will use SMOTE to oversample from the minority technique [23]. The SMOTE method will oversample all attack classes to ensure that they have the same number of rows as the most populous attack class. Under-sampling of the majority class method is not employed since it would result in the loss of a large amount of data.

Step 5: Using the classification Algorithms and cross-validation: We used the classification algorithms discussed in Sect. 4.1 and the k-fold cross-validation technique on these classification algorithms This procedure minimizes the risks against overfitting and identifies the optimal collection of hyperparameters.

As seen in Fig. 1, we utilised the standard 5-fold cross-validation and then partitioned the data into five subsets referred to as folds.

Furthermore, we train and evaluate the machine learning model using 5-fold cross-validation. This process should be repeated five times more until each K-fold serves as the test set. Finally, we use the classification measure discussed in Sect. 4.2 to evaluate the proposed model's performance. For the k-fold cross-validation procedure, we take the average of our recorded scores. These average scores represent our model's performance metrics.

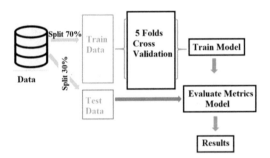

Fig. 1. 5-folds cross-validation strategy.

7   Simulation

Binary Classification. For binary classification, as seen in Tables 3, 4 and 5, our suggested model findings indicate that NB has the shortest process time but does not provide the best metric outcomes. Unlike KNN, RF and SVM provide great metric results, particularly for fridge, garage doors, and GPS trackers, but need the least processing time. CART, on the other hand, has the best metric result for all IoT sensors, especially for thermostats, motion lights, and Modbus. Additionally, it has an excellent time process.

Table 3. The evaluation metrics results for binary classification of the fridge and garage door

Datasets	Models	Accuracy	precision	recall	F1-score	AUC-ROC	Test time
Fridge sensor	LR	0.995354	0.999246	0.991197	0.995205	0.999937	0.072165
	LDA	0.938651	1.000000	0.873878	0.932685	0.996122	0.047687
	KNN	1.000000	1.000000	1.000000	1.000000	1.000000	0.145996
	CART	0.999972	1.000000	0.999942	0.999971	0.999971	0.025332
	NB	0.987126	1.000000	0.973533	0.986587	0.999887	0.014890
	SVM	1.000000	1.000000	1.000000	1.000000	1.000000	0.648848
	RF	0.999972	1.000000	0.999942	0.999971	1.000000	0.392164
Garage door	LR	1.000000	1.0	1.000000	1.000000	1.0	0.042886
	LDA	0.965508	1.0	0.961881	0.980568	1.0	0.026137
	KNN	1.000000	1.0	1.000000	1.000000	1.0	0.154180
	CART	1.000000	1.0	1.000000	1.000000	1.0	0.010799
	NB	0.998623	1.0	0.998478	0.999238	1.0	0.008986
	SVM	1.000000	1.0	1.000000	1.000000	1.0	0.423376
	RF	1.000000	1.0	1.000000	1.000000	1.0	0.114898

Table 4. The evaluation metrics results for binary classification of GPS sensor and Modbus

Datasets	Models	Accuracy	precision	recall	F1-score	AUC-ROC	Test time
GPS sensor	LR	0.998595	0.996672	0.999881	0.998274	0.999999	0.122840
	LDA	0.970392	0.992721	0.933998	0.962454	0.998453	0.052854
	KNN	1.000000	1.000000	1.000000	1.000000	1.000000	0.108358
	CART	0.999976	1.000000	0.999940	0.999970	0.999970	0.042280
	NB	0.934629	0.864219	0.995588	0.925258	0.997830	0.016914
	SVM	1.000000	1.000000	1.000000	1.000000	1.000000	2.131937
	RF	1.000000	1.000000	1.000000	1.000000	1.000000	0.703196
Modbus	LR	0.999362	0.999486	0.999074	0.999280	0.999998	0.059349
	LDA	0.992165	0.982616	1.000000	0.991231	0.999987	0.045753
	KNN	0.973717	0.972720	0.967798	0.970247	0.996619	0.059847
	CART	1.000000	1.000000	1.000000	1.000000	1.000000	0.028879
	NB	0.991710	0.981624	1.000000	0.990726	0.999993	0.011975
	SVM	0.995081	0.995873	0.993004	0.994434	0.999928	2.963834
	RF	1.000000	1.000000	1.000000	1.000000	1.000000	0.621191

Table 5. The evaluation metrics results for binary classification of light_motion, thermostat and weather datasets.

Datasets	Models	Accuracy	precision	recall	F1-score	AUC-ROC	Test time
Light Motion	LR	0.998981	0.999553	0.999042	0.999298	0.999995	0.054809
	LDA	0.949215	1.000000	0.929976	0.963710	0.999984	0.038904
	KNN	0.999861	1.000000	0.999808	0.999904	1.000000	0.182159
	CART	1.000000	1.000000	1.000000	1.000000	1.000000	0.012555
	NB	0.997361	0.998277	0.998085	0.998181	0.999979	0.012671
	SVM	0.998704	0.999170	0.999042	0.999106	0.999991	1.792915
	RF	0.999954	1.000000	0.999936	0.999968	1.000000	0.222028
Thermostat	LR	0.985000	1.0	0.956413	0.977714	0.999992	0.077189
	LDA	0.947916	1.0	0.848653	0.918123	0.999729	0.046920
	KNN	0.995434	1.0	0.986731	0.993321	0.997734	0.517705
	CART	1.000000	1.0	1.000000	1.000000	1.000000	0.030779
	NB	0.964023	1.0	0.895456	0.944838	0.999131	0.019945
	SVM	0.985526	1.0	0.957941	0.978511	0.999992	6.036044
	RF	0.999917	1.0	0.999759	0.999879	1.000000	0.502481
	LR	0.985000	1.0	0.956413	0.977714	0.999992	0.077189
Weather	LR	0.999253	0.998179	1.000000	0.999088	0.999999	0.090064
	LDA	0.982065	0.995002	0.961018	0.977711	0.999777	0.072898
	KNN	0.999952	1.000000	0.999882	0.999941	1.000000	0.111353
	CART	0.999928	0.999882	0.999941	0.999912	0.999930	0.025719
	NB	0.966757	0.924903	1.000000	0.960984	0.999964	0.017554
	SVM	1.000000	1.000000	1.000000	1.000000	1.000000	2.648572
	RF	0.999952	0.999941	0.999941	0.999941	1.000000	0.842818

Table 6. The evaluation metrics results for multi-class classification of the fridge, garage door, GPS and Modbus datasets.

Datasets	Models	F1-score	AUC-ROC	Test time
Fridge sensor	LR	0.95285714	1.000000	6.60396957
	LDA	0.94387755	0.99889665	0.0867672
	KNN	0.99982507	0.99989796	1.19085169
	CART	1.000000	1.000000	4.33804512
	NB	1.0	1.0	0.0369017

(continued)

Table 6. (*continued*)

Datasets	Models	F1-score	AUC-ROC	Test time
	SVM	nan	nan	nan
	RF	0.999971	1.000000	0.392164
Garage door	LR	0.99887755	1.000000	13.91549373
	LDA	0.99887755	1.000000	0.17120194
	KNN	1.000000	1.000000	4.54301476
	CART	1.000000	1.000000	3.19970441
	NB	1.000000	1.000000	0.03989363
	SVM	nan	nan	nan
	RF	1.000000	1.000000	3.68578005
GPS sensor	LR	0.998274	0.999999	0.122840
	LDA	0.962454	0.998453	0.052854
	KNN	1.000000	1.000000	0.108358
	CART	0.999970	0.999970	0.042280
	NB	0.925258	0.997830	0.016914
	SVM	1.000000	1.000000	2.131937
	RF	1.000000	1.000000	0.703196
Modbus	LR	0.99908163	1.000000	11.09529757
	LDA	0.99911565	1.000000	0.17060351
	KNN	0.99928571	0.99957143	0.57856083
	CART	1.00	1.00	5.97901511
	NB	1.00	1.00	0.03487062
	SVM	nan	nan	nan
	RF	1.000000	1.000000	7.25255418

Table 7. The evaluation metrics results for multi-class classification of Light_Motion, Thermostat and Weather datasets.

Datasets	Models	F1-score	AUC-ROC	Test time
Light Motion	LR	0.92576531	0.98973015	19.20963979
	LDA	0.92405612	0.99068966	0.16256261
	KNN	0.93793367	0.96485423	15.65115356
	CART	0.93971939	0.99264295	6.34001136
	NB	0.92739796	0.99326296	0.04186344

(*continued*)

Table 7. (*continued*)

Datasets	Models	F1-score	AUC-ROC	Test time
Thermostat	SVM	nan	nan	nan
	RF	0.94209184	0.99466067	8.68648648
	LR	0.99991254	1.000000	7.45151281
	LDA	0.97542274	1.000000	0.10870838
	KNN	0.99997085	0.99998299	1.74333715
	CART	1.000000	1.000000	2.71826553
	NB	1.000000	1.000000	0.03490448
Weather	SVM	nan	nan	nan
	RF	1.000000	1.000000	3.3241117
	LR	1.000000	1.000000	7.05299664
	LDA	0.98895408	0.99984098	0.18192601
	KNN	0.9990051	0.99943149	0.4717381
	CART	0.999912	0.999930	0.025719
	NB	1.000000	1.000000	0.04391789
	SVM	nan	nan	Nan
	RF	1.000000	1.000000	6.9339807

Multi-class Classification. For multi-class classification, the results are depicted in Tables 6 and 7. It is clear from the two tables that lda has the lowest metric results. Contrary to CART, KNN and Random Forest have the best metric scores, but they do not have good results for time processing. On the other hand, NB outperforms all other classifiers methods for majorities of sensors, and it has the best time processing.

8   Conclusion

This paper presents an enhanced non-combined method for analysing multiple IoT sensor datasets. This method is based on supervised machine learning techniques and a cross-validation approach. The findings indicate that CART approaches perform better in binary classification than other machine learning techniques Additionally, for multi-class classification results indicate that Naive Bayes has the highest accuracy and processing speed. As a future direction, we will focus on the security of Cloud Computing Connectivity to combat and mitigate IoT cyber threats against physical IoT sensors using artificial intelligence and machine learning. Future research might be done to enhance the performance of other recent works.

References

1. Gowda, V.D., et al.: Internet of Things: Internet revolution, impact, technology road map and features. Adv. Math. Sci. J. **9**(7), 4405–4414 (2020)
2. Yousefnezhad, N., Avleen, M., Kary, F.: Security in the product lifecycle of IoT devices: a survey. J. Netw. Comput. Appl. 102779 (2020)
3. Rondon, L.P., et al.: Survey on enterprise Internet-of-Things systems (E-IoT): a security perspective. Ad Hoc Netw. **125**, 102728 (2022)
4. Guo, G.: A Machine learning framework for intrusion detection system in IoT networks using an ensemble feature selection method. In: 2021 IEEE 12th Annual Information Technology, Electronics and Mobile Communication Conference (IEMCON), pp. 0593–05992021). https://doi.org/10.1109/IEMCON53756.2021.9623082
5. Ahmad, Z., et al.: Network intrusion detection system: A systematic study of machine learning and deep learning approaches. Trans. Emerg. Telecommun. Technol. **32**(1), e4150 (2021)
6. Kilincer, I.F., Ertam, F., Sengur, A.: Machine learning methods for cyber security intrusion detection: Datasets and comparative study. Comput. Netw. **188**, 107840 (2021). https://doi.org/10.1016/j.comnet.2021.107840
7. Sarker, I.H.: CyberLearning: effectiveness analysis of machine learning security modeling to detect cyber-anomalies and multi-attacks. Internet of Things **14**, 100393 (2021). https://doi.org/10.1016/j.iot.2021.100393
8. Ma, X., Cheng, X.: Detection and analysis of network intrusion data set based on KNN algorithm. World Sci. Res. J. **7**(6), 118–123 (2021)
9. Kaushik, R., Singh, V., Kumar, R.: Multi-class SVM based network intrusion detection with attribute selection using infinite feature selection technique. J. Discr. Math. Sci. Cryptog. **24**(8), 2137–2153 (2021)
10. Khan, M.A., et al.: Voting classifier-based intrusion detection for IoT networks. In: Saeed, F., Al-Hadhrami, T., Mohammed, E., Al-Sarem, M. (eds.) Advances on Smart and Soft Computing: Proceedings of ICACIn 2021, pp. 313–328. Springer Singapore, Singapore (2022). https://doi.org/10.1007/978-981-16-5559-3_26
11. Wester, P., Fredrik, H., Robert, L.: Anomaly-based intrusion detection using tree augmented naive bayes. In: 2021 IEEE 25th International Enterprise Distributed Object Computing Workshop (EDOCW). IEEE (2021)
12. Alshamy, R., et al.: Intrusion detection model for imbalanced dataset using SMOTE and random forest algorithm. In: International Conference on Advances in Cyber Security. Springer, Singapore (2021)
13. Noureen, S.S., et al.: Anomaly detection in the cyber-physical system using logistic regression analysis. In: 2019 IEEE Texas Power and Energy Conference (TPEC). IEEE (2019)
14. Shen, Z., Yuhao, Z., Weiying, C.: A bayesian classification intrusion detection method based on the fusion of PCA and LDA. Secur. Commun. Netw. **2019** (2019)
15. Rhohim, A., Vera, S., Muhammad Arief, N.: Denial of service traffic validation using K-fold cross-validation on software defined network. eProc. Eng. **8**(5) (2021)
16. Moustafa, N.: New generations of Internet of Things datasets for cybersecurity applications based machine learning: TON_IoT datasets. In: Proceedings of the eResearch Australasia Conference, Brisbane, Australia (2019)
17. Alsaedi, A., Moustafa, N., Tari, Z., Mahmood, A., Anwar, A.: TON_IoT telemetry dataset: a new generation dataset of IoT and IIoT for data-driven intrusion detection systems. IEEE Access **8**, 165130–165150 (2020). https://doi.org/10.1109/ACCESS.2020.3022862
18. Pooja, T.S., Purohit, S.: Evaluating neural networks using Bi-Directional LSTM for network IDS (intrusion detection systems) in cyber security. Glob. Transit. Proc. **2**(2), 448–454 (2021)

19. Ferrag, M.A., et al.: Deep learning-based intrusion detection for distributed denial of service attack in agriculture 4.0. Electronics **10**(11), 1257 (2021)
20. Khan, A., Chase, C.: Detecting attacks on IoT devices using featureless 1D-CNN. In: 2021 IEEE International Conference on Cyber Security and Resilience (CSR). IEEE (2021)
21. Park, S.H., Hyun, J.P., Young-June, C.: RNN-based prediction for network intrusion detection. In: 2020 International Conference on Artificial Intelligence in Information and Communication (ICAIIC). IEEE (2020)
22. Swarnalatha, G.: Detect and classify the unpredictable cyber-attacks by using DNN model. Turkish J. Comput. Math. Educ. (TURCOMAT) **12**(6), 74–81 (2021)
23. Gulowaty, B., Ksieniewicz, P.: SMOTE algorithm variations in balancing data streams. In: Yin, H., Camacho, D., Tino, P., Tallón-Ballesteros, A.J., Menezes, R., Allmendinger, R. (eds.) Intelligent Data Engineering and Automated Learning – IDEAL 2019: 20th International Conference, Manchester, UK, November 14–16, 2019, Proceedings, Part II, pp. 305–312. Springer International Publishing, Cham (2019). https://doi.org/10.1007/978-3-030-33617-2_31

Shape Trajectory Analysis Based on HOG Descriptor for Isolated Word Sign Language Recognition

Sana Fakhfakh[✉] and Yousra Ben Jemaa

L3S Laboratory ENIT, El Manar University Tunis, Tunis, Tunisia
{sana.fakhfakh,yousra.benjemaa}@enis.tn

Abstract. Sign language is based principally on hands gestures. To have a robust recognition system, each hand modality must be correctly presented using both motion and shape descriptors. This paper proposes an enhanced isolated word sign language recognition system based on hands trajectories analysis able to solve the most challenges such as signer's interchangeability and speed variation. In this context, Histogram of Oriented Gradients (HOG) and Support Vector Machine (SVM) are introduced. The performance of our proposed system is tested on public databases (RWTH-Boston-50 and RWTH-Boston-104) with signer-independent condition and outperformed the recent existing works.

1 Introduction

In recent years, the impact of advanced technology on deaf and hearing impaired community is important. In fact, many systems are proposed and tried to offer an automatic and efficient communication tools able to recognize the different signs used by the deaf community. These systems treat alphabets [1], isolated [2–12] or continuous signs [13, 14]. This work deals only with isolated word signs which solely one sign gesture is introduced [10]. Consequently, we focused only on features which are principally linked to hand characteristics [16].

In this context, many researches are interested in the hand shape and hand motion cues together as the best gestures characteristics processing [2, 3, 5–7, 10–12]. Many challenges are fixed in this context like signer independent (SI) conditions [2, 3, 12], signer' interchangeability [12], speed variation [12], sensor based [5, 11], dimensional reduction [6], etc. Due to the words gestures complexity and variability, the existing IWSL systems proposed to use gloves [7], Kinect camera [11], etc. as acquisition devices to facilitate features extraction steps and avoid environmental and gestures dynamic constraints which are expensive devices. Further, many techniques are introduced with vision-based systems in order to make user more freely [15] but different restrictions are fixed in this case like using only the dominant hand or working with a limited vocabulary [17]. Others propose to apply tracking techniques [18, 19] as a solution to describe words gestures dynamic as kalman and particular filters. But, the similarity of some trajectories makes the recognition step of similar word gestures very hard [11, 12] even with the application of powerful classification techniques like CNN [12]. This is due to the absence of spatial data analysis step after the extraction of gesture motion description.

Each word sign can be defined as a group of paths represented by a trajectories matrix. These trajectories take into account the gesture starting conditions which guarantee environmental robustness. In addition, a shape trajectory analysis step is proposed in order to introduce the semantic trajectories relationship. This offers the possibility to avoid temporal data and to introduce only spatial information which ensures the speed variation and signer interchangeability robustness. Finally, the new extracted features are trained and tested in order to recognize all words signs gestures. The main idea of this work is presented in Fig. 1.

Fig. 1. Proposed isolated word sign language recognition system

The paper is organized as follows: Sect. 2 describes our proposed process to extract hand motion trajectories. Shape trajectory analysis step by HOG descriptor as shape analysis technique is detailed in Sect. 3. In Sect. 4, we present the experimental and the comparative study. Finally, Sect. 5 concludes this paper.

2   Trajectories Matrix Extraction

Hand motion and shape cues are considered together in this work when extracting all words gestures trajectories as proposed in our work [10]. Therefore, two levels are proposed as presented in Fig. 2.

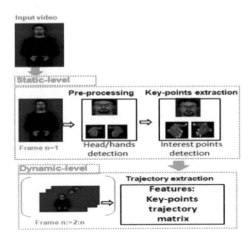

Fig. 2. Trajectories matrix extraction process [10]

2.1 The Static-Level

The static-level is applied in the first frame of word gesture video, when the hand is generally opened and the occlusion condition is totally absent, in order to extract 17 key points from the head and the two hands regions. The choice of these points is based on deaf nervous system functioning which totally related to finger flexion and spatial movement analysis. So we propose it carefully without using state of the art face and hand detector. These points are detected using a non intrusive method without acquisition devices and environmental conditions like background, clothing, dominant hand, etc. This level occupied two principal stages:

- Region of Interest detection (ROI): occupied to extract head and hands regions. In this case, three steps are introduced. First, skin color detection is done based on color space YCbCr which is robust to lighting conditions. Second, closing and opening techniques are introduced to eliminate the obtained small regions after skin color extraction step. Only the three biggest segmented regions is reserved which are linked to head and hands regions. Viola and Jones technique is applied also in order to identify face region and grantee a performing ROI localization stage. The two rest regions are considered as hands objects. Also, a hand detection step is applied as proposed in our work also [1] in order to take in to consideration to short-sleeved shirt condition after ROI detection. This grantee a good key point extraction step.
- Key points extraction: 17 interest points are proposed which are:

  - The three gravity centers related to head, left and right hands positions. The detection of gravity center step is based on the Eq. (1).

$$C_x = \frac{\sum_{i=1}^{n} X_i}{n} \qquad (1)$$

    Where n presents the number of pixels X_i presented in the extracted skin region.
  - The ten fingers tip positions
  - The four points from wrist line hand extremity related to left and right hands. These 14 proposed points related to hand fingers tip and hand wrist line extremities are extracted by searching the deepest points. In this case we propose a hand region convexity analysis. First, a hand convex hull must be calculated in order to found all convexity defects, which defined as the area ratio related of gesture contour and convex hull based on the Eq. (2):

$$\delta = \frac{\text{Contour Area}}{\text{Hull Area}} \qquad (2)$$

    We propose to extract all convexity defects using Open CV function, which provides an easily starting, ending and deepest points detection. Each deepest point D_p existing in each detected convexity defect presents the starting point S_p in the next convex defect and the ending point E_p in the current convex defect calculated by the Eq. (3):

$$Dp = \frac{Sp + Ep}{2}. \qquad (3)$$

In this case, the 7 proposed points are totally linked to the 7 deepest detected points due to each gesture starting having generally an open hand.

2.2   The Dynamic-Level

The dynamic-level is occupied to extract all trajectories linked to the 17 proposed key points using particular filter as tracking technique. Finally a key point trajectory matrix (KPTM) is birthed related to each word gesture as features vector. In order to abolish redundancy and decrease time processing, an elimination step of some consecutive frames is also introduced. All details about the trajectories matrix extraction are presented in our work [10]. Figure 3 presents examples of extracted trajectories.

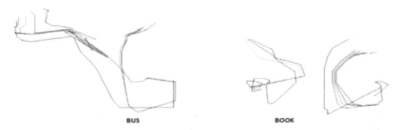

Fig. 3.   Examples of extracted trajectories

3   Shape Trajectory Analysis

In order to characterize shape trajectories, many descriptors can be used such as: SIFT [20], Local Binary Patterns (LBP) [21], and HOG [22, 23]. HOG is widely used and proves its performances in many shape recognition context [24]. HOG was proposed by Dalal and Triggs [25] in order to introduce principally fine and discriminate data based on gradient features representation. Different principal steps were proposed in this context:

1) Computing the intensity gradient magnitude and the orientation for all image pixels.
2) Dividing images in different grid of cells.
3) Computing the histogram of gradients for each obtained cell.
4) Combining all obtained HOGs.

So, the principal idea related to HOG descriptor is that the local shape can be described by the distribution of the gradient intensity or direction of the contours. So, trajectory signature image is divided into 8 × 8 cells. For each cell, HOG features are calculated. Finally, a combination of all extracted features is introduced as final descriptor. All proposed HOG details are presented in Fig. 4.

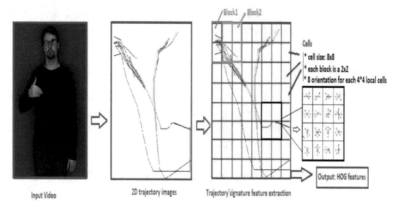

Fig. 4. HOG features for trajectory' signature.

4 Experimental Results

4.1 Dataset

In this paper, we propose to evaluate our proposed approach using RWTH-Boston-50[1] and RWTH-Boston-104[2]. The two proposed challenging databases are collected by using one camera placed in the front and three others in the laterals, with different sign execution speeds and sizes, in order to treat the signer's interchangeability and the speed variation conditions. Two females and one male singer with different clothes are used to this task and partitioned in the test and train data taking into consideration the signer independent condition.

4.2 Test Protocol

In this paper, we use RWTH-Boston-50 for the test and RWTH-Boston-104 training corpus for the train. In addition, solely 15 words signs, which introduce both right and left hands when applying gesture, are introduced. The chosen words are: book, toy, ariv1, hmwrk, bdown, give, box, have, house, write, people, friend, future, leg and movie. In addition, correctly classified recognition CCR is introduced as classification metric. In this case, an extraction of the proposed 15 word gestures from the RWTH-Boston-104 sentences training corpus becomes a necessity. The extraction step is based on the RWTH-Boston-104 ground truth data which introduces each start and end sign position.

To introduce our proposed approach based principally on shape trajectory analysis step, we propose to augment the shape trajectories variations possibility in order to have more powerful recognition system. We propose 3 principal transformations steps. First, we apply randomly for each image sign trajectories 7 transformations based on adding

---

[1] rwth-aachen Homepage, https://www-i6.informatik.rwthaachen.de/aslr/database-rwth-boston-50.php.

[2] rwth-aachen Homepage, https://www-i6.informatik.rwthaachen.de/aslr/database-rwth-boston-104.php.

different geometries variations like scale, rotation, brightness, zoom, cropping random region, adding noise, contrast, etc. Second, we propose to introduce inversion image step in order to take into account when signs are applied using right or left hand. Third, we introduce the first and second presented steps with and without RRF step to have in the total 240 images for each word sign gesture. Finally, 3600 images, both train and test data are generated and classified using SVM classifier.

4.3   Results

In this section, we present an evaluation step for our proposed approach based on accuracy metrics and SVM classifier. Table 1 presents the obtained results applied on RWTH Boston-104 and RWTH-Boston-50 databases.

Table 1. Our proposed approach result.

Approach	CCR
Proposed descriptor	92.27%

As presented in Table 1, a 92.27% CCR value is obtained. This interesting result proves the importance of introducing a spatial data analysis step using HOG for all trajectories describing each isolated sign gesture. Also, to more validate our proposed approach we introduce a cross-validation stage by splitting our proposed dataset randomly into five folds. A 91.72% (± 3) classification accuracy is obtained. This proves the innovation incorporated by our paper coming with the use of HOG descriptor to represent the trajectory pattern of hand's interest point.

4.4   Comparative Study with Existing Work

To highlight the robustness of our proposed approach, we compare it to existing one using deep learning as shape analysis technique [12]. Like our proposed approach, this work takes into account the motion and the shape of the hand to describe gestures. The most important that this work introduces the most recent techniques based on deep learning concepts. Two convolutions Neural Network (CNN) are applied simultaneous to the left and the right combined to Particle filter technique in order to grantee a good hand tracking phase. Thus, as vector characteristics, the hand energy is calculated and applied. In addition, the same test protocol presented in Sect. 4.2 is introduced in work [12]. The overview of [12] is presented in Fig. 5.

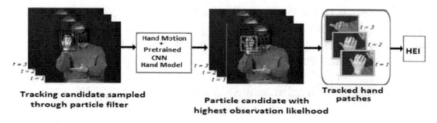

Fig. 5. Overview of the work [12]

Results are presented in Table 2 showing an amelioration of 3% with our approach.

Table 2. Performance comparison for the two approaches: RWTH database

Approach	CCR
Our approach	92.27%
[12]	89.33%

This proves widely our system robustness to:

- All geometry variations as translation, rotation and scale.
- Environment variations as: background, contrast, lighting, colors, user clothes, noise, etc.
- Sign execution speed and signers' interchangeability conditions.
- In addition, this amelioration proves:
- The pertinence of the proposed trajectories which powerfully describe the hand motion and pose in each time.
- The performance of introducing HOG descriptor as feature for shape trajectory analysis process.
- The robustness of spatial data representation step which proves the existence of a gesture' trajectories relationship.

5   Conclusion and Future Works

In this paper, we propose a new approach for isolated word sign language recognition based on shape trajectory analysis. We propose to define all extracted trajectories as a signature related to each word sign. So, a trajectory' signature analysis step is introduced based on HOG descriptor. Also, shape trajectory analysis step is proved when using CNN concept in our recent accepted paper [26]. This offers the possibility to eliminate all time concept related to the dynamic data and just introduces the spatial data in trajectories characterization step. The obtained results prove our idea performance and our system robustness. In addition, our approach outperforms the existing work [12]. In the future, other methods can be applied to more improve shape trajectory analysis step like introducing geodesic distance as elastic metric and taking advantage of the geometry space concept [27].

References

1. Fakhfakh, S., Jemaa, Y.B.: Hand and wrist localization approach for features extraction in Arabic sign language recognition. In: IEEE/ACS 14th International Conference on Computer Systems and Applications (AICCSA), pp. 774–780. Hammaet-Tunisia (2017)
2. Naftel, A., Khalid, S.: Motion trajectory learning in the dft coefficient feature space. In: Fourth IEEE International Conference on Computer Vision Systems, p. 47 (2006)
3. Bhuyan, M.K., Bora, P.K., Ghosh, D.: Trajectory guided recognition of hand gestures having only global motions. In: World Academy of Science Engineering and Technology, pp. 753–764 (2008)
4. Wang, H., Stefan, A., Moradi, S., Athitsos, V., Neidle, C., Kamangar, F.: A system for large vocabulary sign search. In: Kutulakos, K.N. (ed.) ECCV 2010. LNCS, vol. 6553, pp. 342–353. Springer, Heidelberg (2010). https://doi.org/10.1007/978-3-642-35749-7_27
5. Boulares, M., Jemni, M.: 3d motion trajectory analysis approach to improve sign language 3d-based content recognition. In: Procedia Computer Science, vol. 13, pp. 133–143 (2012)
6. Lin, W.-Y., Hsieh, C.-Y.: Kernel-based representation for 2d/3d motion trajectory retrieval and classification. In: Pattern Recognition, vol. 46, pp. 662–670 (2013)
7. Mohandes, M., Deriche, M.: Arabic sign language recognition by decisions fusion using dempster-shafer theory of evidence. In: Computing, Communications and IT Applications Conference, pp. 90–94 (2013)
8. Geng, L., Ma, X., Wang, H., Gu, J., Li, Y.: Chinese sign language recognition with 3d hand motion trajectories and depth images. In: Intelligent Control and Automation (WCICA)11th World Congress on IEEE, pp. 1457–1461 (2014)
9. Pu, J., Zhou, W., Zhang, J., Li, H.: Sign language recognition based on trajectory modeling with hmms. In: Tian, Q., Sebe, N., Qi, G.-J., Huet, B., Hong, R., Liu, X. (eds.) MMM 2016. LNCS, vol. 9516, pp. 686–697. Springer, Cham (2016). https://doi.org/10.1007/978-3-319-27671-7_58
10. Fakhfakh, S., BenJemaa, Y.: Gesture recognition system for isolated word sign language based on key-point trajectory matrix. Comput. Sistem. **22**(4), 1415–1430 (2018)
11. Sidig, A.I., Mahmoud, S.A.: Trajectory based Arabic sign language recognition. Int. J. Adv. Comput. Sci. Appl. **9**(4) (2018)
12. Lim, K.M., Tan, A.W.C., Lee, C.P., Tan, S.C.: Isolated sign language recognition using convolutional neural network hand modelling and hand energy image. In: Multimedia Tools and Applications, pp. 1–28 (2019)
13. Holden, E.J., Lee, G., Owens, R.: Automatic recognition of colloquial Australian sign language. In: Seventh IEEE Workshops on Applications of Computer Vision (WACV/MOTION 2005), vol. 1, pp. 183–188 (2005)
14. Starner, T., Pentland, A.: Real-time American sign language recognition from video using hidden Markov models. In: Proceedings of International Symposium on Computer Vision - ISCV, pp. 265–270 (1995)
15. Cooper, H., Holt, B., Bowden, R.: Sign language recognition. In: Visual Analysis of Humans: Looking at People, pp. 539–562 (2011)
16. Agrawal, S.C., Jalal, A.S., Tripathi, R.K.: A survey on manual and non-manual sign language recognition for isolated and continuous sign. Int. J. Appl. Pattern Recogn. **3**, 99–134 (2016)
17. Zaki, M.M., Shaheen, S.I.: Sign language recognition using a combination of new vision based features. Pattern Recogn. Lett. **32**, 572–577 (2011)
18. Stenger, B.: Template-based hand pose recognition using multiple cues. In:Computer Vision – ACCV, pp. 551–560 (2006)
19. Roussos, A., Theodorakis, S., Pitsikalis, V., Maragos, P.: Hand tracking and affine shape-appearance handshape sub-units in continuous sign language recognition. In: Kutulakos,

K.N. (ed.) ECCV 2010. LNCS, vol. 6553, pp. 258–272. Springer, Heidelberg (2012). https://doi.org/10.1007/978-3-642-35749-7_20

20. Lowe, D.G.: Distinctive image features from scale-invariant keypoints. Int. J. Comput. Vis. **60**(2), 91–110 (2004)
21. Wang, L., Wang, R., Kong, D., Yin, B.: Similarity assessment model for Chinese sign language videos. IEEE Trans. Multim. **16**(3), 751–761 (2014)
22. Koller, O., Forster, J., Ney, H.: Continuous sign language recognition: towards large vocabulary statistical recognition systems handling multiple signers. Comput. Vis. Image Understand. **141**, 108–125 (2015)
23. Dalal, N., Triggs, B.: Histograms of oriented gradients for human detection. In: IEEE Computer Society Conference on Computer Vision and Pattern Recognition (CVPR 2005), vol. 1, pp. 886–893 (2005)
24. Melaugh, R., Siddique, N., Coleman, S.A., Pratheepan, Y.: Feature selection, reduction and classifiers using histogram of oriented gradients: how important is feature selection (2018)
25. Dalal, N., Triggs, B.: Histograms of oriented gradients for human detection. In: IEEE Computer Society Conference on Computer Vision and Pattern Recognition (CVPR 2005) (2005)
26. Fakhfakh, S., Jemaa, Y.B.: Deep learning shape trajectories for isolated word sign language recognition, accepted. Int. Arab J. Inf. Technol. **19**(4) (2022)
27. Ben Tanfous, A., Drira, H., Ben Amor, B.: Coding Kendall's shape trajectories for 3D action recognition. In: Proceedings of IEEE Computer Vision and Pattern Recognition, pp. 2840–2849 (2018)

How Australians Are Coping with the Longest Restrictions: An Exploratory Analysis of Emotion and Sentiment from Tweets

Kawser Irom Rushee[1(✉)], Md Shamsur Rahim[2], Andrew Levula[2], and Mehregan Mahdavi[2]

[1] American International University-Bangladesh, Dhaka, Bangladesh
rushee@aiub.edu
[2] Sydney International School of Technology and Commerce, Sydney, Australia
{shamsur.r,andrew.l,mehregan.m}@sistc.nsw.edu.au

Abstract. Australia is one of the nations having the most extended international border closure and lockdown because of the COVID-19 pandemic. This provides a unique opportunity to explore and understand the emotion and sentiment in the tweets posted by Australians. To utilise this opportunity, tweets from Twitter were collected since the beginning of the pandemic till the 30th of October 2021. Search queries were generated to get COVID-19 and lockdown-related tweets that returned any tweets with the relevant tags. After collecting the tweets, several text pre-processing techniques were applied. Later, sentiment analysis and emotion detection were performed on the pre-processed tweets. Lastly, results were aggregated together and the findings were discussed. Findings from this study suggested that sentiments and emotions fluctuated depending on time and region. The understanding of people's sentiment and emotion towards lockdown presented in this paper may help the policymakers in decision making in future especially with the new variant (Omicron) of COVID-19.

1 Introduction

The impact of COVID-19 immensely affected people worldwide. People around the world are physically affected with the disease and a large number of people died and are still dying. The fear and precocious mentality of people made them isolate to avoid contracting the virus. One of the most common initiatives to keep people safe from COVID-19 was the lockdown, which affected people emotionally, financially and in many other ways [1]. Since the beginning of the pandemic this social distancing, isolation affected people's mental health condition and many people got insomnia, anxiety and depression disorders [1]. Levula et al. [22] highlighted that a lack of social engagement and extended periods of social isolation decreases ones mental health and the effects differ significantly across different life stages. There is hardly any country which is not affected by COVID-19. Australia is one of the countries that had the most extended lockdown period and it affected people mentally and financially. Moreover, many anti-lockdown movements were also seen during the lockdown in Australia. Protests are not the only means for expressing people's opinions, emotions, and feedback regarding the lockdown; social media have become a common platform to express people's feelings.

Twitter is one of the most popular social media platforms. People can share their thoughts, good news, bad news, or any kind of emotions through tweets with a lot of people by a single tweet. People use twitter to get connected with their family, friends, colleagues, and also followers. People tweet to state their thoughts as text to share with others. They also post photos, videos, and website links. Moreover. people can also search for any topic available to access in twitter. During this pandemic, many people use twitter to express their opinions and emotions which makes twitter a gold mine to extract novel insights from the tweets to understand human nature during the pandemic.

The volume of tweets increased during lockdown as people turned to social media as an outlet to express their thoughts, situations and emotional states. They posted for seeking help, prayer, giving information, and for other forms of social and emotional support. Many insights might be hidden in tweets which are not known to people. Analysing tweets related to the pandemic might give some new insights or establish any hypothesis truly related to health conditions and thus analysis of tweets might benefit policy and decision makers.

Since Australia was one of the countries that had the most extended lockdown period. Australia is still disconnected from the rest of the world regarding travelling, and some states just got out of prolonged lockdown such as Melbourne at the time of writing this paper. Besides, different states and territories in Australia have faced different periods and different levels of lockdown. As a lot of people use twitter in Australia, it makes an interesting case to study the tweets originated from different Australian states and territories to understand how people's sentiment and emotions fluctuate over the prolonged travel restrictions and lockdowns.

To analyse the tweets originated from Australia, at first tweepy API was used to collect tweets that contains specific keywords since the beginning of the pandemic (20th March 2020) till 30th October 2021. After collecting ~ 30k tweets, text processing was carried out to make the tweets analytic ready. Once the tweets were ready to be analysed, we applied sentiment analysis and emotion detection techniques. Later, the results were visualised and interpreted using different visualisation tools.

This paper presented a comprehensive and exploratory analysis of the sentiments and emotions from the tweets related to lockdown in Australia. Although, there are existing studies that address similar issues, to the best of authors' knowledge, no studies investigate the change in sentiment and emotion depending on the originating states and territories of a specific country or region and also not focused specifically on lockdown in covid. As the states and territories in Australia went through different periods of lockdowns, therefore, there is a research gap to understand how people expressed themselves in social media like Twitter in different regions within a country. This paper analysed the long-term duration data, data from the beginning of the covid and lockdown situation to the recent lockdown, which is not unfold in existing research. This paper addressed this research gap and explored how people's sentiment and emotions evolved over time depending of the regions.

This paper has been structured as follows: Sect. 2 presents the critical analysis of the related works, Sect. 3 discusses the methodology of the study, Sect. 4 describes results and findings and finally Sect. 5 concludes the paper.

2   Related Works

A good number of studies have been conducted to analyse the sentiment and detect emotions from tweets. Depending on the targeted geography, these studies can be divided into two categories: (1) country-specific, and (2) global. The studies under the country-specific category aimed to analyse the sentiment and emotion from tweets for a specific country. On the other hand, studies under the global category, collected tweets around the world and then analysed the tweets for sentiment analysis and emotion detection. A critical analysis of the relevant research is presented in this section.

Under the category of country-specific, research was carried out to examine the sentiment and emotion from tweets from a specific country. Vibha et al. [2] performed sentiment analysis and emotion detection on the tweets collected from India. Pokharel [3] performed a similar study for Nepal. Pastor [4] performed only sentiment analysis on the tweets originated from Luzon, Philippines. Lastly, Zhou et al. [5] performed sentiment analysis on the tweets generated from New South Wales, Australia. One of the common limitations of theses studies include not analysing the different regions separately. In addition, these studies only analysed the tweets generated in 2020, they have not included the tweets till the end of 2021 (Table 1).

Table 1. Critical analysis of the studies related to sentiment analysis and emotion detection from tweets

Reference	Sentiment?	Emotion?	Targeted Geography	Size of the data set	Data set duration
[2]	Yes	Yes	India	24000	25-03-2020 to 28-03-2020
[3]	Yes	Yes	Nepal	615	21-05-2020 to 31-05-2020
[6]	Yes	Yes	Global	50,200,500	15th April 2020, from 24th April 2020 to 25th April 2020, and from 13th May 2020 to 14th May 2020
[7]	Yes	N/A	Global	92,646 and 85,513	N/A
[8]	Yes	Yes	Global	20,325,929	28 January 2020 to 9 April 2020
[4]	Yes	N/A	Luzon, Philippines	~2500	from the date of lockdown in the Luzon area until the 3rd week
[5]	Yes	N/A	NSW, Australia	94 million	1st January 2020 to 22nd May 2020

The studies under the global category collected the tweets from around the world and then performed sentiment analysis and/or emotion detection. Nemes & Kiss [6] analysed twitter data collected globally for sentiment analysis and emotion detection using RNN. In a similar study, Lwin et al. [8] used CrystalFeel algorithm for emotion detection and performed sentiment analysis on a global scale. Bhat et al. [7] performed sentiment analysis only on tweets collected throughout the world. The main limitation of the studies is they have not analysed different regions separately as emotion and sentiment could vary depending on the location.

3 Methodology

The methodology followed in this study to perform sentiment analysis, emotion detection and topic detection can be divided into three consecutive steps: (1) data collection and preparation; (2) experiment, and (3) result analysis. In the first step, tweets were collected from twitter based on keywords and locations, and different preprocessing tasks were carried out. In step 2, experiment was conducted to determine sentiment, emotion, and topics. Finally, in step 3, results generated from the previous step were analysed. Figure 1 illustrates the three steps of the methodology, and they are discussed further in the following subsections.

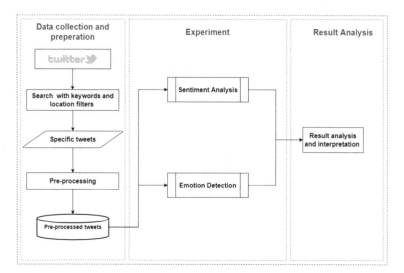

Fig. 1. Steps in the methodology of the study.

3.1 Data Collection and Preparation

Dataset for the analysis was from primary data source and qualitative data. Tweets were collected from Twitter by using search queries to get COVID-19, lockdown-related tweets from different states of Australia. The aim was to collect tweets, time and location from the beginning of the pandemic, 20th March 2020 till 30th October 2021, that returned tweets relevant to the search query tags. Tweets were filtered at the time of data collection as we ignored retweets and collected tweets for only English Language. Moreover, only those tweets were collected which have the location of any place in Australia only. Dataset was made with the relevant tweets, time of the tweet and the location. Dataset was collected by using Tweepy API [9] and the total instance number of the dataset was 29810 for the exploratory analysis.

Text Pre-processing

Text pre-processing techniques were applied to clean tweet text using re library which includes regular expression operations. Any type of punctuation (comma, apostrophe, quotation, question, exclamation, brackets, braces,@,https//, hashtag, parenthesis, dash, hyphen, ellipsis, colon, semicolon, and underscore) were removed using regular expression and mapped with the tweets to get cleaned text of that tweet. Texts were also converted into lower case and added in the dataset.

Tokenization

After performing initial pre-processing, the pre-processed tweet text was used for tokenization. NLTK library [10] was used and tokenization was done for each pre-processed tweet using word_tokenize method. After tokenization individual words were found for each tweet and the tokenization attribute was added to our dataset.

Stopword Removal

Words that are commonly used in the language but do not have much impact are called stopwords (the, in, at, that, which, on, etc.). After tokenization the tokenized words for each tweet were processed for stopword removal. Wordcloud libraray's STOPWORD class was used for this purpose and after the process an attribute was added to our dataset named nonstop as it consisted of words without stopwords. It helped us to analyze our dataset with meaningful words.

Stemmer

Stemmer was used to find the most meaningful form of a word using the attribute added by stopwords removal. Words which were denoting the same meaning but in different forms were made into one form after the process of stemming, like "Testing", "Tester" these words converted into "Test", etc. The NLTK libarary's [10] PorterStemmer() class was imported to use the stem method for stemming words in each tweet. After the stemming process stemmed attribute was added to our dataset containing stemmed word for each tweet.

3.2 Experiment

Sentiment Analysis

Sentiment analysis is a subpart of NLP (Natural Language Processing) that retrieves polarity, subjectivity, and affective state of any given text. In this paper sentiment analysis was done on the pre-processed tweet texts. Tweet text data were analysed using the TextBlob class of textblob library [11] and sentiment polarity and subjectivity were calculated. Then SentimentIntensityAnalyzer class of nltk.sentiment.vader library and vader_lexicon package was used to identify each tweet's sentiment using the polarity_scores method. Vader (Valence Aware Dictionary for Sentiment Reasoning)) library used both qualitative and quantitative methods to produce a high-standard sentiment lexicon that is especially conformed to social media platform's texts [12]. A list of lexical features (e.g., words) labelled according to their semantic as either positive or negative is called sentiment lexicon [13, 14]. In Vader lexical features were combined with consideration for five generalizable rules that assimilate grammatical and syntactical conventions that used in human expression or strengthen sentiment intensity. Moreover, Vader retained and improved the benefits of traditional sentiment lexicons without large-scale training and also suitable for texts for social media platforms like Twitter [12]. Sentiment value calculated as positive, negative or neutral. After sentiment analysis the dataset was evaluated and it consisted of seven attributes tweet, time of tweet, location, cleaned text, sentiment, polarity and subjectivity.

Emotion Detection

Emotion prediction was done using emotion classification of Ekman's six basic emotions (joy, fear, sadness, surprise, disgust, anger) [15]. EmotionPredictor class [15] was used to classify each tweet's emotion. Emotion detection was done by trained recurrent neural network (RNN) models, which can predict emotions from English tweets. The EmotionPredictor class used word and character-based recurrent and convolutional neural network. It is suitable for Twitter texts and performed better over bag of words and latent sematic indexing models [15]. Emotion was detected for all the pre-processed tweets and after that process our dataset got another attribute- tweet emotion.

4 Results and Discussion

Once the sentiments and emotions from the tweets were identified, results were analysed to understand the findings. To visualise the results and interpret them, we used two visualisation software: (1) Microsoft Power BI, and (2) Tableau. Firstly, we plotted the number of different sentiments and emotions. Figures 2 and 3 depict the number of different sentiments and emotions found after analysing the tweets.

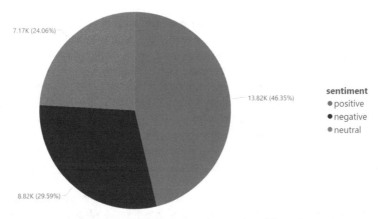

Fig. 2. The number of different sentiments found from the tweets.

From Fig. 2, we can see almost half (46.35%) of the total tweets were positive in nature, followed by negative (29.59%) and neutral (24.06%) sentiment. This finding indicates that Australians were mostly positive about the lockdown during the period of this pandemic. However, from Fig. 1, it can be also observed that negative sentiment is the second most sentiment (8.82K). This indicates that even though most of the tweets are positive, there a significant number of negative tweets too regarding the lockdown. After calculating the number of different types of sentiment, next we calculated the number of each emotion.

In this study, we extracted 6 emotions (i.e., joy, fear, surprise, sadness, anger, disgust) from the tweets related to the lockdowns in Australia. Figure 3 represents the number of different emotion types.

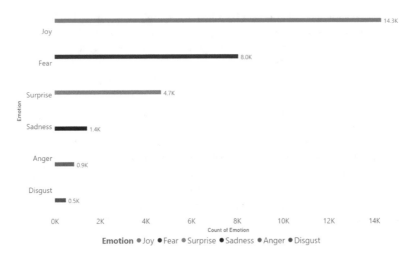

Fig. 3. The number of different emotions found from the tweets.

From Fig. 3, we can see joy is the most common sentiment (14.3K) extracted from the tweets, followed by fear (8.0K), surprise (4.7K), sadness (1.4K), anger (0.9K) and disgust (0.5K). This finding is aligned with the previous finding from the number of sentiments as positive sentiment and emotion of joy are the highest in their category; followed by negative sentiment and emotion of fear. Another important finding is there are more tweets that express fear compared to anger (angry in later part of the lockdown). This finding highlighted that people were mainly afraid of the lockdown rather that angry. However, it would be interesting to observe whether the levels of fear and anger changed over time depending on states or territories. To the authors' knowledge, no previous studies have performed sentiment and emotion analysis for different regions within a country.

As the sentiments and emotions of the tweets may fluctuate over time and it may differ depending on the originating states or territories, we plotted the count of each sentiment and emotion for each state and territory. After plotting the sentiment and emotion, we looked for any spike, in particular sentiment and emotion, and drilled in further to find out the exact date range and tried to understand the reason behind the spike by matching states and territories government announcements from the official website of Parliament of Australia [16] and Wikipedia. In addition, for sentiment, we also observed whether a particular type of sentiment crossed another type of sentiment or not.

After plotting the emotion and sentiment for the tweets originated from New South Wales (NSW), from Fig. 4, we can see that the types of emotion and sentiment varies depending on the time. In terms of emotion, the first peak of emotion was observed on 23 March 2020 when the NSW Premier announced new restrictions that temporarily shut down non-essential services and businesses. The top three emotions were joy, fear, and surprise in that day. On the other hand, for sentiment, number of neutral tweets were higher than positive tweets. Between 31 August 2020 and 6 September 2020, it can be observed that the number of tweets containing negative sentiment was higher than positive sentiment for the first time since the beginning of the pandemic. Based on the announcements from NSW Health [17], it can be observed that NSW Health urged people to wear mask while on public transport which might have caused this increase in negative sentiment. Another peak between 14–20th December 2020 was occurred because of the stay-at-home directions reintroduced in the northern beaches' suburbs in NSW [18].

Since 18 June 2021, a new variant of COVID-19 (delta) cluster was detected and on 25 June 2021, four Sydney local government areas (LGAs) were under lockdown. This incident caused several spikes in the next couple of weeks. For the week of 16 August 2021–22 August 2021, it can be observed that the number of tweets with negative sentiment outnumbered positive sentiment tweets. It could be caused from reducing the 10-km rule to allowing people within 5 kms from their home for shopping, exercise and outdoor recreation [19]. In terms of the emotions of the tweets, fear was the dominating emotion through the pandemic for NSW.

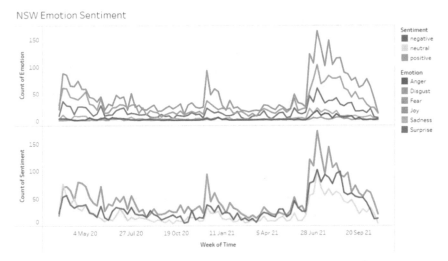

Fig. 4. Count of emotion and sentiment from the tweets for New South Wales

Victoria was the hardest hit state among all states in Australia. Since the beginning of the pandemic, there were six incidents of lockdown. From Fig. 5, we can see that, the first peak (30 March 20206 April 2020) happened when the first lockdown was put in place from March 30 to May 12, 2020 [20]. The second peak occurred between 6 July 2020 and 12 July 2020 when the second lockdown was introduced (July 8 to October 27, 2020). These observations clearly indicates the relationship between the time of lockdown and change in the number of sentiments and emotions. However, one interesting pattern can be observed for the week of 16 August 2021 to 22 August 2021. For this week, the difference between the number of positive and negative sentiment from tweets was the highest for any state, in any time. At that time, Melbourne was going through another lockdown that lasted from August 5–October 25, 2021. In terms of the emotions of the tweets, fear was the dominating emotion through the pandemic for Victoria.

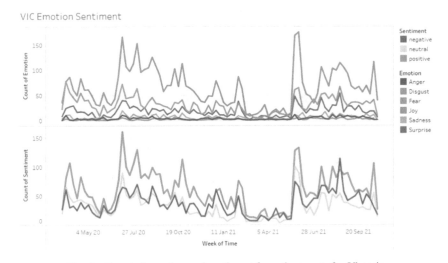

Fig. 5. Count of emotion and sentiment from the tweets for Victoria

In the case of Queensland and South Australia, both states showed similar patterns of peaks that are associated with lockdowns. For Queensland, from Fig. 6(a), more peaks can be observed compared to South Australia as it went through a couple of snap lockdowns. Otherwise, all the observations are similar to NSW and VIC except that, for South Australia the sentiment towards lockdown remained negative for a couple of weeks towards the end of the pandemic.

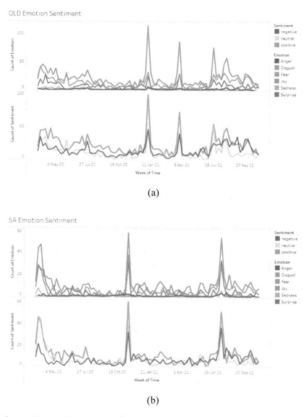

Fig. 6. Count of emotion and sentiment from the tweets for (a) Queensland; (b) South Australia

For Northern Territory, and Tasmania, the number of tweets related to lockdown was very low as these two went through 9 and 0 day of lockdown respectively [21]. Figure 7(a) and (b) represent the count of emotion and sentiment from tweets for Northern Territory, and Tasmania. From Fig. 7(c), we can see that, there is a sudden spike in the negative sentiment from tweets for ACT. After drilling down to the tweets, we observed that the tweets were related to the possibility of a super spreading event from a rally of Donald Trump Jr, the ex-president of the USA. This finding suggests, even though the tweets could have originated from a specific region, it does not mean it would be related to the region.

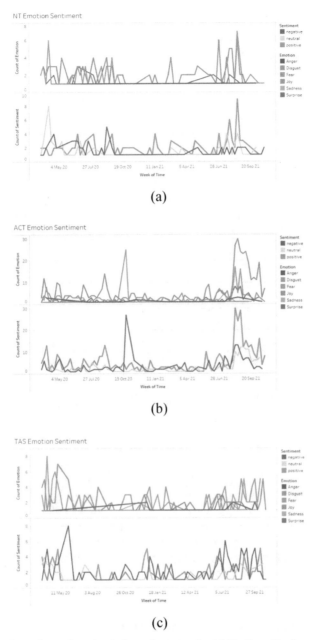

Fig. 7. Count of emotion and sentiment from the tweets for (a) Northern Territory; (b) Australian Capital Territory; and (c) Tasmania.

Finally, we generated a word cloud from the cleaned text to understand what are the key terms that people are using to communicate via Twitter regarding the lockdowns. From Fig. 8, we can see that Australia, lockdown, covid, Melbourne, Sydney, quarantine, travel, work, school, were some of the widely used words.

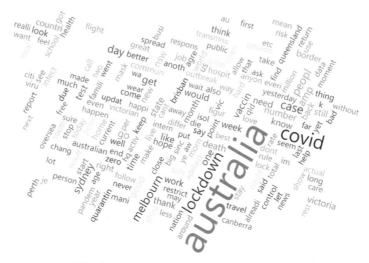

Fig. 8. Word cloud generated from the tweets.

Based on the above findings, we can conclude that there exists a direct relationship between the level of sentiments and emotions and nature of the events that occurred at that time. However, it is also important to observe the different emotional states and sentiment that were expressed across the different states or territories during the pandemic. One of the key findings the level of fear that was expressed by people during the lockdown periods. This has significant implications on the mental health of citizens of these states and territories and it raises policies concerns because of the extended lockdowns and lack of social engagement with members of their communities. The long-term implications of these lockdowns across different life stages will also need to be evaluated and considered in future studies to determine if there are any significant differences and to identify possible mitigation and policy strategies.

5   Conclusion

Australia is one of the countries with the toughest and longest restrictions related to COVID-19. Different states and territories in Australia have faced different period and level of lockdowns. To reveal how emotion and sentiment fluctuates over time depending on the regions regarding lockdown, tweets are required to be analysed separately depending on the origin. In this study, we collected tweets since 20th March 2020 till 30 October 2021 related to COVID-19 and lockdowns with location information. After collecting the tweets, pre-processing techniques were applied to make the tweets analytics ready. Once the tweets were ready, sentiment analysis and emotion detection were performed on the pro-processed tweets. Later, the results were visualised and interpreted to understand the reasons behind the interesting patterns. The study concluded that the sentiment and emotion of tweets remained favourable in most of the times, however, there were occurrences when the sentiment and emotion were unfavourable depending on the state and region. This finding indicates that location has an impact on the overall sentiments and emotions.

Further studies can be conducted to understand other covid related phenomena (e.g., depression) depending on the location and how this would differ across different life stages. In addition, topic detection can be performed to understand different topics nationwide.

References

1. Marroquín, B., Vine, V., Morgan, R.: Mental health during the COVID-19 pandemic: effects of stay-at-home policies, social distancing behavior, and social resources. Psychiatry Res. **293**, 113419 (2020)
2. Barkur, G., Vibha, G.B.K.: Sentiment analysis of nationwide lockdown due to COVID 19 outbreak: evidence from India. Asian J. Psychiatry **51**, 102089 (2020)
3. Pokharel, B.P.: Twitter sentiment analysis during covid-19 outbreak in Nepal. SSRN 3624719 (2020)
4. Pastor, C.K.: Sentiment analysis of Filipinos and effects of extreme community quarantine due to coronavirus (COVID-19) pandemic. SSRN 3574385 (2020)
5. Zhou, J., Yang, S., Xiao, C., Chen, F.: Examination of community sentiment dynamics due to COVID-19 pandemic: a case study from a state in Australia. SN Comput. Sci. **2**, 1–11 (2021)
6. Nemes, L., Kiss, A.: Social media sentiment analysis based on COVID-19. J. Inf. Telecommun. **5**, 1–15 (2021)
7. Bhat, M., Qadri, M., Noor-ul-Asrar Beg, M.K., Ahanger, N., Agarwal, B.: Sentiment analysis of social media response on the Covid19 outbreak. Brain Behav. Immun. **87**, 136 (2020)
8. Lwin, M.O., et al.: Global sentiments surrounding the COVID-19 pandemic on Twitter: analysis of Twitter trends. JMIR Public Health Surveill. **6**, e19447 (2020)
9. https://docs.tweepy.org/en/stable/client.html#tweepy.Client
10. NLTK. https://www.nltk.org/
11. Loria, S.: Textblob documentation. Release **0.15**(2), 269 (2018)
12. Hutto, C., Gilbert, E.: Vader: A parsimonious rule-based model for sentiment analysis of social media text. In: Proceedings of the International AAAI Conference on Web and Social Media (2014)
13. Liu, B.: Sentiment analysis and subjectivity. Handb. Nat. Lang. Process. **2**, 627–666 (2010)
14. Indurkhya, N., Damerau, F.J.: Handbook of natural language processing. Comput. Linguist. **37**, 395–397 (2011)
15. Colnerič, N., Demšar, J.: Emotion recognition on Twitter: comparative study and training a unison model. IEEE Trans. Affect. Comput. **11**, 433–446 (2018)
16. Parliamentary Library. https://www.aph.gov.au/About_Parliament/Parliamentary_Departments/Parliamentary_Library/pubs/rp/rp2021/Chronologies/COVID-19StateTerritoryGovernmentAnnouncements
17. NSW Health. https://www.health.nsw.gov.au/news/Pages/20200831_00.aspx
18. Roxburgh, A., et al.: Adapting harm reduction services during COVID-19: lessons from the supervised injecting facilities in Australia. Harm Reduct. J. **18**, 20 (2021)
19. https://www.health.nsw.gov.au/news/Pages/20210816_01.aspx

20. Boaz, J.: Melbourne passes Buenos Aires' world record for time spent in COVID-19 lockdown. ABC News (2021)
21. Infogram.   https://infogram.com/number-of-days-in-lockdown-by-australian-state-1h7z2l 8xmznyx6o?live
22. Levula, A., Harré, M., Wilson, A.: The association between social network factors with depression and anxiety at different life stages. Commun. Ment. Health J. **54**, 842–854 (2018)

COVID-19 Article Classification Using Word-Embedding and Extreme Learning Machine with Various Kernels

Sanidhya Vijayvargiya[1]([⊠]), Lov Kumar[1], Aruna Malapati[1],
Lalita Bhanu Murthy[1], and Aneesh Krishna[2]

[1] BITS-Pilani, Hyderabad, India
{f20202056,lovkumar,arunam,bhanu}@hyderabad.bits-pilani.ac.in
[2] Curtin University, Perth, Australia
A.Krishna@curtin.edu.au

Abstract. The impact of the COVID-19 pandemic on the socially networked world cannot be understated. Entire industries need the latest information from across the globe at the earliest possible. The business world needs to cope with a very volatile market due to the pandemic. Businesses need to be swift in sensing potential profit opportunities and be updated on the changing consumer demands. Technological advances and medical procedures that successfully deal with COVID-19 can help save lives on the other side of the world. This seamless passage of crucial information, now more than ever, is only possible through the networked world. There are on average 821 articles published online on COVID-19 a day. Manually going through around 800 articles in a day is not feasible and highly time-consuming. This can prevent the industries and businesses from getting to the relevant information in time. We can optimize this task by applying machine learning techniques. In this work, six different word embedding techniques have been applied to the title and content of the articles to get an n-dimensional vector. These vectors are inputs for article classification models that employ Extreme Learning Machine (ELM) with linear, sigmoid, polynomial, and radial basis function kernels to train these models. We have also used feature selection techniques like the Analysis of Variance (ANOVA) test and Principal Component Analysis (PCA) to optimize the models. These models help to filter out relevant articles and speed up the process of getting crucial information to stay ahead of the competition and be the first to exploit new market opportunities. The experimental results highlight that the usage of word embedding techniques, feature selection techniques, and different ELM kernels help improve the accuracy of article classification.

Keywords: COVID-19 · Data imbalance · Feature selection · Word embedding

1 Introduction

Article Classification or Text Classification is crucial to help reduce time and effort exerted by humans. By combining word embedding techniques with machine

L. Barolli et al. (Eds.): AINA 2022, LNNS 451, pp. 69–81, 2022.
https://doi.org/10.1007/978-3-030-99619-2_7

learning classifiers, article classification models can convert text-based data to vectors and categorize articles on suitable topics. Article classification can be particularly helpful in going through the articles published during COVID-19 and sorting out the ones related to business. Businesses have faced the acute effects of the pandemic [1]. Therefore, they must go through the relevant articles to stay on par or ahead of the industry.

This research is essential to allow businesses to become aware of new market opportunities at the earliest and plan long-term strategies after COVID-19 [2]. This paper aims to optimize the time for industries most affected by COVID-19, like businesses, to find relevant news. This news can help them provide a timely response to the ever-changing state of the world due to the pandemic. In previous research, different data mining techniques were used to classify articles. However, there was a lack of emphasis in these works on how COVID-19 has affected the classification of the articles. Two problems are faced when classifying articles.

- **Word Embedding:** The article or text classification models are built on the title and content of the data. The problem lies in converting this text data to numerical data to apply machine learning techniques. In this work, six different word embedding techniques such as Global Vectors for Word Representation (GLOVE)[1], Continuous Bag of Words Model (CBOW)[2], Skipgram (SKG)[3], Google news word to vector(w2v)[4], FastText (FTX)[5], and term frequency, inverse document frequency (TFIDF) are applied on both the content and the title of articles. These techniques help us represent the word not just as a number but as a vector in n-dimensional space.
- **Number of Features:** The number of features input to the model impacts its predictive ability. Certain features obtained from word-embedding techniques can be redundant or not have a significant impact on the dependent variable. Such features can lead to lower accuracy of the prediction models. In this study, we have used two different feature selection techniques to eliminate irrelevant and redundant features that hinder the model's predictive ability.

This paper is arranged as follows: Past literature available on article classification and various word embedding techniques is reviewed in Sect. 2. Section 3 describes the experimental dataset collection along with various methodologies used. Section 4 discusses the research methodology employed in this work through an architecture framework. Experimental results and their analysis are presented in Sect. 5. Section 6 describes the comparative analysis of models developed using different word-embedding techniques, sets of features, and various machine learning models. Finally, Sect. 7 concludes the reported work and gives future research directions.

---

[1] https://nlp.stanford.edu/projects/glove/.

[2] https://towardsdatascience.com/nlp-101-word2vec-skip-gram-and-cbow-93512ee24 31.

[3] https://towardsdatascience.com/nlp-101-word2vec-skip-gram-and-cbow-93512ee24 31.

[4] https://code.google.com/archive/p/word2vec/.

[5] https://FastText.cc/.

2 Related Work

Researchers have used various techniques to classify texts more accurately. Previously, various classification algorithms alike Naive-Bayes, maximum entropy, RCNN have been studied and their key conclusions are discussed in this section. K. Nigam et al. proposed the use of maximum entropy for text classification [3]. Maximum entropy is used for text classification by estimating the conditional distributions of the class variables given in the document. The paper compared maximum entropy with Naive Bayes, and while maximum entropy showed to be significantly better sometimes, it was also considerably worse other times. On its best performance, it produced an accuracy of 78.8%. The authors concluded that this model might be sensitive to poor feature selection.

H. Lodhi et al. proposed an approach to classify documents based on string kernels [4]. For modestly sized datasets, in comparison with standard word feature space kernels, it showed positive results. For larger documents, the paper introduced an efficient technique for good approximations. On the Reuters dataset, word kernel and contiguous n-gram kernel outperformed this model in most cases. The authors hypothesized that the outstanding results on smaller datasets suggest that the kernel is conducting something akin to stemming, hence providing semantic links between words that the word kernel must regard as separate, based on the inconsistent results. In large datasets with enough data to learn the importance of the terms, this impact is less noticeable.

A common classification algorithm is Naive Bayes. McCallum and Nigam [5] compared the two approaches in which Naive Bayes is used- a multivariate Bernoulli model and a multinomial model. A Bayesian network with no dependencies between words and binary word features is known as a multivariate Bernoulli model. A unigram language model with integer word counts is known as a multinomial model. The Bernoulli model performed better with small vocabulary sizes, but the multinomial model performed better at larger vocabulary sizes and reduced error by 27% on average. On the Newsgroups dataset, the multinomial model achieves an accuracy of 85%, and the Bernoulli model achieves an accuracy of 74%. The multinomial model is almost uniformly better than the multivariate Bernoulli model.

A critical step of text classification is word embedding. The Facebook AI research team of Joulin et al. [6] compared their text classifier FastText(FTX) with other deep learning classifiers based on accuracy and other tests. They trained FastText on more than 1 billion words in less than 10 min using a standard CPU and classified 0.5 million sentences among 312k classes. FastText with bigram achieves an accuracy of 96.8% on Sogou and 98.6% on DBpedia. Overall the accuracy is slightly better than char-CNN and char-CRNN and a bit worse than VDCNN. The difference between the speed of FastText compared to neural networks increases with the size of data, going up to at least 15,000x. FastText is able to perform on par with recently proposed models made with deep learning while being much faster.

This research work has uniquely focused on classifying COVID-19 articles. Unlike previous works, Extreme Learning Machine with various kernels has been employed to build the classification models. ELM has the advantage of providing higher scalability and is less computationally complex than other classification methods. Article classification using only the title of articles is considerably less computationally expensive than using the article's content. It can be used on certain occasions based on the accuracy required. The impact on accuracy when classifying solely based on the title or the content of an article has also been addressed in this paper. The conclusions derived from this paper using ELM for classification can be compared to the above-mentioned text classification techniques to understand the advantages and disadvantages of ELM and justify the novelty of this research.

3   Study Design

This section presents the details regarding various design settings used for this research.

3.1   Experimental Dataset

In this paper, the proposed models are validated using a COVID-19 article dataset. This dataset was collected from Kaggle[6]. The COVID-19 Public Media Dataset is a resource containing more than 350,000 online articles scraped over the span of one year from January 1 to December 31, 2020. It possesses articles in English from more than 60 resources from various domains. This dataset is too large to wholly analyze by converting it to numerical data using word embedding techniques. Thus, the data analyzed was limited to the months of March, April, and May. This data was then split into four categories- finance, business, general, and tech. These act as the output classes of the article classification model. The distribution of the articles from these categories over the months of March, April, and May is displayed in Fig. 1. The category with the highest number of articles is business, with 17,015 in March, 22,270 in April, and 26,952 in May. The dataset was split into title and content to train two different sets of models, one with its input being the title of the articles and the other with its input being the content.

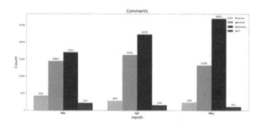

Fig. 1. Data-Sets

6 https://www.kaggle.com/jannalipenkova/covid19-public-media-dataset.

3.2 Word Embedding

The dataset consists of title and content, both of which can be used to classify articles. In this work, six different word embedding techniques, including Google news word to vector(w2v), Continuous Bag of Words Model (CBOW), Global Vectors for Word Representation (GLOVE), term frequency, inverse document frequency(TFIDF), Skip-gram(SKG), and fastText (FTX) are applied on both title and content extracted from the dataset [7]. These techniques aid in representing the word as a vector in n-dimensional space rather than merely a number. These vectors are used as input to build models that accurately categorize articles.

3.3 Feature Selection Techniques

Upon converting the text to vectors, we use these vectors of n-dimension as inputs to the models. The selection of essential feature vectors also impacts the performance of the models. This study used two feature selection techniques, the Analysis of Variance(ANOVA) and Principal Component Analysis(PCA), to remove irrelevant and redundant features. We then compared the performance of the models with the original set of features to the performance of the models with the refined set of features after ANOVA or PCA.

ANOVA test is a tool that helps split an observed aggregate variability in a dataset into systemic and random factors. The systemic factors have a significant impact on the dependent variable, whereas the random factors do not. ANOVA test helps analyze the significance of the independent variables' impact on the dependent variables. These tests reject the null hypothesis or accept the alternate hypothesis [8].

PCA is a technique for lowering the dimensionality of such datasets which improves the interpretability, and minimizes data loss. We only select the top principal components to get reduced dimensional data. While there is some information loss, we are trading accuracy for simplicity. Sometimes, the variables are highly correlated in such a way that they contain redundant information [9].

3.4 Extreme Learning Machine

ELM (Extreme Learning Machines) are feed-forward neural networks. ELM does not require gradient-based back-propagation to find models [10]. A sequential learning algorithm is followed, which does not require retraining when new data is added and is very efficient when dealing with non-stationary real-time datasets. It is a single hidden layer feed-forward neural network. ELM randomly allocates hidden nodes, generates biases and input weights of hidden layers, and finds the output weights using least-squares algorithms. Instead of back-propagation like a single hidden layer neural network, it performs matrix inverse and does it only once. Compared to traditional neural networks, It aids in the alleviation of sluggish training speeds and over-fitting issues. ELM is based on empirical risk minimization theory, and it only requires one iteration to learn. It has improved generalization, resilience, as well as a quick learning rate. We have used ELM with four different kernels- linear (LINK), sigmoid (SIGK), polynomial (POLYK), and radial basis function (RBFK) [11].

4   Research Methodology

In this work, six different word embedding methods have been applied to the title and content of the articles for feature extraction. Two feature selection techniques have been employed, and these selected features are used as input to develop models for classifying articles. The models are trained with four different kernels of Extreme Learning Machine. A detailed overview of our proposed work is given in Fig. 2.

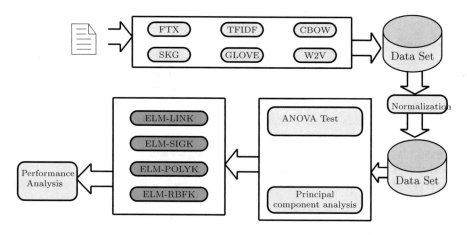

Fig. 2. Research framework

The proposed framework is a multi-step process consisting of feature extraction from text data using word embedding techniques, removing irrelevant and redundant features, and finally developing prediction models using four different kernels of Extreme Learning Machine. The dataset taken is reduced to March, April, and May. The categories of articles taken are limited to the four most common categories in the dataset. The data is reduced to remove outlier categories that do not have sufficient data. There are two types of models that are built. The first category is trained on March data and tested on April data. The second type of model is trained on April data and tested on May data. The dataset is separated into title and content to analyze separately. Then, the six word embedding techniques are applied to the datasets. Furthermore, feature selection techniques are applied to the data, but we also preserve the data with the original set of features. The resulting data is trained with the four different kernels of ELM to get the classification models. The models are then analyzed using accuracy, AUC, and compared with box plot and mean rank.

5 Empirical Results and Analysis

In this study, we have applied six different word embedding techniques, two feature selection techniques, and Extreme Learning Machine (ELM) with four different kernels to classify articles into four different categories. Each word embedding technique is applied to the title and content of the articles separately. The dataset was split into two based on the month of publication of the article. The first set of models was trained on data from March and tested on the data from April, whereas the other was trained on data from April and tested on data from May. Therefore, a total of 288 [2 datasets (1 content + 1 title) * 6 word-embedding techniques * 2 month-split (trained on March data + trained on April data) * 3 sets of features * 4 kernels with ELM] distinct prediction models are built in this study. The predictive ability of these trained models is evaluated with Area Under Curve(AUC) and accuracy performance values, as shown in Table 1.

Table 1. AUC values for extreme learning with various kernels

	Title								Content							
	MAR-APR				APR-MAY				MAR-APR				APR-MAY			
	All Features															
	SIGK	RBFK	LINK	PLYK	SIGK	RBFK	LINK	PLYK	SIGK	RBFK	LINK	PLYK	SIGK	RBFK	LINK	PLYK
TF-IDF	0.8	0.82	0.81	0.84	0.81	0.86	0.82	0.85	0.92	0.94	0.94	0.96	0.92	0.94	0.94	0.96
CBOW	0.66	0.75	0.67	0.78	0.67	0.77	0.68	0.79	0.84	0.89	0.85	0.89	0.83	0.89	0.84	0.9
SKG	0.57	0.73	0.67	0.73	0.58	0.75	0.68	0.75	0.83	0.84	0.84	0.85	0.83	0.84	0.84	0.86
GLOVE	0.78	0.86	0.81	0.86	0.78	0.86	0.81	0.87	0.86	0.89	0.88	0.9	0.86	0.89	0.88	0.91
W2V	0.82	0.85	0.83	0.86	0.81	0.86	0.83	0.86	0.88	0.9	0.89	0.91	0.88	0.9	0.9	0.92
FTX	0.59	0.65	0.64	0.66	0.62	0.66	0.65	0.67	0.69	0.72	0.72	0.74	0.69	0.72	0.72	0.73
	Anova Features															
TF-IDF	0.78	0.81	0.79	0.84	0.79	0.82	0.8	0.85	0.92	0.94	0.94	0.96	0.92	0.94	0.94	0.96
CBOW	0.66	0.75	0.67	0.77	0.67	0.77	0.67	0.79	0.84	0.89	0.85	0.89	0.82	0.89	0.84	0.9
SKG	0.58	0.73	0.67	0.73	0.58	0.75	0.68	0.75	0.83	0.84	0.84	0.85	0.83	0.84	0.84	0.86
GLOVE	0.78	0.86	0.81	0.86	0.78	0.86	0.81	0.87	0.86	0.89	0.88	0.9	0.86	0.89	0.88	0.91
W2V	0.82	0.85	0.83	0.85	0.81	0.86	0.83	0.86	0.88	0.9	0.89	0.91	0.88	0.9	0.9	0.92
FTX	0.59	0.65	0.64	0.66	0.62	0.66	0.66	0.67	0.69	0.72	0.72	0.74	0.69	0.72	0.72	0.73
	PCA Features															
TF-IDF	0.82	0.8	0.82	0.82	0.83	0.8	0.83	0.84	0.91	0.91	0.91	0.92	0.91	0.91	0.91	0.92
CBOW	0.73	0.75	0.73	0.77	0.73	0.76	0.74	0.79	0.87	0.88	0.88	0.89	0.87	0.88	0.88	0.89
SKG	0.64	0.7	0.71	0.73	0.65	0.7	0.73	0.74	0.83	0.84	0.84	0.85	0.84	0.85	0.85	0.86
GLOVE	0.84	0.85	0.84	0.85	0.85	0.85	0.85	0.86	0.87	0.88	0.88	0.89	0.88	0.88	0.88	0.89
W2V	0.84	0.85	0.84	0.85	0.84	0.85	0.84	0.85	0.88	0.89	0.89	0.89	0.89	0.89	0.89	0.9
FTX	0.55	0.59	0.64	0.62	0.6	0.6	0.66	0.62	0.71	0.69	0.72	0.73	0.71	0.68	0.72	0.73

The high value of AUC as shown in Table 1 confirms that the developed models can satisfactorily classify the articles into various categories. The AUC values are split column-wise, where the first eight columns correspond to the models that have ML techniques applied to the title of the article. In contrast, for the remaining eight columns, the ML techniques have been applied to the content of the article. Both of these sets of eight columns are further split into sets of four columns. Mar-Apr columns represent the AUC of models trained on March data and tested on April data, whereas Apr-May columns represent the AUC of the models trained on April data and tested on May data. The data also has three equal row-wise segments, where the segments correspond to the data with All Features, ANOVA test features, and PCA features. For each mode of feature selection, AUC values of all combinations of six word embedding techniques and four ELM kernels are presented.

5.1   Comparative Analysis

This section compares the models developed using different word embedding techniques, feature selection techniques, and different kernels of ELM. The comparison is based on statistics like Area Under Curve(AUC), and a visual representation of the comparative performance is provided using box plots. These box plots are further analyzed using descriptive statistics such as Mean, Median, Max, Min, Q1(25%), and Q3(75%). In this paper, we have also used the Friedman test [12] to corroborate the results. The goal of the Friedman test is to accept or reject the following hypothesis.

- **Null hypothesis** - The different ML techniques used do not produce a significant impact on the performance of the article classification models.
- **Alternate Hypothesis** - The different ML techniques do produce a significant impact on the performance of the article classification models.

The Friedman test was carried out with a significance level of $\alpha = 0.05$ with degrees of freedom as 5 for word embedding techniques, 3 for kernel comparison of ELM, 2 for feature selection, 1 for months, and 1 for title and content.

5.2   Word-Embedding

In this study, six different word embedding techniques such as Global Vectors for Word Representation(GLOVE), term frequency-inverse document frequency (TF-IDF), Skip-gram (SKG), Google news word to vector(W2V), Continuous Bag of Words(CBOW), and FastText (FTX) have been used to compute the numerical vector of the title and content of the articles. AUC has been used as the primary metric to compare the predictive ability of models built using different word embedding techniques.

Box-Plot: Word-Embedding: Figure 3a visually depicts the performance value, i.e., AUC of different word embedding techniques in box plot diagrams. From Fig. 3a, we can deduce that the models developed using the word vectors computed by TF-IDF and W2V have a better predictive ability than other models. However, as compared to other techniques, the models produced with FTX have low predictive ability. The models created with W2V have a mean AUC of 0.868, a maximum AUC of 0.915, and a Q3 AUC of 0.894, which means that 25% of the models created with W2V have an AUC of 0.894. These conclusions are supported by the accuracy data of these models, which show that kernels using W2V and TF-IDF outperform kernels using alternative word-embedding strategies for both title and content.

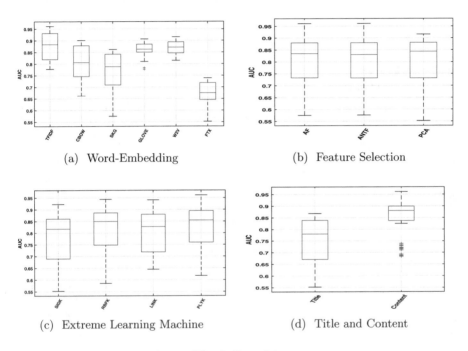

(a) Word-Embedding   (b) Feature Selection

(c) Extreme Learning Machine   (d) Title and Content

Fig. 3. Box-plot

Friedman Test: Word-Embedding: In this study, Friedman Test is also applied to the AUC values to better compare the predictive ability of the models developed using various word embedding techniques. The goal of the test is to accept or reject the null hypothesis. The null hypothesis is, "the article classification models developed using different word-embedding techniques do not have a significant difference in their predictive ability." The Friedman test was carried out with a significance level of $\alpha = 0.05$ with degrees of freedom as 5. Table 2a containing the mean ranks for the various word embedding techniques. The lower the mean rank, the better the models' performances are. W2V has the lowest mean rank of 1.833, while FTX has the highest mean rank of 5.917. TF-IDF with a mean rank of 1.896 is marginally outperformed by W2V. TF-IDF's high performance could possibly be due to the fact that news articles share a lot of common phrases and words. TF-IDF through its system of assigning weights to each term can effectively prevent common words used in articles from having an impact on the classification better than other word-embedding techniques.

5.3   Feature Selection

In the proposed research, we use two different types of feature selection techniques- ANOVA test and PCA, and in a third set of models, we use all of the original features for training predictive models for article classification. These feature selection techniques have been applied to both the title and content datasets.

Table 2. Friedman test

(a) Word-Embeddin

Technique	Mean rank
TFIDF	1.896
CBOW	4.000
SKG	4.917
GLOVE	2.438
W2V	1.833
FTX	5.917

(b) Feature Selection

Technique	Mean rank
AF	1.615
ANTF	2.177
PCA	2.208

(c) ELM Kernels

Technique	Mean rank
SIGK	3.93
RBFK	2.29
LINK	2.70
PLYK	1.06

Box-plot: Feature Selection: Figure 3b suggests that all feature selection techniques resulted in high performance, with PCA having the highest median of AUC. All three have a very close mean AUC of 0.807 for all features(AF), 0.805 for Anova test features, and 0.807 for PCA. AF has a min of 0.574 AUC, max of 0.961 AUC, a median of 0.834 AUC, Q1 of 0.73 AUC, and Q3 of 0.879 AUC. AF and ANOVA seem to give similar results as per the box plot.

Friedman Test: Feature Selection: We have also employed the Friedman test on the AUC values of the models, built using three different set of features, to compare the different feature selection techniques based on their predictive ability. The null hypothesis, which is to be accepted or rejected based on the Friedman test, is "the article classification models developed using different sets of features do not have a significant difference in their predictive abilities." The Friedman test was carried out with a significance level of $\alpha = 0.05$ with degrees of freedom as 2. Table 2b presents the mean ranks of the three sets of features. The mean ranks from the Friedman test help judge how different techniques under study perform compared to each other. Lower mean ranks correspond to better performance in comparison to others. The models trained using the set of original features have the lowest mean rank of 1.615, followed by ANOVA with 2.177, and finally PCA with 2.208. The mean ranks show that the models perform better when all features are included, and the use of feature selection techniques like PCA and ANOVA test only regresses the performance of the models.

5.4   Extreme Learning Machine

The study makes use of four different kernels for Extreme Learning Machine to classify articles. These kernels are applied with different word-embedding techniques and on both the title and the content datasets. In this work, we have used the Sigmoid kernel(SIGK), the Radial Basis Function kernel(RBFK), the Linear kernel(LINK), and the Polynomial kernel (PLYK).

Box-plot: Extreme Learning Machine: Figure 3c helps visualize statistics such as Mean, Median, Min, Max, Q1, and Q3 of AUC with the different kernels used with ELM. The results show that the model trained using the Polynomial Kernel has the overall best performance with a mean AUC of 0.830. The sigmoid kernel performs the worst in comparison to other kernels, with a mean AUC of 0.779. The models developed using the Polynomial kernel achieve 0.961 max AUC, 0.617 min AUC, 0.855 median AUC, 0.759 Q1, and 0.894 Q3. RBFK and LINK have worse AUC and performance than PLYK but significantly better than SIGK.

Friedman Test: Extreme Learning Machine: The Friedman test is also applied to the AUC values of the different kernels to help statistically compare the performance of models built using different ELM kernels. The test aims to accept or reject the null hypothesis. The null hypothesis for this test is, "the article classification models developed using the different kernels of Extreme Learning Machine do not have a significant difference in their predictive abilities." The Friedman test was carried out with a significance level of $\alpha = 0.05$ with degrees of freedom as 3. The mean rank of different kernels after applying the Friedman test is present in Table 2c. From Table 2c, we can observe that the mean rank of 1.069 of POLYK is the lowest among all. The sigmoid kernel has a mean rank of 3.931 is the highest and thus symbolizes that the sigmoid kernel is the worst out of the four.

5.5   Title and Content

In this paper, there are two different types of models based on the data used for training the models. One set is trained using the title of the article, and the other set is trained using the content of the article.

Box Plots: Title and Content: Figure 3d shows the visual depiction of the predictive ability of the models trained on the title compared to that of the models trained on content. The models trained on content outperformed the models trained on title in every AUC box plot metric. The models trained on content had a mean AUC of 0.856, median AUC of 0.880, 0.961 max AUC, 0.682 mean AUC, 0.837 Q1, and 0.899 Q3.

Friedman Test: Title and Content: The Friedman test is applied to the AUC values of the models trained on title and content data to help compare the performance of the two sets of models. The purpose of this test is to accept or reject the null hypothesis, which is, "whether the models trained on title or content of the article have significant impact on its performance." The Friedman test was carried out with a significance level of $\alpha = 0.05$ with degrees of freedom as 1. The models trained on the content of the article have the lowest mean rank of 1, whereas the models trained on the title of the articles have a mean rank of 2.

6 Conclusion

In this paper, using different variations of word embedding techniques, feature selection techniques, and different Extreme Learning Machine kernels, we build models to classify COVID-19 articles based on content and their title. Unlike existing research, this work focuses on classifying COVID-19 specific articles in different categories to help businesses get to crucial information and news faster. To represent the word as a vector in n-dimensional space, we used six distinct word-embedding techniques. Three sets of features selected using feature selection techniques and four distinct kernels of the Extreme Learning Machine are used to assess these methods' prediction potential. Finally, using AUC and accuracy performance metrics, the prediction ability of these models is evaluated and compared. The main conclusions we derived were the following:

- The high value of AUC confirms that the developed models can satisfactorily classify the articles into various categories and speed up the process of classification for businesses.
- The models developed by W2V have the best classification accuracy, with TF-IDF coming close.
- The feature selection techniques regressed the performance of the models. The models performed best with their original features.
- The models which used Extreme Learning Machine with polynomial kernel significantly outperformed the other kernels.

References

1. Donthu, N., Gustafsson, A.: Effects of COVID-19 on business and research (2020)
2. Krishnamurthy, S.: The future of business education: a commentary in the shadow of the COVID-19 pandemic. J. Bus. Res. **117**, 1–5 (2020)
3. Nigam, K., Lafferty, J., McCallum, A.: Using maximum entropy for text classification. In: IJCAI 1999 Workshop on Machine Learning for Information Filtering, vol. 1, pp. 61–67. Stockholom, Sweden (1999)
4. Lodhi, H., Saunders, C., Shawe-Taylor, J., Cristianini, N., Watkins, C.: Text classification using string kernels. J. Mach. Learn. Res. **2**(Feb), 419–444 (2002)
5. McCallum, A., Nigam, K., et al.: A comparison of event models for Naive Bayes text classification. In: AAAI 1998 Workshop on Learning for Text Categorization, vol. 752, pp. 41–48. Citeseer (1998)
6. Joulin, A., Grave, E., Bojanowski, P., Mikolov, T.: Bag of tricks for efficient text classification. arXiv preprint. arXiv:1607.01759 (2016)
7. Lai, S., Liu, K., He, S., Zhao, J.: How to generate a good word embedding. IEEE Intell. Syst. **31**(6), 5–14 (2016)

8. Cuevas, A., Febrero, M., Fraiman, R.: An anova test for functional data. Comput. Stat. Data Anal. **47**(1), 111–122 (2004)

9. Wold, S., Esbensen, K., Geladi, P.: Principal component analysis. Chem. Intell. Lab. Syst. **2**(1–3), 37–52 (1987)

10. Huang, G.-B., Zhu, Q.-Y., Siew, C.-K.: Extreme learning machine: theory and applications. Neurocomputing **70**(1–3), 489–501 (2006)

11. Bergman, S.: The Kernel Function and Conformal Mapping, vol. 5. American Mathematical Society (1970)

12. Zimmerman, D.W., Zumbo, B.D.: Relative power of the Wilcoxon test, the Friedman test, and repeated-measures anova on ranks. J. Exp. Educ. **62**(1), 75–86 (1993)

An Improved Ant Colony Optimization Based Parking Algorithm with Graph Coloring

Marco Agizza, Walter Balzano[(✉)], and Silvia Stranieri

University of Naples Federico II, Naples, Italy
marco.agizza@outlook.com, walter.balzano@gmail.com,
silvia.stranieri@unina.it

Abstract. In an era which is going towards the spread of smart cities and intelligent transportation systems, vehicular ad hoc networks are an interesting framework to propose innovative solutions. One of the most tedious problems for drivers in a urban environment is the parking process. Indeed, drivers looking for an available parking slot keep being the main cause of traffic congestion, which also involves a high stress and air pollution level. In this work, we provide a smart parking solution, aiming at a higher context awareness for drivers, by relying on a well known optimization problem, the ant colony. By choosing an opportune criterion to update the pheromone, we push drivers to choose possibly uncrowded paths, ending up with a solution which guarantees a fair node distribution with respect to the available parking slots.

1 Introduction

Nowadays, the Internet of Things (IoT) [1] is becoming more and more integral part of our lives. Just think of autonomous vehicles [2], smart cities [3,4], autonomous vehicular networks [5], and also environmental preservation [6,7]. This trend is supposed to continue if we just have a look at the study reported in [8], where they predict that, by the year 2050, almost the 65% of population will be living in the city. This phenomenon would clearly increase the number of vehicles along the city roads, by highlighting the relative problems, consequently: starting from the traffic congestion, the gas emission [9], to the high demand of parking space. In this context, we strongly believe that a mechanism to favor a higher network awareness for drivers should be employed to face the high congestion risk in cities.

In this work, we want to focus on the parking problem: it is very often the case that people give up on a show, or a dinner because of the difficulty in finding a free parking slot. Moreover, the crowd due to drivers looking for a parking space has also severe consequences on environmental pollution, as well as on drivers stress level.

The parking problem has already been investigated by researchers from several perspectives: some centralized solutions [10] have been provided to assist

L. Barolli et al. (Eds.): AINA 2022, LNNS 451, pp. 82–94, 2022.
https://doi.org/10.1007/978-3-030-99619-2_8

users to get parking information by relying on a web application, but also decentralized approaches, like [11], that aim at maximizing the autonomy of drivers, without any intermediaries, which in our opinion is more suitable for the problem we are facing.

For this reason, in this work we study the solution to a well known optimization problem, based on the behavior of real ants in nature, the ant colony optimization problem (ACO, for short), with the goal of applying such a solution to the parking problem in a decentralized manner. In particular, our approach takes inspiration from ACO, but we follow an opposite direction: indeed, while in ACO ants are attracted from paths with a higher amount of pheromone, which is the chemical substance released from an ant that followed the same path previously, in our scenario the pheromone acts as a repulsive for drivers, so to avoid crowded situations.

More specifically, anytime a driver follows a path P, the associated pheromone will be updated so to make other drivers understand that P could be a potentially crowded path. In this way, anytime a driver has to choose a path, by following the pheromone it is guaranteed that the less crowded one is selected. The final result is a context-aware and self-organizing network [12], characterized by an even distribution of vehicles among the available parking slots, with a lower gas emission due to the multiple tours that drivers usually have to perform to find a space.

2   Related Works

Smart parking is one of the most challenging application of vehicular ad hoc networks (VANETs, for short), see [13] for a survey. They are networks in which vehicles act like nodes and their interconnections are given by the signal strength between a pair of nodes. They can communicate via broadcasting directly, if they are close enough with respect to the communication range needed, or via multihop by using other vehicles or road infrastructures as bridge. In this field, there are multiple open issues, some of which have been largely investigated in this Journal, such as [14], where vehicular network connectivity is studied examining the impact of mobility dependent metrics, or even [15] where a routing protocol in VANETs is proposed, by using node features such as position, density, and link quality.

These networks are a powerful and promising framework to apply artificial intelligence aiming at reaching intelligent cities, among which smart parking.

The literature is made of several solutions aimed at reducing the traffic congestion, such as [16] where the importance of the context is clearly highlighted, but also [17] where authors provide a fog-computing-based mechanism to avoid congestion in vehicular networks, and also solutions specifically applied at smart parking. Several authors rely on ACO to find a solution for the parking problem: for instance, in [18], they apply ACO to solve the parking problem, but in a totally different manner, since they focus on autonomous vehicles. See [19] for a survey on ACO. In our work, we deal with human drivers and we want to facilitate their parking process. Also authors in [20] uses ACO: indeed, our aim is to improve the solution they provided, by also considering the actual vehicle distribution during the parking problem, which is not taken into account. Precisely, we assume a similar environment classification by keeping the spirit of the solution provided, but with a higher precision description of the connection between nodes, and also proposing an opportune implementation and simulation with different numbers of vehicles. ACO provides, in our opinion, an interesting way of approaching the solution of a problem, as it follows the biological behavior of living beings: indeed, it has not only been used for parking applications, but also in robotics, such as in [21], where it is applied to avoid collisions with an obstacle while programming the movement of a robot.

There are also different types of solutions, interesting as well, like the one provided in [22]. These kinds of solutions are based on game theory aspects, and they focus on the effort needed for the driver to find a free parking slot. These are much more theoretic solutions, with respect to ours, which instead provide an actual implementation with a real-world-scenario proposal, and relative simulation results.

3 Preliminaries

In this section, we are going to recall some concepts our work is based on. First of all, it is necessary to remind the definition of the Ant Colony Optimization problem and the way a solution is computed.

3.1 ACO

As precisely explained in [19], Ant Colony Optimization was introduced in 90's as a nature-inspired metaheuristic for the solution of hard combinatorial optimization problems. It follows the behavior or real ants looking for food: in the beginning, they visit the neighborhood of their nest at random, they find some food source and get it back to the nest. During the trip, ants release a pheromone trail to guide other ants towards the food.

One of the main component of ACO is the pheromone model, which is the probabilistic model defining the pheromone parameters used to select the solution for each iteration. Precisely, at any iteration ants probabilistically construct solutions to the combinatorial optimization problem under consideration, exploiting the pheromone model. Then, some solutions are used to update the trail level before the next iteration starts. Such an updating is performed through an update rule, which in many ACO algorithms is a variation of the following formula. [20]:

$$\tau_i^j = (1 - \rho) * \tau_i^j + \Delta\tau_i^j \tag{1}$$

where ρ is the evaporation rate needed to guarantee an opportune convergence of the algorithm, and $\Delta\tau_i^j$ is the current pheromone value.

4   Model Definition

In this section, we are going to describe the model on which our solution has been built. The environment is classified as a directed weighted graph $G = (V, E)$, where the set of nodes V is a set of parking regions, and their connections are represented through the edges in E. An example can be seen in Fig. 1. In our setting, we assume that each driver, that has to park his car, has a starting region that is known, and a destination region towards which he wants to get as close as possible. Hence, the graph configuration depends on a fixed destination region for any driver taking part in the parking process.

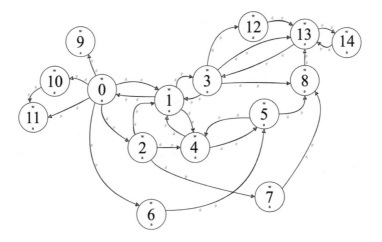

Fig. 1. Parking regions graph.

As shown in Fig. 1, to each parking region of the model, and each edge connecting nodes, some properties are provided:

- For nodes:
  - w: the distance to walk to reach the destination region;
  - a: the number of available parking spots in the region.
- For edges
  - d: the distance to travel by road to reach the destination node of the arc from the source one;
  - p: the probability with which each vehicle will visit the destination node of the arc from the source one; since it is a probability, it is a value between zero and one. As in the standard ACO, this parameter works as the pheromone but, differently form ACO, it has a repulsive power, rather than an attractive one.

5   An Ant Colony Optimization Parking Algorithm

The Ant Colony Optimization (ACO) is a genetic algorithm that allows you to solve the Traveling Salesman Problem (TSP) that requires to find the minimum costing path to reach a specific node in the graph visiting every node at most once. In the ACO, every agent participates individually in the construction of the solution, building, iteratively, different candidate solutions for the TSP that converge, step by step, in a single one thanks to a shared memory system which allows each agent to share with the others information relating to their experience about the choice of a certain direction rather than another [23,24]. According to the rules of the original ACO, the solution to the problem is theoretically excellent, but it is not if we consider other values of the real world as the traffic congestion: after some time every agent will follow the same path because it brings to food and the number of agents on the path is not important due to the fact that there are no lanes, traffic lights and, at most, the ants will walk on each other to reach the destination.

In other words, the original ACO algorithm finds the shortest route to reach a destination but it also causes the formation of traffic jams and requires a higher time for each agent to reach the destination, that would take us away from achieving our goal. For this reason, we provide a variation of the standard ACO algorithm, bu starting with a graph coloring to improve the context-awareness of the vehicular network.

5.1   Graph Coloring

In this section, we provide a coloring mechanism to avoid drivers choosing the wrong edges. When we speak about wrong edges, we mean edges that push the driver irreversibly away from the destination, rather than getting him closer to it. Since, at the very beginning of the execution, there is no pheromone yet to inform drivers of which edges should be picked and which should not, a graph coloring is needed to prevent wrong choices that would lead to a bad exploration of the graph. If we consider that the start node is the 0 one and the destination node is the 8 one, at the beginning of the execution of the algorithm, without considering the graph coloring, the vehicle will have to do the following evaluation to choose the node to move to, by maximizing the ratio $\frac{a}{d*w}$. Such a ratio is:

- $\frac{4}{1.6*1.16} = 2.16$ for the node 1,
- $\frac{3}{1.2*1.5} = 1.67$ for node 2,
- $\frac{7}{2*3} = 1.17$ for node 6,
- $\frac{16}{2.4*3.4} = 130.56$ for node 9,
- $\frac{21}{0.8*2.8} = 9.38$ for node 10,
- $\frac{4}{1.3*3.34} = 0.92$ for node 11.

With these values, the node 9 would be chosen, but it would bring the vehicle irreversibly far from the destination.

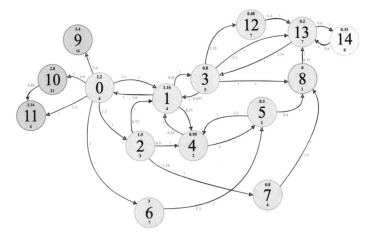

Fig. 2. Colored parking regions graph assuming that the destination is the node 8.

In order to avoid such a problem, the graph is previously colored as in Fig. 2, so that every agent is aware of which node can bring it to the destination, or can get him closer to it, at every step. Precisely, we define a parameter θ, which indicates the maximum distance that is reasonable to walk from the parking slot to the destination:

- The red nodes are those that do not lead to the destination node, and hence they should be avoided;
- The green nodes are the ones that brings to the destination or to a parking region within the distance θ;
- The yellow is associated to those nodes v having only one outgoing arc, which is directed only to a node having v among its adjacent nodes. Such a configuration might be source of annoying loops, and for this reason we impose that yellow nodes should be visited only once by each vehicle, unless a parking slot is made available in the region: in this last case, a further visit is allowed to complete possibly the parking process.

Clearly, as in the standard version of ACO, in the beginning, the graph does not provide a meaningful pheromone information, indeed it has value 1 on every edge. In this phase, the edge chosen by the driver, among the available ones with the same amount of pheromone, is the one that maximizes the ratio $\frac{a}{d*w}$, where a, d, and w are defined in Sect. 4.

5.2 The Pheromone Model

In the proposed algorithm, in order to avoid situations where all vehicles follow the same path, the pheromone causes a repulsion instead of attraction and it is modified as described below:

- decrease of p value: when a vehicle travels on an arc, the p value of the entering arcs of the node to which it has moved are decreased by:

$$\frac{(d * w)}{(a + 1)} * 10^{-1,9} \tag{2}$$

  The decreased value is deliberately lower to the increased one in order to avoid losing information too quickly.
- increase of p value: when a vehicle leaves a region or when it parks in a region, the p value of the entering arcs of the node is increased by:

$$\frac{a}{(d * w)} * 10^{-2} \tag{3}$$

  The increase is proportional to the available parking spots in the region in which the driver has parked or has leaved and inversely proportional to the distance to travel by road and the one to walk to reach the driver destination node;
- total evaporation: a the end of each step (every vehicle has made a movement) an update of p of each arc is made so that those arches that have been visited previously attractive again:

$$\frac{1}{d * 1000} \tag{4}$$

  this p increase on each visited edge is inversely proportional to the length of the edge in order to disadvantage the choice of longer ones.

where the value of w for the destination node is not equal to 0 but to 0.02 considering that also a parking slot in the destination region is not the place where the driver is headed.

5.3 Constraint Relaxation

In a realistic scenario in which there are few available parking slots, it is not always possible to guarantee that all vehicles will reach the destination or an acceptable node (according to the distance θ) without visiting a node they have already visited; for this reason, a process of relaxation of the constraint on the θ value is needed in order to facilitate the choice of the region in which allocating the car.

More precisely, the relaxation of the constraint on the acceptable walking distance θ, which consists in increasing the distance θ by 0.5, is applied every time a vehicle visits a node that has already been visited previously; in this way, for example, also the red nodes could be considered for allocation (desperate times call for desperate measures).

Through such a trick, unwanted loops are handled, and it is guaranteed that for each driver a solution will be eventually found.

5.4   The Algorithm

As well explained in [20], every algorithm that follows ant colony optimization is based on the repetition of "Initialization", "Construction" and "Updating" until a "Termination condition" is satisfied.

In this section, we are going to precisely illustrate the Algorithm 1 we have built, where:

- D is the drivers population in which every driver has a position, a θ and a list of positions history to keep track of the nodes he has already visited;
- E is the set of edges of the parking regions graph.

After coloring the graph and initializing the drivers population D, as long as there is a driver who has not parked, each driver will always move, step by step, through the edges that maximize the pheromone or the ratio $\frac{a}{d*w}$ ($findNextNode$). Every movement of the drivers results in a decrease and an increase in the pheromone on the edges:

- increase of pheromone: the incoming edges pheromone of a parking region is increased when a driver park or leaves that parking region;
- decrease of pheromone: the incoming edges pheromone of a parking region is decreased when a driver move into it;

Each position taken by the driver is collected in a list so that, if a node is visited more than once by the same driver, the relaxation on θ constraint can take place.

6   Valuation

In this section, we want to provide some simulation results from the execution of the provided algorithm, as well as analyzing the behavior of the algorithm by changing the number of drivers, or by changing the destination node.

Algorithm 1. optAco

1: Color the graph
2: **for** each d in D **do**
3:     *set d → pos equal to start_node*
4:     *set d → θ equal to 0.4*
5: **end for**
6: **repeat**
7:     **for** d in D **do**
8:         **if** d not parked **then**
9:             *n_node ← findNextNode(d → pos)*
10:            **if** *n_node ≥ 0* **then**
11:                *increaseIncomingEdgesPh(d → pos)*
12:                *moveToNode(d, n_node)*
13:                *decreaseIncomingEdgesPh(d → pos)*
14:                **if** *d has already visited n_node* **then**
15:                    *relax Walk Distance Constraint(d)*
16:                **end if**
17:                **if** *n_node → w ≤ d → θ ∧ n_node → a ≥ 0* **then**
18:                    *park d in n_node*
19:                    *increaseIncomingEdgesPh(d → pos)*
20:                **end if**
21:            **end if**
22:        **end if**
23:    **end for**
24:    **for** e in E **do**
25:        $Pheromone_e - = Evaporation$
26:    **end for**
27: **until** *every d parked*

6.1   Simulation Results

Considering the graph in Fig. 2, if the starting node is 0, the driver population has a cardinality of 15, the path of nodes of the graph covered by each agent to reach the destination is shown in the Table 1.

If we consider that at the step 16 parking spots is left and available again in node 14, path of the drivers is shown in Table 2

6.2   Benchamarks

We also perform some observations on the behavior of our algorithm, by starting from the Fig. 2, to check the number of steps needed to complete the parking process, by first varying the number of drivers (Table 3), then the destination nodes (Table 4).

The columns of Table 3 represent the number of drivers, while the rows are the number of steps needed on average to complete the parking process. The columns of Table 4 represent the destination node, while the rows are the number of steps needed on average to complete the parking process.

Table 1. Driver path with always 0 available parking spots in 14

	m_1	m_2	m_3	m_4	m_5	m_6	m_7	m_8	m_9	m_10
D_1	0	1	3	8						
D_2	0	2	7	8	13					
D_3	0	1	3	8	13					
D_4	0	2	7	8	13					
D_5	0	1	3	8	13					
D_6	0	2	7	8	13					
D_7	0	1	3	8	13					
D_8	0	2	7	8	13					
D_9	0	1	3	8	13	3				
D_{10}	0	2	7	8	13	3	8	13	3	
D_{11}	0	1	3	8	13	14	13	3		
D_{12}	0	2	7	8	13	3	8	13	3	
D_{13}	0	1	4	5						
D_{14}	0	2	7	8	13	3	12	13	14	13
D_{15}	0	1	4	1	4					

Table 2. Driver path with a new available parking spots in node 14 from step 16

	m_1	m_2	m_3	m_4	m_5	m_6	m_7	m_8	
D_1	0	1	3	**8**					
D_2	0	2	7	8	**13**				
D_3	0	1	3	8	**13**				
D_4	0	2	7	8	**13**				
D_5	0	1	3	8	**13**				
D_6	0	2	7	8	**13**				
D_7	0	1	3	8	**13**				
D_8	0	2	7	8	**13**				
D_9	0	1	3	8	13	**3**			
D_{10}	0	2	7	8	13	3	8	13	**3**
D_{11}	0	1	3	8	13	14	13	**14**	
D_{12}	0	2	7	8	13	3	8	13	**3**
D_{13}	0	1	4	**5**					
D_{14}	0	2	7	8	13	3	12	13	**3**
D_{15}	0	1	4	1	**4**				

Table 3. Step needed varying the number of drivers

	5	10	15	20	25
Average number of steps	3.8	4.2	5.07	5.7	6.56

Table 4. Step needed varying destination node

	3	8	12
Average number of steps	3.1	4.2	3.7

7   Conclusions

In big urban areas the congestion due to drivers looking for parking is a very big deal. Considering the increasing number of people which is going to populate the cities in the next years, an intervention on such an unease should be taken into account, both for traffic management reasons, but also for drivers comfort and environment pollution reasons.

In this work, we provide an adaptation of the known ant colony optimization, not commonly applied in this field, to favour a higher context-awareness in vehicular ad hoc networks, by using an opportune graph coloring mechanism, ending up with a self-organized network able to indicate to drivers where they should allocate their cars. Such an ability is obtained through an ad hoc pheromone updating, aiming at make more attractive those paths that are not crowded, and less attractive those that have been previously chosen by other drivers.

Simulation results and tests show that the vehicles distribution obtained through the proposed algorithm is fair, and it allows avoiding crowd and congested paths.

Acknowledgments. This paper has been produced with the financial support of the Justice Programme of the European Union, 101046629 CREA2, JUST-2021-EJUSTICE, JUST2027 Programme. The contents of this report are the sole responsibility of the authors and can in no way be taken to reflect the views of the European Commission.

References

1. Greengard, S.: The Internet of Things. MIT Press (2021)
2. Wiseman, Y.: Autonomous vehicles. In: Research Anthology on Cross-Disciplinary Designs and Applications of Automation, pp. 878–889. IGI Global (2022)
3. Ghazal, T.M., et al.: IoT for smart cities: machine learning approaches in smart healthcare-a review. Future Internet **13**(8), 218 (2021)
4. Lapegna, M., Stranieri, S.: DClu: a direction-based clustering algorithm for VANETs management. In: Barolli, L., Yim, K., Chen, H.C. (eds.) Innovative Mobile and Internet Services in Ubiquitous Computing. IMIS 2021. Lecture Notes in Networks and Systems, vol. 279, pp. 253–262. Springer, Cham (2021). https://doi.org/10.1007/978-3-030-79728-7_25

5. Alsarhan, A., Al-Ghuwairi, A.R., Almalkawi, I.T., Alauthman, M., Al-Dubai, A.: Machine learning-driven optimization for intrusion detection in smart vehicular networks. Wirel. Pers. Commun. **117**(4), 3129–3152 (2021)

6. Di Luccio, D., et al.: Coastal marine data crowdsourcing using the internet of floating thingsithe results of a water quality model. IEEE Access **8**, 101209–101223 (2020)

7. Romano, D., Lapegna, M.: A GPU-parallel image coregistration algorithm for InSar processing at the edge. Sensors **21**(17), 5916 (2021)

8. Statista: Proportion of population in cities worldwide from 1985 to 2050 (2021)

9. Pope Iii, C.A., et al.: Lung cancer, cardiopulmonary mortality, and long-term exposure to fine particulate air pollution. Jama **287**(9), 1132–1141 (2002)

10. Shahzad, A., Choi, J., Xiong, N., Kim, Y.-G., Lee, M.: Centralized connectivity for multiwireless edge computing and cellular platform: a smart vehicle parking system. Wirel. Commun. Mob. Comput. **2018** (2018)

11. Singh, P.K., Singh, R., Nandi, S.K., Nandi, S.: Smart contract based decentralized parking management in ITS. In: Lüke, KH., Eichler, G., Erfurth, C., Fahrnberger, G. (eds.) Innovations for Community Services. I4CS 2019. Communications in Computer and Information Science, vol. 1041, pp. 66–77. Springer, Cham (2019). https://doi.org/10.1007/978-3-030-22482-0_6

12. Amato, F., Casola, V., Gaglione, A., Mazzeo, A.: A semantic enriched data model for sensor network interoperability. Simul. Model. Pract. Theory **19**(8), 1745–1757 (2011)

13. Yousefi, S., Mousavi, M.S., Fathy, M.: Vehicular ad hoc networks (VANETs): challenges and perspectives. In: 2006 6th International Conference on ITS Telecommunications, pp. 761–766. IEEE (2006)

14. Naskath, J., Paramasivan, B., Aldabbas, H.: A study on modeling vehicles mobility with MLC for enhancing vehicle-to-vehicle connectivity in VANET. J. Ambient Intell. Hum. Comput. **12**(8), 8255–8264 (2021)

15. Ghaffari, A.: Hybrid opportunistic and position-based routing protocol in vehicular ad hoc networks. J. Ambient Intell. Hum. Comput. **11**(4), 1593–1603 (2020)

16. Sepulcre, M., Gozalvez, J., Härri, J., Hartenstein, H.: Contextual communications congestion control for cooperative vehicular networks. IEEE Trans. Wirel. Commun. **10**(2), 385–389 (2010)

17. Yaqoob, S., Ullah, A., Akbar, M., Imran, M., Shoaib, M.: Congestion avoidance through fog computing in internet of vehicles. J. Ambient Intell. Humaniz. Comput. **10**(10), 3863–3877 (2019). https://doi.org/10.1007/s12652-019-01253-x

18. Wang, X., Shi, H., Zhang, C.: Path planning for intelligent parking system based on improved ant colony optimization. IEEE Access **8**, 65267–65273 (2020)

19. Dorigo, M., Blum, C.: Ant colony optimization theory: a survey. Theor. Comput. Sci. **344**(2–3), 243–278 (2005)

20. Balzano, W., Stranieri, S.: ACOp: an algorithm based on ant colony optimization for parking slot detection. In: Barolli, L., Takizawa, M., Xhafa, F., Enokido, T. (eds.) Web, Artificial Intelligence and Network Applications. WAINA 2019. Advances in Intelligent Systems and Computing, vol. 927, pp. 833–840. Springer, Cham (2019). https://doi.org/10.1007/978-3-030-15035-8_81

21. Viet, N.H., Vien, N.A., Lee, S.G., Chung, T.C.: Obstacle avoidance path planning for mobile robot based on multi colony ANT algorithm. In: First International Conference on Advances in Computer-Human Interaction, pp. 285–289. IEEE (2008)

22. Mamandi, A., Yousefi, S., Atani, R.E.: Game theory-based and heuristic algorithms for parking-IoT search. In: 2015 International Symposium on Computer Science and Software Engineering (CSSE), pp. 1–8. IEEE (2015)

23. Amato, F., Casola, V., Mazzeo, A., and Romano, S.: A semantic based methodology to classify and protect sensitive data in medical records. In: 2010 Sixth International Conference on Information Assurance and Security, pp. 240–246. IEEE (2010)
24. Amato, F., Casola, V., Mazzocca, N., Romano, S.: A semantic approach for fine-grain access control of e-health documents. Log. J. IGPL **21**(4), 692–701 (2013)

A Review About Machine and Deep Learning Approaches for Intelligent User Interfaces

Antonino Ferraro[1(✉)] and Marco Giacalone[2]

[1] University of Naples Federico II, via Claudio 21, 80125 Naples, Italy
`antonino.ferraro@unina.it`
[2] LSTS, Vrije Universiteit Brussel, 4B304, Pleinlaan 2, 1050 Brussels, Belgium
`marco.giacalone@vub.ac.be`

Abstract. The last few years have seen a huge explosion in the use of *Machine Learning* (ML)-based approaches, particularly *Deep Neural Networks* (DNNs) in a variety of fields, to solve complex prediction problems, or in industry to provide a very effective predictive maintenance system for equipment, or in the field of image manipulation and computer vision. In addition, recent publications have contributed to the evolution of *Intelligent User Interfaces* (IUIs) through DNN-based approaches. This paper aims to share a recent overview of published work on the development of IUIs, initially through ML techniques and then, analyze only those based on DNN models. The ultimate goal is to provide researchers with concrete support to be able to develop IUI projects and to be able to inform them about the latest developments on *Artificial Intelligence* (AI) models used in this field.

Keywords: Intelligent User Interface · Intelligent systems · Application fields · Machine Learning · Deep Learning

1 Introduction

Nowadays there is a growing demand for automation of tasks or processes, this is due to many factors, first of all the exponential growth of the Internet and related technologies that led to the increase of information that must be used as input to process an output. The large amount of data to be managed has driven the evolution of intelligent systems for their processing, since for a human being to manage them has become burdensome and complicated. Therefore, the discipline of data science has emerged to exploit the large amount of data available to gain advantages on their correct interpretability by combining the notions of statistics, distributed systems, databases and data mining [1,18]. A popular approach for data analysis by data scientists is ML and other *Artificial Intelligence* (AI) techniques such as *Deep Learning* (DL).

Already now it is evident that many intelligent and interactive systems would not work without an approach based on artificial intelligence, i.e. using ML or

L. Barolli et al. (Eds.): AINA 2022, LNNS 451, pp. 95–103, 2022.
https://doi.org/10.1007/978-3-030-99619-2_9

DL techniques, going to motivate how these techniques contribute to greater development efficiency, in particular we can reflect on the evolution of innovative user interfaces in various fields thanks to the use of these new methodologies [2], for example in predictive maintenance in the industrial environment [3], in forecasting and future estimation of stock markets [4], in data filtering systems [6], in medical diagnostics [7] and in cybersecurity and attack prevention [8].

AI technology enables the creation of IUIs that were previously not possible, as they may be able to make decisions independently or in part, this brings different paradigms of human-computer interaction. Moreover, IUIs based on ML in addition to the interest of the research world, has found open field in the industrial and business scenes, just think of the widespread intelligent video recommender systems that provide users with suggestions for the next videos to be viewed or even recommender systems for online shopping, to optimize the volume of sales, recommending interesting products to visitors of a website. This article aims to provide researchers with a recent overview of User Interface Systems based on AI, in particular ML and DNNs. In particular we want to emphasize articles that exploit these paradigms for the implementation of IUI Systems.

The article organized as follows: Sect. 2 presents an overview of user interface systems that make the most use of Machine Learning and Deep Learning patterns. Section 3 discusses the most recent research articles that have proposed AI-based IUIs, specifically analyzing their methodology. Finally, Sect. 4 concludes the paper by doing a summary analysis of the state of research in this field.

1.1   Application Trends for Intelligent User Interfaces

In recent years IUIs are very common in our lives, this is because they have brought immense benefits not only in terms of fruition of content or services for end users but also in terms of optimization of business processes. The purpose of this paragraph is to report real examples of user interfaces that we use every day and then appreciate their usefulness.

Large amounts of information are problematic to manage, this in any application domain, just think of online stores that have thousands of items for sale. It was necessary to develop algorithms that would critically filter certain items based on the user, then dynamically. Problems of this kind have emerged also in the field of audio and video streaming, in news portals or in social networks [17,23,30]. There are several criteria for filtering and presenting information to an end user including collaborative filters, content-based filters, and filters based on demographic and contextual factors.

Interactive technologies have invaded the market and are now an integral part of everyday life, the use of natural language has become an input for many intelligent interfaces, just think of the questions we ask on search engines through the microphone of a smartphone or the use of voice assistants in the home or interaction with chat bots for after-sales services or requests for assistance. This is a demonstration of how much IUIs have evolved through Human-Computer Interaction (HCI) and Natural Language Processing (NLP).

Finally we can see how text management has become fundamental, the effective interpretation of writing input by a user interface allows for better performance, such as if you want to search for something on the web in the shortest possible time, in keyword recognition or in general for information extraction. NLP and ML have a fundamental role in Intelligent Text Input (e.g. smartphone keyboards that predict what you want to write). Results of this work will be exploited for the design and implementation of the recommender system used in the European Project CREA2 [5,11,25].

Fig. 1. Practical applications of Intelligent User Interfaces (IUIs).

2   IUIs Overview

IUIs are created to improve the dialogue between a human user and a machine that represents or executes a set of patterns to satisfy requests. Moreover, the design of intelligent interfaces draws on different heterogeneous disciplines (e.g. software engineering, AI, psychology, HCI and others) and their integration has resulted in the evolution and release of practical and efficient user interfaces. But in order to get to this point IUIs had to be able to reason about the information of the users using it, meant as a correct interpretation (and also anticipation) of the requests to be satisfied and be able to "learn" what the user wants to get at that moment, analyzing not only what is his static knowledge base but also a dynamic knowledge given by the surrounding world. A strong evolutionary thrust is given mainly to the use of ML techniques, specifically several models are used in IUIs to provide them with learning capabilities including deep neural networks, Bayesian networks, decision trees, decision models based on fuzzy logic etc. The goal of these approaches is to make user interfaces more user-friendly, improve the interaction with them but also provide support to the interface developer.

3   IUIs Classification

In this section we aim to discuss how intelligent interfaces have evolved with the contribution of ML and DL. Specifically we will go over what are the latest

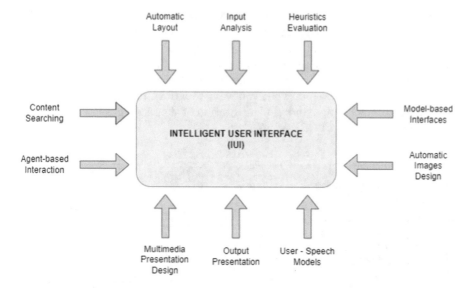

Fig. 2. Targets of an Intelligent User Interface (IUI).

articles in the literature (*Scopus* - from 2021). In particular from the analysis of the literature we can see that the major contributions have been inherent in three macro-categories:

– Chatbots and NLP;
– Adaptive UIs;
– Intelligent Input Text Optimization.

The use of natural language is the main interaction between people and the dialogue could offer new opportunities. The same thing happens with computer systems, the use of natural language with a user interface offers huge advantages to improve human-computer interaction. Chatbots have a fundamental role as a link between a human user and a computer system, it can be defined as a software solution designed to simulate a human conversation: a non-human contact that uses AI algorithms to return a structured dialogue to the end user.

There are many challenges in managing huge amounts of information; often in some domains there is a need to restrict the data output for the end user, just think of streaming music or video and social content [26, 27], obviously through dynamic criteria that vary based on the user of the UI.

Modern IUIs try to predict the behavior of the user using it, based on this assumption methods have been introduced to understand user intentions, specifically probabilistic methods have been introduced. A practical context is modern smart smartphone keyboards, here HCI models have been used to obtain optimal design layouts based on the device in use and statistical models for word correction and prediction.

In the following subsections we will analyze the most recent published articles divided according to the AI techniques used: ML-based or specifically with DNN-based schemes, that are used for the development of recent IUIs.

3.1   Machine Learning-Based Papers

We explore in detail the recently published papers that have proposed a methodology based on Machine Learning, excluding solutions with deep neural networks, which we will address in a dedicated way later.

Regarding Chatbots and NLP, the work of Akkineni et al. [9] discusses the development of a chatbot in an educational context, specifically to support college students. These type of systems have also been proposed in the medical field, including that of Nayak et al. [10] to be able to answer the questions inherent in COVID-19. Also in the field of psychology, papers have been submitted for help with depression, such as the proposal by Liu et al. [7].

About Adaptive User Interfaces, in the automotive domain, Wolf et al. [12] presented a ML-based framework to optimize interactions with the car's onboard interface to avoid inattentions while driving, meanwhile [13] shows a model-based environment for the generation of context-aware graphical user interfaces at runtime. Table 1 summarizes the published papers based on ML techniques.

Table 1. Articles based on ML models (excluding DNN models), evaluated by category, model used, and application context.

Article	Category	Model	Application
[9]	Chatbots and NLP	Retrieval and Generative-based	Educational
[10]	Chatbots and NLP	Natural-language understanding	Health
[12]	Adaptive UIs	Decision Tree Classifier	Automotive
[13]	Adaptive UIs	Statistics models	Health
[14]	Intelligent Input Text	SVM, XG boost	Health

3.2   Deep Neural Network-Based Papers

Now we will focus on recently published articles that have proposed solutions with methodology based on artificial neural networks and we will see how the spread of Deep Learning has also strongly involved the development of intelligent interfaces.

Rocha et al. [15] proposed a chatbot to facilitate self-reading of energy meters, through different ML techniques, including Retina Net networks, while [21] implemented a health-bot using a Recurrent Neural Network (RNN) with the purpose of self-monitoring health.

In the field of text optimization, the article [32] presented a text classification method for tourism questions based on a Long Short Term Memory (LSTM), instead [8] presented a SQL-Injection recognizer based on a Multilayer Perceptron (MLP) model. Table 2 summarizes the published papers based on DNN techniques.

Table 2. Articles based on DNN models, evaluated by category, model used, and application context.

Article	Category	Model	Application
[15]	Chatbots and NLP	Retina Net	Industrial
[20]	Chatbots and NLP	CNN, LSTM, Transformers	Educational
[21]	Chatbots and NLP	Sequential model	Health
[22]	Chatbots and NLP	CNN	Health
[24]	Adaptive UIs	Knowledge Layers (KL)	UI Optimization
[28]	Intelligent Input Text	BERT	Fauna
[29]	Intelligent Input Text	LSTM, BLSTM, GRU	Environmental disaster
[31]	Intelligent Input Text	DLSTA	Human emotion identification
[32]	Intelligent Input Text	LSTM	Tourist trips
[8]	Intelligent Input Text	MLP	Computer security

4 Conclusions

In this paper, we proposed a brief analysis on papers published on *Scopus* since 2021 inherent to Intelligent User Interfaces. The goal is to provide researchers with the latest developments in this field. Particular attention has been paid to papers that have used Machine Learning models, first in a general way, then dwelling on those that have proposed models based on Deep Neural Networks. The articles have been classified by category, model used and application context. We can conclude by stating that ML models have spread massively also in the world of UIs and are used in a variety of fields, such as Health, Industrial, Tourism, Educational, Natural Phenomena, Computer Security and Automotive, this is because these models have allowed interfaces to evolve and be very efficient in a person's daily life, before this was practically impossible.

Acknowledgments. This paper has been produced with the financial support of the Justice Programme of the European Union, 101046629 CREA2, JUST-2021-EJUSTICE, JUST2027 Programme. The contents of this report are the sole responsibility of the authors and can in no way be taken to reflect the views of the European Commission.

References

1. Amato, F., Cozzolino, G., Moscato, F., Moscato, V., Picariello, A., Sperli, G.: Data mining in social network. In: De Pietro, G., Gallo, L., Howlett, R.J., Jain, L.C., Vlacic, L. (eds.) Intelligent Interactive Multimedia Systems and Services, pp. 53–63. Springer, Cham (2019). https://doi.org/10.1007/978-3-319-92231-7_6

2. La Gatta, V., Moscato, V., Postiglione, M., Sperlì, G.: CASTLE: cluster-aided space transformation for local explanations. Exp. Syst. Appl. **179**, 115045 (2021)
3. Petrillo, A., Picariello, A., Santini, S., Scarciello, B., Sperlí, G.: Model-based vehicular prognostics framework using Big Data architecture. Comput. Ind. **115**, 103177 (2020). ISSN 0166-3615. https://doi.org/10.1016/j.compind.2019.103177
4. Xu, H., Chai, L., Luo, Z., Li, S.: Stock movement prediction via gated recurrent unit network based on reinforcement learning with incorporated attention mechanisms. Neurocomputing **467**, 214–228 (2022). ISSN 0925-2312. https://doi.org/10.1016/j.neucom.2021.09.072
5. Giacalone, M., Salehi, S.: CREA: an introduction to conflict resolution with equitative algorithms. In: Romeo, F., Dall'Aglio, M., Giacalone, M., Torino, G. (eds.) Algorithmic Conflict Resolution (2019)
6. Jeyaraj, S., Raghuveera, T.: A deep learning based end-to-end system (F-Gen) for automated email FAQ generation. Exp. Syst. Appl. **187**, 115896 (2022). ISSN 0957-4174. https://doi.org/10.1016/j.eswa.2021.115896
7. Liu, H., Peng, H., Song, X., Xu, C., Zhang, M.: Using AI chatbots to provide self-help depression interventions for university students: a randomized trial of effectiveness. Internet Interv. **27**, 100495 (2022). ISSN 2214-7829. https://doi.org/10.1016/j.invent.2022.100495
8. Jothi, K.R., Balaji, B.S., Pandey, N., Beriwal, P., Amarajan, A.: An efficient SQL injection detection system using deep learning. In: 2021 International Conference on Computational Intelligence and Knowledge Economy (ICCIKE), pp. 442–445 (2021). https://doi.org/10.1109/ICCIKE51210.2021.9410674
9. Akkineni, H., Lakshmi, P.V.S., Sarada, L.: Design and development of retrieval-based chatbot using sentence similarity. In: Nayak, P., Pal, S., Peng, S.-L. (eds.) IoT and Analytics for Sensor Networks: Proceedings of ICWSNUCA 2021, pp. 477–487. Springer, Singapore (2022). https://doi.org/10.1007/978-981-16-2919-8_43
10. Nayak, N.K., Pooja, G., Kumar, R.R., Spandana, M., Shobha, P.: Health assistant bot. In: Shetty, N.R., Patnaik, L.M., Nagaraj, H.C., Hamsavath, P.N., Nalini, N. (eds.) ERCICA 2020, pp. 219–227. Springer, Singapore (2022). https://doi.org/10.1007/978-981-16-1338-8_19
11. Giacalone, M., Salehi, S.S.: Online dispute resolution: the perspective of service providers. In: Romeo, F., Dall'Aglio, M., Giacalone, M., Torino, G. (eds.) Algorithmic Conflict Resolution (2019)
12. Wolf, J., Wiedner, M., Kari, M., Bethge, D.: HMInference: inferring multimodal HMI interactions in automotive screens. In: 13th International Conference on Automotive User Interfaces and Interactive Vehicular Applications, pp. 230–236. Association for Computing Machinery, New York (2021). https://doi.org/10.1145/3409118.3475145
13. Mezhoudi, N., Vanderdonckt, J.: Toward a task-driven intelligent GUI adaptation by mixed-initiative. Int. J. Hum. Comput. Interact. **37**(5), 445–458 (2021). https://doi.org/10.1080/10447318.2020.1824742
14. Wawer, A., Chojnicka, I., Okruszek, L., et al.: Single and cross-disorder detection for autism and schizophrenia. Cogn. Comput. **14**, 461–473 (2022). https://doi.org/10.1007/s12559-021-09834-9
15. Rocha, C.V.M., et al.: A chatbot solution for self-reading energy consumption via chatting applications. J. Control Autom. Electr. Syst. **33**(1), 229–240 (2022). https://doi.org/10.1007/s40313-021-00818-6

16. Qureshi, K.N., Ahmad, A., Piccialli, F., Casolla, G., Jeon, G.: Nature-inspired algorithm-based secure data dissemination framework for smart city networks. Neural Comput. Appl. **33**(17), 10637–10656 (2020). https://doi.org/10.1007/s00521-020-04900-z

17. Canonico, R., et al.: A smart chatbot for specialist domains. In: Barolli, L., Amato, F., Moscato, F., Enokido, T., Takizawa, M. (eds.) WAINA-2020, pp. 1003–1010. Springer, Cham (2020). ISSN 21945357. ISBN 9783030440374. https://doi.org/10.1007/978-3-030-44038-1_92

18. Khan, M., et al.: Enabling multimedia aware vertical handover management in Internet of Things based heterogeneous wireless networks. Multimedia Tools Appl. **76**(24), 25919–25941 (2017)

19. Lv, Z., Piccialli, F.: The security of medical data on internet based on differential privacy technology. ACM Trans. Internet Technol. **21**(3), 1–18 (2021)

20. Saadna, Y., Boudhir, A.A., Ben Ahmed, M.: An analysis of ResNet50 model and RMSprop optimizer for education platform using an intelligent chatbot system. In: Ben Ahmed, M., Teodorescu, H.-N.L., Mazri, T., Subashini, P., Boudhir, A.A. (eds.) Networking, Intelligent Systems and Security: Proceedings of NISS 2021, pp. 577–590. Springer, Singapore (2022). https://doi.org/10.1007/978-981-16-3637-0_41

21. Aggarwal, H., Kapur, S., Bahuguna, V., Nagrath, P., Jain, R.: Chatbot to map medical prognosis and symptoms using machine learning. In: Khanna, K., Estrela, V.V., Rodrigues, J.J.P.C. (eds.) Cyber Security and Digital Forensics: Proceedings of ICCSDF 2021, pp. 75–85. Springer, Singapore (2022). https://doi.org/10.1007/978-981-16-3961-6_8

22. Jain, A., Yadav, K., Alharbi, H.F., Tiwari, S.: IoT & AI enabled three-phase secure and non-invasive COVID 19 diagnosis system. Comput. Mater. Continua (CMC) **71**(1), 423–438 (2022). https://doi.org/10.32604/cmc.2022.020238

23. Amato, A., Cozzolino, G., Giacalone, M.: Opinion mining in consumers food choice and quality perception. In: Barolli, L., Hellinckx, P., Natwichai, J. (eds.) 3PGCIC-2019, pp. 310–317. Springer, Cham (2020). https://doi.org/10.1007/978-3-030-33509-0_28

24. Zouhaier, L., Hlaoui, Y.B.D., Ayed, L.B.: A reinforcement learning based approach of context-driven adaptive user interfaces. In: 2021 IEEE 45th Annual Computers, Software, and Applications Conference (COMPSAC), pp. 1463–1468 (2021). https://doi.org/10.1109/COMPSAC51774.2021.00217

25. Giacalone, M., Loui, R.P.: Dispute resolution with arguments over milestones: changing the representation to facilitate changing the focus. Jusletter IT IRIS **2018**, 167–174 (2018)

26. Amato, F., Moscato, V., Picariello, A., Sperli, G.: KIRA: a system for knowledge-based access to multimedia art collections. In: Proceedings of the IEEE 11th International Conference on Semantic Computing, ICSC 2017, pp. 338–343 (2017). Art. no. 7889559. ISBN 9781509048960. https://doi.org/10.1109/ICSC.2017.59

27. Amato, F., et al.: Challenge: processing web texts for classifying job offers. In: Proceedings of the 2015 IEEE 9th International Conference on Semantic Computing, IEEE ICSC 2015, pp. 460–463. Institute of Electrical and Electronics Engineers Inc. (2015). Art. no. 7050852. ISBN 9781479979356. https://doi.org/10.1109/ICOSC.2015.7050852

28. Edwards, T., Jones, C.B., Corcoran, P.: Identifying wildlife observations on Twitter. Ecol. Inf. **67**, 101500 (2022). ISSN 1574-9541. https://doi.org/10.1016/j.ecoinf.2021.101500

29. Eligüzel, N., Çetinkaya, C., Dereli, T.: Application of named entity recognition on tweets during earthquake disaster: a deep learning-based approach. Soft. Comput. **26**(1), 395–421 (2022). https://doi.org/10.1007/s00500-021-06370-4

30. Castiglione, A., Cozzolino, G., Moscato, V., Sperli, G.: Analysis of community in social networks based on game theory. In: Proceedings of the IEEE 17th International Conference on Dependable, Autonomic and Secure Computing, IEEE 4th Cyber Science and Technology Congress, DASC-PiCom-CBDCom-CyberSciTech 2019, pp. 619–626. Institute of Electrical and Electronics Engineers Inc. (2019). Art. no. 8890406. https://doi.org/10.1109/DASC/PiCom/CBDCom/CyberSciTech.2019.00118

31. Guo, J.: Deep learning approach to text analysis for human emotion detection from big data. J. Intel. Syst. **31**(1), 113–126 (2022). https://doi.org/10.1515/jisys-2022-0001

32. Luo, W., Zhang, L.: Question text classification method of tourism based on deep learning model. Wirel. Commun. Mob. Comput. **2022** (2022). Article ID 4330701, 9 pages. https://doi.org/10.1155/2022/4330701

A Survey on Neural Recommender Systems: Insights from a Bibliographic Analysis

Flora Amato[1], Francesco Di Cicco[2], Mattia Fonisto[1(✉)], and Marco Giacalone[3]

[1] University of Naples Federico II, 80125 Naples, Italy
{flora.amato,mattia.fonisto}@unina.it
[2] University of Turin, 10124 Turin, Italy
francesco.dicicco@unito.it
[3] Vrije Universiteit Brussel, 1050 Brussels, Belgium
marco.giacalone@vub.be

Abstract. In recent years, deep learning has gotten a lot of attention, notably in fields like Computer Vision and Natural Language Processing. With the growing amount of online information, recommender systems have shown to be an effective technique for coping with information overload. The purpose of this article is to provide a comprehensive overview of recent deep learning-based recommender systems. Furthermore, it provides an experimental assessment of prominent topics within the latest published papers in the field. Results showed that explainable AI and Graph Neural Networks are two of the most attractive topics in the field to this day, and that the adoption of deep learning methods is increasing over.

Keywords: Recommender systems · Deep learning · Research trend analysis

1 Introduction

A recommender system's purpose is to make relevant suggestions to a group of users for contents that they might be interested in. As the Internet continues to expand at extremely rapid speed, effective recommender systems for filtering through the flood of contents are becoming increasingly crucial. In general, suggestions are made based on the user's preferences, item attributes, previous interactions between the user and the item, and other factors such as demographic information. The architecture of these recommenders is mainly determined by the domain and the specific characteristics of the data available. Collaborative Filtering (CF), Content-Based Filtering (CBF), and Hybrid Filtering (HF) systems are the three basic types of recommendation models [1].

Deep learning has had a significant impact on recommendation architectures, providing more chances to increase recommender performance. Recent deep learning-based recommender systems have the ability to overcome the limitations of traditional models and produce high-quality recommendations: they can capture non-linear interactions and enable the codification of increasingly complex data sources including textual and visual data. Multilayer Perceptron (MLP), Autoencoder (AE), Convolutional Neural Network (CNN), Recurrent Neural Network (RNN) and Attention Module (AM) are the basic techniques on top of which neural recommendation models are built [2].

The purpose of this study is to examine some literature on deep learning-based recommender systems and to give a bibliographic analysis of most recent works published on the subject. It gives a broad overview for readers to easily grasp the world of deep learning-based recommendation, as well as a basic guideline for selecting deep neural networks to tackle specific recommendation tasks.

To summarize, the paper is structured as it follows:

- In Sect. 2, we present an overview of recommendation models based on conventional and deep learning approaches;
- In Sect. 3, we do a bibliographic study of the most recent articles published on the subject of recommenders;
- In Sect. 4, we discuss current trends and give an insight on the possible future directions of the field.

2 Conventional and Deep Learning Approaches to Recommendation

From now on, we will refer to the entities a system recommends as 'items' and the entities a system recommends to as 'users'.

The generic recommendation problem can be summarised as estimating the ratings for items that have not yet been seen by a user. Intuitively, this estimate is based on the user's evaluations of other things as well as some other data, such as demographic information. Once the ratings for the items that have yet to be rated are estimated, a recommender can recommend the ones with the highest estimated ratings to the user.

The following is a more formal formulation of the suggestion problem: let U represent the set of all users, and V represent the set of all potential items to recommend. Let r be a utility function that quantifies the usefulness of item v to user u, such as $r : U \times V \rightarrow R$, where R is a totally ordered set. Then, for each user $u \in U$, a recommender wants to find an item $v' \in V$ that maximizes the usefulness for that user:

$$\forall u \in U, \quad v'_u = \arg\max_{v \in V} r(u, v)$$

The usefulness $r(u, v)$ of an item v to user u is frequently referred as a 'rating' in recommender systems. The main challenge with these architectures is that the utility function r is specified on a subset of the $U \times V$ space rather than its entirety, which implies that r must be extended to the whole space somehow [3].

2.1 Conventional Approaches

Based on how suggestions are generated, recommender systems are typically categorized into the following types:

- CF recommender systems make recommendations to the user based on the preferences of other users with similar ratings, by exploiting the commonality between user similar ratings;
- CBF recommender systems generate recommendations based on similarities between items similar to those the user liked in the past, by exploiting the descriptive characteristics of items;
- HF recommender systems combines multiple approaches together overcoming disadvantages of one approach by compensating with the other.

Each item and each user are mapped to an embedding vector in a shared embedding space $E = \mathbb{R}^d$ in both content-based and collaborative filtering approaches. The embedding space is often low-dimensional, such that d is significantly lower than $\|U\|$ and $\|V\|$, and captures some latent structure of the item or user set. In the embedding space, similar elements cluster together. A similarity metric, such as cosine similarity, dot product, or Euclidean distance, defines the concept of closeness.

Matrix Factorization (MF) is one conventional collaborative filtering approach. Let Y be the user-item rating matrix $Y \in R^{m \times n}$, where m is the number of users and n is the number of items. Each entry $y_{i,j} \in R$ defines the preference of i-th user for j-th item, while $y_{i,j} = 0$ indicates that preference of i-th user for j-th item is not available. MF aims at determining two embeddings $\bar{U} \in R^{m \times d}$ and $\bar{V} \in R^{n \times d}$ such that $Y \approx \bar{U}\bar{V}^T$.

The idea behind these approaches is to find a way to generate user and item embeddings in a shared space such that their dot product, or some other similarity metric, is a good approximation of user ratings [4].

2.2 Neural Approaches

Deep learning is defined by the ability to learn deep representations from data, which entails learning several levels of representations and abstractions. Moreover, deep neural networks are composite, since they can combine several neural building blocks into a single architecture. This is especially useful when dealing with content-based recommendations as contents are usually multi-modal, such as web contents. CNNs and RNNs, for example, exploit the underlying structure in vision and human language [5], respectively. Traditional recommender algorithms may also be described as neural differentiable networks, as shown in [6]. As a result, the conventional approaches become less appealing.

The existing models are classified in the following subsections depending on the types of deep learning building blocks used.

Autoencoders. One of the issues that need to be tackled for recommender systems is the sparsity of the ratings input matrix, especially in CF recommender systems. There have been works that tried to face this issue with matrix factorization techniques, such as SVD [7] or Regularized Alternative Least Square algorithm [8]. The problem with these methods is that they are linear and do not allow to capture subtler latent factors. Similar or even better results are achieved with neural networks since they are able to catch and naturally represent the non-linearities missed by the above techniques.

The AutoRec model in [9] is one of the first attempts to use autoencoders for CF recommender systems. The results obtained in terms of RMSE show that it was able to outperform previous state-of-the-art solutions, such as the Local Low-Rank Matrix Factorisation (LLORMA) in [10] and the Restricted Boltzmann machine for CF (RBM-CF) in [11].

In [12] an autoencoder is fed with sparse rating inputs in order to reduce their dimensionality and then reconstruct them into dense rating vectors. In particular, Stacked Denoising Auto-Encoders (SDAE) are used to perform CF. To reconstruct the matrix of ratings in order for it to not be sparse, two autoencoders are defined: the *Uencoder*, that takes as input the sparse vector of the user and returns as output a dense vector for the same user, and the *Vencoder*, that does the same with the item vector. Therefore, given a sparse matrix Y of the user ratings, where y_{ij} is the rating the user i has given to item j, the goal is to predict an estimate for the missing ratings $\hat{y}_{ij}$.

Another solution to the same problem is proposed in [13] with the combined use of matrix factorization and Marginalized Denoising Auto-Encoders (mDA), which is a variant of the SDAE used in [12] with a lower computational cost of training with respect to their stacked version. The approach proposed is a hybrid model that takes advantage of both the rating matrix and side information. The latent factors of the first one are captured by making use of matrix factorization, while the side information latent factors are caught thanks to the use of the mDA. The Auto-Encoders can also be stacked in order to gain performance for different training tasks.

CNNs. With convolution and pool operations, CNNs are able to handle unstructured multimedia input. As a result, the majority of CNN-based recommendation models use CNNs to extract features from a variety of unstructured sources, such as photos and videos. It is also possible to apply CNNs directly to feedback matrices when working with collaborative filtering approaches.

In [14], the authors presented a CNN-based recommender model to solve the ACM Multimedia 2016 Grand Challenge user-tag prediction problem. This model integrated both the visual information and the user context in a unified network. The intuition behind this architecture is that an image can have multiple annotations based not only on the visual content, but also on the current context, such as time and location. The network is made up of two branches: the first one handles visual information through a pre-trained AlexNet, while the second one handles the context information through a MLP called ContextNet.

In [15], the authors proposed a CNN-based recommender system to solve the image recommendation problem. They elaborated a two branches network where two images and one user preferences are firstly mapped into the same embedded space and then these embeddings are used to compute some distances. What distinguishes this architecture is the Comparative Deep Learning (CDL) technique used to train the model, which consists of computing the distances between one source against two targets embeddings, one 'positive' and one 'negative', for calculating a triplet loss over the difference between these two. This technique assumes that the distance between the source and a 'positive' target should be less than the one between the source and a 'negative' target.

In [16], the authors tackle the multimedia recommendation problem by exploiting the attention mechanisms when applied to items, such as videos, and their components, such as frames of a video. The idea behind this recommender is that there exists item and component implicitness within users' implicit feedback, such as viewing a video, that blur users' real preferences. The implicit feedback necessitates attention on the collection of items on both levels in order to better describe users' preferences. The proposed network is called Attentive Collaborative Filtering (ACF) and it is based on two attention modules: the component-level attention module, which uses a CNN feature extraction network to learn to select informative components of multimedia items, and the item-level attention module, which learns to score item preferences.

In [17], the authors developed a Convolutional Neural Collaborative Filtering (ConvNCF), employing CNNs within a CF system in a different way than usual. CF traditionally measures the user-item affinity with the inner product in the embedding space. While inner product is excellent in capturing the overall structure in user-item interaction data, the prediction function's expressive potential is limited by its simplicity and linearity. The majority of neural recommender models rely on neural networks to optimize either the user or item representation, as seen above, however this does not let the prediction function capture the correlations among embedding dimensions. ConvNCF works as follows: it first computes the outer product above user and item embedding to explicitly capture pairwise correlations between embedding dimensions, and then it employs a convolution neural network (CNN) above the matrix generated by outer product to learn correlations in a hierarchical manner.

RNNs. The way most recommender systems capture the relationship between users and items over time is usually very limited, and non-existent in early works of collaborative filtering recommender systems, where the user was seen as static and as if its tastes did not evolve in time. Recurrent Neural Networks have been adopted in order to address this limitation, and improve predictions over both short and long-term profiles of collaborative filtering methods.

A first non-neural attempt to incorporate temporal information regarding users' tastes has been made in [18]. The time-aware factor model applied is compared with other factor models, Singular Value Decomposition (SVD) and SVD++ [19], that did not take into account the temporal dimension of the users. The timeSVD++ factor model showed better results with respect to its counterparts for different factor dimensionalities.

Later works focused also on the evolution in time of items and the interactions between users and items. These works made extensive use of RNNs in particular.

In [20], the existing relationship between users and items is looked as in coevolution through time. The RNN is first employed in order to capture this coevolving aspect of users' and items' latent features. The algorithm used is named DeepCoevolve and it models the evolution of users' and items' embeddings by adopting two update functions. The embeddings are updated whenever an interaction occurs between a user and an item.

In [21], the focus is on another aspect useful to recommender systems: the difference between short and long-term recommendations. The results show that RNNs are well-suited to be used in collaborative filtering (CF) recommender systems. In the work was used the GRU architecture [22] with the Hinge objective function, since its complexity is not linear in the number of items as it is the case with the Categorical Cross-Entropy (CCE) which is usually used in language modeling. In order to improve the long-term predictions, given the RNNs' inclination for short-term predictions, three ways were explored: random dropout of items from a training sequence, shuffling the training sequences, selecting multiple targets to be positive in order to produce a loss function that favors long-term predictions. This last approach was possible due to the use of the Hinge loss rather than the CCE. The three approaches were tested on the well-known Netflix dataset recording the ratings of the users' on a catalog of films, and all three led to an improvement on long-term predictions, with the combination of dropout and shuffling being the best.

3   Bibliographic Analysis

The quantity of research papers on recommender systems has grown tremendously in recent years, indicating that deep learning, which is a hot topic in research nowadays, is definitely becoming prevalent in recommender systems. Figure 1 shows the number of publications on the subject of recommender systems from 2017 to 2021.

We conducted a bibliographic analysis over the 1000 most recent papers on the subject of recommender systems as they are published on the arXiv repository. We scraped the arXiv website in order to get the .pdf files, and then extracted the text from them in a format that was suitable for analysis. For this particular task, we made use of a slightly modified version of the repository published by [23].

In an attempt to assess what are the most recent research directions on the subject of recommender systems, we used the Latent Dirichlet Allocation (LDA) algorithm [24] in order to find the most relevant topics in the papers gathered. LDA is a generative probabilistic model of a corpus that estimates the probability of a document to belong to a topic, where a topic is a collection of documents that share semantically similar words.

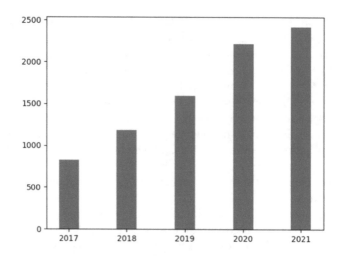

Fig. 1. Number of recommender systems' publications per year

The optimal `n_components` parameter for the LDA procedure [25] has been found by a trial and error approach. We looked at both the t-distributed Stochastic Neighbor Embedding (t-SNE) [26] as well as the top-K words for each clustering obtained by using a different number of topics (`n_components`) in the LDA.

Starting from Fig. 2b, the clusters seem to overlap between each other, indicating that for a larger number of topics it would generate more intricated clusters. The reason for this could be found in the inter-disciplinary scope of some papers that refer to all the different topics, as well as bad clustering in general. A good example of an overlapping topic is provided by the clusters within Fig. 2b, where topic 4 is the most intertwined one. By looking at the top-K words and relevant papers characterizing topic 4, we assess that one possible label for it is 'AI Fairness', which makes sense with respect to its inherent cross-domain applications. By the same logic, we assign labels to all other topics as shown in Table 1, except for topic 3 for which it was not possible to infer a label due to an incorrect sampling.

We chose to describe the results obtained by clustering for 5 topics, since it was possible to get clusters that are well-separated as well as to identify a strong multidisciplinary topic. Following our intent of identifying the most prominent topics in the recommender systems field, we propose topics 1 and 4 as the most interesting and transversal ones. In fact, there is a growing interest in the field of explainable AI [39] and in the adoption of Graph Neural Networks (GNNs) that crosses several fields of research, including the one object of this study.

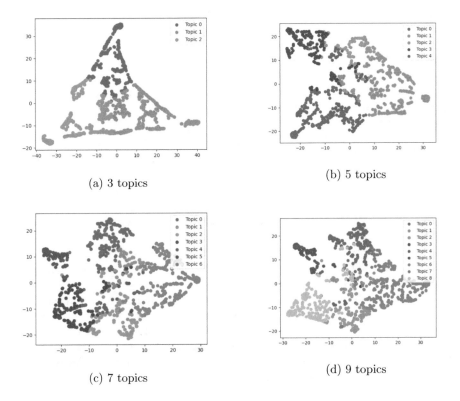

(a) 3 topics

(b) 5 topics

(c) 7 topics

(d) 9 topics

Fig. 2. *t*-SNE for different number of topics

Table 1. Table describing topics in Fig. 2b

Topic	Label	Top-K words	Relevant papers
0	Probabilistic approaches	Condition, determine, variance, minimum, estimate, bound, error, standard, linear, parameter	[27–29]
1	Graph Neural Networks	Capture, feature, batch, attention, dimension, representation, prediction, vector, matrix, layer	[30–32]
2	Computer Vision and Data Visualization	Classification, label, machine, description, content, identify, accuracy, domain, extract, language	[33–35]
3	N.A.	Effect, application, event, documentation, technology, control, communication, access, management, development, environment, support	N.A.
4	AI Fairness	Reward, reinforce, question, agent, person, prefer, experience, world, policy, feedback, decision	[36–38]

4   Conclusions

The goal of the study is to provide a broad review of recommender systems, from classical approaches to the most recent and state-of-the-art deep learning algorithms. Furthermore, in an attempt to suggest valuable directions for future research, we conducted a bibliographic analysis of the most recently published papers in the field.

We found out that there is a growing interest in adopting deep learning methods, such as GNNs, and that there could be further interest in adding explainability to it.

To the best of our knowledge, there is no survey in the field of recommender systems that employed probabilistic generative modeling to discover the main research interests for further work on the basis of recently published papers.

In order to improve performances, the model used for clustering the papers could be changed into a neural one. In addition, the construction of a labeled dataset would make it possible to use supervised learning instead of the unsupervised approach used in this study.

The extrapolation of texts from academic .pdf files is very limited due to the lack of a well-established open-source tool. Moreover, for the bibliographic analysis, the limited sample of papers used could induce a bias issue towards topic extrapolation. An obvious improvement would be to enlarge the dataset, but better tools for scraping an online archive of papers and extrapolating texts from .pdf files are needed for this to be feasible.

Acknowledgments. This paper has been produced with the financial support of the Justice Programme of the European Union, 101046629 CREA2, JUST-2021-EJUSTICE, JUST2027 Programme. The contents of this report are the sole responsibility of the authors and can in no way be taken to reflect the views of the European Commission.

References

1. Melville, P., Sindhwani, V.: Recommender systems. Encyl. Mach. Learn. **1**, 829–838 (2010)
2. Zhang, S., Yao, L., Sun, A., Tay, Y.: Deep learning based recommender system: a survey and new perspectives. ACM Comput. Surv. (CSUR) **52**, 1–38 (2019)
3. Adomavicius, G., Tuzhilin, A.: Toward the next generation of recommender systems: a survey of the state-of-the-art and possible extensions. IEEE Trans. Knowl. Data Eng. **17**, 734–749 (2005). https://doi.org/10.1109/tkde.2005.99
4. Laishram, A., Sahu, S.P., Padmanabhan, V., Udgata, S.K.: Collaborative filtering, matrix factorization and population based search: the nexus unveiled. In: ICONIP (2016). https://doi.org/10.1007/978-3-319-46675-0_39
5. Amato, F., Casola, V., Mazzeo, A., Romano, S.: A semantic based methodology to classify and protect sensitive data in medical records. In: 2010 6th International Conference on Information Assurance and Security, pp. 240–246 (2010)
6. He, X., et al.: Neural collaborative filtering. In: Proceedings of the 26th International Conference on World Wide Web, pp. 173–182 (2017)

7. Koren, Y., et al.: Matrix factorization techniques for recommender systems. IEEE Comput. **42**, 30–37 (2009). https://doi.org/10.1109/mc.2009.263

8. Zhou, Y., Wilkinson, D.M., Schreiber, R., Pan, R.: Large-scale parallel collaborative filtering for the Netflix prize. In: AAIM (2008). https://doi.org/10.1007/978-3-540-68880-8_32

9. Sedhain, S., Menon, A.K., Sanner, S., Xie, L.: AutoRec: autoencoders meet collaborative filtering. In: WWW (2015). https://doi.org/10.1145/2740908.2742726

10. Lee, J., Kim, S., Lebanon, G., Singer, Y., Singer, Y.: Local low-rank matrix approximation. In: ICML (2013)

11. Salakhutdinov, R., Mnih, A., Hinton, G.E.: Restricted Boltzmann machines for collaborative filtering. In: ICML 2007 (2007). https://doi.org/10.1145/1273496.1273596

12. Strub, F., Mary, J.: Collaborative filtering with stacked denoising autoencoders and sparse inputs. In: NIPS 2015 (2015)

13. Li, S., Kawale, J., Fu, Y.: Deep collaborative filtering via marginalized denoising auto-encoder. In: CIKM (2015). https://doi.org/10.1145/2806416.2806527

14. Rawat, Y.S., Kankanhalli, M.S.: ConTagNet: exploiting user context for image tag recommendation. In: ACM International Conference on Multimedia (2016). https://doi.org/10.1145/2964284.2984068

15. Lei, C., Liu, D., Li, W., Zha, Z.-J., Li, H.: Comparative deep learning of hybrid representations for image recommendations. In: 2016 IEEE Conference on Computer Vision and Pattern Recognition (CVPR) (2016). https://doi.org/10.1109/cvpr.2016.279

16. Chen, J., et al.: Attentive collaborative filtering: multimedia recommendation with item- and component-level attention (2017). https://doi.org/10.1145/3077136.3080797

17. Du, X., et al.: Modeling embedding dimension correlations via convolutional neural collaborative filtering. arXiv arXiv:1906.11171 (2019)

18. Koren, Y.: Collaborative filtering with temporal dynamics. In: KDD (2009). https://doi.org/10.1145/1557019.1557072

19. Koren, Y.: Factorization meets the neighborhood: a multifaceted collaborative filtering model. In: KDD (2008). https://doi.org/10.1145/1401890.1401944

20. Dai, H., Wang, Y., Trivedi, R., Song, L.: Deep coevolutionary network: embedding user and item features for recommendation. arXiv arXiv:1609.03675 (2016)

21. Devooght, R., Bersini, H.: Long and short-term recommendations with recurrent neural networks. In: UMAP (2017). https://doi.org/10.1145/3079628.3079670

22. Cho, K., van Merriënboer, B., Bahdanau, D., Bengio, Y.: On the properties of neural machine translation: encoder-decoder approaches. In: SSST@EMNLP (2014). https://doi.org/10.3115/v1/w14-4012

23. Yu, C., Zhang, C., Wang, J.: Extracting body text from academic PDF documents for text mining. arXiv arXiv:2010.12647 [cs.IR] (2020)

24. Blei, D.M., Ng, A.Y., Jordan, M.I.: Latent dirichlet allocation. J. Mach. Learn. Res. **3**, 993–1022 (2001). https://doi.org/10.1016/b978-0-12-411519-4.00006-9

25. Pedregosa, F., et al.: scikit-learn: machine learning in Python. J. Mach. Learn. Res. **12**, 2825–2830 (2011)

26. Van der Maaten, L., Hinton, G.E.: Visualizing data using t-SNE. J. Mach. Learn. Res. **12**, 2825–2830 (2008)

27. Han, C., Rao, N., Sorokina, D., Subbian, K.: Scalable feature selection for (multi-task) gradient boosted trees. arXiv arXiv:2109.01965 [stat.ML] (2021)

28. Abroshan, M., Yip, K.H., Tekin, C., van der Schaar, M.: Conservative policy construction using variational autoencoders for logged data with missing values. arXiv arXiv:2109.03747 [cs.LG] (2021)
29. Giorgio, G., Federico, N., Massimo, R.: Matrix completion of world trade. arXiv arXiv:2109.03930 [econ.G4N] (2021)
30. Shen, Y., et al.: How powerful is graph convolution for recommendation? arXiv arXiv:2108.07567 [cs.IR] (2021)
31. Chen, Y., et al.: Attentive knowledge-aware graph convolutional networks with collaborative guidance for personalized recommendation. arXiv arXiv:2109.02046 [cs.IR]. 12 Amato, Di Cicco, Fonisto, Giacalone (2022)
32. Jung, S., et al.: Global-local item embedding for temporal set prediction. In: 15th ACM Conference on Recommender Systems, September 2021 (2021). https://doi.org/10.1145/3460231.3478844
33. Zhang, Z., Zhang, C., Niu, Z., Wang, L., Liu, Y.: GeneAnnotator: a semi-automatic annotation tool for visual scene graph. arXiv arXiv:2109.02226 [cs.CV] (2021)
34. Zeng, Z., et al.: An evaluation-focused framework for visualization recommendation algorithms. arXiv arXiv:2109.02706 [cs.HC] (2021)
35. Bulathwela, S., Perez-Ortiz, M., Novak, E., Yilmaz, E., Shawe-Taylor, J.: PEEK: a large dataset of learner engagement with educational videos. arXiv arXiv:2109.03154 [cs.IR] (2021)
36. Dori-Hacohen, S., et al.: Fairness via AI: bias reduction in medical information. arXiv arXiv:2109.02202 [cs.AI] (2021)
37. Chaney, A.J.B.: Recommendation system simulations: a discussion of two key challenges. arXiv arXiv:2109.02475 [cs.IR] (2021)
38. Zhang, D., Wang, J.: Recommendation fairness: from static to dynamic. arXiv arXiv:2109.03150 [cs.IR] (2021)
39. Covert, I., Lundberg, S., Lee, S.-I.: Explaining by removing: a unified framework for model explanation. arXiv arXiv:2011.14878 [cs.LG] (2020)

Information Networking and e-Government in United Nations and Europe

Alfonso Marino, Paolo Pariso, and Michele Picariello(✉)

Dipartimento di Ingegneria, Università degli Studi della Campania "Luigi Vanvitelli",
Via Roma, 29, Aversa, CE, Italy
{alfonso.marino,paolo.pariso,michele.picariello}@unicampania.it

Abstract. The purpose of this paper is, theoretically and empirically, to explore how public e-government models are used, analyzing their strengths and weakness. The study intends to analyze the typology of e-government models, their maturity and their implementation, bearing on the Analysis of the international organization UN e-government 2020 report that shows the countries with the biggest implementation of Digital Government. The study aims to know, through a SWOT and correlational analysis, whether they are relevant for productivity and efficiency, effectiveness, inclusion and sustainability, legitimacy and trust. The analysis of knowledge, allowed us to stress the key role of contingent factors around the digital transition process of presidency action. It was also highlighted that there is no correlation with GDP, but it is always hampered by the digital divide.

Keywords: E-government · E-bureaucracy · ICT · Public sector · Digital divide

1 Introduction

The implementation of a digital process through information technologies (ICT) has as its fundamental aim the advance of effectiveness and efficiency: within the government, amplifies the perception of transparency and increases the previous aspects, arriving to ferry the complete concept of presidency in e-government. The serious change needed for the digital transition has been called into question by the appearance of the COVID-19 pandemic [15]. At this particular time in history, where social distancing has been one of all the cornerstones of the fight against the spread of the virus, the employment of tools for the general public sector supported digital identity, shared infrastructure and customary services, has proved to be crucial. That's why the employment of digital technologies and data [1] strategically, has allowed - to some realities - to adapt to the pandemic crisis with agility and speed. This phenomenon has become a part of the strategic vision of the state at the very best level, to confirm more efficiency and transparency within the public sector, and to produce better, cheaper and faster services to citizens and businesses. Increasingly, people's expectations of services, also because the way and quality of its delivery, are changing, and the government must meet these needs and expectations in United Nation (UN) and European Union (EU) [2] areas.

2 Literature Review

The standard and reliability of public services [3] include a major impact on the integrity and trust [7] of public sector organizations [4]. In contrast, poorly designed and implemented services that don't embed citizen demands at their core may undermine the flexibility of the presidency to see societal welfare and people's wellbeing [5]. The adoption of digital technologies [6] and the application of information can transform the interior processes and operations of the state and, consequently, how they design and deliver public services. Digital technologies enable greater efficiency, agility and responsiveness in governments, allowing them to react quickly and even anticipate people's needs. E-government is geared toward bringing greater sectoral efficiencies through the adoption of digital technologies, making existing procedures and public services more cost and time-effective. Governments embraced digital technologies with a view to improving public services, but their approach often lacked coherence and sustainability across different sectors and levels, which is significant to making synergies for integrated, seamless and proactive service delivery. Numerous factors contribute to the maturity of eGovernment, all of which are thanks to the expansion of technology and therefore the complexity of human life. The complexity of presidency actions must even be met if an e-government process is adopted. [8] The foremost important factors within the must create maturity in eGovernment are Widespread Global Network Growth, Network-Based Growth, Reduce costs, Rising public expectations.

Currently, many case studies have relative success in their application, to continue the implementation of Digital Government strategies [9]. The most important obstacle is to form these applications and their architectures reliable, customizable and particularly effective. This goal reduces, within the wake of previous experiences, the possibilities of failure of subsequent Digital Government architectures [10, 11] precisely because they need to be considered "native" within the whole government structural methods. Having a conscious and knowledgeable design of the project of transition to the digital action of the presidency, allows you to master the legislative, semantic, logical and organizational aspects [12]. The choice of a particular architecture of the Digital Government allows an accurate description of the decision-making process that underlines the importance to spot some strategic steps in such process, namely the identification of the varieties of Digital Government currently in use; the characteristics of Digital Government architecture and therefore the associated challenges; the basic architecture building blocks found in Digital Government architectures.

At this time, the questions we ask ourselves are which models ensure a good and efficient transition process from standard government processes to Digital Government? Does the Digital Divide [16] affect the transition process? Additionally, is GDP a limiting factor for the advancement and development of the technological process?.

3 Methodology

In order to gauge research questions as conceptual background, the strategy has been wont to examine different models of e-government. To do this, we have used printed and digital bibliographic resources that also sit down with global cases. Because of the immense bibliographical availability, we've got surveyed different areas of the maturity of the e-government, realizing a comparative classification with other studies. Our attention was focused on Journal ad Reference Papers written in English and published by 2019 to 2021.

4 Empirical Results

From the listed categories, it emerges that we have five relations between government and stakeholders (C2G, B2G, G2N, G2C, and G2E). The first, C2G, refers to direct contact with the private citizen for private use, like the renewal of hunting and fishing licenses, payment of taxes, infringements, state taxes, and requests for reimbursement. The B2G model refers to the services required by entrepreneurs, businesses and corporations, for commercial purposes (profit or non-profit) like - for instance - applications for commercial licenses, application for a brand new workforce through employment agencies. Finally, from G2N to G2E, through the G2G, everyone works on two-way communication between government and government agencies or employees, to enhance services within the state or outside it. Therefore, all this can be possible at a time when there is cooperation pushed between multiple actors: government organizations, individuals and corporations, take a key role inside and out of doors the general public sector to make public value. Having, therefore, a holistic approach makes completely far more understandable and feasible. The ecosystem must be in a position to supply growth, predisposed to vary and technological-social innovation. It should be home-grown, advantage local knowledge, also inclusive, and make sure that any transformation is geared toward creating equal opportunities for all people to access reliable and quality services. Meanwhile, it should be collaborative since providing integrated digital services requires a high degree of coordination among ministries and agencies and new mindsets in government and society. Additionally, learn by people-centric approaches to service delivery and program management, addressing concrete problems and desires experienced by different groups in society. At the systemic level, a holistic approach to digital government transformation requires building deep capabilities and capacities. [13] Political commitment at the very best levels of a presidency is crucial, as could be a clear vision of the aim of state transformation guided by a group of core values that are aligned with the 2030 Agenda for Sustainable Development. Identifying the right process to define an e-government model is vital and to try to do so the process of implementing digital government transformation and highlighting the key pillars of a method and implementation plan is used as a capacity development tool to spot the weather and steps needed to maneuver the digital government transformation process forward. The whole process is iterative so that it is improved and is schematizable in four procedural steps:

Set true analysis to assess digital transformation capacity gaps and opportunities; Articulate a shared vision of state transformation and the way digital technologies are accustomed achieve societal goals;

Use a technique and a digital government implementation road map within which key pillars are identified;

Put monitoring and evaluation mechanisms in situ to gather feedback that ought to then be wont to inform the following rounds of situation analysis, strategy development and implementation.

We want certain key points.

It is necessary, at this time, to proceed with an analysis that permits in an exceedingly more specific way the points on which to intervene within the "governmental machine" and which are the basic aspects and the motivations for the digital transformation. The government has to know this level of information in IT, DATA, computer science (AI), cybersecurity, privacy and other critical aspects, to work out the number of basic skills within the country. In this way, we are able to determine the minimum level of competence and professionalism necessary to be able to implement the complete digitization process.

In this process, emerges how complex it's to go to define a development model of eGovernment and every one the method activities associated with it, thanks to a range of political, economic, and technological reasons, and therefore the completion of related projects and activities. That is why some e-government experts [14] have decided to draw a transparent vision for the e-government by designing a scientific model and by clarifying the implementation steps, public sector agents will help to more accurately assess the progress of the implementation. It should be noted that these models of eGovernment are compared for various degrees of maturity and by a variety of country, where applied.

5 Discussion

Starting from Methodology and research [17] questions above mentioned the discussion highlight that some countries, who show the possible impact of Digital Government Transformation, conducted as part of the research. The idea is to show a real aspect of what has been introduced in the previous paragraph in terms of productivity and efficiency; effectiveness, inclusion and sustainability; legitimacy and trust.

The idea of having an overall government vision that takes into account the needs of stakeholders, bringing together the results in a single e-government portal, represents a clear sign of the success of government policies, where there is a single point of access, data collection, request tax and legal documentation, as well as having the possibility of total customization of the electronic portfolio. All the countries on the list have a comprehensive legal and regulatory framework for e-government that establishes guidelines relating to digital identity, personal data and online information. To this is added specific national strategy for new technologies such as artificial intelligence, deep machine learning and block chain. Interoperability is absolutely an integral part of the project, to ensure an increase in the quality of the related services, ensuring that the entire government process and its model are rethought. Another aspect that emerges is the positive correlation between GDP per capita and the value of EGDI. Countries with a higher gross domestic product, also by more advanced technological progress, can exploit these two aspects as an advantage for the digital transition.

The exchange of information between government and stakeholders should be fair and supported by regular transitions of information flows. This two-way exchange strengthens a positive trend with the public, even sharing technical, economic and/or financial information. The circumstance not to be neglected is that this data can be "machine ready" that is to be readable even by machine format and not only "human interface.

6 Conclusions

The conclusions of this paper is that at the theoretically level e-government is developing through ICT and digital platform, at the same time, the research as put on evidence some organizational bottlenecks (I.e. Digital Divide). This analysis was performed both UN and EU geographic areas. These geographic areas are relevant for productivity and efficiency, effectiveness, inclusion and sustainability, legitimacy and trust. A continuously focus on the transition of UN and UE areas, is strategic for accumulation knowledge, linked to a better insight of the phenomenon.

References

1. Juana-Espinosa, S., Luján-Mora, S.: Open government data portals in the European Union: considerations, development, and expectations. Technol. Forecast. Soc. Change **149**, 119769 (2019). https://doi.org/10.1016/j.techfore.2019.119769, ISSN 0040-1625
2. Marino, A., Pariso, P., Picariello, M.: The influence of digital levers on the growth of social contexts and economic development within European countries. In: 2021 The 5th International Conference on E-Commerce, E-Business and E-Government, pp. 141–146, April 2021
3. Marino, A., Pariso, P., Picariello, M.: E-government and Italian local government: managerial approach in two macro areas to improve manager's culture and services. In: Barolli, L., Woungang, I., Enokido, T. (eds.) Advanced Information Networking and Applications. AINA 2021. Lecture Notes in Networks and Systems, vol. 227, pp. 107–116. Springer, Cham (2021). https://doi.org/10.1007/978-3-030-75078-7_12
4. OECD: Government at a Glance 2019, OECD Publishing, Paris (2019). https://www.oecd-ili brary.org/governance/government-at-a-glance-2019_8ccf5c38-en. Accessed 08 April 2021
5. Welby, B.: The impact of digital government on citizen well-being. OECD Working Papers on Public Governance, No. 32, OECD Publishing, Paris (2019). https://doi.org/10.1787/24b ac82f-en. Accessed 08 April 2021
6. Stratu-Strelet, D., Gil-Gómez, H., Oltra-Badenes, R., Oltra-Gutierrez, J.V.: Critical factors in the institutionalization of e-participation in e-government in Europe: technology or leadership? Technol. Forecast. Soc. Change **164**, 120489 (2020). https://doi.org/10.1016/j.techfore. 2020.120489
7. Porumbescu, G.A.: Linking public sector social media and e-government website use to trust in government. Gov. Inf. Q. **33**(2), 291–304 (2016)
8. Ghorbanizadeh, V., Radsaz, H., Abbaspour, J.: Fragmentation of barriers to the establishment of e-Government in Iran. Q. J. Inf. Technol. Manag. Stud. Sec. Year **39**(8), 1–32 (2014)
9. Martin, N.J., Gregor, S.D., Hart, D.: Using a common architecture in Australian e-Government: the case of smart service Queensland. In: ICEC 2004 (2004)
10. Tambouris, E., Kaliva, E., Liaros, M., Tarabanis, K.: A reference requirements set for public service provision enterprise architectures. Softw. Syst. Model. **13**(3), 991–1013 (2012). https://doi.org/10.1007/s10270-012-0303-7
11. Hornnes, E., Jansen, A., Langeland, Ø.: How to develop an open and flexible information infrastructure for the public sector? In: Wimmer, M.A., Chappelet, J.-L., Janssen, M., Scholl, H.J. (eds.) EGOV 2010. LNCS, vol. 6228, pp. 301–314. Springer, Heidelberg (2010). https://doi.org/10.1007/978-3-642-14799-9_26
12. EU European Commission: EIRA—Joinup. European Commission (DG Informatics)—ISA2 Programme (2019). https://joinup.ec.europa.eu/solution/eira/release/v300. Accessed 12 July 07

13. International Telecommunication Union: New initiatives to support digital literacy for seniors in Singapore. Digital Inclusion Newslog. http://digitalinclusionnewslog.itu.int/2018/12/22/new-initiatives-to-support-digitalliteracy-for-seniors-in-singapore/. Accessed 22 Dec 2018

14. Fath-Allah, A., Cheikhi, L., Al-Qutaish, R., Idri, A.: E-Government maturity models: a comparative study. Int. J. Softw. Eng. Appl. **5**, 71–91 (2014). https://doi.org/10.5121/ijsea.2014.5306

15. Marino, A., Pariso, P.: The global macroeconomic impact of covid-19: four European scenarios. Acad. Strateg. Manag. J. **20**(2), 1–21 (2021)

16. Marino, A., Pariso, P.: From digital divide to e-government: re-engineering process and bureaucracy in public service delivery. Electron. Gov. Int. J. **16**(3), 314–325 (2020)

17. Marino, A., Pariso, P.: Human resource management in public transports: organizational typologies and research actions. VINE J. Inf. Knowl. Manag. Syst. (2021). https://doi.org/10.1108/VJIKMS-01-2021-0006

A Microservices Based Architecture for the Sentiment Analysis of Tweets

Beniamino Di Martino[1,2], Vincenzo Bombace[1], Salvatore D'Angelo[1], and Antonio Esposito[1(✉)]

[1] Department of Engineering, University of Campania "Luigi Vanvitelli", Caserta, Italy
{beniamino.dimartino,salvatore.dangelo,antonio.esposito}@unicampania.it,
vincenzo.bombace@studenti.unicampania.it
[2] Department of Computer Science and Information Engineering, Asia University, Taichung, Taiwan

Abstract. Sentiment Analysis techniques have been largely applied to Tweets, newsgroups and Social Networks in general, with several applications in sociological studies. Users tend to comment and express their opinions much more genuinely on Social Networks, as if their natural filters were somehow lifted. In particular, complaints regarding malfunctions of specific services are often filed in form of public comments or Tweets, on the official accounts of the Service providers. In some cases, people just express dissatisfaction regarding services on their own accounts, and use hashtags to better identify the specific topic they are referring to. In this paper, a framework for the analysis of Tweets is proposed, with the specific objective to identify malfunctioning of essential services, such as water, electrical, gas or public illumination. Since the number of comments and Tweets to analyse is considerable, a microservices based architecture, with Docker containers and Kafka queues, has been created. This allows to define a scalable and parallelizable architecture, whose characteristics can be adapted to the number of Tweets to be analysed, which are in turn treated as a continuous data streaming.

1 Introduction

The application of Sentiment Analysis techniques to comments reported on Social Networks, especially Twitter, has vastly increased in the past years. It is quite common for data analysts to use Twitter's data to experiment with their new solutions and to implement new Natural Language Processing (NLP) techniques and algorithms. Commercial and political uses of such techniques are probably the most common applications: brands often try to understand the customers' feelings towards their products, in order to improve their commercials and reach wider audiences, while Political Parties try to catch people's opinions and to adapt to the public reactions in order to gain more consent.

While the application of Sentiment Analysis techniques on Tweets is not a novelty, there is still room for improvement, especially as regards the implementation of flexible, scalable and easy to use frameworks. In this paper a Container

© The Author(s), under exclusive license to Springer Nature Switzerland AG 2022
L. Barolli et al. (Eds.): AINA 2022, LNNS 451, pp. 121–130, 2022.
https://doi.org/10.1007/978-3-030-99619-2_12

based framework for the Sentiment Analysis of Tweets is presented. The use of Microservices and containers to implement such a framework is the focus of the work, as it provides the scalability and flexibility that most of the existing, centralized solutions, do not [6]. Edge computing solutions that distribute the computation over remote nodes and servers are available [4], but they are not always feasible or are far too complicated for certain tasks.

Section 2 of this paper describes the general design of the framework, while the implementation details are reported in Sect. 3, with screens of the realised framework in use. Section 4 closes the paper with remarks and pointers to future developments.

2   The Container Based Framework

The architecture used to realize the Twitter analysis framework described in this paper, is based on a microservices structure, a design model created through independent components that carry out specific processes. What distinguishes the microservices-based architecture from traditional monolithic approaches is the breakdown of the application into its basic functions. Each function, identified with a service, can be designed, developed and implemented independently. Therefore, independent services can work without expressly relying on other existing components, and will continue to fulfill their objectives even if the whole environment in which they run changes, provided that they continue to receive the requested input.

In monolithic architectures, all processes are closely linked to each other and run as a single service. This means that adding a new functionality, or improving an existing one, becomes much more complex in a monolithic system. This complexity limits experimentation and makes it more difficult to implement new ideas. Monolithic architectures represent an additional risk to application availability, since the presence of numerous dependent and closely related processes increases the impact that an error in a single process has on the whole framework.

A microservices-based architecture is built from independent components that run each application process as a service. These services communicate through a well-defined interface, that in most uses lightweight APIs, but that can also rely on communication queues. Each service performs only one function, but it must do it efficiently and efficaciously; each service can be updated, distributed and resized to respond to the request for specific functions of an application, without interfering with the other existing microservices.

Given these premises, the reasons to choose a microservices based architecture for the development of the framework presented in this work are justified by a series of non-functional requirements that the architecture must meet:

- **Agility**. Since the world of Sentiment Analysis is rapidly evolving, with new techniques becoming available continuously, changes in the framework are to be expected. Development teams need to act in small, well-defined contexts so they can work more independently and quickly. This reduces the development cycle times.
- **Scalability and Flexibility**. Microservices allow you to scale each service independently to meet the functionality required for the specific application. In the case of the presented framework, more resources can be allocated to a single task, for example to accelerate the acquisition of Tweets or the analysis of ingested ones.
- **Simplicity and Distribution**. Microservices support continuous integration and continuous delivery, so new algorithms can be tested more easily, and it is possible to eventually restore previous settings when something goes wrong without important consequences on the framework.
- **Technological freedom**. Microservices-based architectures do not necessarily apply a single approach to the entire application. Development teams have the freedom to choose the best tools to solve their specific problems. As a result, the teams building microservices can choose the best tool for each particular objective or goal their component has been designed for. In our development, Flask [5] and Python are the most used technologies, but nothing prevents to use Javascript and/or Java to implement the very same components.
- **Reusability of the code**. Dividing the software into small and well-defined modules allows the complete reusability of the produced code.
- **Resilience**: The independence of services increases the resilience of an application in case of errors. In a monolithic architecture, an error in a single component could have repercussions on the entire application. With microservices, applications can fully handle service failures by isolating functionality without blocking the entire framework.

2.1 Design of the Framework

The methodology applied in the present study takes in consideration data extracted from Twitter, and made available by registering to the Twitter Developer portal, which gives access to a free API. The API is free to use, but it limits the number of calls that a developer can do per day.

The Tweets examined in this study will not be downloaded randomly, but they will be selected by hashtag, which will thus define a precise context for the text contained within the Tweets.

The goal is therefore to use the information contained in the Tweets to carry out a sentiment analysis, using a microservices architecture, where each service is contained within a limited container, and will play a well-defined role.

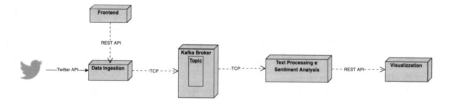

Fig. 1. The components of the container based framework

As shown in Fig. 1, the framework is constituted by four components:

- a **Frontend** component, which allows users to communicate with the Platform through a users friendly Graphical Interface
- a **Twitter data ingestion** component, which gathers data from the Twitter platform using its native APIs.
- a **Kafka Queue** [7], acting as a Broker, which collects the information gathered by the ingestion components and allows the other containers in the framework to access them at their own pace.
- a **Text Processing and Sentiment Analysis** component which acts as the core of the whole platform, having the objective to actually elaborate the incoming data and to run the sentiment analysis,
- a **Visualization** component that integrates with the graphical interface shown in the front-end, that supports the navigation of the results obtained through the Sentiment Analysis.

The platform is thus modelled after a Big Data pipeline, such the one presented in [3], but it uses components developed within containers that communicate with each other through Rest APIs or Kafka queues. Each of these components runs within a separate container, which allows for the creation of multiple instances of them, running in parallel and thus providing scalability to the whole framework. The container that executes the **Frontend** actually runs a Flask based web interface, whose specific objective is to show a simplified user interface. This interface is used to enter the specific keywords that filter the original Tweets downloaded from the Twitter platform. The keywords are translated into hashtags, that Twitter users generally apply to their comments, in order to provide some kind of context. Since we are interested in reducing the number of Tweets to analyse at the same time, and to specify a particular context on which the analysis can focus, using customized hashtags for filtering seems the most viable solution.

The Frontend starts the research of Tweets, by sending the required hashtags to the Data Ingestion component through a Restful API.

The **Twitter data ingestion** component has been realized through a microservice running within a container. It employs, as the Frontend, a Flask-based service that receives the requested tags from the Fontend, and then uses the APIs provided by the Twitter platform to retrieve a set of Tweets. Once acquired, the Tweets are shown to the framework user as they are, without any kind of pre-processing, and are then sent to the Kafka Queue, by using the *fromTwitter* topic as a collector. The **Kafka queue** acts as a Broker for the incoming Tweets, and thanks to the possibility to use different topics at the same time, it fully supports the scalability of the whole framework. The **Text Processing e Sentiment Analysis** component has been designed as a Python based microservice, executed within a container, which takes the data from the Kafka queue, using the aforementioned "fromTwitter" topic, and implements two separated substeps of the whole processing workflow:

1. Initially, the microservice cleans all the incoming information, operating as a Data Curator, by removing all stop-words, links, symbols (like emojis) and punctuation marks. After this, it applies stemming algorithms to the text, in order to simplify it. Stemming consists in reducing a word, or set of words, to its base form, thus eliminating not important information.
2. The second step consists in the Sentiment Analysis, which evaluates the stemmed text and assigns it a value, representing its polarity. The examined text and the polarity are organized into a Pandas Dataframe, and can be retrieved by any external service through an API, also implemented in Python.

The **Visualization** microservice runs within a different container, and uses Flask to show a user friendly web interface, where the results obtained by the Sentiment Analyser are presented. In particular, the data are retrieved from the Sentiment Analyser container through its exposed APIs, and will be available through two different views:

- A tabular view, in which the Tweets are simply listed with their corresponding polarity value-
- A Map based view, in which Tweets are geo-localized and shown on a World Map, where the polarity of the Tweet is stressed by the color used to represent it on the Map.

For each of these components, Sect. 3 reports more detailed information and screens of their graphical realization.

3   Platform Implementation

In order to implement the architecture presented in Fig. 1 using containers, the Docker technologies has been applied. In particular, a standard *docker-compose.yam* file has been created, containing all the information deemed necessary to create the several Docker containers that compose the final application, together with the network connecting them. In this way, we obtain a multi-container framework, where each container runs a microservice, connected to one another through a shared network and a Kafka queue. The images used to create the Kafka and Apache Zookeeper containers (the latter is needed to correctly manage Kafka) are downloaded from the Docker Hub, and are represented by **wurstmeister/kafka** and **wurstmeister/zookeeper**[1]. These two images have hundreds of pulls from the repository, and also simplify the configuration of the containers, by exposing specific parameters directly in the Docker compose file, which are necessary to set-up the Kafka service.

- **KAFKA_ADVERTISED_HOST_NAME** sets the name of the Kafka queue to which all other containers will be able to connect.
- **KAFKA_ZOOKEEPER_CONNECT** sets address and port of the Kafka container.
- **KAFKA_CREATE_TOPICS** generate the Topics, that are part of Kafka configuration, and are needed to keep and retrieve messages from the Queue.

The other services are obtained by including Python scripts in the configuration file, through different *Dockerfile* dedicated to each container, which are then called by the *docker-compose.yam* file.

As already stated in the description of the design of the application reported in Sect. 2, the Front-end of the whole system has been realized using the Flask technology, which allows for the development of simple, yet efficient and captivating, web pages. In our case, a Python script implementing the Flask web page has been created, which incorporates the possibility to choose specific hashtags on which the research of Tweets can start, and launch the research and the Sentiment Analysis.

Figure 2 reports instead the web page that allows users to choose the hashtags to use to download the Tweets. The interface also allows for the selection of a maximum number of Tweets to analyse, and to start the download and analysis by clicking on a simple button. The Form shown by the Front-end has been implemented using the *wtforms* module, that allows to set specific rules which can refuse hashtags and keywords considered not fit for the research (slang words, swearing or similar). A list of currently examined hashtags is shown on the page, so that the user always knows what the system is looking for each time the research starts.

All the data representing the hashtags and the number of Tweets to downloaded are coded as a JSON string, and sent to the Twitter Data Ingestion container, which exposes a Restful API for this purpose. The API interface is called through the *flask_restful* module.

The Twitter Data Ingestion Container is in charge of the ingestion of Tweets, and of sending the results to the Kafka topic. In order to connect to the Twitter database, it is necessary to subscribe to the developer APIs offered by the Social Media platform, and to use the code that Twitter itself provide. This has already been done in other works, such as [2], that we are using as a baseline. The container has to be configured with a set of parameters, that are used to connect it to the Kafka queue:

- Kafka_broker_url is the address of the Kafka Broker, where the Queue is waiting for the data produced by the ingestion container.

---

[1] http://wurstmeister.github.io/kafka-docker/.

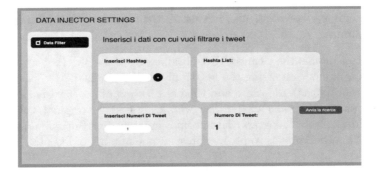

Fig. 2. Homepage of the Frontend

- Value_serializer defines the format of the messages to be sent into the Kafka topic. In this specific case, a JSON string coded with ASCII has been chosen, but other options are available.
- API_version sets the API version used when communicating with Kafka.

The ingestion container works in a very simple way: it runs a *for loop*, cycling through all of the hashtags that have been sent by the Front-end, and it requests the Twitter API to obtain the Tweets that have been labelled with each of the examined hashtags. The API allows also to configure a maximum number of Tweets (previously set via the Fornt-end) to download, to choose the Tweets' language (here *Italian* is the default choice), and to decide if we want to select the most popular Tweets or the most recent ones. For the last choice, in this work a *mixed* mode has been selected, so that all kinds of Tweets are returned by the API.

The collected Tweets are pre-processed, in order to sent clear information to the Sentiment Analysis container. As already explained in Sect. 2, there are some simple elaborations that are done on the Tweets' text.

- Removal of Emoji: this is done through the *re* Python module, by defining an *emoji pattern*, which can recognize the ASCII codes used for emoji, emoticons, symbols pictographs, and even flags.
- Removal of punctuation marks: this task is simply obtained by exploiting the *punctuation* method of the *string* module from Python. *Stemming*: the stemming task is not implementable through standard Python libraries, but it can be realized by exploiting external modules, such as the ones provided by NLTK. The library contains a good number of Natural Language Processing modules, among which several stemmers. For the purpose of this papers, the **Snowball Stemmer** has been employed, as it also provides rules an algorithms to support the Italian language, which is not completely covered by similar instruments.

Once the Tweets have been downloaded and cleaned, they are included into a Pandas dataframe, that is then sent to the Kafka Queue. The dataframe contains vectors of numbers or of strings, having all the same length, and grouped into named columns:

- **Tweets** contains the cleaned Text obtained from the Tweets;
- **Len** is a number reporting the length of the Tweets field;
- **Date** reports the exact date and time of creation of the Tweet;
- **User** reports the name of the User who generated the original Tweet;
- **Position** contains geographical information regarding the examined Tweet;
- **Like** is another numerical field, expressing the number of likes the Tweet has received

The so built dataframe is sent to the Kafka queue, where it will be collected by the Sentiment Analysis component and visualized. The pre-processed Tweets are collected by the Kafka queue, which the Sentiment Analyser container periodically probes, to check if new data are available.

In order to detect the polarity of the Tweet, a simple library has been exploited: TextBlob[2] allows for the analysis of simple texts, including the ones that are in Italian, and provides an output polarity in the range $[-1; +1]$, with 0 being a neutral score, -1 negative and $+1$ positive. The Sentiment Analyser takes advantage of the stemmatization and punctuation removal operated by the Ingestion container, even if the very same TextBlob offers the same functions. In this way, it would be possible to replace the Sentiment Analysis container, and to choose a different library to calculate the polarisation of Tweets, without the need to re-implement parts of the code. Once TextBlob has calculated the polarity, or the sentiment of the Tweet, a new dataframe is built and sent to the Visualization container. The new dataframe is simply obtained from the previous one, by adding **Sentiment** numerical column just after the **Like** column. The Sentiment Analyser sends the data to visualize through an API, and then the Visualizer creates the views, according to option chosen by the final user. As already stated in the Design Sect. 2, there are two available views.

The first one, shown in Fig. 3, is quite simple and straightforward: the Tweets are just listed, with all the information contained in the dataframe retrieved from the Kafka queue. This surely is the simplest view possible, but it is not very easily navigable, and geographical information are not properly used. The view shown in Fig. 4 is much more interesting, as it visualizes each Tweet according to the specific geographical location from which it has been created, according to the Twitter API. Red pointers show negative Tweets, while Blue pointers are used for positive comments. When a Tweet is neutral, a Grey pointer is used on the map. While the map refers to the whole world, since we are focusing on Tweets in Italian, most of them will come from Italy o other Italian speaking countries, such as Switzerland. Of course, some can be also seldom generated in non Italian speaking countries, by tourists for example: the map shown in Fig. 4 shows such a situation, where a negative Tweet in Italian has been registered between the United Arab Emirates and Oman. The visualization of the map has been obtained via the **folium** library, which shows a World Map where

---

[2] https://textblob.readthedocs.io/en/dev/.

Tweet serializzati

	tweets	len	date	position	likes	sentiment
0	SpaceX Tesla Ferrari Stellantis come superare le grandi crisi i vari italiani e le aziende che lavorano con El	110	2021-09-24 10:38:10	World	251	0
1	Ed ora i nazisti del vaxino si toglieranno il camice bianco per indossare quello verde green Hanno una Tesla e ti	113	2021-10-03 08:41:44	Roma, Lazio	1	0
2	Non so per quale motivo ogni giorno mi chiedo il perch la la mia auto preferita Tesla	85	2021-10-03 08:34:52		0	0
3	tuoa hio gazzler Import Tesla	29	2021-10-03 08:21:47	Mombasa	0	0
4	Inoltre le batterie elettriche scariche inquinano molto molto di piu e smaltirle sara quasi impossib	100	2021-10-03 08:02:54	Rome	1	0
5	RT Altro trimestre da record per TeslaAltro risultato incredibile per Tesla Rilasciati oggi i dati	98	2021-10-03 07:56:18		0	0
6	Il suo non sar forse un nome particolarmente noto al grande pubblico ma che il patrimonio di KoGuan Leo fosse da	112	2021-10-03 07:56:00		0	0
7	Tesla segna ottime vendite nell ultimo trimestre nonostante la crisi dei chip diglicsofia	89	2021-10-03 07:45:41	Milano	0	0
8	RT Grande successo in India per gli scooter elettrici Ola S La versione base costa l equivalente di euro con k	113	2021-10-03 07:41:56	middle of nowhere, r. odonico	0	0
9	Tesla segna ottime vendite nell ultimo trimestre nonostante la crisi dei chip	77	2021-10-03 07:36:32		1	0
10	Cos avrebbe fatto De Ligt se non fosse diventato un calciatore La risposta del difensore molto autoironica	106	2021-10-02 08:00:00		702	0
11	Bianconeri da dove griderete ForzaJuve ToroJuve LIVE su IT	58	2021-10-02 14:34:31	Allianz Stadium	609	1
12	Nedved a Dazn S per il Toro la partita dell anno Per noi una partita importante una delle giocate in	101	2021-10-02 15:24:55		818	0
13	RT Locatelli a DAZN La dedica a tutta la mia famiglia juventina a casa Saranno felici	85	2021-10-03 09:06:14	Coglionazzolandia	0	0
14	RT MATCHDAY Roma vs Empoli Serie A giornata ottobre Stadio Olimpico Ore Visibile su	90	2021-10-03 09:05:39	a snoxall, s aprochrxm	0	0
15	Ok diciamo pure che lo scrivono solo media rivolti ai subumani per Ravezzani ods ecc	84	2021-10-03 09:05:34	A Sud	0	1
16	Ciao Camilla se esci dall applicazione la prossima volta che apriral l app di DAZN ti verranno c	96	2021-10-03 09:05:03		0	0
17	RT Alle siamo in diretta con NapoliMilan Guarda la gara su Sky DAZN Prime Video Follo	87	2021-10-03 09:03:26	Milano	0	0
18	Hey HELP ITA sto cercando di utilizzare la chat ma devo parlare con un umano Mezz ora che sto aspettando Co	107	2021-10-03 09:03:19	Milan, Italy	0	0
19	e su diretta IT e Mihajlovic ex Lazio dai	45	2021-10-03 09:03:11	Firenze	0	0

Fig. 3. List of examined Tweets

markers can be placed. The Twitter API, at least for free developers, does not provide latitude and longitude, but only the name of the place where the Tweet was generated. However, by using the **geopy.geocoders**[3]Python module, that converts the name of geographical points in such coordinates, to be used on a map. Since not all users give Twitter the permissions to locate them, there are may situations in which this information is not provided by the API at all, so the corresponding comments cannot be displayed on the map, and the tabular visualization of Fig. 3 is needed.

Fig. 4. Geo-localized map of examined Tweets

4 Conclusion and Future Works

The work presented in this paper represents a first step to the realization of a solid container based Sentiment Analysis framework, which can ensure high scalability and availability to final users. The possibility to modify and substitute the containerized components independently represents an important advantage

---

[3] https://geopy.readthedocs.io/en/stable/.

for developers, especially when new technologies need to be tested. Also, since the microservices based architecture allows to scale the single components independently, it is theoretically possible to better balance the used resources and to concentrate on specific tasks that, at an unknown moment in time, can request more computational power than usual. In the future, further research will be carried out on the Sentiment Analyser: first of all, tests must be run to verify the actual accuracy of the employed library, which has been taken as it is, without applying any kind of optimization; second, it is necessary to identify solutions that can be applied to posts and comments that are much longer than standard Tweets, for which the currently used libraries could not represent the best choice. Another aspect to be considered is the possibility to integrate the whole framework with container oriented platforms, such as Kubernetes [1], which could be used to better manage the workload for the containers, an aspect that has not been taken in consideration in this work.

Acknowledgments. This project has received funding from the European Union's Horizon 2020 research and innovation program through the NGI ONTOCHAIN program under cascade funding agreement No 957338.

References

1. Bernstein, D.: Containers and cloud: from LXC to Docker to Kubernetes. IEEE Cloud Comput. **1**(3), 81–84 (2014)
2. Di Martino, B., Colucci Cante, L., Graziano, M., Enrich Sard, R.: Tweets analysis with big data technology and machine learning to evaluate smart and sustainable urban mobility actions in Barcelona. In: Barolli, L., Poniszewska-Maranda, A., Enokido, T. (eds.) Complex, Intelligent and Software Intensive Systems. CISIS 2020. Advances in Intelligent Systems and Computing, vol. 1194, pp. 510–519. Springer, Cham (2020). https://doi.org/10.1007/978-3-030-50454-0_53
3. Di Martino, B., et al.: A big data pipeline and machine learning for a uniform semantic representation of structured data and documents from information systems of Italian ministry of justice. Int. J. Grid High Perform. Comput. (IJGHPC) (2021, in press)
4. Di Martino, B., Venticinque, S., Esposito, A., D'Angelo, S.: A methodology based on computational patterns for offloading of big data applications on cloud-edge platforms. Future Internet **12**(2), 28 (2020)
5. Grinberg, M.: Flask Web Development: Developing Web Applications with Python. O'Reilly Media, Inc. (2018)
6. Jaramillo, D., Nguyen, D.V., Smart, R.: Leveraging microservices architecture by using Docker technology. In: SoutheastCon 2016, pp. 1–5. IEEE (2016)
7. Thein, K.M.M.: Apache Kafka: next generation distributed messaging system. Int. J. Sci. Eng. Technol. Res. **3**(47), 9478–9483 (2014)

Container-Based Platform
for Computational Medicine

Gennaro Junior Pezzullo[1(✉)], Beniamino Di Martino[2,3], and Marian Bubak[4,5]

[1] Department of Engineering, University of Rome "Campus Bio-Medico",
Rome, Italy
`gennaro.pezzullo@unicampus.it`
[2] Department of Engineering, University of Campania "Luigi Vanvitelli",
Caserta, Italy
`beniamino.dimartino@unicampania.it`
[3] Department of Computer Science and Information Engineering,
Asia University, Taichung City, Taiwan
[4] Sano Centre for Computational Medicine, Krakow, Poland
`m.bubak@sanoscience.org`
[5] ACC Cyfronet AGH, Krakow, Poland
`bubak@agh.edu.pl`

Abstract. In recent years the concept of "containerization" thanks to commercial software development has become a very promising paradigm also for e-science. In this paper, we present how this paradigm may significantly facilitate the creation of platforms for advanced scientific simulations. We present how with currently available containerization technologies, you can design a platform for simulation in the field of computational medicine with the same functionality as cloud system called Atmosphere, developed in 2011-14 ACK Cyfronet AGH for the Virtual Physiological Human community. The Atmosphere, based on the virtualization concept, provides support thanks to an intuitive interface that performs workflows on-demand cloud computing as well as access to cloud storage. After careful analysis of the Atmosphere structure, a redesign of the platform was subsequently carried out using container-based technologies, in particular Kubernetes. After architectural remodeling based on analysis of requirements and containerization possibilities, an implementation has followed through a Maven project in Java using the Kubernetes API. All this was followed by a validation phase to verify the actual operation.

1 Introduction

The main goal of the research presented here was to check to what extent the development of virtualization and containerization technologies [1] facilitates the creation of scientific computing platforms using cloud computing and HPC resources [2] (the infrastructures needed to work with big data such as computing power, databases, networking or storage can be provided in minutes within the

L. Barolli et al. (Eds.): AINA 2022, LNNS 451, pp. 131–140, 2022.
https://doi.org/10.1007/978-3-030-99619-2_13

cloud [3]). Such platforms are very much needed in many scientific simulations, especially in computational medicine. Multiple simulation runs are performed, some of them are interrupted in the run after being judged not to be promising. These platforms should enable the repeatability of calculations and easy storage and comparison of results. Most of the applications that implement scientific simulations have the form of a workflow, therefore these platforms should enable the performance of these workflows in an optimal way. The following research on the "Porting of a Cloud Platform for Medical Web Services to Container Technology" can be divided into five main parts:

- The first item is an analysis of a project developed by a team of Polish researchers for the VPH community which takes the name of "Atmosphere". This platform is able to provide support to doctors thanks to an intuitive interface that performs workflows on-demand [4,5]. Moreover, it is developed with cloud computing, cloud storage and specific virtualization technologies;
- the second step is a redesign of the Atmosphere Platform and its changes both in the implementation and in the workflow;
- the third part is an analysis of the progress of the state of the art, also referring to the main advantages in the use of container technology, instead of virtualization.
- the fourth part focuses on the actual practical implementation and validation of a part of the redesigned Atmosphere Platform;
- Finally, the last phase is related to the conclusion of this work.

2 Descriptions of the Functionality Provided by Atmosphere

In general, the final goal achieved with Atmosphere is the resolution of problems related to the supply of distributed computing resources oriented towards medical research [6]. During the design phase, it is important to understand how to redistribute existing applications on distributed resources so that every application always runs with the most appropriate resource for the context [7]. Moreover, a basic requirement is that the user must never realize that any kind of infrastructural change is taking place. The redesigned platform sees within the entire ecosystem three separate users who are respectively:

- Developers: they create the services, they know software development, but not necessarily the cloud,
- End Users: who are those who use the services,
- Administrators: cloud experts.

The developers' main task is to create a service. To achieve their goal, they need to contact the VPH-Share cloud platform which creates a virtual machine with the required hardware resources and shares an instance of this. The developers can log into the machine and install their software. Then, they may instruct the VPH-SCP to upload the service to the cloud and at this point, it is available for the end-user. On the other side, the end-user has the possibility to select all the available services on the cloud. When he selects a service VHP-Share Cloud Platform spawns an instance of this and updates the end-user with all the access information. At this point, the user can interact with the service. When he ends his work he can shut down the service and make computational resources free [8].

3 New Architecture Based on Containerization Concept

The basic idea of this redesign is to use the concept of the container instead of virtual machines. The change will bring benefits in terms of performance because a virtual machine to be enabled needs a complete OS installation, which affects the computer's memory and other resources [9]. And more, this causes a waste of resources in the various stages of the software life and the virtual machine can easily imagine restraining the ideal setting of multi-cloud where moving an application should be easy and fast. In particular, the idea is to develop the system in the following way: after expanding the service and its related image, it is published on a sharing platform (such as docker HUB). If it is required, the image is downloaded, a container is generated with the service inside and a container instance is created. The graph has been divided into 3 parts and to each part, a number has been assigned in order to spell out the activities, too.

3.1 Design Providing Functionality for Administrators

The first actor is the administrator who has to:

- configure the folder and path where images are downloaded,
- configure the Kubernetes machine.

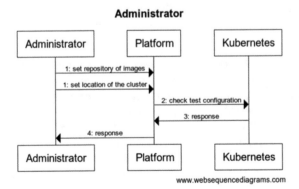

Fig. 1. The figure shows the sequence diagram relating to the administrator describing the operations necessary to configure the tool.

The administrator basically takes care of initializing the machine. In particular, he configures the image repository on the platform (which for example can be DockerHub) and sets the location of the cluster. After doing this, the next step is a validation run on the platform" to verify that everything is working correctly. In the event that the validation result is positive, the administrator's work is finished. Otherwise, it is necessary to repeat the procedure starting from the first point until the validation is successful Fig. 1. After that, Kubernetes is online.

3.2   Design Providing Functionality for Developers

The second actor is the developer who basically:

- deals with creating a service,
- shares it on a hub.

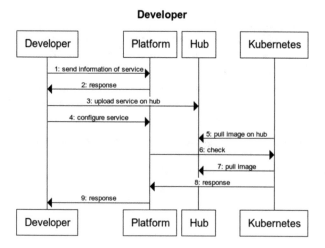

Fig. 2. The figure shows the sequence diagram relating to the developer describing the operations necessary to develop and load service on the platform

The first operation of the developer is to send to the platform all the information relating to the services, such as what the service does if you need credentials and everything related to the documentation of the service. At this point, once the platform has been checked, it sends a response to the developer which can be positive or negative. If the answer is positive, the developer loads the service on the hub and configures it on the platform. Next, before targeting the service as "available" on the platform, a service check is performed by pulling from the hub and the developer is informed about the actual status of the service. If the check is successful, the service is made available to all users connected to the platform. If not, it means that something either in the upload on the hub or in the development of the service was not successful. It is, therefore, necessary to check the service and only then start again from point 3 in Fig. 2.

3.3   Design Providing Functionality for End-Users

Finally, the end-user can make a request to use a particular service. Only after the request is accepted, the image associated with the service is downloaded and an instance of this is created. The third and last step is the one performed by the end-user.

The end user:

- makes his request,
- if he receives a confirmation message, he uses the services,
- he receives another confirmation message when he asks to terminate the service.

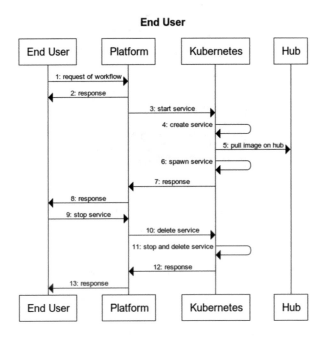

Fig. 3. The figure shows the sequence diagram for the end user describing the operations necessary to use a service

The front end in this case does not change. While the back end is completely different, the operation is as follows: the user request the platform to run a service. Depending on which service is used (even if it is not explicitly shown in the graph), the service may or may not require credentials.

Once the request is made, a confirmation or an error message is given. In case of confirmation, Kubernetes is used to create a service, pull the associated image from the hub and finally the service is spawned. In the end, a confirmation message is sent to the user to tell him that he can start using the service. If something went wrong instead an error message will appear to the end-user and the service will not be spawned. From here, the next step is done at the end of using a service. After finishing, the user sends a request to the platform to terminate the service. At this point, Kubernetes is ordered to stop and delete the service and once done, a confirmation message is sent to the user (Fig. 3).

4   Container-Based Platform

In the Atmosphere project, virtual machines were used to initialize the services; as we know, these VMs necessarily require the installation of an operating system to work, as it is shown in Fig. 4

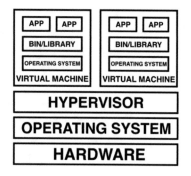

Fig. 4. Architecture of virtual machines: the figure describes the main components needed to start virtual machines. In this architecture, applications need an operating system to function.

To work this operating system requires:

- Computational resources,
- Memory resources.

In addition to this, it slows down the ideal multi-cloud scenario [10,11] where moving an application should be quick and easy. Instead of virtualizing all this set of resources, it would certainly be more convenient to abstract only that part of the environment necessary for the applications to make them work and the relative instructions. In this scenario, the use of containers is useful. In fact, from an architectural point of view, these containers are located on a physical server, on which an operating system runs and shares its kernel [12]. In this way the containers do not individually need an OS to be started (Fig. 5).

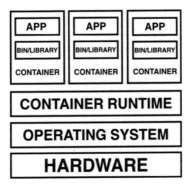

Fig. 5. Architecture of Containers: the figure describes the main components needed to start containers. In this architecture, applications do not need an operating system to function

After these considerations we can affirm that an approach based on the use of containers rather than on the use of virtual machines brings the following advantages:

- Simplified deployment: in fact, the use of containers simplifies the deployment of any application as it is "packaged" into a single component that can be distributed and configured with a single command line. This means that it is no longer necessary to configure the execution environment,
- Quick availability: by virtualizing only the components necessary to run the application, instead of the entire machine, startup times are significantly reduced compared to those of a virtual machine,
- Wide portability: a great advantage in the life cycle of software is the ability to create and replicate the various components and containers allowing you to do it quickly and easily. This leads to the creation of development, proof of concept and production environments without the need to operate configurations,
- More granular control: the containers can incorporate both an entire application or a single component. This approach allows developers to divide resources into microservices, thereby providing greater control over application runnability, reuse of individual components, and improvement of the performance of the entire infrastructure,
- Lightweight: A big selling point of containers is that they are much lighter than virtual machines in terms of hardware resources. This is because they do not need a dedicated operating system. As a result, they are much faster to be activated and deactivated.

5   Implementation and Proof of Concept

In developing the program, a top-down approach has been preferred. Therefore, functions have been first developed individually and then, only later, merged.

In general, the main objective has been to reproduce the part relating to the end-user. It is also important to specify that in this implementation we have used a limited version of Kubernetes, which is called Minikube [13], installed on Docker in a container. Maven projects have initially been developed [14], then the dependency of Kubernetes in the pom.xml file has been inserted. The created program, therefore, allows you to perform the following operations:

- View active deployments,
- Ability to spawn a deployment,
- Ability to delete a deployment.

Everything has been automated using Java code and can be executed with simple commands via the terminal. The former commend does not require input parameters. It connects via commands provided by the Kubernetes documentation to Minikube shows us the currently active deployments. As for the possibility of spawning a deployment, for greater convenience, it has been decided to create a document containing all the yml files, each with a unique code. When this unique code is called from the terminal, the respective image of the program is created and deployed. Finally, as regards elimination, the operating logic is very similar to spawning since the unique code is used here, too, but in this case, it is necessary to deactivate the container. With a simple if/else, the user is given the opportunity to choose what to do. Of course, in a future implementation, a simpler interface could be used, such as a web app. The three sub-programs have been validated individually to verify effective operation. All of them returned positive results in their operation. To validate the actual reliability of the program, all the operations have been first performed on the local terminal using kubectl commands, after by the program developed in Java and, in the end, the results have been always completely equivalent. It is possible to view the code from this GitHub repository [15]. Drawing the conclusions, we can say that, thanks to the first sub-programs, we can check that everything has gone well. With this code we have reproduced all tasks in Fig. 3.

6   Conclusion

We started with an existing software called Atmosphere. In Sect. 2 we have analyzed it and extrapolated its most important characteristics. In Sect. 3 a redesign of the Atmosphere application has been then carried out using container technologies. In Sect. 4 we analyzed the main differences between containers and virtual machines and the state of the art of the project. Finally, in Sect. 5 one of the main parts was implemented with Java code. In particular, that concerning the back-end of the "end-user". Tests have been carried out to monitor the actual functioning of what has been produced. Surely, among the future implementations, the most relevant could be to create the implementation of the developer, administrator and a front-end interface as it was with Atmosphere to ensure that the end-user can be interfaced. We consider also an extension of this new platform with the semantic techniques [16,17] to compose in automatic way containers for specific simulation services.

Acknowledgements. The authors would like to thank their colleagues from ACC Cyfronet Marek Kasztelnik, Maciej Malawski, Jan Meizner and Piotr Nowakowski for their help in understanding the structure of the Atmosphere platform and for valuable discussions and suggestions regarding the Kubernetes. The work was carried out during the GJP's stay at AGH under the Erasmus+ program. MB research was partially supported by EU H2020 grant Sano 857533 and IRAP FNP project.

References

1. Aversa, R., Branco, D., Di Martino, B., Venticinque, S.: Container based simulation of electric vehicles charge optimization. In: Barolli, L., Woungang, I., Enokido, T. (eds.) AINA 2021. LNNS, vol. 227, pp. 117–126. Springer, Cham (2021). https://doi.org/10.1007/978-3-030-75078-7_13

2. Di Martino, B., Cretella, G., Esposito, A.: Cloud portability and interoperability. In: Cloud Portability and Interoperability. SCS, pp. 1–14. Springer, Cham (2015). https://doi.org/10.1007/978-3-319-13701-8_1

3. Pop, F., Kołodziej, J., Di Martino, B. (eds.): Resource Management for Big Data Platforms. CCN, Springer, Cham (2016). https://doi.org/10.1007/978-3-319-44881-7

4. Aarestrup, F.M., et al.: Towards a European health research and innovation cloud (HRIC). Genome Med. **12**(1), 1–14 (2020)

5. Martí-Bonmatí, L., et al.: Primage project: predictive in silico multiscale analytics to support childhood cancer personalised evaluation empowered by imaging biomarkers. Eur. Radiol. Exp. **4**(1), 1–11 (2020)

6. Kasztelnik, M., et al.: Support for taverna workflows in the VPH-share cloud platform. Comput. Methods Prog. Biomed. **146**, 37–46 (2017)

7. Meizner, J., Nowakowski, P., Kapala, J., Wojtowicz, P., Bubak, M., Tran, V., et al.: Towards exascale computing architecture and its prototype: services and infrastructure. Comput. Inform. **39**(4), 860–880 (2020)

8. Nowakowski, P., et al.: Cloud computing infrastructure for the VPH community. J. Comput. Sci. **24**, 169–179 (2018)

9. Bubak, M., et al.: The EurValve model execution environment. Interface Focus **11**(1), 20200006 (2021)

10. Malawski, M., Gajek, A., Zima, A., Balis, B., Figiela, K.: Serverless execution of scientific workflows: experiments with HyperFlow, AWS Lambda and Google cloud functions. Futur. Gener. Comput. Syst. **110**, 502–514 (2020)

11. Gerhards, M., Sander, V., Živković, M., Belloum, A., Bubak, M.: New approach to allocation planning of many-task workflows on clouds. Concurr. Comput. Pract. Experience **32**(2), e5404 (2020)

12. Yadav, A.K., Garg, M.L., Ritika: Docker containers versus virtual machine-based virtualization. In: Abraham, A., Dutta, P., Mandal, J., Bhattacharya, A., Dutta, S. (eds.) Emerging Technologies in Data Mining and Information Security. Advances in Intelligent Systems and Computing, vol. 814, pp. 141–150. Springer, Singapore (2019). https://doi.org/10.1007/978-981-13-1501-5_12

13. Kubernetes documentation. https://kubernetes.io/

14. Maven documentation. https://maven.apache.org/

15. Github repository. https://github.com/gennarojuniorpezzullo1/Atmosphere_Container_Implementation.git

16. Bellini, E., Cimato, S., Damiani, E., Di Martino, B., Esposito, A.: Towards a trust-worthy semantic-aware marketplace for interoperable cloud services. In: Barolli, L., Yim, K., Enokido, T. (eds.) CISIS 2021. LNNS, vol. 278, pp. 606–615. Springer, Cham (2021). https://doi.org/10.1007/978-3-030-79725-6_61
17. Di Martino, B., Gracco, S.A.: Semantic techniques for IoT sensing and eHealth training recommendations. In: Barolli, L., Yim, K., Enokido, T. (eds.) CISIS 2021. LNNS, vol. 278, pp. 627–635. Springer, Cham (2021). https://doi.org/10.1007/978-3-030-79725-6_63

Digital Twins for Autonomic Cloud Application Management

Geir Horn[(⊠)], Rudolf Schlatte, and Einar Broch Johnsen

Department of Informatics, University of Oslo, P.O. Box 1080 Blindern,
0316 Oslo, Norway
Geir.Horn@mn.uio.no, {rudi,einarj}@ifi.uio.no

Abstract. Cloud applications are distributed in nature, and it is challenging to orchestrate an application across different Cloud providers and for the different capabilities along the Cloud continuum, from the centralized data centers to the edge of the network. Furthermore, optimal dynamic reconfiguration of an application often takes more time than available at runtime. The approach presented in this paper uses a concurrent simulation model of the application that is continuously updated with real-time monitoring data, optimizing, and validating deployment reconfiguration decisions prior to enacting them for the running applications. This enables proactive decisions to be taken for a future time point, thereby allowing ample time for the reconfiguration actions, as well as realistic Bayesian estimation of the application's time variate operational parameters for the optimization process.

1 Introduction

A Cloud application is intrinsically a set of components distributed over different locations and infrastructures, hence it becomes challenging to maintain the application performance and stability over time as it may require constant attention of dedicated DevOps engineers. This challenge is not new and *Autonomic Computing* has been proposed as a solution [17]. The core concept is the Monitor, Analyse, Plan, Execute—with Knowledge (MAPE-K) [14] feedback loop continuously monitoring essential application metrics, and then plan and adapt the application to the application's current execution context. The concept has been used to build application management platforms for mobile computing [9], ubiquitous computing [11], and Cross-Cloud computing [12].

A rational decision is in psychology and economy understood as the choice maximizing the *utility* of the decision maker [10]. Autonomic computing approaches therefore assume that the DevOp engineer's decisions can be replaced by a *utility function* that balances the different concerns and trade-offs implicit in the decision [18]. However, research has shown that it is hard for a DevOps engineer to formulate the utility function [8]. This barrier may be even higher as the number of configuration choices increases when the application can be deployed on heterogeneous computers that may be severely restricted in capacity, or when it is possible to deploy the application components in variants targeting different hardware accelerators like Graphics Processing Units (GPUs), Tensor Processing Units (TPUs), and Field-Programmable Gate Arrays (FPGAs).

L. Barolli et al. (Eds.): AINA 2022, LNNS 451, pp. 141–152, 2022.
https://doi.org/10.1007/978-3-030-99619-2_14

An alternative to model the utility is to simulate the candidate deployments and pick the one that best satisfies the high-level goals and intentions of the DevOps engineer. This approach resembles Data Farming, *i.e.*, simulation experiments where the model parameters are varied across the simulations to compare the simulation output with measurements from the real system, thereby identifying the unknown underlying system parameters [13]. Data farming has been used similarly to make better decisions by complementing and improving the quality of the data mining of the observed 'big data' [24]. However, autonomic Cloud application management based on the MAPE-K loop continuously monitors the operational parameters of the application and its execution context. This allows the simulation model to be causally connected with the real word through the measurements. Hence, the model becomes a Digital Twin (DT) [3] of the running application. Instead of executing multiple simulations to identify unknown system parameters, as one would do for data farming, the novelty here is to execute multiple simulations for various reconfiguration options using the DT model as a part of making optimized decisions for the application configuration for the application's current execution context.

The optimized autonomic Cloud application management and the concept of a DT are discussed in Sect. 2 as a background to understand the problem and its complexity. A prerequisite for the proposed DT approach to work is that the executable modelling language used to implement the DT is accurately representing the salient application characteristics. The Abstract Behavioral Specification (ABS) language fulfills the DT requirements, and Sect. 3 introduces the ABS language and a generic ABS model for Cloud applications. Section 4 briefly discusses the context of our proposed solution, and Sect. 5 concludes the paper.

2   The Cloud Application and Its Digital Twin

2.1   Cloud Application Modelling

Cloud computing is first and foremost a *business model* where the infrastructure necessary to execute an application will be rented on demand. Permanently renting resources will in the long run be more costly than owning the resources, and hence this implies that Cloud computing is for applications that are infrequently needed, or to cover intermittent peak loads of permanently running applications. It is not necessary to manage automatically short-lived applications. Hence, the type of applications considered here are the ones that are running for a significant time with variable resource needs over the execution time and where the resources needed permanently should be owned by the application owner with the variable needs for resources covered by rented resources offered by the Cloud providers.

A Cloud application is therefore a *distributed* application: It is minimally divided between the part running on the private infrastructures and a variable part running in the Cloud. The application can be seen as an orchestration of *software components*, where the term *component* is understood as an abstraction for an application building block [20]. The application components can be either software modules, applications, or packages; web services or Cloud platform functions; or data sources or sinks.

Any computing system is fundamentally about data transformation. This property transfers to the individual application components. Each component collects some input data, does some manipulation of the data, and produces some form of output data. These three phases can be seen as conducted in a strict sequence, and each phase has a starting time and a duration. This simple data flow delay model is illustrated in Fig. 1. It is assumed that the autonomic application management system monitors four metrics for each component: The component's start time and the duration of the three phases. Note that this fundamental view also applies for stream data processing modules where the module's functionality should be further refined into component functions processing a part of the data input flow producing a part of the output flow. Finally, the three phases are also valid irrespective of the programming paradigm and philosophy used to create the component.

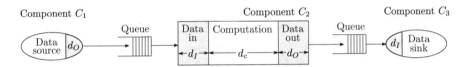

Fig. 1. The simple data flow model where there are intrinsic stochastic delays in each component caused by data input, d_I, data computation, d_c, and data output, d_O. Each component has individual probability distributions for these delays. Delays in the virtual work queues are included in the downstream component's input delay.

Hence, without lack of generality the application can be modelled as a set of components, each fulfilling the three phases of the data transformation. The time taken in each of the three phases will consequently depend on the data location and data size, and on the vertical scalability parameters like the number of cores given to a component or the amount of memory it has. In addition, it is necessary to model the data flow of the application since the input data for a component comes from one or more other computing components or data sources, and the output data flows into other computing components or data sinks. This is illustrated in Fig. 2. Horizontal scalability, *i.e.* how many instances there are of a particular component type, will affect the overall application completion time, the *makespan*, but the number of components will not affect how long it takes one component to finish its data transformation phases. There are already many options for describing this topology model of the application covering the type of components, the application's data flow, the components' computational requirements and constraints, their scalability constraints, and the application's DevOps engineers' deployment and operational goals [5].

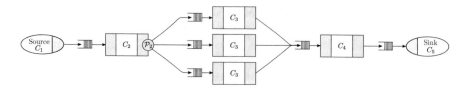

Fig. 2. Data flows where data is split on multiple instances of a component type, here C_3, require policies $\mathcal{P}$ on the upstream components deciding how the data is forwarded to the instances of the downstream component type(s). Popular policies are *round-robin*, *random*, or *broadcast* where the same data goes to all downstream instances.

Assuming that such a domain specific model of the managed application is available, it can be converted automatically into a DT simulation model representing the three fundamental phases of computation for each component, and the component's output-to-input connectivity. In general, all duration parameters are stochastic. In the best case, the DevOp engineers may have an idea for the *prior* duration time distributions to be included in the domain specific model. Otherwise, one may assume some standard distributions, typically Gaussian for their analytic tractability and well understood use in non-linear regression [23]. It is however difficult for the application's DevOp engineers to know the parameters for these duration distributions. For instance the duration of the computation phase will depend on data size and possibly also on the content of the data, and so it is hard to estimate the mean duration and the variance of the computation time *a priori*.

The DT component model will therefore maintain secondary probability distributions for the parameters of the duration distributions used in its three execution phases. These secondary distributions may better be defined by the application's DevOp engineers. The parameters of these secondary *prior distributions* will be recursively refined based on the monitored metric values from the running application using the Markov Chain Monte Carlo (MCMC) method, and in the case no prior secondary distribution is given it can be estimated from available monitoring information using an empirical Bayes approach [6].

The scalability parameters of a DT component will be set by the optimizer of the autonomic application management platform. Some of these *vertical* scalability parameters will directly affect the performance of the component, *e.g.*, the number of cores or the amount of memory or the use of a hardware accelerator. Changing the component performance will directly affect the secondary distributions. Thus, the DT component model will maintain and update one set of secondary distributions per hardware configuration option available for the component.

The location of the components will affect the parameters for the two communication phases of the component model. Consider a computing component reading data from a data component. In this case the duration of the data input phase will be shorter if the data component is located in same region of the same Cloud provider as the computing component. The minimum duration will be when both components are located in the same data centre, but there is no way for the application management to control the Cloud provider's allocation of loads to its data centres within a given region.

Hence, even for the same region of the same Cloud provider the mean duration of the data input phase will be a *random variate*, and this variation may be orthogonal to the variation caused by the size of the data set exchanged. Thus, the DT model should also maintain and update location specific secondary distributions for the parameters of the communication delay distributions.

2.2 Digital Twins

A DT is a digital replica of an underlying system, often called the Physical Twin (PT). The DT is connected to its PT in real-time through continuous data streams such as sensor measurements at different locations and by other ways of collecting data. This turns the DT into a *live replica* of the PT, with the purpose of providing insights into its behaviour, and clearly distinguishes a DT from, *e.g.*, a standard simulation model.

A DT is commonly seen as an architecture with three layers: the *data layer* with, *e.g.*, Computer Aided Design (CAD) drawings and sensor data, an *information layer*, which turns these raw data into structured data, and an *insight layer*, which applies different analysis and visualization techniques to the structured data. The analysis techniques of the insight layer can be classified as follows: The DT is typically able to compute an approximation of how the PT acts in a given scenario (*simulation* or "what-happened" scenarios), or to estimate how the PT will behave in the future based on historical and current data (*prediction* or "what-may-happen" scenarios). By configuring the parameters of the different models, the DT may analyse the consequence of different options on future behaviour (*prescription* or "what-if" scenarios).

In the context of Cloud application management, we aim for a DT which can offer prescription. The data layer of the DT will initially consist of information about the configuration of the heterogeneous Cloud environment in which the application will execute, the resource profile of different locations such as bandwidth, memory and processing capacity, and the data flow topology of the application. The DT will continuously add information to the data layer about the software components to be orchestrated, including the duration of executing the different phases of the data transformation associated with a software component on a given location in the various Cloud environments. Thus, the main technology needed to implement the data layer of a digital twin is a solution for timed data streams, such as a time-series data base.

The information layer will assimilate data from the data layer into a model which combines the application topology with the current configuration of the heterogeneous Cloud environment, transforming recorded or default durations to location-independent execution costs and gradually improving the precision of costs by learning from previous executions of the application workflow. Thus, the information layer combines information about the application topology and the resources provided at different locations in the current configuration of the Cloud environment. A challenge in the construction of DTs is to find a representation at the information layer which supports the required level of analysis. For a DT supporting Cloud application management, we propose to represent the information layer as an executable model in which different deployment decisions may be easily expressed and efficiently explored through simulation. The model should have a formal semantics which makes it transparent for the Cloud application management how to configure the parameters and understand the results from the executable model. This turns the DT into a tool for optimized Cloud application management.

2.3   Optimized Cloud Application Management

Each application component, $C \in \mathbb{C}$, has a set of requirement attributes, $\mathbb{A}_C$. These typically specifies the resource necessary for the component to perform as expected. Thus, a requirement attribute can be the number of cores useful for the component ranging from a minimum number to a maximum number, or it can be the amount of memory, or the Cloud providers that can be used, or the geographic location of the hosting data centre, or the number of copies or instances of the component the application can successfully exploit. The requirement attributes can be continuous or discrete. Each value of a requirement attribute value, $a_{C,i}$, is taken from the attribute's *domain*, $a_{C,i} \in \mathbb{A}_{C,i}$, for all the requirement attributes $i = 1, \dots |\mathbb{A}_C|$.

A *configuration* of a component is an assignment of values to all its requirement attributes. The different ways a component C can be configured is its *variability space*, which is the Cartesian product of its attribute domains, $\mathbb{V}_C = \mathbb{A}_{C,1} \times \cdots \times \mathbb{A}_{C,|\mathbb{A}_C|}$. The configuration of the Cloud application is a flattened vector, c, of the configurations of *all* the components of the application. The variability space of the application is therefore the Cartesian product of the variability spaces for its components, $\mathbb{V} = \mathbb{V}_1 \times \cdots \times \mathbb{V}_{|\mathbb{C}|}$.

The ultimate goal of the DevOps engineers or the autonomic platform managing the application will be to find *the best* configuration, $c^* (t_k)$ for the application's current *execution context*, $\boldsymbol{\theta}(t_k)$, which is a vector of the metric values of the monitored application at the time point t_k. The traditional applications of autonomic computing assumes that the goals and preferences of the application's DevOops engineers can be captured as a *utility function* from the variability space to the unit interval, $U : c \in \mathbb{V} \mapsto [0, 1]$. Finding the best configuration given the current execution context means, most likely, solving a non-linear mixed continuous-discrete optimization problem

$$c^* (t_k) = \underset{c(t_k) \in \mathbb{V}}{\operatorname{argmax}} U \left(c(t_k) \mid \boldsymbol{\theta}(t_k), \boldsymbol{\phi} \right) \tag{1}$$

where $\boldsymbol{\phi}$ is a set of fixed parameters for the utility function family. This problem must be solved subject to a set of deployment constraints whenever there is a new execution context, *i.e.*, whenever any of the monitored values changes. The metric values can change abruptly and frequently, for instance as application users come and go, leaving little time for the optimization of the configuration before the next context change happens. The latter point is remedied by checking only the constraints, or the Service-Level Objectives (SLOs), of the optimization problem when a new measurement arrives, and then stay with the existing configuration if it remains feasible under the altered execution context.

The Bayesian approach of the DT defined in Sect. 2.1 allows the optimization to be done entirely based on the SLOs defined for the application. The same solver used for the mathematical programme (1) can be used. Searching for an optimal configuration, it will generate a sequence of candidate application configurations, $c(t_k)$, feasible for the application's current execution context. One may then run a set of parallel simulations using the DT application model for each candidate configuration where the components randomly draw delay times from the delay distributions using the Maximum Likelihood Estimates (MLEs) of all delay distribution parameters. Alternatively, the secondary distributions, updated on each measurement from the running application, allow precise confidence intervals to be given for the parameters of the delay distributions. The simulation can be executed using these worst case limits for the distribution parameters. For both alternative ways of selecting the delay distribution parameters, one computes the relevant application performance metrics indicating the goodness of the overall application performance indicators under the chosen set of distribution parameters for a candidate configuration from the ensemble of its DT simulations.

Evaluating the SLOs for the computed application performance metric values will indicate if a configuration candidate is likely to remain feasible, or if a new and better configuration must be deployed for guaranteeing the continued feasibility of the application. Only the feasible configuration candidates will be retained and scored according to some high level goals set by the DevOps engineers like 'least deployment cost' or 'minimal reconfiguration from the currently running configuration', and the candidate with the highest score can be selected as the next deployment configuration.

The approach can be illustrated with a small data farming application with three components: A dispatcher component sending off data parallel jobs to a set of worker components, and one result processing component receiving the output of the workers, see Fig. 2. All jobs must be completely processed by a given deadline. Assume that the execution time delay distribution of a job on a worker is Gaussian, $N(\mu, \sigma)$. At the 95% confidence level, the delay value will be in the interval $\mu \pm 1.96\sigma$. From the confidence intervals of the secondary distributions at the same level of significance one has that $\mu \in [\mu^-, \mu^+]$ and $\sigma \in [\sigma^-, \sigma^+]$, and so the worst case bound on the execution time of one job at one worker is $\mu^+ + 1.96\sigma^+$. However, as multiple workers are processing jobs in parallel, there could be a queue of results for the downstream component causing additional global delays, and possibly this component could end up as a bottleneck in the application. For this particular example, one could calculate the queuing delays using queueing theory, but given that the arrival distribution of the results on the last component is unknown and dependent on the processing delay distributions of the worker, it is in general impossible to model this system analytically and assess if the deadline is met for a given application configuration. As the autonomic application management must work for any application, simulating the DT is the only way to collect realistic performance statistics for the managed application.

3   Abstract Behavioral Specification Model as the Digital Twin

3.1   The Abstract Behavioral Specification Language

The ABS language combines implementation-level specifications with verifiability, high-level design with executability, and formal semantics with practical usability. ABS is a concurrent, object-oriented, modelling language with a functional layer with algebraic datatypes and side-effect-free functions. Of particular interest for this paper is *Real-Time ABS* [16], which additionally features a time and resource model: going beyond purely behavioural specifications, ABS models can specify time elapsed on a logical clock at certain points in the program logic, as well as resource usage when executing certain code paths. This resource usage will again influence the observed logical time for model execution. Instead of a full introduction to the language, which can be found, *e.g.*, in the language manual[1], this section maps ABS language features to the concepts of Cloud application modelling discussed in Sect. 2.1.

ABS as a language is based on the Actor model [2], with actors executing concurrently and communicating via asynchronous method calls. Method call results are returned via Future variables, which the caller can synchronize on. This naturally models the components of a Cloud application and their synchronous and asynchronous communication patterns.

Actors in ABS possess state in the form of fields bound to values. Computations on values (numbers, strings, boolean, actor references, futures, and user-defined algebraic data types) are expressed in a purely functional sub-language that is amenable to formal analysis. Depending on the desired level of detail of the Cloud application model, we can consider only the control flow of the data transformation steps mentioned in Sect. 2.1, or include manually abstracted or real computations in the model. In case the Cloud model has different, data-dependent control flow that needs to be modeled, typically some form of data will be included in the model. Multiple instances of components, and dynamic creation and tear-down of components, can be modelled by creating and releasing actors modelling these components. It is straightforward to model auto-scaling resource pools of such components.

As mentioned, the time consumed by a component's computation can be modelled via the logical clock implemented in Real-Time ABS. Elapsed time can be specified as constant or dependent on actor state, *e.g.*, as a function of the length of an input parameter to the computation. Logical time only elapses when explicitly specified; otherwise, computations are modelled to run "infinitely fast".

To model the deployment of the Cloud application components on machines with different resources, ABS implements a language feature called *Deployment Component* [16]. A deployment component serves as a location for one or more actors, and has a set of resources like computation speed, network bandwidth, memory, number of cores that it distributes among its actors. An actor can explicitly specify a resource need like computation or bandwidth for a step in its computation; that operation will take a certain amount of logical time depending on the amount of available resources.

Finally, ABS implements a *Model Application Programming Interface (API)* that allows to access the state of dedicated objects and call methods on these objects via Hypertext Transfer Protocol (HTTP) requests from outside the running model. Object state and method call results are returned in the JavaScript Object Notation (JSON) data format. For simple visualizations and interactions with a running model, the model can serve HyperText Markup Language[2] (HTML) and JavaScript to a web browser and react to requests from that browser session.

---

[1] https://abs-models.org/manual/.

[2] https://html.spec.whatwg.org/.

3.2   Modelling Cloud Applications with ABS

The ABS features are closely aligned with the Cloud application modelling concepts, so converting those models into ABS will lead to understandable code, even when done automatically or semi-automatically. All of data-dependent delays, deployment-dependent delays and data transfer delays are directly represented at the ABS language level.

The component types shown in Fig. 1 can be implemented as actors. Each component's queue will be modelled explicitly inside the ABS actor, to ensure the First In is the First Out (FIFO) semantics and to make the queue length and other attributes available in the model.

Simple component parameterization, *e.g.*, parameters for random distributions of data and computation sizes, can be implemented via local actor state. These parameters can be changed at run time by a dedicated monitor component, and accessed and stored from outside via the Model API during and after model execution.

More complex behavioral differences, *e.g.*, the data flow policies shown in Fig. 2, can be implemented either via ABS traits as mix-ins of methods into class definitions or, in case policies should be adaptable at run time, via a method changing its behavior depending on local state. Again, policy state can be accessed from outside the model via the Model API.

The Cloud application topology can be implemented via ABS deployment components that model the spatial arrangement of components on machines with varying attributes. Some additional modeling or implementation work might be required if it becomes necessary to model complex machine topologies with non-trivial cost functions for transferring data between two given machines.

Data that needs to be persisted, *e.g.*, initial and computed random distribution parameters for each component, can be exported via the Model API. ABS includes read-only support for the SQLite database engine; this can be used for bulk initialization of components at model run-time. Behavior or parameter changes during a model run can be triggered via the Model API.

4   Discussion

There are many approaches available for a Cloud operator to schedule the incoming workload [1, 19]. It is important to note that the Cloud application management discussed in this paper takes an *application centric view*. In other words, there are no restrictions on new Cloud resources other than budget and time to acquire the needed virtual resources. The application components are allocated to the available virtual resources ignoring how the application's virtual resources are *scheduled* on the physical data centre hardware by the involved Cloud operator(s).

This paper has discussed the modelling of the DT using Bayesian estimation of the stochastic parameters of the involved distributions. The resulting DT model can be seen as resembling application modelling using a Bayesian Network (BN) [7]. BNs assume binary node states organized as a Directed Acyclic Graph (DAG), and have recently been used for applications with DAG dependencies to schedule the application components on a fixed number of heterogeneous resources [22]. The closest related approach to the ideas presented here is probably the modelling of planning problems as a dynamic BN in time and delay variables with direct sampling of the simulated BN to estimate the overall plan completion time (makespan) [4].

However, autonomic Cloud application management is not only about executing the application as quickly as possible, and the DT allows constrained optimization for multiple objectives on carefully selected adequate resources. The use of the ABS modelling language allows the extension of the DT refining the application model if needed to capture all details for validating various candidate application configurations and probabilistically simulating their performance indicators; some related case studies are listed in [25].

5   Conclusion

Constrained autonomic Cloud application management requires the optimization of the application configuration. This paper has presented the vision of using a Digital Twin (DT) based on the Abstract Behavioral Specification (ABS) language as a simulation model to assess the feasibility of various application configuration candidates given the application's Service-Level Objectives (SLOs). The proposed DT approach avoids the need for the cumbersome utility function definition normally required for autonomic computing, and allows better expressiveness for the application model thereby enhancing the optimization of the application's deployed configuration. This promising approach will hopefully soon be implemented in an autonomic Cloud application management platform to demonstrate its benefits for real world applications.

Acknowledgements. This work has received funding from the European Union's Horizon 2020 research and innovation programme under grant agreement No. 871643 MORPHEMIC (http://morphemic.cloud) *Modelling and Orchestrating heterogeneous Resources and Polymorphic applications for Holistic Execution and adaptation of Models In the Cloud.*

References

1. Arunarani, A.R., Manjula, D., Sugumaran, V.: Task scheduling techniques in cloud computing: a literature survey. Fut. Gener. Comput. Syst. **91**, 407–415 (2019). https://doi.org/10.1016/j.future.2018.09.014
2. Agha, G.A.: ACTORS - A Model of Concurrent Computation in Distributed Systems. MIT Press (1990)
3. Barricelli, B.R., Casiraghi, E., Fogli, D.: A survey on digital twin: definitions, characteristics, applications, and design implications. IEEE Access **7**, 167653–167671 (2019). https://doi.org/10.1109/ACCESS.2019.2953499
4. Beaudry, E., Kabanza, F., Michaud, F.: Planning for concurrent action executions under action duration uncertainty using dynamically generated Bayesian networks. In: Proceedings of the International Conference on Automated Planning and Scheduling, pp. 10–17 (2010)
5. Bergmayr, A., et al.: A systematic review of cloud modeling languages. ACM Comput. Surv. (CSUR) **51**(1), 22:1–22:38 (2018). https://doi.org/10.1145/3150227
6. Carlin, B.P., Louis, T.A.: Bayesian Methods for Data Analysis. Chapman and Hall/CRC (2008). https://doi.org/10.1145/3150227

7. Darwiche, A.: Modeling and Reasoning with Bayesian Networks. Cambridge University Press (2009). ISBN 978-0-521-88438-9. https://doi.org/10.1017/CBO9780511811357. https://www.cambridge.org/core/books/modeling-and-reasoning-with-bayesian-networks/8A3769B81540EA93B525C4C2700C9DE6

8. Floch, J., et al.: Playing MUSIC – building context-aware and self-adaptive mobile applications. Softw. Pract. Exp. **43**(3), 359–388 (2013). https://doi.org/10.1002/spe.2116

9. Geihs, K., et al.: A comprehensive solution for application-level adaptation. Softw. Pract. Exp. **39**(4), 385–422 (2009). https://doi.org/10.1002/spe.900

10. Gilboa, I.: Rational Choice. MIT Press (2010). https://doi.org/10.1002/spe.900

11. Hallsteinsen, S., et al.: A development framework and methodology for self-adapting applications in ubiquitous computing environments. J. Syst. Softw. **85**(12), 2840–2859 (2012). https://doi.org/10.1016/j.jss.2012.07.052

12. Horn, G., Skrzypek, P.: MELODIC: utility based cross cloud deployment optimisation. In: Proceedings of the 32nd International Conference on Advanced Information Networking and Applications Workshops (WAINA), pp. 360–367. IEEE Computer Society (2018). https://doi.org/10.1109/WAINA.2018.00112

13. Horne, G., Schwierz, K.-P.: Summary of data farming. J. Syst. Softw. **5**(1), 8–27 (2016). https://doi.org/10.3390/axioms5010008

14. IBM: An architectural blueprint for autonomic computing, 3rd edn. White Paper, p. 34. IBM (2005). https://doi.org/10.1016/j.jss.2012.07.052

15. Johnsen, E.B., Hähnle, R., Schäfer, J., Schlatte, R., Steffen, M.: ABS: a core language for abstract behavioral specification. In: Aichernig, B.K., de Boer, F.S., Bonsangue, M.M. (eds.) FMCO 2010. LNCS, vol. 6957, pp. 142–164. Springer, Heidelberg (2011). https://doi.org/10.1007/978-3-642-25271-6_8

16. Johnsen, E.B., Schlatte, R., Tapia Tarifa, S.L.: Integrating deployment architectures and resource consumption in timed object-oriented models. J. Syst. Softw. **84**(1), 67–91 (2015). https://doi.org/10.1016/j.jlamp.2014.07.001

17. Kephart, J.O., Chess, D.M.: The vision of autonomic computing. J. Syst. Softw. **36**(1), 41–50 (2003). https://doi.org/10.1109/MC.2003.1160055

18. Kephart, J.O., Das, R.: Achieving self-management via utility functions. J. Syst. Softw. **11**(1), 40–48 (2007). https://doi.org/10.1109/MIC.2007.2

19. Kumar, M., Sharma, S.C., Goel, A., Singh, S.P.: A comprehensive survey for scheduling techniques in cloud computing. J. Netw. Comput. Appl. **143**(12), 1–33 (2019). https://doi.org/10.1016/j.jnca.2019.06.006

20. Lau, K.-K., Wang, Z.: Software component models. J. Syst. Softw. **33**(10), 709–724 (2007). https://doi.org/10.1109/TSE.2007.70726

21. Pezoa, F., Reutter, J.L., Suarez, F., Ugarte, M., Vrgoč, D.: Foundations of JSON schema. In: Proceedings of the 25th International Conference on World Wide Web, WWW 2016, pp. 263–273 (2016). https://doi.org/10.1145/2872427.2883029

22. Muhuri, P.K., Biswas, S.K.: Bayesian optimization algorithm for multi-objective scheduling of time and precedence constrained tasks in heterogeneous multiprocessor systems. Appl. Soft Comput. **92**(12), 106274 (2020). https://doi.org/10.1016/j.asoc.2020.106274

23. Rasmussen, C.E., Williams, C.K.I.: Gaussian Processes for Machine Learning. MIT Press (2005). https://doi.org/10.7551/mitpress/3206.001.0001

24. Sanchez, S.M., Sánchez, P.J.: Better Big Data via data farming experiments. In: Tolk, A., Fowler, J., Shao, G., Yücesan, E. (eds.) Advances in Modeling and Simulation. SFMA, pp. 159–179. Springer, Cham (2017). https://doi.org/10.1007/978-3-319-64182-9_9

25. Schlatte, R., Johnsen, E.B., Kamburjan, E., Tapia Tarifa, S.L.: Modeling and analyzing resource-sensitive actors: a tutorial introduction. In: Damiani, F., Dardha, O. (eds.) COOR-DINATION 2021. LNCS, vol. 12717, pp. 3–19. Springer, Cham (2021). https://doi.org/10.1007/978-3-030-78142-2_1
26. Tao, F., Zhang, H., Liu, A., Nee, A.Y.C.: Digital twin in industry: state-of-the-art. J. Syst. Softw. **15**(4), 2405–2415 (2019). https://doi.org/10.1109/TII.2018.2873186

Opportunities and Advantages of Cloud Migration of a Smart Restaurant System

Beniamino Di Martino, Luigi Colucci Cante[⊠], and Nicla Cerullo

University of Campania "L. Vanvitelli", Caserta, Italy
{beniamino.dimartino,luigi.coluccicante}@unicampania.it,
nicla.cerullo@studenti.unicampania.it

Abstract. The restaurant sector, despite being increasingly developing nowadays, is still very ineffective due to the low use of technology. The latter has introduced significant improvements in this sector through the simple use of applications, cloud computing, NFC sensors and IoT. For this reason, it was proposed to design a system that would allow automated and intelligent control and management of a restaurant using IoT for product safety in the kitchen and booking of parking spaces, tables and meals through an interactive Menu. So, something that would provide ease of organisation in the process of ordering food, regularly checking and monitoring its quality to maintain proper hygiene. In addition, there will also be the possibility to collect information about the availability of parking spaces, which will be detected through infrared proximity sensors, which will avoid a lack of organisation. These improvements in restaurant management have greater results through the use of cloud services, in fact in the article, the migration of a restaurant application to the cloud is proposed using the services and patterns offered by Amazon. The use the latter allows an increase in performance, availability and security of the entire system.

1 Introduction

The large-scale deployment of resources provided as a service has led to the progress of a real process of "cloudification" of legacy systems. As a result, many industries are benefiting from the advantages and opportunities provided by cloud computing to improve the performance of their own applications. In recent years, a revolution has also been happening in the restaurant industry that aims to overcome inefficiencies in the sector by introducing Cloud services and IoT technologies. New devices, such as monitoring the availability of parking spaces, or the introduction of a system for booking parking spaces, tables and meals through an interactive menu, have allowed the introduction of the Smart Restaurant concepts. The use of IoT technologies, and the use of cloud services provided by cloud providers, are the fundamentals of smart restaurants.

In this paper, the migration of an application for the intelligent management of restaurants is proposed using cloud services and cloud patterns offered by Amazon. The cloud topology obtained is modeled through the **OpenTosca**

© The Author(s), under exclusive license to Springer Nature Switzerland AG 2022
L. Barolli et al. (Eds.): AINA 2022, LNNS 451, pp. 153–162, 2022.
https://doi.org/10.1007/978-3-030-99619-2_15

tool [1], which uses the **TOSCA** (*Topology and Orchestration Specification for Cloud Applications*) standard to model a cloud application. TOSCA is now the de-facto modeling language for service designs in many companies. By adopting a model-driven standard like TOSCA, it becomes easier for these organizations to create an application-centric playground, where any application instance can be deployed faster, using the same set of configurations and with high reliability. This paper is organized as follows. In the Sect. 2, existing works on legacy systems' cloud migration are analyzed. Section 4 describes the architecture of intelligent restaurant management application proposed, highlighting what weaknesses could be improved thought a cloud migration. Section 5 shows the agnostic version of the cloud architecture, which leverages the cloud agnostic patterns proposed in [3] to perform the cloud migration process. Then, the vendor-specific version of the cloud architecture is proposed, leveraging the AWS services offered by Amazon. The modeling of the vendor specific architecture with the Open-Tosca tool is also proposed. Finally, the paper concludes in Sect. 6.

2   Related Works

The "cloudification" of existing systems is an activity that has grown exponentially in recent years. As a result, a lot of research work in the last few years has been aimed towards migrating whole components of on-premise applications on cloud.

The work [4] proposes a generic framework that can be used to migrate processes from a legacy delivery model to a cloud delivery model. The framework examines all of the organisation's governing policies, laws and other internal or external factors that model its operations.

In the work [7] an architecture for cloudification of legacy applications is proposed, which consists of three parts: a Web portal, a SaaS service supermarket and a SaaS application development platform.

The work [2] investigates the need for academic institutions to provide new services, in a cloud model, to be used either in teaching or research activities. One of the main decisions to be addressed is related to the cloud model to adopt (private, public or hybrid), and what the mixing of functionalities to use for the hybrid one, so two different methodologies (Cost/features and Semantic-based) are been experimented in order to identify the best suited cloud model to adopt for a specific problem.

The work [5] proposes the study of cloud migration of an monitoring platform providing accurate quality and performance indicator in order to monitor an internal company's products. A comparative analysis of the best cloud service and deployment model has been performed. After analyzing the different cloud solutions and challenges, will follow the studies of the design, development and migration of the application mentioned above.

The work presented in this paper differs from the existing approaches, as it directly applies cloud migration techniques based on cloud computing patterns and uses existing services provided by cloud vendor as Amazon.

3 Background

Amazon Web Services, Inc. (AWS) is a subsidiary of Amazon providing on-demand cloud computing platforms and APIs to individuals, companies, and governments, on a metered pay-as-you-go basis. These cloud computing web services[1] provide a variety of basic abstract technical infrastructure and distributed computing building blocks and tools. One of these services is **Amazon Elastic Compute Cloud (EC2)**, which allows users to have at their disposal a virtual cluster of computers, available all the time, through the Internet. AWS's virtual computers emulate most of the attributes of a real computer, including hardware central processing units (CPUs) and graphics processing units (GPUs) for processing; local/RAM memory; hard-disk/SSD storage; a choice of operating systems; networking; and pre-loaded application software such as web servers, databases, and customer relationship management (CRM).

In addition to services, AWS also provides a set of patterns, which are called **AWS Cloud Design Patterns (CDP)** [6], that are a collection of solutions and design ideas for using AWS cloud technology to solve common systems design problems. To create the CDPs, AWS's members reviewed many designs created by various cloud architects, categorized them by the type of problem they addressed, and then created generic design patterns based on those specific solutions. This beta Cloud Design Pattern web site is the culmination of work by many different architects including **Ninja of Three (NoT)** who shared their expertise and experience in building cloud solutions.

The AWS Architecture Center also provides some reference architectures[2] prepared by industry experts that help cloud architects design secure, high-performance, resilient and efficient infrastructures for a wide set of applications and workloads.

4 Smart Restaurant System Architecture

The native application architecture is described in this section. This application involves two main actors:

- *Customer*: once registered on the application, through a Login process he can choose to make a reservation of a parking space. Then, he can choose to: i) book a table by selecting the number of seats; ii) order any product of each category available (appetizer, first course, second course, dessert, fruit, drink); iii) make the payment of the bill;

---

[1] https://docs.aws.amazon.com/index.html.
[2] https://aws.amazon.com/it/architecture/.

- **Restaurant Owner**: he will be able to manage the application, in particular by inserting, whenever he needs it, special offers or by eliminating, adding or modifying products in the menu.

In order to model the physical aspects of a system, but also to visualize the organization and relationships between its components, a Component Diagram has been created. It is shown in Fig. 1.

The diagram is made up of the following components:

- **Mobile Application**: interface module with all the various components, made available to all users;
- **Registration & Authentication**: module that allows authentication and registration and will be connected to a specially dedicated database (Registration & Authentication Database) containing information about each user;
- **Parking System**: module used to collect and disseminate information on the availability of parking spaces, which will be detected through infrared proximity sensors and whose reservation is made through an NFC sensor;
- **Table booking System**: module that manages the booking of tables in the restaurant through the use of NFC sensors;
- **Online ordering system**: module that manages orders for products that will be made through the use of an NFC sensor;
- **Food quality management system**: module that deals with the management of food quality, in particular the control through a temperature sensor and color;
- **Payment system**: module that allows the customer to interface with an external payment system, which has a relative database (Payment Database);
- **Mobile App Database**: module for the management and storage of data relating to payments, authentications, reserved parking spaces and tables and orders placed.

The component diagram shown in Fig. 1 follows the architectural pattern **Model-View-Controller (MVC)** so as to separate the various functionalities between different blocks. In particular, the *Model* allows the implementation of the reservation of parking spaces, tables, products and many other features in the application; the *View* represents the interface with all the various components made available to users; the *Controller* represents a meeting point between the graphical interface and the data: for example, when the user needs to change the information of the profile, the Controller notifies the Model so that it implements the requests coming from the user.

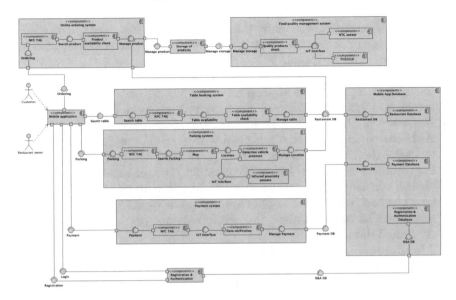

Fig. 1. Component diagram of smart restaurant system.

5   Cloud Migration of Smart Restaurant System

5.1   Agnostic Cloud Architecture

As a starting point for this work we were inspired by an AWS reference architecture that models a generic application that connects multiple restaurants using IoT and Machine Learning[3]. Following the suggestions of the reference architecture it has been possible to map some agnostic Cloud Computing patterns into the system architecture elements. Figure 2 shows this mapping and it is described below:

- *Online ordering system, Table booking system, Parking system*: mapped with the *Elastic Queue* as it allows to control possible high accesses to table and product bookings that would slow down the system. Therefore it manages these queues by adjusting the number of instances of the application components that manage these requests. This avoids possible bottlenecks.
- *Parking system*: mapped with *Exactly-once Delivery* as it avoids a high number of messages relating to the booking of parking spaces by the same customer which would lead to an overload by the same identifier. To avoid this, filtering is applied to the messages.
- *Payment system, Registration & Authentication system*: mapped with *At-least-once Delivery* as it ensures the receipt of any requests for payment or access to the application. Thanks to this pattern, if the confirmation of receipt of a message is not delivered after a certain period of time, the

---

[3] https://d1.awsstatic.com/architecture-diagrams/ArchitectureDiagrams/connected-restaurants-using-iot-ai-ml-ra.pdf?did=wpcardtrk=wpcard.

message is returned. Therefore, thanks to such retransmission, it is ensured that, for each message retrieved by a recipient, an acknowledgement is sent to the sender of the message.

- **Mobile Application**: mapped with **User Interface Component** as it provides an input for the application and a link between asynchronous communication with application components and synchronous user access.
- **Restaurant Database, Payment Database and Registration & Authentication Database**: mapped with **Key-Value Storage** to be able to model the three databases as NoSQL allowing the storage of data relating to orders, payments and authentication through excellent management of their consistency, but above all scalability. It is an extremely flexible pattern and can be adapted to the changing needs of different customers. It is recommended to use it given the presence of not particularly complex and structured data.
- **Monitoring**: inserted in order to be able to monitor the system and make decisions regarding the quantity of requests received. It was mapped with **Watchdog** as it allows high reliability since it monitors the system for any anomalies in the components and in this case also provides for appropriate resolution.

Finally, the entire architecture has been mapped with the following patterns: i) **Elastic Load Balancer** to make sure that we have an automatic scalability of the application components of such distributed application; ii) **Private Cloud** to provide a high level of security and privacy for the user that can, however, reduce its elasticity, the ability to provide a pay-per-use pricing model and above all guarantee a secure connection; iii) **Unpredictable Workload** to align the numbers of resources to the workload that can change according to the seasons, as there can be periods when the user has a high peak and others when it is stable. Thus, the load is unpredictable as there may be numerous accesses due to weather phenomena, offers for limited periods or sudden interest from the public due to word of mouth.

5.2   AWS Cloud Architecture

Starting from the Agnostic Cloud Computing Patterns, it was possible to carry out a mapping with Vendor Specific Cloud Computing Patterns. In particular, AWS (Amazon Web Services), which offers a large number of services, was taken into account. Table 1 shows a comparison between vendor patterns taken into consideration and corresponding agnostic patterns is shown.

While, the services related to the previous patterns are as follows: *Amazon EC2, SQS* (Simple Queue Service), *Amazon API Gateway, Dynamo DB, Amazon S3* (Simple Storage Service), *Amazon Cognito, KMS* (Key Management Service), *CloudWatch, Auto Scaling, AMI* (Amazon Machine Images), *Elastic Load Balancing, AWS Lambda, AWS IoT Core, AWS IoT Device Defender, AWS IoT Device Management*.

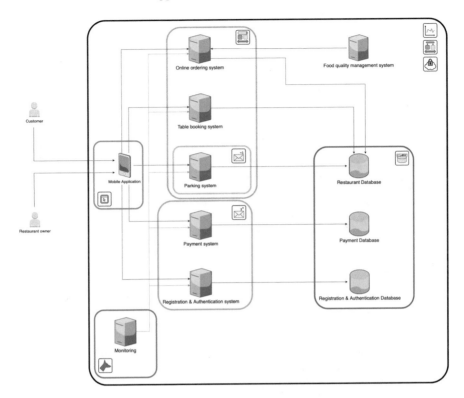

Fig. 2. Agnostic architecture of smart restaurant system.

Table 1. Vendor patterns vs agnostic patterns

AWS pattern	Agnostic pattern
Scale out	Elastic load balancer
Priority queue	Elastic queue
Storage index	Key-value storage
Monitoring integration	Watchdog

All these services was used in the AWS architecture shown in Fig. 3, which has the following workflow: Users and Restaurant Owner access the Mobile Application running on an **EC2** instance through an authentication, also running on an EC2 instance, in which the **AWS Cognito** service is deployed, through which Customer and Restaurant Owner can register by entering username and password and in which data encryption is guaranteed through **KMS**. From an API Gateway, which is able to handle numerous requests at the same time, it was possible to route the requests towards the private resources of the VPC. The Ordering process, performed on an EC2 instance, is connected to the EC2 instance that manages the quality of the products through the **AWS Lambda**

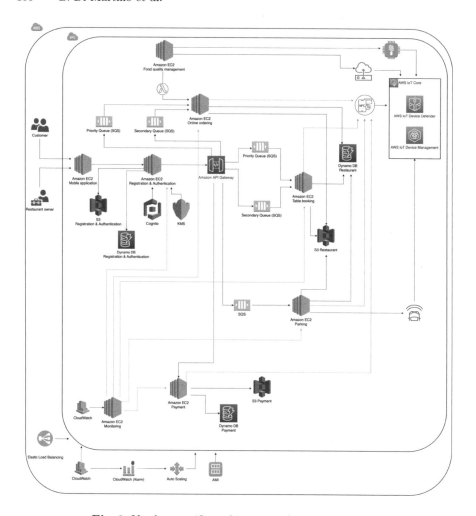

Fig. 3. Vendor specific architecture of smart restaurant.

service, since the latter allows to facilitate and speed up the search for "quality approved" foods, also managing their scalability and performance.

The EC2 instances of Registration & Authentication, Online Ordering, Table booking, Parking and Payment, in order to have scalability, security and high performance, will have the data saved in a Simple Storage Service (S3) and that of the meta-data in the Dynamo DB. The latter, moreover, with the exception of Registration & Authentication, are connected to **AWS IoT Core**, as they are connected to IoT devices. The latter, in turn, is connected to **AWS IoT Device Defender** and **AWS IoT Device Management** for security, device access protection and integrity monitoring. Parking's EC2 instance is managed by a queue in order to have fault tolerance, microservices scalability and lossless message storage and reception. On the other hand, the Online Ordering and

Table booking EC2 instances are modelled by means of two **SQS** queues with different priorities: the "Priority", having a higher priority, is intended for users as there will certainly be more requests than the Restaurant owner which, for this reason, is associated with the "Secondary".

The Monitoring component was executed on an EC2 instance and has a **CloudWatch** connected to it as it allows monitoring and optimising the use of resources. This service is actually applied to the entire VPC (Virtual Private Cloud), as well as the Elastic Load Balancing, Auto Scaling and **Amazon Machine Images** (AMI) services in order to have automatic scalability, reliable security and load balancing.

Finally, Fig. 4 shows the AWS architecture realised in OpenTOSCA, in which some components have been added: i) Apache server (*Apache-2.4*) hosted (hostedOn) on *Ubuntu*; ii) various Python scripts (*Python_3*), realised in Python, which implement the functionalities of the single nodes to which they are connected; iii) some IoT devices used (*NFC-tag, TemperatureSensor, Color-sensor, InfrarredProximitySensor*). This architecture has been modelled in TOSCA as it allows the deployment and management of Cloud applications to be automated. It makes it possible to describe the structure of an application, modelled as a topological model, the dependencies of the components and the necessary infrastructure resources in a portable way.

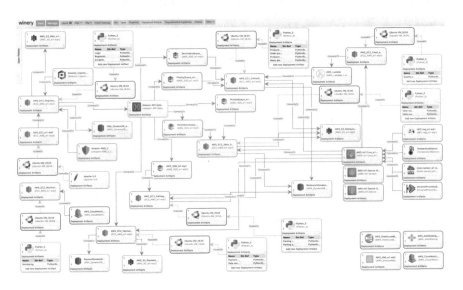

Fig. 4. Vendor specific architecture of smart restaurant modeled with OpenTOSCA.

6   Conclusion

In this paper the migration of a restaurant application to the cloud is proposed mapping some agnostic cloud computing patterns into on-premise system. Two different architectures are provided: an agnostic architecture and a vendor specific architecture using services and cloud computing patterns offered by Amazon. Different benefits have been gained from this migration to the cloud, such as the introduction of automatic scalability, enhanced security of sensitive data and greater load balancing.

In future work, this migration process could be enriched with services and patterns provided by other cloud provider, such as Azure, Google, IBM, etc. In this way a comparison between the different solutions could be obtained.

Acknowledgements. The work described in this paper has been supported by the Project VALERE "SSCeGov - Semantic, Secure and Law Compliant e-Government Processes".

References

1. Binz, T., et al.: OpenTOSCA – a runtime for TOSCA-based cloud applications. In: Basu, S., Pautasso, C., Zhang, L., Fu, X. (eds.) ICSOC 2013. LNCS, vol. 8274, pp. 692–695. Springer, Heidelberg (2013). https://doi.org/10.1007/978-3-642-45005-1_62
2. Cantiello, P., Di Martino, B., Mastroianni, M., Cante, L.C., Graziano, M.: Towards a cloud model choice evaluation: comparison between cost/features and ontology-based analysis. Int. J. Grid Utility Comput. (2022, article published/in press). http://hdl.handle.net/2122/15035
3. Fehling, C., Leymann, F., Retter, R., Schupeck, W., Arbitter, P.: Cloud Computing Patterns. Springer, Vienna (2014). https://doi.org/10.1007/978-3-7091-1568-8
4. Fratila, L.-A.: Enterprise architecture and corporate governance a cohesive approach towards cloud migration in the banking industry. Int. J. Econ. **VIII**(5) (2020)
5. Gianno, F.: Development and cloud migration of booking flow monitoring tool. Ph.D. thesis, Politecnico di Torino (2020)
6. Young, M.: Implementing Cloud Design Patterns for AWS. Packt Publishing Ltd., Birmingham (2015)
7. Yu, D., et al.: A practical architecture of cloudification of legacy applications. In: 2011 IEEE World Congress on Services, pp. 17–24. IEEE (2011)

Analysis of Techniques for Mapping Convolutional Neural Networks onto Cloud Edge Architectures Using SplitFed Learning Method

Beniamino Di Martino[✉], Mariangela Graziano, Luigi Colucci Cante, and Datiana Cascone

University of Campania "L. Vanvitelli", Caserta, Italy
{beniamino.dimartino,mariangela.graziano,
luigi.coluccicante}@unicampania.it,
datiana.cascone@studenti.unicampania.it

Abstract. The Convolutional Neural Network is a machine learning algorithm of increasing interest in recent years for its use in computer vision. Today, there are a lot of applications in safe driving, object recognition, person identification and in healthcare. On the other hand, many devices do not have the computational power to support a deep neural network and, moreover, a machine learning algorithm requires a training set of considerable size for optimization, which is continuously updated and common to multiple users. The shared data relating to the images, can generate security problems to the system by falling within the field of data privacy. Local regulations, such as the GDPR in Europe, provide for high levels of security, in particular data defined "sensitive", such as biometrics and health data. Using this data in a shared environment can lead to a data breach, not sharing it degrades CNN's performance. In this article we will illustrate the mechanisms of subdivision of a convolutional neural network between edge devices, with limited computational power and a public cloud platform. The distribution of the computation aims at convolution of the neural network and at preserving system security. In particular, an example of distribution will be illustrated using the tree-computation pattern on SplitFed Learning architecture.

1 Introduction

Artificial intelligence is today a science that is increasingly closer to companies and people [5]. In recent times, applications that use machine learning algorithms are more and more used by ordinary people or unskilled employees [4]. Machine vision is the most required algorithms, it simulates one of the five senses of man, maybe the most important. Different applications use these techniques in various fields, some of these are: i) safe self-driving [3]; ii) intelligent video surveillance systems[1]; iii) tools to predict complex diseases (such as retinopathy

---

[1] https://www.smartbuildingitalia.it/wp-content/uploads/2019/11/mattia-bastianini-.pdf.

L. Barolli et al. (Eds.): AINA 2022, LNNS 451, pp. 163–172, 2022.
https://doi.org/10.1007/978-3-030-99619-2_16

in diabetic patients [9]) or in some well-known video games [7]. Devices that require these algorithms, usually, do not have sufficient computing power to ensure high performance and speed of execution. The use of a remote computing platform, made available by a public cloud platform [2], can solve this problems. However, in specific cases, this solution is not enough. In this paper we show how a machine learning algorithm, as CNN, can is splitted on distributed nodes of network of Cloud Edge Architecture. We use a SplitFed Learning model [12], already known in literature, to split neural network and the computational pattern "Tree Computation" [6], to distributed the computation in order to optimize speed up and data security level of system. The mapping of machine learning algorithm and Tree Computation Pattern performs model show in [6]. This paper is organized as follows. In Sect. 2 all the notions useful to understand this work are reported. Section 3 illustrates a description of Federate Learning focusing on Split Learning and SplitFed Learning techniques. Section 4 contains a description of computational patterns used in Federate Learning. Section 5 proposes an implementation of CNN in the Federate Learning. Finally, the paper concludes with Sect. 6.

2   Background and Related Works

Artificial Neural Network (ANN or NN) [14,15] is a machine learning algorithm, a mathematical model inspired by the human cerebral cortex. It is composed of groups of artificial neurons, connected to each other through one or more interconnection layers. A particular model of ANN is the Convolutional Neural Network (CNN), that is specializes in image recognition. The goal of a CNN is to classify pictures based on some features. Like an human brain, makes visual-conceptual associations, makes simplifications and then recognizes objects or people. This algorithm can be seen as a chain of stages, in each of them the neurons are organized in groups. The different stages are show in Fig. 1. The Input Stage receives the data vector representing the pixels of the input image; Convolutional Stage extracts the features and identify patterns; Pooling Stage reduces the dimensionality of the image (increasing abstraction); Fully Connected Network performs image classification and finally Output Stage presents the result.

Fig. 1. Stages of convolutional neural network

In a CNN there can be more convolutional stages and more pooling stages, greater is a number of this, greater is the complexity of the network, better is ability to recognize images. CNN, like any machine learning algorithm, must be trained to perform a task correctly. For this reason it must receive a lot of labeled data (training set), which is used to learn the patterns. If it receives a larger data set, it will be more intelligent. We have analyzed several papers in which the applications of a convolutional neural network are often used by devices with high computing capacity that can reside both locally and on the cloud platform. In this configuration the most used computational pattern is **Producer/Consumer**. The work [13] proposes an approach for privacy-preserving federated learning employing an SMC protocol based on functional encryption. To evaluate this approach the authors use a federated learning process to train a CNN on the MNIST dataset. In the article [8] the authors have reformulated the Federate Learning approach as a group knowledge transfer training algorithm, called FedGKT to address the resource-constrained reality of edge devices. This algorithm is used to train small CNNs on edge nodes and periodically transfer their knowledge by knowledge distillation to a large server-side CNN. It reduced demand for edge computation, lower communication bandwidth for large CNNs, and asynchronous training, all while maintaining model accuracy. Training a convolutional neural network, however, remains an unsolved phase, requiring high communication costs, energy consumption and security problem.

3   Federate Learning

First defined by Google in 2017 [1], Federated Learning (FL-also known as collaborative learning), is a technique that allows different participants to build a common and robust model. It is used to train an algorithm of machine learning, part of the algorithm is implemented on the edge device while the remainder is performed in a server. Simplifies edge-side computation.

3.1   Split Learning

Split learning is a new technique that allows you to train a machine learning algorithm while preserving the privacy of data [11]. It introduces a subdivision, between client and server, in the execution of the training phase. As we can see in Fig. 2, the complete model (W) is divided into two parts: client side (W^C) and server side network portion (W^S). The client trains the client-side model up to a level called "Cut Layer". The server receives some parameters, called "Smashed data" and runs only the server-side model. Training of the complete model is performed by sequential (forward/backward) propagation without data exchange.

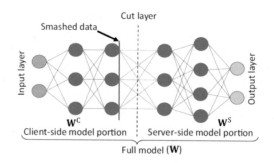

Fig. 2. Split Learning's model

3.2   SplitFed Learning

SplitFed Learning (SFL) is the fusion of Federated Learning (FD) and Split Learning (SL) techniques. In this methodology, the client side model is run as in FL, while the complete model is trained by dividing it on the client side and server side as in SL to exploit the advantages of privacy and computation. In this architecture it is possible introduce an additional worker node, called Fed Server, dedicated to synchronizing the model between the various clients participating in the training and preserve local model of CNN [12].

4   Computational and Architectural Patterns for Federate Learning

Design Patterns represent a general, reusable and portable solution to a problem that commonly occurs within a given context. This definition can be extended to other categories of patterns, sayed Computational Patterns, which represent a computational solution to a problem in distributed network. One of this, **Tree Computation Pattern**, organizes the computation according to a hierarchical tree. Starting with the root node, the original problem is splitted into simpler sub-problems, until a set of computationally executable tasks is identified. Then, the partial results are aggregated and the solution of the complex problem is obtained. Figure 3 summarizes the scheme of this pattern.

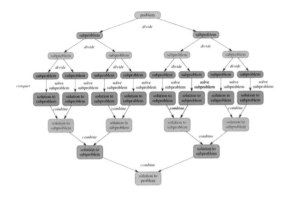

Fig. 3. Tree computation pattern

5 An Implementation Proposal of CNN in Federate Learning

This article focuses on the analysis of the optimization of the distribution of a CNN on Cloud Edge architecture. The elements on the edge are characterized by low computing power and are mobile devices, a server supports the training and initialization phases in order to reduce interactions between two part. On the cloud side, the public platform provides with the most expensive computational part of CNN (deep learning). The distribution of the computation between the nodes occurs through the use of a method seen in the paper [6] where the application deployment proceeds for the following steps:

STEP 1: application of a computational pattern to chosen algorithm;
STEP 2: decomposition of pattern according to architectural patterns;
STEP 3: pattern guided deployment in specific scenario.

In first step, given the high interest and complexity of a CNN, this type of application has been selected. The distribution of the computation will take place according to the SplitFed Learning technique which involves three or more actors, who collaborate in the formation of the model. In Fig. 4 is shown the sequence diagram. The first actor is the Fed Server which starts communication sending a local model to client, this model contains weights and activation function for each layer of CNN. The client starts his forward propagation, on its local data, until the cut layer. The forward propagation continues on cloud platform which perform deep layers, receiving only the smashed data. Cloud side evaluates the Loss function, corrects parameters and sends them back to the client (backward propagation). This loop proceeds until an acceptable Loss value is obtained in client side. Once the ideal values for parameters have been obtained, these are transmitted to Fed Server which makes an average, Fed Average [10]. Finally, Fed Server updates the model, which will be more and more accurate as more clients participate in the formation of the global model.

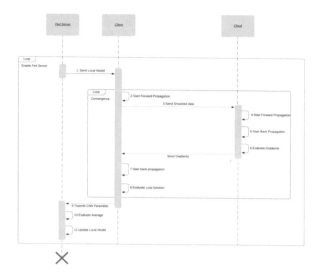

Fig. 4. Sequence diagram of CNN in SplitFed Learning

So, the sequence diagram reported in Fig. 5 shown the interaction in Tree Computation pattern as reported in Sect. 4. It represents this information in the following way: there are three actors who work together to reach the goal, this are: Root, Level1 and Level2. Root is the initial problem and the Levels the partial computational solutions. Root assigns a sub-problem to Level1 which, cannot determine the solution, splits it into a smaller problem and assigns it to Level2 which calculates its part of the sub-problem. That is aggregated and transmitted back to Level1 which in turn aggregates, calculates the solution and sends back to the Root.

Finally, in Fig. 6 a possible mapping between the sequence diagrams of the CNN and the Tree Computation is represented.

In second step, see Fig. 7, our application has been distributed on an N-tier architecture, where the tiers that compose it are: **Data Ingestion Layer** component which detects and receives the data, carries out the initialization phase of the algorithm, part of the data acquisition layer; **Resource Management Level** is a system interface to the data source; **Business Logic Layer** it is the central part of the project, includes everything related to data processing and all the functions of the application. Furthermore, in N-Tier diagram, Camera represents a generic image input sensor, it sends data to a memory element, Local Data. The Local Model contains the custom model of each device while the Global Model is a component of the Fed server, calculates and averages the client models.

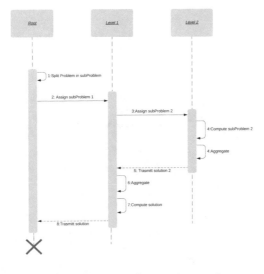

Fig. 5. Sequence diagram of tree computation pattern

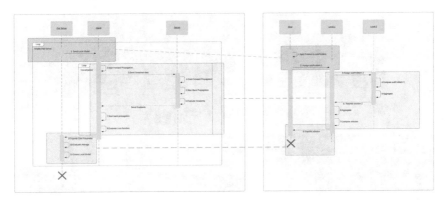

Fig. 6. Mapping of SplitFed Learning onto tree computation pattern

In this last step a possible deployment on generic architecture is illustrated. In Fig. 8, on the edge-site we have client device, this represent a generic element of a heterogeneous system. Fed Server can be implemented with a low computational component (computes the average of some parameters) and a small storage capability (saves model's parameters).

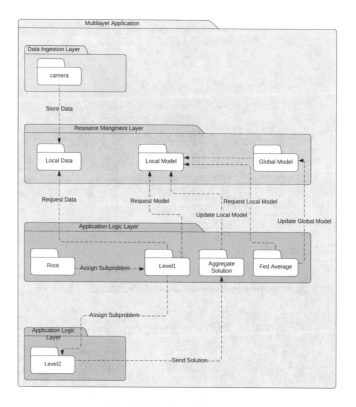

Fig. 7. N-tier architecture

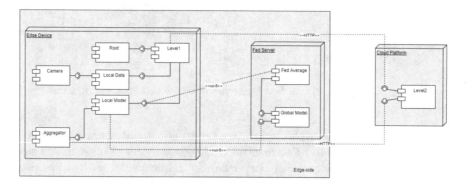

Fig. 8. Deployment diagram

6   Conclusion

In this paper an application of Tree Computing Pattern to CNN on a distributed cloud network has been proposed. This configuration could be extended to multiple levels of computation depth and used in enterprise that use heterogeneous devices. A business such as government or a hospital needs adequate data confidentiality but at the same time may require devices with real-time response. The optimization of the IT distribution among the various nodes of the network can standardize the response quality of the CNN exploiting the computing capacity of the most powerful devices. The large number of network elements can improve the quality of the overall model. In the next works we will develop the deployment on the remote platform for generic applications and specific vendors, comparing the performances. The division into tasks of both edge and cloud computing makes the application scalable and portable in cloud platform. This paper is intended to be only a starting point for the creation of cloud-distributed applications of convolutional neural networks.

Acknowledgements. The work described in this paper has been supported by the Project VALERE "SSCeGov - Semantic, Secure and Law Compliant e-Government Processes".

References

1. Beaufays, F.S., Chen, M., Mathews, R., Ouyang, T.: Federated learning of out-of-vocabulary words (2019)
2. Cantiello, P., Di Martino, B., Mastroianni, M., Cante, L.C., Graziano, M.: Towards a cloud model choice evaluation: comparison between cost/features and ontology-based analysis. Int. J. Grid Util. Comput. (2022, article published/in press). http://hdl.handle.net/2122/15035
3. Chishti, S.O.A., Riaz, S., BilalZaib, M., Nauman, M.: Self-driving cars using CNN and Q-learning. In: 2018 IEEE 21st International Multi-Topic Conference (INMIC), pp. 1–7. IEEE (2018)
4. Di Martino, B., Colucci Cante, L., Graziano, M., Enrich Sard, R.: Tweets analysis with big data technology and machine learning to evaluate smart and sustainable urban mobility actions in Barcelona. In: Barolli, L., Poniszewska-Maranda, A., Enokido, T. (eds.) CISIS 2020. AISC, vol. 1194, pp. 510–519. Springer, Cham (2021). https://doi.org/10.1007/978-3-030-50454-0_53
5. Di Martino, B., Cascone, D., Colucci Cante, L., Esposito, A.: Semantic representation and rule based patterns discovery and verification in eProcurement business processes for eGovernment. In: Barolli, L., Yim, K., Enokido, T. (eds.) CISIS 2021. LNNS, vol. 278, pp. 667–676. Springer, Cham (2021). https://doi.org/10.1007/978-3-030-79725-6_67
6. Di Martino, B., Esposito, A.: Applying patterns to support deployment in cloud-edge environments: a case study. In: Barolli, L., Woungang, I., Enokido, T. (eds.) AINA 2021. LNNS, vol. 227, pp. 139–148. Springer, Cham (2021). https://doi.org/10.1007/978-3-030-75078-7_15
7. Goericke, S.: Using convolution neural networks to develop robust combat behaviors through reinforcement learning. Ph.D. thesis, Naval Postgraduate School (2021)
8. He, C., Annavaram, M., Avestimehr, S.: Group knowledge transfer: federated learning of large CNNs at the edge. arXiv preprint arXiv:2007.14513 (2020)

9. Kwasigroch, A., Jarzembinski, B., Grochowski, M.: Deep CNN based decision support system for detection and assessing the stage of diabetic retinopathy. In: 2018 International Interdisciplinary PhD Workshop (IIPhDW), pp. 111–116. IEEE (2018)

10. Lu, Y., Fan, L.: An efficient and robust aggregation algorithm for learning federated CNN. In: Proceedings of the 2020 3rd International Conference on Signal Processing and Machine Learning, pp. 1–7 (2020)

11. Thapa, C., Chamikara, M.A.P., Camtepe, S., Sun, L.: SplitFed: when federated learning meets split learning. arXiv preprint arXiv:2004.12088 (2020)

12. Thapa, C., Chamikara, M.A.P., Camtepe, S.A.: Advancements of federated learning towards privacy preservation: from federated learning to split learning. In: Rehman, M.H., Gaber, M.M. (eds.) Federated Learning Systems. SCI, vol. 965, pp. 79–109. Springer, Cham (2021). https://doi.org/10.1007/978-3-030-70604-3_4

13. Truex, S., et al.: A hybrid approach to privacy-preserving federated learning. In: Proceedings of the 12th ACM Workshop on Artificial Intelligence and Security, pp. 1–11 (2019)

14. Wang, S.C.: Artificial neural network. In: Interdisciplinary Computing in Java Programming, pp. 81–100. Springer, Cham (2003). https://doi.org/10.1007/978-1-4615-0377-4_5

15. Zhang, Z.: Artificial neural network. In: Multivariate Time Series Analysis in Climate and Environmental Research, pp. 1–35. Springer, Cham (2018). https://doi.org/10.1007/978-3-319-67340-0_1

In-cloud Migration of a Custom and Automatic Booking System

Beniamino Di Martino, Mariangela Graziano(✉), and Serena Angela Gracco

University of Campania "L. Vanvitelli", Caserta, Italy
{beniamino.dimartino,mariangela.graziano}@unicampania.it,
serenaangela.gracco@studenti.unicampania.it

Abstract. Nowadays the advancement of technologies such as the Internet of Things and Machine Learning has allowed the development of increasingly complex systems to simplify and automate countless different processes. On the other hand, the affirmation of Cloud Computing, and therefore the delivery of different services through the Internet as tools and applications like data storage, servers, databases, networking, and software has made it possible to reach even higher levels in terms of cost savings, increased productivity, speed and efficiency, performance, and security.

This paper illustrates the practical example of the possible On-Cloud migration of an On-premise application, using the services and patterns offered by Amazon. The application in exam has the purpose of automatically managing access and booking of a room, and possibly a parking lot, in a tourist complex. Since the original project did not foresee mechanisms of autoscaling of the computational resources, there is no monitoring mechanism or database redundancy, the usage of a Cloud approach was necessary to make possible the automatic scaling of resources based on the unpredictable load of requests from the users of the platform and to achieve performance, security and availability increase of the whole system.

1 Introduction

The accommodation industry is currently undergoing many transformations as a result of the of data explosion, social media advertising and a general trend to travel further and further away. These changes have obviously impacted by the way a hotel, or accommodation system in general, manages its customers, customer requests and therefore the need to develop increasingly complex, automated and highly scalable reservation management systems.

A hotel reservation system is a complex automated tool that schedules the dates and duration of a guest's stay, records payments, and the most advanced systems also allow customers to select extras such as reserving parking, customising room services such as temperature and light level, or other goodies to put in their room upon arrival. In a scenario where mobile bookings on travel websites are increasing at a faster rate than desktop bookings and an increasing percentage of them come from smart devices, cloud computing has shown the potential to meet these demands by providing remote access and mobile optimisation, intuitive and web-optimised interface, speed of deployment and scalability.

In this paper, a possible application of Cloud Computing in the hotel industry is proposed. This paper is organised as follows. In the Sect. 2, existing works on possible applications of cloud computing technologies and possible strategies for migrating applications from different industries and software companies to the cloud are analysed. Section 3 gives an overview of the architecture of the non-cloud booking system, with all its many functionalities and analysing all the improvements that a migration to the cloud could bring. Section 4 illustrates the cloud agnostic architecture applicable to the proposed system and then gives an overview of the concrete services and cloud models offered by Amazon applied to the project using the OpenTosca modelling tool that can model networks in a standardised way, improve automation, enable portability and overcome interoperability issues more easily. Finally, the paper concludes with Sect. 5, in which the results of the proposed work are discussed along with a comparison with a different cloud provider and some suggestions for future developments.

2   Background and Related Work

Many software companies and several industry sector were migrating to the Cloud in order to take advantage [21] of this technical paradigm such as elasticity, and reduced operational costs. The impact analysis of a potential migration to Cloud is reported in [1,2]. A comprehensive analysis of strategies and methods for cloud migration and service migration in a new computing paradigm were reported in [13,25]. In recent years, the potential and advantages of cloud computing has prompted many to consider and invest in a possible migration of existing applications to the cloud, involving several sectors, which have adopted a migration to cloud computing to improve their performance. In [7] the authors investigated the need for academic institutions to provide new services, in a cloud model, to be used either in teaching or research activities. The works [11,20] both outlined some considerations on the application of cloud computing to the healthcare sector and described the benefits of migrating to cloud computing in the public health sector for India and Valle del Cauca. Another interesting work focused on the usage of Cloud technology in the restaurant sector and accommodation industry. The article [15] proposed a cloud reservation system that improved the reservation system for a restaurant chain.The benefits that cloud computing can provide to the hotel industry were detailed in [5,18,19] , which specifically showed how the migration to cloud computing had an extremely relevant impact on the way information and IT resources were managed by the hotel industry. The research [24] proposed a cloud computing model for managing budget hotels to optimise operational costs, increasing process efficiency and enhancing customer satisfaction using a case study of hotel chains in Thailand. The work [23], investigated the overall adoption of Cloud Computing technology in the hotel industry, its challenges and benefits. The research conducted by the Grand Thornton company [12] focuses on the adoption of the cloud technology for one of the most significant IT spend items for hotels - the property management system (PMS) - which enables the automation of all front desk activities and acts as the'back bone'of the hotel.

3   Off-Cloud Booking System Architecture

The Property Management System in examination aims to automatically manage the access and reservation of a room, and possibly a parking space, in a tourist complex. After signing up and logging in to the mobile application or the system's website, the user can start looking for a room in the facility of interest and then access the customisation system by choosing services such as adjusting the temperature, lights or serving coffee and hot drinks in the room and then possibly booking the room. Once the user accesses the facility, an automatic system directs him/her to the nearest car park, if chosen during the booking process, and to the room booked. The intelligent distance control system will continuously monitor in real time the crowding situation in the most crowded areas of the facility, alerting customers via an intelligent screen located outside the monitored areas. In order to get a clearer view of what the parts of the system are and how they interact, it was necessary to go into more detail and create the UML Component Diagram shown in Fig. 1.

Different colors were used to highlight the application structure following the Three-Tier software architectural pattern [17]: **green** for the **Data layer**, **orange** for the **Application layer** and finally **purple** for the **Presentation layer**. Analyzing the various components in more detail, we find the **System front end** component that deals with interfacing with the user providing many interaction such as authentication provided by the **Login** and **Register** interfaces exposed by the **Authenticator** system. The interfaces **Select services**, **Search for room** and **Pay for reservation** allow the research for a room, the setting of the various customization services and the parking service, and then to finalize the booking process thanks to the functionalities offered by the third-party **Payment system** component. The **Smart IoT room services system** component takes care of making effective all the customizations required by the user by controlling all IoT systems located in the rooms of tourist facilities, using the input data coming from different sensors processed by a big data pipeline (Data management system). The **Smart parking reservation system** component manages the incoming data by the parking sensors that detect whether the parking space is occupied or not, the data entered by the user at the time of booking and, on the basis of these, manages to identify and indicate the user the best near parking lot at check-in. The **Smart distances manager system**, exploiting the data coming from cameras located in the most crowded and frequented places of the structure, analyzes first the distances between customers, then, by means of the component **AI Data interpreter**, detects too close distances and alerts customers thanks to smart screens located at the entrances

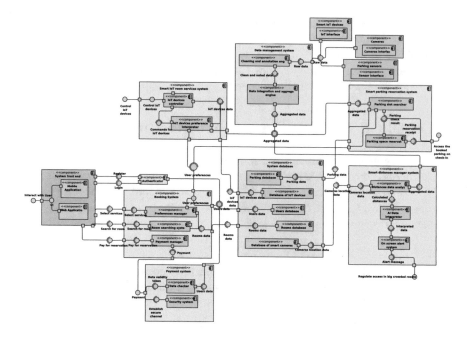

Fig. 1. UML Component Diagram

of these areas. Finally, the **System database** component encompasses all the structures of storage of user data, rooms, location of parking, sensors and cameras. This architecture thus described is not suitable for high demand loads from users, because it does not yet provide autoscaling mechanisms [4] of computational resources. For the time being, no mechanism is foreseen monitoring, nor database redundancy.

4   In-Cloud Migration of on Premise Custom and Automatic Lodge Booking System

4.1   Agnostic Cloud Architecture

Cloud Migration [8,14] is that IT process [3] with the goal of moving data and applications to a Cloud environment. Migration can take place from an on-premise data center to the public cloud, or from one Cloud platform to another.

The starting point for the cloud migration of the application in analysis was to locate the agnostic cloud patterns deemed more suitable particularly to achieve the automatic on-demand provisioning of computational resources, the load balancing and the interfacing with external services. Patterns are a widely used concept in computer science to describe good solutions to reoccurring problems in an abstract form. Such conceptual solutions can then be applied in concrete use cases regardless of used technologies, such as software, middleware, or programming languages.

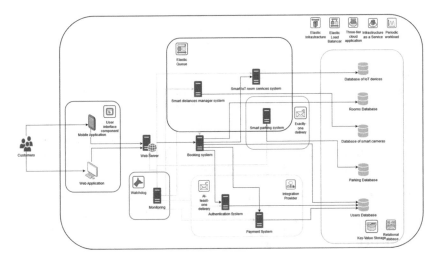

Fig. 2. Agnostic Cloud Architecture

It was thus useful to see such concept implemented also in the Cloud area focusing on the application of design Patterns[1] for Cloud Services [9]. As can be noted in Fig. 2, the **Elastic Infrastructure, Elastic Load Balancer, Infrastructure as a Service** and **Periodic Workload** have been mapped to the entire application [16] adjusting so its capacity automatically and dynamically to maintain stable and predictable performance while minimizing costs. The **Elastic Queue** pattern has been mapped with the components **Smart Parking System, Smart Distances Manager** and **Smart IoT Room Preferences** to manage the queues of requests to the various services of the system in order to regulate the number of component instances handling requests. Finally, the **Watchdog** pattern has been mapped to the monitoring part of the application while for the storage part, the **Key value** and **Relational Database** patterns were used.

4.2   AWS Cloud Architecture

Amazon's vendor-specific patterns [10, 22] are analyzed in the following section along with required services, underlining the correspondence with some of the agnostic patterns described previously.

4.2.1   AWS Cloud Patterns
The main AWS Cloud Patterns used are listed below:

- **Scale Out Pattern**: The approach according to which several servers are inserted in parallel with identical specifications for handling high traffic loads

---

[1] https://www.cloudcomputingpatterns.org/.

is known as "Scale out". In fact, Scale Out provides multiple virtual servers to run and that a Load Balancer is used to distribute the load to each independent virtual server. Depending on the specific system, in fact, the traffic could drastically increase over weeks, days or even hours. The AWS cloud service therefore makes it easy to dynamically change the number of virtual servers that manage processing, making it possible to adapt to such sudden changes in the traffic volume. A combination of three services is used: a load balancing (**Elastic Load Balancer**), the monitoring tool (**CloudWatch**) and the automatic scale-out service (**Auto Scaling**), to easily structure a system that can be scaled automatically depending on the load.

- **Job Observer Pattern**: As mentioned, AWS has a system known as Auto Scaling that can scale up or down automatically the number of **EC2** instances and working in conjunction with a tool resource monitoring known as **CloudWatch**, to increase or decrease EC2 instances in compliance with monitored value. **CloudWatch** in this case can monitor the number of messages in the **Simple Queue System (SQS)** queues, an additional service provided by AWS. The Job Observer Pattern then uses SQS to handle job requests, **Auto Scaling**, and CloudWatch to structure a system capable of automatically increasing or decreasing the number of servers batch based on the number of messages (job requests) within a queue.

- **Monitoring Integration Pattern**: AWS provides a monitoring service that is unable to observe operation internal virtual server (operating system, middleware, applications and so on), it is required have an independent monitoring system. The AWS **Cloud Monitoring** monitoring service provides an API, which allows you to use the independent monitoring system to perform centralized control, including the Cloud side, to get information.

Other pattern used were the **Storage Index Pattern** to make more efficient the key-value storage databases and the **Read Replica Pattern** to increase the performance of the processes reading data even with a large number of database instances is created.

4.2.2   AWS Services

Having first identified all the useful AWS Patterns that it was possible to implement, it was then mandatory to find all the services that could help to realize all the functionalities required for the system. Some of the most relevant AWS services are here presented:

- **Amazon EC2**: Amazon Elastic Compute Cloud (EC2) is a web service which provides secure and scalable compute capacity in the cloud. It is designed to make cloud computing at Web scale easier for developers. Via the intuitive Web service interface of Amazon EC2 it is possible obtain and configure the capacity in a simple and immediate way. The user has complete control of its own IT resources, which can be run in the Amazon's highly efficient processing environment.

- **Amazon CloudWatch**: Amazon CloudWatch is a monitoring and observability service created to provide concrete data and analysis, for monitoring applications, respond to changes in system-wide performance, optimize resource utilization and get a unified view of operational integrity. CloudWatch collects monitoring and operational data in the form of logs, parameters and events, providing a unified view of AWS resources, of applications and services running on AWS and on local servers.

- **Amazon SageMaker**: Amazon SageMaker helps developers and data scientists to prepare, build, train and implement quickly high-quality machine learning models combining a wide range of features created ad hoc for the ML. Amazon SageMaker is a comprehensive service that accelerates innovation with tools created specifically for each stage of ML development, including labeling, data preparation, feature design, detection of statistical bias, autoML, training, optimization, hosting, explainability, monitoring and workflows. This tool was used in combination with **AWS Glue ETL**, **Amazon Kinesis Firehose** and **Amazon Simple Storage Service** to build a pipeline for the data coming from the smart cameras and sensors to build a ML model to compute the distances among people in the different places of the building and possibly predict a different time to come back in case the place is overcrowded.

Figure 3 illustrated the system architecture created using the OpenTosca[2] tool[6], underlying all the **AWS patterns** taken into consideration along with necessary complementary **AWS services**.

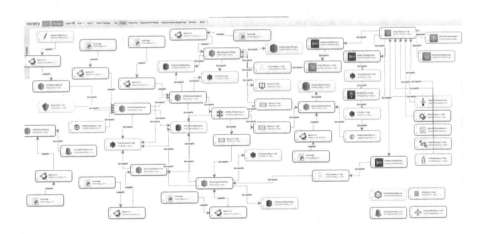

Fig. 3. AWS Cloud Architecture using OpenTosca and Winery

---

[2] https://www.opentosca.org/.

Moving on to analyze the **Topology Template** created, through a **Apache server** hosted *(hostedOn)* on **Ubuntu**, in its own right once hosted on an EC2 instance, users access the Booking System through a Registration or Authentication process, performed on EC2, and represented with the **AWS_EC2_ Registration & Authentication** node on which it is hosted (hostedOn) **Ubuntu (Ubuntu-VM_18.04)**. Also connected to the node in question are *(ConnectsTo)* **Amazon_Cognito** and **AWS KMS** that allow user registration, scalability and encryption some data. To this node is connected *(ConnectsTo)* then the service **AWS_EC2_BookingSystem** which takes care of managing user requests and route them to the appropriate services. Then through an **Amazon API Gateway**, which is able to handle numerous requests at the same time, it was possible to route the latter to each **EC2 instance**, each carried over to an **EC2** node: **AWS_EC2_SmartIoTRoomServices**, **AWS_EC2_SmartParking**, **AWS_EC2_SmartDistancesManager**, **AWS_EC2_Payment**.

AWS_EC2 nodes mentioned will have related data and metadata saved in *(ConnectsTo)* an **S3 (Amazon-S3)** and **Dynamo DB (AWS_DynamoDB)** or on an **RDS service** following respectively the **Storage Index** and **Read Replica patterns**. Focusing on the **EC2 SmartIoTRoomService**, it can be seen that it is connected *(ConnectsTo)* to the **AWS IoT Events service** and to the **Lambda service** that defines the detector model based on user preferences managed by EC2; in the case of Smart Parking system there is the Lambda service that analyzes the data coming from sensors and those of users to locate the best free nearest parking. The **Kinesis Firehose** service was used in all cases for normalization of data coming from cameras, parking sensors and IoT systems. However, the output data is used in the case of video cameras for artificial intelligence training thanks to the **Amazon Sage Maker service**. The latter makes thus possible to predict times when the facilities centers are less crowded to better distribute the clientele or identify the suitable free parking space closest to the time of arrival of the customer.

5   Conclusions and Future Works

This article examines a possible migration process of an accommodation management application to the cloud. A cloud agnostic architecture is first provided and then a cloud vendor specific one is described highlighting the cloud patterns and cloud services provided by Amazon that can be applied, along with the possible improvements that such migration to the cloud would bring. Finally, after the cloud migration, we have found that the application now can handle with minimal effort unpredictable workloads with an automatic on demand resources provisioning while maintaining costs low, preserving security and making monitoring of the resources way easier. Amazon provides many different services suitable for every customer possible need, providing a detailed description of each of them, APIs and documentation along with a simple and highly intuitive user interface. It is also possible to make this cloud migration using other cloud providers such as IBM. The differences found with AWS services are minimal.

From the point of view of infrastructure and platforms, both vendors offer customization of computing resources, scalability and base monitoring that already represents an efficient cloud migration. Although IBM with its services does not define specific patterns, but generic adoption patterns, it is still possible to map these services with agnostic patterns and make a three-way comparison between the two cloud vendor specific and agnostic solutions. We have found that services like EC2, SQS, CloudWatch or SageMaker could easily be replaced with IBM services like **IBM Cloud Virtual Servers**, **IBM MQ**, **IBM Cloud Monitoring** and **IBM Watson Machine Learning**. It would be interesting in the future to analyse the proposed model by considering other cloud providers in order to achieve the best solution.

Acknowledgments. The work described in this paper has been supported by the Project VALERE "SSCeGov - Semantic, Secure and Law Compliant e-Government Processes".

References

1. Alonso, J., Orue-Echevarria, L., Escalante, M., Gorroñogoitia, J., Presenza, D.: Cloud modernization assessment framework: analyzing the impact of a potential migration to cloud. In: 2013 IEEE 7th International Symposium on the Maintenance and Evolution of Service-Oriented and Cloud-Based Systems, pp. 64–73. IEEE (2013)
2. Amin, R., Vadlamudi, S.: Opportunities and challenges of data migration in cloud. Eng. Int. **9**(1), 41–50 (2021)
3. Banijamali, A., Pakanen, O.-P., Kuvaja, P., Oivo, M.: Software architectures of the convergence of cloud computing and the internet of things: a systematic literature review. Inf. Softw. Technol. **122**, 106271 (2020)
4. Banker, G., Jain, G.: A literature survey on cloud autoscaling mechanisms. IJEDR **2**, 3811–3817 (2014)
5. Bilgihan, A., Okumus, F., Nusair, K.K., Kwun, D.J.W.: Information technology applications and competitive advantage in hotel companies. J. Hosp. Tour. Technol. **2**(2), 139–153 (2011)
6. Binz, T., et al.: OpenTOSCA – a runtime for TOSCA-based cloud applications. In: Maglio, P.P., Weske, M., Yang, J., Fantinato, M. (eds.) Proceedings of the 8th International Conference on Service-Oriented Computing, ICSOC 2010, San Francisco, CA, USA, 7–10 December 2010, pp. 692–695. Springer, Heidelberg (2010). https://doi.org/10.1007/978-3-642-45005-1_62
7. Cantiello, P., Di Martino, B., Mastroianni, M., Cante, L.C., Graziano, M.: Towards a cloud model choice evaluation: comparison between cost/features and ontology-based analysis. Int. J. Grid Util. Comput. (2022)
8. Cretella, G., Di Martino, B.: An overview of approaches for the migration of applications to the cloud. In: Caporarello, L., Di Martino, B., Martinez, M. (eds.) Smart Organizations and Smart Artifacts, pp. 67–75. Springer, Cham (2014). https://doi.org/10.1007/978-3-319-07040-7_8
9. Dai, J., Huang, B.: Design patterns for cloud services. In: Agrawal, D., Candan, K.S., Li, W.-S. (eds.) New Frontiers in Information and Software as Services, pp. 31–56. Springer, Heidelberg (2011). https://doi.org/10.1007/978-3-642-19294-4_2

10. Di Martino, B., Cretella, G., Esposito, A.: Cloud services composition through cloud patterns: a semantic-based approach. Soft. Comput. **21**(16), 4557–4570 (2016). https://doi.org/10.1007/s00500-016-2264-1
11. Rodríguez, C.A.G., Delgado, A.R.A.: Advantages of migration to the cloud computing in the public health sector of Valle del Cauca (2019)
12. Grant Thornton, UK: Emerging clouds in hotel technology (2016). https://www.grantthornton.co.uk/
13. Kaisler, S., Money, W.H.: Service migration in a cloud architecture. In: 2011 44th Hawaii International Conference on System Sciences, pp. 1–10. IEEE (2011)
14. Kratzke, N.: A brief history of cloud application architectures. Appl. Sci. **8**(8), 1368 (2018)
15. Lo, C.-Y., Lan, T.-S., Lin, C.-T., Kuo, Y.-T.: The development and construction of a cloud reservation service system for a western-style restaurant chain. Afr. J. Bus. Manage. **6**(11), 4066–4078 (2012)
16. Pozdniakova, O., Mazeika, D.: Systematic literature review of the cloud-ready software architecture. Baltic J. Mod. Comput. **5**(1), 124 (2017)
17. Ralph, M., Simon, B., Lehmann, S., Steuer, D.: Software architecture for modular, extensible and reusable signal processing components. In: Proceedings of Advances in Automation, Multimedia and Video Systems, and Modern Computer Science, pp. 304–308 (2001)
18. Sakthivelmurugan, V., Vimala, R., Aravind Britto, K.R.: Star hotel hospitality load balancing technique in cloud computing environment. In: Peter, J.D., Alavi, A.H., Javadi, B. (eds.) Advances in Big Data and Cloud Computing: Proceedings of ICBDCC18, pp. 119–126. Springer, Singapore (2019). https://doi.org/10.1007/978-981-13-1882-5_10
19. Schneider, A.: The adaptation of cloud computing by the hotel industry (2012)
20. Singh, M., Gupta, P.K., Srivastava, V.M.: Key challenges in implementing cloud computing in Indian healthcare industry. In: 2017 Pattern Recognition Association of South Africa and Robotics and Mechatronics (PRASA-RobMech), pp. 162–167. IEEE (2017)
21. Tran, V., Keung, J., Liu, A., Fekete, A.: Application migration to cloud: a taxonomy of critical factors. In: Proceedings of the 2nd International Workshop on Software Engineering for Cloud Computing, pp. 22–28 (2011)
22. Varia, J.: Best practices in architecting cloud applications in the AWS cloud. Cloud Comput. Princ. Paradigms **18**, 459–490 (2011)
23. Vella, E., Yang, L., Anwar, N., Jin, N.: Adoption of cloud computing in hotel industry as emerging services. In: Chowdhury, G., McLeod, J., Gillet, V., Willett, P. (eds.) Transforming Digital Worlds, pp. 218–228. Springer, Cham (2018). https://doi.org/10.1007/978-3-319-78105-1_26
24. Wiboonrat, M.: Cloud computing in budget hotel management. In: 2014 IEEE International Conference on Management of Innovation and Technology, pp. 327–332. IEEE (2014)
25. Zhao, J.-F., Zhou, J.-T.: Strategies and methods for cloud migration. Int. J. Autom. Comput. **11**(2), 143–152 (2014)

Anomalous Witnesses and Registrations Detection in the Italian Justice System Based on Big Data and Machine Learning Techniques

Beniamino Di Martino[1,2], Salvatore D'Angelo[1(✉)], Antonio Esposito[1], and Pietro Lupi[3]

[1] Department of Engineering, University of Campania "Luigi Vanvitelli", Caserta, Italy
{beniamino.dimartino,salvatore.dangelo,antonio.esposito}@unicampania.it
[2] Department of Computer Science and Information Engineering, Asia University, Taichung, Taiwan
[3] Quarta Sezione Civile, Tribunale di Napoli, Naples, Italy
pietro.lupi@giustizia.it

Abstract. It is estimated that in 2020 the amount of data produced was about 44 zettabytes for a per capita daily production of about 16 gigabytes. These numbers make us think about how much knowledge is possible to extract from the data produced every day by every single inhabitant of the earth. Supported by the growing diffusion of frameworks and tools for the automatic analysis of Big Data, Machine Learning and Deep Learning, we can try to extract knowledge from all this information and use it to offer a greater definition to human knowledge. In this paper we present two techniques that exploit the knowledge provided by data analysis to identify anomalies in the Italian judicial system, in particular in the civil process. The first anomaly concerns the presence of "serial witnesses", people who lend themselves to provide testimony of the facts occurred in different trial proceedings where places dates and events overlap highlighting a false testimony. The second anomaly relates to "multiple entries" by lawyers with the aim of being able to happen upon a judge "favorable" to the case. The two anomalies presented, but the possibilities are endless, are identified through the definition of Big Data pipelines for data aggregation, information extraction and data analysis.

1 Introduction

In the Italian legal system there are two sections, the civil section and the criminal section.

If we take into consideration only the data published by the Ministry of Justice regarding acts filed electronically, we see that from July 2014 to June 2020, almost 62 million acts were filed. There are over one million active parties and almost 3000 public administrations. Over 38 million digital native documents received by magistrates. Nearly 150 million communications sent and nearly 3 million payments made.

Within all this data there is a great deal of knowledge that is difficult to extract if you do not have the right tools.

A very valid support to the analysis of the so-called Big Data is given by artificial intelligence, specifically by all those techniques of machine learning that go under the name of Machine Learning and Deep Learning. Artificial intelligence provides a very valuable contribution in data analysis, for example, in the case of large volumes of data allows us to delegate complex activities such as pattern recognition and learning to computer-based approaches.

AI makes it possible to cope with data velocity through automatic decision-making systems that bring other decisions, but above all with data variety through the extraction, structuring and understanding of unstructured data.

The spread machine learning techniques and the rising of deep learning and deep neural networks puts even more emphasis on relationships between intelligence based analysis, information mining systems and the data itself.

The availability and access to data is one of the main problems in the application of machine learning and deep learning techniques, so it is very important for the Ministry of Justice in this case but for all institutions to have access to the potential knowledge held.

In this paper we present some techniques for the extraction of knowledge from structured and unstructured data through the definition of Big Data Pipelines, built for extracting, collecting and storing both structured and unstructured data from effective users' "data flows", in order to mine data for extracting new information to support data producers and consumers and their application in the context of the Telematic Civil Process of the Ministry of Justice for the detection of anomalies. We mainly want to address two anomalies:

- serial witnesses
- anomalous registrations

Starting from the work done in [4], we have defined some use cases for the application of big data pipelines and data lake for the extraction of information for the detection of some anomalies that can and have occurred, in the Italian judicial system. The first anomaly we are going to analyze is the identification of serial witnesses within civil trials. In order to understand the reasons for this application of the pipeline, we must first understand in principle how a civil trial is conducted, in particular the part relating to the persons called to testify to the facts which are the subject of the trial. Let's imagine a process related to an accident between two cars, the drivers of these cars do not reach an agreement for an amicable resolution of the incident and decide to proceed by legal means to prove their version of the facts. The two drivers represent the parties and are represented by their respective lawyers. The process begins when the lawyer of one of the parties registers the case at the court of competence, at this moment also begins all the production of data related to the process. During the discussion of the case under consideration, the parties can bring in persons who witnessed the events and can testify to what happened. On the day of the discussion of the case, the lawyer asks the judge to hear the version of the facts presented by the witness, who is identified by the judge and heard. And it is precisely at this moment that the anomaly occurs. It often happens that are "hired" people extraneous to the facts testify falsely in court, and often these people lend themselves to do so for many times. The judge at this time has nothing to support the identification of these people other than his own personal ability to identify that the facts presented do not correspond to the reality of the facts

Another anomaly that we will analyze is the possibility of a lawyer to register the same dispute twice. The anomaly occurs because the system is unable to analyze the content of the dispute and there are no restrictions on the parties involved and their respective lawyers. The system allows the registration of the same dispute with the same parties and the same lawyer. Once the dispute has been registered, the system assigns the dispute to one of the available judges. The anomaly could be exploited to impose a preference on the judge, registering the dispute several times until the judge assigned ex officio is the one preferred by the lawyer.

2   Related Works

The abundance of data made available by the digitalization of Civil Trials, corresponding to the creation of the Telematic Civil Process in Italy, has fueled several research efforts. These works aim to exploit the deluge of information in order to improve the existing digital process and promote the reorganization of Italian justice, with a better reuse of existing resources. The importance of reorganizing the Italian justice has been stressed in several works, such as [8], where the necessity to revise the existing procedures in order to improve the current situation, especially regarding the duration of trials, is explicitly expressed. The work presented in [6] goes into this specific direction, trying to analyze the data from the Telematic Civil Process in order to identify the characteristics of critical Trials, by means of statistical and graph-based analysis.

The complexity of the data coming from the Telematic Civil Process needs a specific organization of the techniques to apply, and the design of ad-hoc frameworks. A first attempt to systematize the analysis of data from the Telematic Civil Process has been carried out in [2], where a complex Big Data pipeline has been proposed to address both structured and not structured sources, and to build a shared Datalake, enhanced with semantic technologies.

Using Big Data technologies to address the analysis of unstructured sources, by organizing and preprocessing them accurately, is not a novel approach, as it has been successfully applied in other such as in [5], where the creation of Datalakes is explicitly addressed. Surveys on Big Data technologies and pipelines [9] suggest how these should be organized to tackle the common issues that generally arise in the analysis of structured and unstructured information.

The work presented in [4] focuses on Artificial Intelligence and Machine Learning techniques to provide the automatic annotation of plain text resources, which was motivated by the necessity to detect sensible information in unstructured documentation, expressed by Chancellor offices. Name Entity Recognition techniques are used in this work, and are exploited to identify all the subjects involved in Trials just by analyzing the textual documentation attached to them.

Starting from these research effort, in the work presented in this paper Name Entity Recognition techniques and Big Data oriented technologies are applied, in combination with Microservices based architectures, to design a system in which entities are recognized and analyzed in order to detect specific anomalies in the registration of testes in Trials, as it will better described in the following sections.

3   The Data Analysis Pipeline

In order to analyze the data incoming from the sources provided by the Ministry of Justice, a Big Data pipeline has been designed, and then implemented through a Microservices based architecture. In this section, the design of the Pipeline will be presented, together with some pointers to implementing software artifacts and libraries that can be used in the various steps constituting it. The Core of the Pipeline and of these work, represented by the implementation of the Anomalies detection algorithms and the Microservices architecture are instead presented in Sect. 4 and 5.

Figure 1 describes the designed Pipeline, which consists of five interconnected stages.

1. The **Data Ingestion** step is necessary to correctly gather the data incoming from the Ministry of Justice. The sources to be taken in consideration come in different formats and are structured in different ways, and the ingestion layer needs to take in consideration all of them. This also means that different technologies must be adopted in the same stage, with different libraries being employed.

   - Unstructured Data are represented by textual documents, either obtained by scanning previously printed Word files, or directly by converting them into PDF. In both cases, the text must be first recognized from the document via an **Optical Character Recognition** process, and only then they can be actually analyzed by the Pipeline. Several OCR libraries and tools are available: Tesseract [11] has been successfully used in several context, and supports several programming language, among which Python, so it can be considered as a good starting point to analyse the unstructured sources. At this stage, no pre-processing is run, as the Ingestion simply works as a collector of data, to be fed to the following steps.
   Most of the textual documents consist of Judicial sentences, documentations provided by testes and parties, or results of the consultation of technical experts and consultants on specific matters.

- Structured Data are represented by information contained in Databases, which can be much more easily accessed via standard means, such as SQL queries, or by reading structured logs produced by the Ministry of Justice systems. At the moment of writing, the Ministry of Justice employs relational databases, which are generally supported by any programming language, having the right connector installed. Logs are much more interesting, as they follow a specific format that the Ministry uses to track all the events in its archives, so specific tools must be developed. Software like Logstash [12] can be employed here, giving the possibility to define the structure of the analyzed logs and to easily transform them for later elaboration.

  Structured data are very useful, as they provide a frame against which unstructured data can be organized, and can be thus exploited to build the Semantic layer used in the next stages of the Pipeline.

2. The **Data Curation** stage is where the initial elaborations on the incoming data are carried out. In particular, the unstructured information acquired by the Ingestion stage are here subjected to a series of transformations, that make them ready for the analysis. In particular, by using Natural Language Processing libraries (NLP), such as Spacy and NLTK [10], the text is cleansed from unnecessary punctuation marks, stopwords and not important words (conjunctions, adverbs, articles) are removed, and Stemming is applied. Stemming is particularly important, as it allows for the reduction of all terms to a basic form, which is much more easy to understand for NLP programs. Libraries used for the Data Curation needs to be specifically set for the Italian language, which is used in the documentation from the Ministry of Justice.

3. The **Data Analysis and Access** stages can be considered as the Core of the whole Pipeline. Here, the cleansed text and the data recovered from the structured sources are analyzed, and **Name Entity Recognition** (NER) is applied in order to determine the presence of testes and parties in the documents. The mere presence of such subjects in the text is of course not sufficient: relationships among people involved in the Trials, locations and times are necessary to carry out the anomalies detection approaches.

  Geolocalization data are obtained by external services, not connected to the Ministry of Justice. These information are retrieved on request, through specific APIs offered by such services (i.e. Google Maps). It is also possible to leverage specific libraries, such as the **geopy.geocoders**[1] Python module, that converts the name of geographical points in map coordinates. These are needed to understand if the locations and times declared by testes and parties in a Trial are consistent with their geographical distance and travel times. Section 4 is more detailed regarding the core aspects of the Pipeline.

---

[1] https://geopy.readthedocs.io/en/stable/.

As also explained in the work presented in [2], all the information extracted from the text is used to populate an Ontology, which becomes a sort of structured and complex vocabulary, where terms are annotated, can be retrieved with SQL like queries, and compared through properties that create relationships among them. Used together with the original texts and documents, such an Ontology can be used to create a Datalake of information, which would also contain the results of the analysis, ready for visualization. The vision of a Juridical Ontology [1,3] that can represent entities and their relations in this domain reflects the necessity to identify relationships among subjects through the NER, also implemented in this stage.

4. In **Data Visualization** the results of the analysis made during the previous stages are shown to users. There are several possible formats to show the results as annotations on the examined texts. The CoNLL-U Format[2] and the BRAT standoff format[3] are notable examples that can be used to easy implement visualization tools for annotated texts. The work described in [7] reports several examples of such annotated visualization, and Fig. 2 shows how a Judicial sentence could be represented after the NER has been applied and the concepts have been identified and connected to Ontology classes and individuals. The Spacy libraries, such as Displacy[4], include visualization modules that allow the correct representation of annotated texts

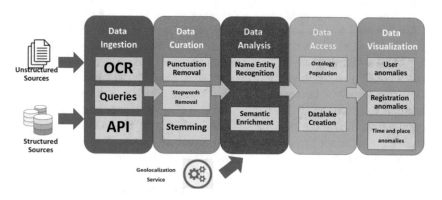

Fig. 1. Big Data pipeline customized for the anomalies detection

---

[2] https://universaldependencies.org/format.html.

[3] https://brat.nlplab.org/standoff.html.

[4] https://spacy.io/usage/visualizers.

Fig. 2. Visualization of annotated text

4   Anomalies Detection

4.1   Identification of "Serial Witnesses"

For the identification of "Serial Witnesses", we use some custom entities extracted with NER: Witness, Date, Location, ID Number. So we can combine all the information we have to create a unique identification of the witness. At this point, two implementations of the anomaly identifier are possible:

- A retrospective analysis of past sentences which analyzes the sentences to derive an average value of testimonies per witness and show all witnesses who deviate from the average. Specifically, the technique used is that of the identification of outliers of which different methods are available in the literature that differ according to the number of features that I want to consider and the distribution of the data on which to apply the technique.
- A real-time analysis that supports the judge at the moment of identification. The system warns the judge if the witness is among the outliers and, therefore, it is a case to which particular attention should be paid. The system could also support the judge in verifying the facts presented by the witness, applying the NER to the testimony written by the clerk and verifying its correspondence with the entities present in the summons presented at the time of registration of the judicial procedure in the system.

4.2   Recognition of Anomalous Registrations

At least two approaches are possible for this other anomaly:

- an analysis of the text present in the documents presented to search for identical documents and present an equality score purely relative to the percentage of identical words found, as most plagiarism systems do.
- the other approach would be based on natural language processing techniques to model an artificial intelligence that takes into account synonyms, alternative expressions and the context.

5   Microservices Architecture

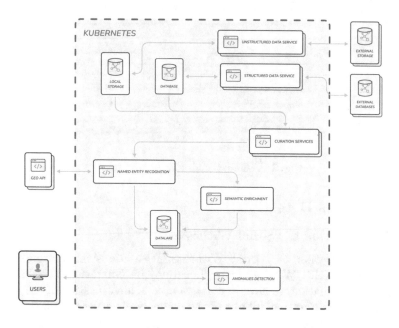

Fig. 3. Microservices architecture for anomalies detection

The entire analysis pipeline was designed for microservices on a kubernetes cluster. In the proof of concept, the cluster is a single node cluster on which all the microservices listed below are run (Fig. 3):

- a service that retrieves unstructured documents converts them into text and saves a txt file with the same name as the original file in shared storage.
- a service that extracts structured data from external databases and saves the extracted and transformed data in an indexed database.
- a service for the data cleansing phase that prepares the data for the next phase.
- a service that implements a deep learning model for the extraction of customized entities from the text, and in the case of places these are also geo-referenced through external geolocalization services.
- once the NER has been applied, the result is also used to enrich a domain ontology through the dedicated service.
- finally the result of all the analyzes and the results are saved in a datalake that will support the anomalies detections phase

6   Conclusion and Future Works

In this paper, the design of a Big Data pipeline for the analysis of Anomalies in Civil Trials has been presented, together with an implementing architecture, based specifically on Microservices. Using Microservice for this purpose increases the scalability, availability and reusability of the components of the framework, in which each element can be seamlessly substituted without affecting the other components and stages of the pipeline.

The whole pipeline focuses on the analysis of both structured and non structured data, and it is specialized in the identification of two different kinds of anomalies: Serial Testes and Erroneous Registrations. Once such anomalies have been identified, textual annotations must be visualized to help laypeople, mostly Judges and Chancellors, in easily identifying the subjects and documents which arise the anomaly alarms. In the present work, the technologies that can be used for the implementation of such anomalies recognition have been suggested, and preliminary methodologies have been addressed.

In future works, we aim at implementing the designed architectural components introduced here, and to apply the Pipeline to a set of sentences and database tables, provided by Italian Courts, which will represent our Case study.

Acknowledgements. The work described in this paper has been supported by the Project VALERE "SSCeGov - Semantic, Secure and Law Compliant e-Government Processes".

References

1. Ceci, M., Gangemi, A.: An OWL ontology library representing judicial interpretations. Semant. Web **7**(3), 229–253 (2016)
2. Di Martino, B., et al.: A big data pipeline and machine learning for a uniform semantic representation of structured data and documents from information systems of Italian ministry of justice. Int. J. Grid High Perform. Comput. (IJGHPC) (2021, in press)
3. Di Martino, B., Marino, A., Rak, M., Pariso, P.: Optimization and validation of eGovernment business processes with support of semantic techniques. In: Barolli, L., Hussain, F., Ikeda, M. (eds.) CISIS 2019. AISC, vol. 993, pp. 827–836. Springer, Cham (2020). https://doi.org/10.1007/978-3-030-22354-0_76
4. Di Martino, B., Marulli, F., Lupi, P., Cataldi, A.: A machine learning based methodology for automatic annotation and anonymisation of privacy-related items in textual documents for justice domain. In: Barolli, L., Poniszewska-Maranda, A., Enokido, T. (eds.) CISIS 2020. AISC, vol. 1194, pp. 530–539. Springer, Cham (2021). https://doi.org/10.1007/978-3-030-50454-0_55
5. Fang, H.: Managing data lakes in big data era: what's a data lake and why has it became popular in data management ecosystem. In: 2015 IEEE International Conference on Cyber Technology in Automation, Control, and Intelligent Systems (CYBER), pp. 820–824. IEEE (2015)
6. Martino, B.D., Cante, L.C., Esposito, A., Lupi, P., Orlando, M.: Temporal outlier analysis of online civil trial cases based on graph and process mining techniques. Int. J. Big Data Intell. **8**(1), 31–46 (2021)

7. Di Martino, B., Marulli, F., Graziano, M., Lupi, P.: PrettyTags: an open-source tool for easy and customizable textual multilevel semantic annotations. In: Barolli, L., Yim, K., Enokido, T. (eds.) CISIS 2021. LNNS, vol. 278, pp. 636–645. Springer, Cham (2021). https://doi.org/10.1007/978-3-030-79725-6_64

8. Massimo Orlando, G.V.: Il controllo di gestione negli uffici giudiziari: il "laboratorio" livorno. Questione Giustizia Trimestrale promosso da Magistratura democratica - Fascicolo 1. Eguaglianza e diritto civile (2020)

9. Oussous, A., Benjelloun, F.Z., Lahcen, A.A., Belfkih, S.: Big data technologies: a survey. J. King Saud Univ. Comput. Inf. Sci. **30**(4), 431–448 (2018)

10. Schmitt, X., Kubler, S., Robert, J., Papadakis, M., LeTraon, Y.: A replicable comparison study of NER software: StanfordNLP, NLTK, OpenNLP, SpaCy, Gate. In: 2019 Sixth International Conference on Social Networks Analysis, Management and Security (SNAMS), pp. 338–343. IEEE (2019)

11. Smith, R.: An overview of the tesseract OCR engine. In: Ninth International Conference on Document Analysis and Recognition (ICDAR 2007), vol. 2, pp. 629–633. IEEE (2007)

12. Turnbull, J.: The Logstash Book. James Turnbull (2013)

A NLP Framework to Generate Video from Positive Comments in Youtube

Hamza Salem[✉] and Manuel Mazzara

Innopolis University, Innopolis, Republic of Tatarstan, Russia
h.salem@innopolis.university, m.mazzara@innopolis.ru

Abstract. Video-sharing sites platforms like YouTube have unique architecture and atmosphere. The comments section is one of the evolution that attracted the users towards expressing opinions and sharing more about videos. Opinions can be used to examine knowledge, user behavior analysis and provide the creator with more ideas to create videos. This paper proposed a novel NLP framework to examine user comments on YouTube and use sentiment analysis to create a short video from positive comments. The results of this study suggest that the framework could be effective to promote the original video using classified community comments. In addition, the results of our implementation indicate that such as framework can be integrated to detect some comments on YouTube and remove negative comments before even posting them.

Keywords: NLP · Short-video · Automation · Video generation · Sentiment analysis

1 Introduction

In 2021, more than two billion people visited YouTube every month [1]. Short-form video platforms such as Tiktok is also one of the leading platforms in the field because they found their niche in short-duration videos, which are attractive for the younger generation. YouTube itself introduced a short-form video service that allows users to create a video of maximum of 15 s, and upload it. From the beginning of 2005 till now, Video comments can be considered as one of the main vectors for social exchange. According to YouTube, more than one billion users interact every week by sharing and commenting on videos.

In this paper, we will show a tool that can take YouTube video URLs to produce short videos from positive comments about it. This paper is organized as follows, Sect. 2 is a brief overview of the literature review for other researchers in the same field. In Sect. 3 we present the approach and the workflow for the tool. Section 4 is an implementation for each step in the workflow. Finally, Sect. 5 present our conclusion and the contribution of our tool.

© The Author(s), under exclusive license to Springer Nature Switzerland AG 2022
L. Barolli et al. (Eds.): AINA 2022, LNNS 451, pp. 193–198, 2022.
https://doi.org/10.1007/978-3-030-99619-2_19

2   Literature Review

The overall aim of this research is to automate the process of generation short video using natural language process (NLP) from YouTube comments [2]. It is then necessary to review the methods that have been used to deal with integrating NLP technology and YouTube as a content source. YouTube is a very rich source for social engagement such as comments. During the last 10 years, researchers used Data sources to train machine learning models using sentiment analysis libraries such as NLTK [3]. There follows a literature review of existing works in this field. The authors in [4] proposed a novel framework based on semantic and sentiment aspects using fuzzy lattice reasoning to meaningful latent-topic detection, by utilizing sentiment analysis of user comments of the Oscar-nominated movie trailers on YouTube. However, this research covers English comments and only for Oscar movies' trailers, not the whole movie.

In another research, [5] the authors investigate the understanding of the perceptions and barriers of end-users towards autonomous vehicles by performing an analysis of reactions to videos about the autonomous vehicle on YouTube. By using the attitudes of the end-users regarding liking, disliking, and commenting patterns. However, the study only analyzed 15 most-viewed videos and their comments. Also, sentiment the lexicon used in this study is based on three established lexicons with the addition of an extra 200 words only.

Several studies, for example [6,7], and [8], have been performed techniques to analyze opinions posted by users about a particular video. The analysis of user comments as a source can be integrated into several applications such as comment filtering, personal recommendation, and user profiling. Also, more recent evidence [9] shows that there are four types of YouTube comments:

1. Short syllable comments.
2. Advertisements of any kind
3. Negative criticism
4. The insane, rambling argument

These comments can be very rich material for data scientists and researchers to develop applications. In this paper, we will pick positive comments using NLP and sentiment analysis to create a short video from positive comments. Such as framework will help Creators to share their communities the positive reaction using YouTube short video platform.

3   Methodology

The setup we used can be found in Fig. 1, which represents the workflow for our framework. Overall, our approach classifies the comments to positive and negative based on sentiment analysis and writes the positive comments to a JSON object that will be an input for Tool to generate the video.

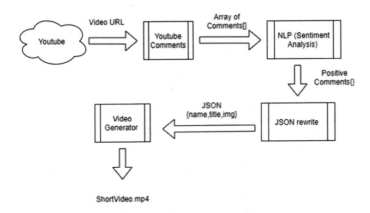

Fig. 1. Generating short video using scraping YouTube comments

As shown in Fig. 1, we initially scrape the top comments section on YouTube video using a Python library Called selenium [10]. The Python Script will export comments and return an array of comments, after the classification process the array will be reshaped to be a JSON object and each object is a video frame that includes (Name, Comment, Image) for each comment. Finally, all frames will be merged together to generate a short video that look like a slide show with background music. The newly generated Video can be used by YouTubers to promote the original video using the Short video feature that exists on YouTube.

4   Implementation

To implement the above methodology, we have built a python script to scrape YouTube comments using the URL of any Video. Code Snippet 1 Shows the code of the scraping process for all top comments and Code Snippet 2 the output of the scraping process for as one comment.

```python
def scrape_comments(url):
    data=[]
    with webdriver.Chrome(ChromeDriverManager().
    install()) as driver:
        wait = WebDriverWait(driver,15)
        driver.get(url)
        driver.execute_script('window.scrollTo(1, 500);')
        #now wait let load the comments
        time.sleep(5)
        driver.execute_script('window.scrollTo(1, 3000);')
        comment_div=driver.find_element_by_xpath
        ('//div[@id="contents"]')
        comments=comment_div.find_elements_by_tag_name
        ('ytd-comment-thread-renderer')
        time.sleep(5)
        print(len(comments))
        for comment in comments:
            comment_text=comment.find_element_by_tag_name
            ('ytd-expander')
            img=comment.find_element_by_tag_name('img')
            name=comment.find_element_by_tag_name('h3')
            if img.get_attribute("src") is not None:
                data.append({'comment':comment_text.text,
                'Name':name.text,
                'Image':img.get_attribute("src")})
    return data
```

Code Snippet 1: Script for scraping YouTube Comments

```json
{
"Comment": "This Combo with Auntie Liz is nice!",
"Name": "FakerUp",
"Image": "https://yt3.ggpht.com/ytc/AKedOLT"
}
```

Code Snippet 2: Comment As JSON Object

Each Comment has comment body and this body is the main input for NLTK sentiment analysis as seen Code Snippet 3.

```python
def is_positive(comment):
    """True if comment has positive compound sentiment,
    False otherwise."""
    return sia.polarity_scores(comment)["compound"] > 0
#{'neg': 0.0, 'neu': 0.616, 'pos': 0.384, 'compound': 0.8168}
```

Code Snippet 3: The Sentiment Intensity Analyzer

By getting the polarity scores function we had the results as a JSON object, for positive comments the 'compound' will be more than 0, and for negative comments will be less than 0. By Using a simple If-statement we have produced an array of positive comments only. These comments will be entered as Array for the next component in the framework that will produce the video. To create a video in Python we have used MoviePy library that can manipulate the video frame by frame [11]. Each Frame will include the comment and name of the account that commented with personal account images. Finally, all frames will be merged together to generate a short video with a music background if needed.

5   Conclusion

In the present paper, we introduced a new Framework to make a short video from any YouTube video with comments using only the positive comments from the community. the proposed method is implemented using several technologies as described in Sect. 4. It can be used to generate a short video to promote the main video as a marketing Framework for YouTubers and the NLP part can be improved to feed other models using the same sentiment analysis. YouTube comments represent how the community interacts with videos and positive feedback will bring more views with the same interest. Also, YouTube is big data set to improve machine learning models based on NLTK too. Ultimately, this Framework should provide another way to promote videos using the community comments. However, We aware that our research may have limitations too. The first is the comments section has a filter from YouTube itself and mostly the top comments are always positive. The second is the other languages that can be supported, so far we have only English as the main language. Future work will look into these limitations and provide a solution that can cover a use case support images in articles and more research about the words in the original text.

References

1. How Many People Use YouTube in 2021? (New Data), January 2021. https://backlinko.com/youtube-users
2. Jones, K.S.: What is the role of NLP in text retrieval? In: Strzalkowski, T. (eds.) Natural Language Information Retrieval. TLTB, vol. 7, pp. 1–24. Springer, Dordrecht (1999). https://doi.org/10.1007/978-94-017-2388-6_1
3. Perkins, J.: Python Text Processing with NLTK 2.0 Cookbook. Packt Publishing Ltd. (2010)
4. Jelodar, H., et al.: A NLP framework based on meaningful latent-topic detection and sentiment analysis via fuzzy lattice reasoning on YouTube comments. Multimedia Tools Appl. **80**(3), 4155–4181 (2021). https://link.springer.com/10.1007/s11042-020-09755-z
5. Das, S., Dutta, A., Lindheimer, T., Jalayer, M., Elgart, Z.: YouTube as a source of information in understanding autonomous vehicle consumers: natural language processing study. Transp. Res. Rec. J. Transp. Res. Board **2673**(8), 242–253 (2019). http://journals.sagepub.com/doi/10.1177/0361198119842110

6. Choudhury, S., Breslin, J.G.: User sentiment detection: a YouTube use case (2010)
7. Cheng, X., Dale, C., Liu, J.: Understanding the characteristics of internet short video sharing: YouTube as a case study. arXiv preprint arXiv:0707.3670 (2007)
8. Siersdorfer, S., Chelaru, S., Nejdl, W., San Pedro, J.: How useful are your comments? Analyzing and predicting YouTube comments and comment ratings. In: Proceedings of the 19th International Conference on World Wide Web, pp. 891–900 (2010)
9. Lynn, T., Endo, P.T., Rosati, P., Silva, I., Santos, G.L., Ging, D.: A comparison of machine learning approaches for detecting misogynistic speech in urban dictionary. In: 2019 International Conference on Cyber Situational Awareness, Data Analytics And Assessment (Cyber SA), pp. 1–8. IEEE (2019)
10. Gundecha, U.: Learning Selenium Testing Tools with Python. Packt Publishing Ltd. (2014)
11. Doberkat, E.-E.: 11. Einfache video-manipulation. In: Python 3. De Gruyter Oldenbourg, pp. 213–224 (2018)

Smart Insole Monitoring System for Fall Detection and Bad Plantar Pressure

Salma Saidani[1]([✉]), Rim Haddad[2], Ridha Bouallegue[3], and Raed Shubair[4]

[1] Innov'Com Laboratory/Sup'Com, National Engineering School of Tunis,
University Tunis El Manar, Tunis, Tunisia
salma.saidani@supcom.tn, salma.saidani@enit.utm.tn
[2] Laval University, Quebec, Canada
rim.haddad@dti.ulaval.ca
[3] Innov'Com Laboratory, High School of communication of Tunis University
of Carthage, Ariana, Tunisia
ridha.bouallegue@supcom.tn
[4] New York University (NYU) Abu Dhabi, Abu Dhabi, United Arab Emirates
raed.shubair@nyu.edu

Abstract. The present work proposes a portable electronic system for the feedback and monitoring of plantar pressure for elderly people in real-time. It is composed of a smart insole, a mobile application, and a cloud server to store data. The smart insole is made up of four resistive pressure sensors for footwear, gyroscopes (GY521) for measuring foot orientation, a Bluetooth HC05 module for transmitting collected data in real time to a laptop, smartphone, and an ATMEGA328P microcontroller. The mobile application was created to provide the patient with instant visual feedback of his actions while wearing the shoes. This technique is useful for monitoring the physical health of elderly persons on a regular basis without interfering with their normal activities.

Keywords: Sensors · Pressure · Mobile application · Real-time · Elderly.

1 Introduction

In the literature of telemedicine, the integration of wearable sensors in health monitoring systems with mobile communications has produced the best-performed results in healthcare services from doctors to patients [1].

Several medical investigations have proven that telemedicine has been adopted to care for patients with chronic diseases [2]. According to the World Health Organization (WHO), there is a high risk that diabetes will become the seventh cause of death in the world by 2030, noticing that the number of deaths each year of people with diabetes is growing. It has around 1.6 million people [3]. Diabetic feet, in particular, necessitate continuous monitoring due to a variety of issues such as peripheral autonomic neuropathy, ulceration, and other microvascular complications such as retinopathy and renal disease [4].

© The Author(s), under exclusive license to Springer Nature Switzerland AG 2022
L. Barolli et al. (Eds.): AINA 2022, LNNS 451, pp. 199–208, 2022.
https://doi.org/10.1007/978-3-030-99619-2_20

To prevent the risk of injury and ulceration to the diabetic foot, it is essential to have a complete examination for the evaluation of the condition of the foot [4]. The current best practice is daily monitoring of the feet [5]. Although the subject's use of the shoe's smart insole can be helpful in detecting or predicting ulceration. Furthermore, according to the WHO, cardiovascular diseases (CVD) continue to be the leading cause of death, accounting for 17.7 million deaths (34.8% of total mortality) each year [6]. Risk factors such as high blood pressure, cholesterol, and impaired kidney function cause more than 80% of cardiovascular disease [7].

Studies have shown that patients with obesity have low intracellular concentrations of magnesium, which can compromise the nutrient's physiological functions [9].

Therefore, this mineral is responsible for protecting against cardiovascular diseases, although changes in the nu trient metabolism of obesity may compromise the functions of this element [8]. In fact, there is a link between cardiovascular diseases and Alzheimer's disease. The root cause of non-inherited Alzheimer's disease (AD) remains unknown despite much research. Since there is strong evidence supporting the hypothesis that cardiovascular disease (CVD) may contribute to the progression of AD, In addition, inflammation accompanies the pathogenesis of AD and CVD and is not only a consequence but also an important factor in disease progression. Also, the risk of developing AD and cardiovascular disease appears to be increased by a wide range of factors, including hypertension, dyslipidemia, and physical activity. Falls for people can present a serious health problem. Every year, the rate of injury and death increases rapidly. The prevention of falls is very important in today's healthcare landscape, where the population is predominantly adult in the world [10]. The risk of falling and being injured increases with age, such as the stability of walking, which degrades due to muscle-skeletal changes. The World Health Organization started to initiate a strategy to reduce the rate of physical inactivity by 10% by 2025 and by 15% in 2030 [11]. While several portable techniques exist, insole-based plantar pressure monitoring systems have surged as a leading technology for monitoring the progression of a patient's chronic disease due to the strong correlation between gait and disease status [12].

Smart insoles equipped with sensors appear as a potential solution to control the gait cycle of patients during their daily activities [13]. These devices are attached to patients to acquire, store, and analyze continuous data related to their health status.

In this paper, we present a mobile application related to a smart shoe insole system to facilitate the physical activity monitoring of the elderly. The remainder of the paper is structured as follows: Sect. 2 presents related works from various studies on health monitoring using wearable sensors. In Sect. 3, we present the analysis of the gait cycle. In Sect. 4, we describe the system realization hardware and software implementation, and in Sect. 5, we conclude and outline.

2   Related Works

Many research studies have discussed the integration of sensors into systems for patients' health control during their daily activities. In general, the concept of these systems is based on various tasks such as detecting data generated by sensors, analyzing information provided by various sensors, and sending alerts in the case of emergency.

All of these systems can be classified based on the number of sensors, the type of sensor, and the utility of their use. James Coates et al. [5] propose a shoe sensing device prototype comprised of 42 sensors: temperature, humidity, acceleration, rotation, bioimpedance, skin temperature, accelerometer, rotation force, humidity, and GSR. The sensors are installed in the following locations: Sensors for force and temperature are placed over the calcaneus (heel), great toe, first metatarsal, and fifth metatarsal. GSR is detected on the fifth metatarsal, with bioimpedance located in the middle of the foot. The power system is powered by a 3.7 V 900 mAh 2.5 h battery that is replaced every two hours. Many types of information are detected in real-time using this system, such as acceleration, rotation, galvanic skin response, environmental temperature, humidity, force, skin temperature, and bioimpedance signals. Sunarya, Unang et al. [14], present a smart shoe equipped with sensors for gait analysis to correct the walking posture and avoid injuries. All of the sensors are placed in the following positions: Eight pressure sensors and two Bosch BMI160 sensors were installed on the outsoles of the shoes. Each Bosch BMI160 sensor is made up of a three-axis accelerometer and a three-axis gyroscope. There were a total of 12 sensors mounted on the outsoles of both the left and right shoes. Ivanov, Kamen et al. [15]. Studied the possibility of using a multimodal smart insole for identity recognition to identify the normal and pathological gait of patients walking outdoors. They used a control module attached to the frontal part of the shoe with an BMI160 inertial sensor for their experiment. Via cable, the module connects a sole composed of nine force sensors attached to a thin, flexible printed circuit board. In this system, the sensors are placed in different anatomical zones of the foot, as follows: one in the big toe (T1), five in the metatarsal heads (M1-M5), one in the midfoot (MF1), and two sensors in the heel (LH1, MH1). Gioacchini, Luca et al. [16] developed a smart insole prototype composed of pressure sensors to control the physical activity of the elderly. Two types of sensors are evaluated in the design of the insole: the FSR402, and the FlexiForce Sensor A301. For the calibration-free insole, the sensor should act as a switch. In fact the A301 Flexiforce sensor is not very sensitive to low weight variation, therefore the FSR 402 is more suitable for the design of smart insole. Ren, Dian et al. [17] represent a smart shoe insole system for plantar pressure measurement using random forest algorithm modules to recognize the behavior and activities of participants.

In fact, smart shoes are composed of seven pressure sensors and extract 167 plantar pressure data features that could identify nine different daily life activities. The sensors are located on the heel, lateral midfoot, center of the midfoot, and lateral sides of the foot. They permit real-time recording during normal activities. Sundarsingh, Esther Florence et al. [18] created a smart shoe system with three force-sensitive resistors placed at plantar pressure points to detect and prevent falls in elderly people with nervous problems. This system is made up of sensors, a microcontroller, and an accelerometer that are all housed in a small module. A threshold-based fall detection algorithm is being used. In the event of a fall, an alert is triggered, and the patient's location is sent to the Internet of Things device via the Telegram application. Li et al. [19] aim to identify falls, so they proposed a threshold algorithm and support vector machine with a mobile phone. Shahzad and Kim [20] desire to detect a fall using a smartphone and accelerometer signals produced during daily activities.

Most previous studies with commercial applications used high-density sensors that covered the entire plantar area. However, in the earlier studies, they used an important number of sensors and hardware, which increased the complexity of the system, and they used electrical cables to connect the in-shoe sensors and the data acquisition system placed around the waist or the ankle, which caused a little weight on the ankle. Therefore, it caused discomfort for the elderly during their exercise, especially if they had diabetes and foot pain. The proposed model uses only four pressure sensors placed at plantar pressure points, which do not add to the device complexity in terms of hardware integration or complex circuit mounting on the shoe and, on the other hand, have a relatively lower material cost. In fact, it is able to collect large amounts of valid data in real-time during indoor or outdoor activities.

3   Gait Analysis

The study of the gait cycle is necessary to define the different pressure points on the foot, as it helps in determining the location of sensors in the shoe insole. A gait cycle is a repetitive pattern that requires steps and strides [21]. It can be classified into the stance phase and swing phase [22]. As shown in Fig. 1, the stance phase accounts for 60% of the total gait cycle during which the foot is in contact with the ground, while the swing phase accounts for 40% of the total gait cycle beginning with the toe-off of the foot and lasting the entire time that the foot is in the air [23,24]. The stance phase is further divided into five phases as follows: initial contact, loading response, midstance, terminal stance, and pre-swing. The swing phase is subdivided into three phases: initial swing, mid swing, and terminal swing [25].

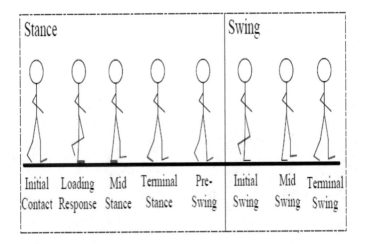

Fig. 1. Gait cycle [24].

4   Hardware and Software Design

4.1   Hardware Descriptions

The proposed system consists of a smart shoe insole prototype. It requires the use of hardware modules, namely: four FSR402 (Force Sensor Resistor) to measure the physical pressure and weight, the ECG sensor to measure the electrical activity of the heart, Gyroscope Gy-521 to measure acceleration, orientation, and the inclination of the foot relative to the ground, an ARDUINO Card to control and retrieve data, an ATMEGA328P microcontroller, electric wires to connect different components together, Bluetooth HC05 to send data to the laptop, and a Secure Digital Card (SD card) to save all patient data. This equipment is interconnected via wires to create a functional prototype of the insole as shown in Fig. 2.

In this system, the plantar pressure data is collected using the Force FSR402, which is cheap, very thin and easy to handle on the insole, single-area and double-wire force resistor, as shown in Fig. 3. FSR402 is a force-sensitive resistor with a round sensing area. It is composed of a solid-thick polymer film and it varies its resistance with pressure applied [26].

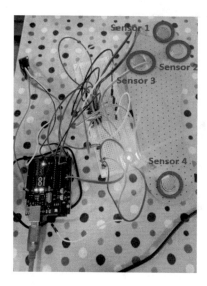

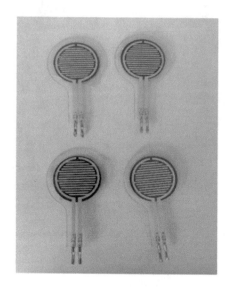

Fig. 2. Smart insole prototype.

Fig. 3. FSR402 sensors.

4.2 Software Implementation

For our proposed smart insole system, we used MIT App Inventor to create an Android mobile application. In fact, for real-time storage, we used the cloud, and then we used Firebase as a platform to implement our cloud database. The object can have complete visibility of his state of health regularly and in real-time. For this reason, the smart insole is a very recommended solution for all patients, especially the elderly, so that they feel better supported during their activities.

During the authentication phase, the user must enter his username and password in order to access our application's main interface. As shown in Fig. 4, if the user makes a mistake when entering one of his coordinates, an error message indicating the authentication failure will appear, and the user will be able to re-enter their information.

Once the authentication has been successfully completed and the user is connected to the insole via Bluetooth, he will be directed to the menu interface of our mobile application named "Followed Feet", which has nine buttons, including one for logging out and eight others, allowing the user to access a set of physiological information such as: Body Mass Index (BMI), Heart Rate, Pedometer interface, Water consumption reminder, and waist to hip rate as shown in Fig. 5.

To ensure a healthy lifestyle for patients and reliable monitoring, first the patient must choose their physical activity to start the measurement. For the BMI calculator interface, it depends on the height of the meter and the weight in kg. Obesity is detected when the BMI value exceeds a certain threshold, as shown in Figs. 6 and 7.

Fig. 4. Login interface.

Fig. 5. Smart insole menu.

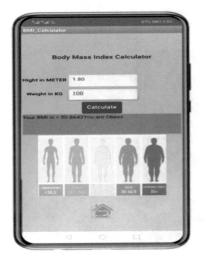

Fig. 6. BMI for obese patients.

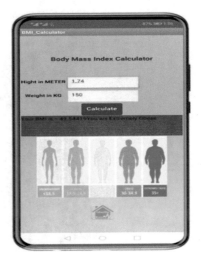

Fig. 7. BMI for extremely obese patients.

As shown in Fig. 8, the WHR interface calculator's value exceeds the threshold value of 0.95, indicating a high risk of metabolic complications such as diabetes and heart disease. During the physical activities, the patient can count his steps, distance in meters, and heart rate pulse as shown in Fig. 9.

Indeed, to avoid any risk, our application can detect any change in the patient's health status, as shown in Fig. 10. In the event of an emergency, such as a fall, a notice is sent out with the patient's specific position.

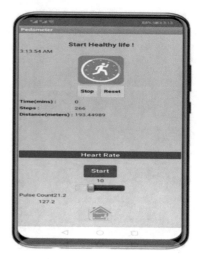

Fig. 8. WHR calculator interface.   **Fig. 9.** Heart rate after activity exercise.

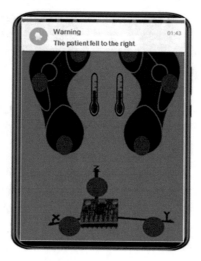

Fig. 10. Fall alert.

5   Conclusion

In this paper, we present our Android mobile application and our prototype of the smart insole system for plantar pressure monitoring. Our proposed system consists of an orthopedic insole equipped with FSR402 sensors to detect all physiological parameters in real-time and a mobile application to improve the quality of monitoring. The main purpose of this project is to help the elderly during their daily activities to prevent diseases and falls. The preliminary tests of our system were performed in the Innov'Com Laboratory at the Higher School of

Communications of Tunis (SUP'COM). The first results demonstrate the efficiency of our system in controlling and detecting bad plantar pressure. We intend to conduct a comparative research of our system and various algorithms and machine learning approaches in the future to establish their reliability in fall detection.

Acknowledgment. We are grateful for the support of the Department of Electrical and Computer Engineering at the New York University of Abu Dhabi (NYU).

References

1. Ping, W., Jin-Gang, W., Xiao-Bo, S., Wei, H.: The research of telemedicine system based on embedded computer. In: Proceedings of the 27th Annual International Conference of the Engineering in Medicine and Biology Society (IEEE-EMBS 2005), Shanghai, China, pp. 114–117. IEEE, January 2006
2. Gani, A., Gribok, A.V., Lu, Y., Ward, W.K., Vigersky, R.A., Reifman, J.: Universal glucose models for predicting subcutaneous glucose concentration in humans. IEEE Trans. Inf. Technol. Biomed. **14**(1), 157–165 (2010)
3. Sonou, A., et al.: Absolute cardiovascular risk of women using hormonal contraception in Porto-Novo. Cardiovasc. J. Afr., **2** (2018)
4. Bowling, F.L., Rashid, S.T., Boulton, A.J.: Preventing and treating foot complications associated with diabetes mellitus. Nat. Rev. Endocrinol. **11**(10), 606–616 (2015)
5. Coates, J., Chipperfield, A., Clough, G.: Wearable multimodal skin sensing for the diabetic foot (2016)
6. World Health Organization: Hearts: technical package for cardiovascular disease management in primary health care (2016)
7. Cortesi, P.A., et al.: Trends in cardiovascular diseases burden and vascular risk factors in Italy: the Global Burden of Disease study 1990–2017. Eur. J. Prev. Cardiol. **28**(4), 385–396 (2021)
8. Dos Santos, L.R., et al.: Cardiovascular diseases in obesity: what is the role of magnesium? Biol. Trace Elem. Res., 1–8 (2021)
9. Leszek, J., et al.: The links between cardiovascular diseases and Alzheimer's disease. Curr. Neuropharmacol. **19**(2), 152–169 (2021)
10. Sterling, D.A., O'Connor, J.A., Bonadies, J.: Geriatric falls: injury severity is high and disproportionate to mechanism. J. Trauma Injury Infect. Crit. Care **50**(1), 116–9 (2001)
11. Hegde, N., Bries, M., Swibas, T., Melanson, E., Sazonov, E.: Automatic recognition of activities of daily living utilizing insole-based and wrist-worn wearable sensors. IEEE J. Biomed. Health Inform. **22**(4), 979–988 (2018)
12. Chen, J.L., et al.: Plantar pressure-based insole gait monitoring techniques for diseases monitoring and analysis: a review. Adv. Mater. Technol., 2100566 (2021)
13. Saidani, S., Haddad, R., Bouallegue, R., Shubair, R.: A new proposal of a smart insole for the monitoring of elderly patients. In: Barolli, L., Woungang, I., Enokido, T. (eds.) AINA 2021. LNNS, vol. 226, pp. 273–284. Springer, Cham (2021). https://doi.org/10.1007/978-3-030-75075-6_22
14. Sunarya, U., et al.: Feature analysis of smart shoe sensors for classification of gait patterns. Sensors **20**(21), 6253 (2020)

15. Ivanov, K., et al.: Identity recognition by walking outdoors using multimodal sensor insoles. IEEE Access **8**, 150797–150807 (2020)

16. Gioacchini, L., Poli, A., Cecchi, S., Spinsante, S.: Sensors characterization for a calibration-free connected smart insole for healthy ageing. In: Goleva, R., Garcia, N.R..C., Pires, I.M. (eds.) HealthyIoT 2020. LNICST, vol. 360, pp. 35–54. Springer, Cham (2021). https://doi.org/10.1007/978-3-030-69963-5_3

17. Ren, D., Aubert-Kato, N., Anzai, E., Ohta, Y., Tripette, J.: Random forest algorithms for recognizing daily life activities using plantar pressure information: a smart-shoe study. PeerJ **8**, e10170 (2020)

18. Sundarsingh, E.F., Saraswathi, V., Sankareshwari, S., Sona, S.: Fall detection smart-shoe enabled with wireless IoT device. Circuit World (2020)

19. Li, X., Nie, L., Xu, H., Wang, X.: Collaborative fall detection using smart phone and Kinect. Mob. Netw. Appl. **23**(4), 775–788 (2018)

20. Shahzad, A., Kim, K.: FallDroid: an automate smart-phone-based fall detection system using multiple kernel learning. IEEE Trans. Ind. Inform. **15**(1), 35–44 (2019)

21. Mariani, B., Hoskovec, C., Rochat, S., Bula, C., Penders, J., Aminian, K.: 3D gait assessment in young and elderly subjects using foot-worn inertial sensors. J. Biomech. **43**(15), 2999–3006 (2010)

22. Saidani, S., Haddad, R., Bouallegue, R.: A prototype design of a smart shoe insole system for real-time monitoring of patients. In: 2020 6th IEEE Congress on Information Science and Technology (CiSt), pp. 116–121. IEEE, June 2021

23. Hynes, A., Kirkland, M. C., Ploughman, M., Czarnuch, S.: Comparing the gait analysis of a Kinect system to the Zeno Walkway: preliminary results. Developing (2019)

24. Yu, J., Gao, W., Jiang, W.: Foot pronation detection based on plantar pressure measurement. In: Journal of Physics: Conference Series, vol. 1646, no. 1, p. 012041. IOP Publishing, September 2020

25. Saboor, A., et al.: Latest research trends in gait analysis using wearable sensors and machine learning: a systematic review. IEEE Access **8**, 167830–167864 (2020)

26. Xi, X., Jiang, W., Lü, Z., Miran, S.M., Luo, Z.Z.: Daily activity monitoring and fall detection based on surface electromyography and plantar pressure. Complexity (2020)

A Recommendation Method of Health Articles Based on Association Rules for Health Terms Appeared on Web Documents and Their Application Systems

Trinh Viet Thong[1], Kosuke Takano[2(✉)], and Kin Fun Li[3]

[1] Course of Information and Computer Sciences, Graduate School of Engineering, Kanagawa Institute of Technology, 1030 Shimo-ogino, Atsugi 243-0202, Kanagawa, Japan
s1985005@cco.kanagawa-it.ac.jp
[2] Department of Information and Computer Sciences, Kanagawa Institute of Technology, 1030 Shimo-ogino, Atsugi 243-0292, Kanagawa, Japan
takano@ic.kanagawa-it.ac.jp
[3] Department of Electrical and Computer Engineering, Faculty of Engineering, University of Victoria, 3800 Finnerty Road, Victoria, BC V8P 5C2, Canada
kinli@uvic.ca

Abstract. This paper presents a recommendation method of health articles based on association rules for health terms appeared on Web documents. On the Web search for health information, we sometimes find it difficult to decide which Web articles are relevant to our symptoms and reliable information. The proposed method classifies health articles and make scores on each article based on association rules, and then, recommends health articles with high support values according to user's symptoms. In the experiment, we evaluate the feasibility of our proposed method.

1 Introduction

The internet, particularly the web, has become a ubiquitous part of our life, so that most people have the access to and are comfortable with suitable health information [1]. On the other hand, as the number of people in need of medical diagnosis increases, the time to wait for a doctor's consultation is increasing. Therefore, the consultation time of doctors for each diagnosis tends to be shorter. As a result, the diagnosis may end without sufficient explanation of the symptoms, and the mental anxiety about the diagnosis and treatment may remain. In addition, it is difficult for the people to accurately grasp their own symptoms and convey it to the doctor without using specialized knowledge and dedicated equipment. In such situations, due to the spread of health information on the website as described above, many patients are accessing to health and medical information on the website to seek about their health status [2]. For this reason, Health Literacy, which selects and utilizes necessary information from health and medical information is becoming an important skill for maintaining and improving health.

© The Author(s), under exclusive license to Springer Nature Switzerland AG 2022
L. Barolli et al. (Eds.): AINA 2022, LNNS 451, pp. 209–219, 2022.
https://doi.org/10.1007/978-3-030-99619-2_21

However, finding useful health information from the vast amount of information on the website is difficult. For example, health information posted on personal blogs on the Web and SNS (Social Network Service) websites contains numerous anonymous comments and opinions from other patients and the quality of the information is also varied. In this case, the treatment may be useful for some people, but it may not be useful for many others. As a result, patients may try to self-treat with information and methods that are not suitable for their condition. In addition, when seeking health information on the Web, multiple Web health articles with different causes and treatment methods may be found for the user's symptoms. In such cases, it is difficult for users to determine which Web health articles are appropriate for their symptoms and reliable information.

In this study, in order to solve such problems, we propose a method for recommending health articles based on association rules for health terms appeared on web documents. The feature of the proposed method is that useful score of health articles calculated based on the association rules. The recommendation score of a health article is calculated based on both the useful score of the health article and the similarity between a word set representing user's health symptom and the health article.

2 Rerated Work

In the medical field, understanding of health, selecting and using essential information from medical and health information (health care and health information), is an important skill to maintain and improve people's health. Health Understanding is a technical term meaning "the ability to collect, understand, evaluate and use information about health and medical care" [2]. When a patient's health knowledge is high, treatment effectiveness tends to be higher than in patients with low health proficiency. Additionally, it has been confirmed that people with low health levels have a higher risk of hospitalization [3, 4]. In this way, understanding of health is one of the factors influencing disparities in health. This is a concept being researched in Japan and abroad, focusing on the medical field [5].

The Web can be used as the first source of health information to gain knowledge about one's own health condition before it can be diagnosed by a specialist. From the information on the Web, a patient can searche for their own health information and treatments, and decide a method of treatment by his/herself to some degree. However, it is difficult to find useful health information from the vast amount of health information on the Web, since huge amount of data is accumulated on the Web. The method proposed by Ueda et al. extracts frequently-appered words from a large number of a cumulative e-mail set by applying TF-IDF to discover valuable information. In their method, the Association Rules is applied as a knowledge acquisition algorithm [6].

3   Proposed Method

3.1   Overview

The proposed method extracts the Association Rule from the frequency of occurrence of health terms contained in articles about health articles on the Web to suggest articles about health. Then, our method calculates the usefulness score of health articles based on the rule weight derived from the association rule. The recommendation score of a health article is calculated based on the usefulness score of the health article with the similarity between the word set representing the user's health symptoms and the health article. Here, in this study, we determine the usefulness of health articles on the Web from the following four perspectives.

1. Easy-to-understand description (S)
2. Causes of the described health (C) symptoms
3. Preventive approach (P) for the described health symptoms
4. Treatments (T) for the described health symptoms

3.2   Extraction of Association Rules

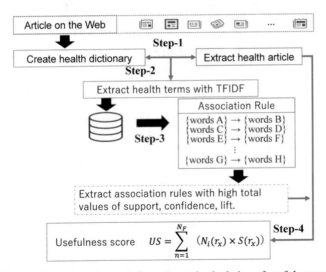

Fig. 1. Extraction of association rules and calculation of useful scores

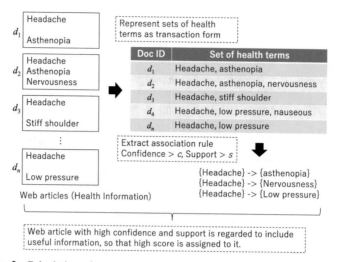

Fig. 2. Calculation of association rules for health information in Web articles

By considering a web health article as a transaction, and each health term in the web health article as an item, the Association Rules for the health terms appeared in the web health article are extracted. Figure 1 shows the process of extracting the Association Rules. In addition, Figure 2 shows the process of calculation of association rules for health information in Web articles.

Step-1: Extraction of health terms in Web article

A dictionary of health terms D is created from n terms $h_1, h_2, ..., h_n$. Terms $w_1, w_2, ..., w_n$ are extracted from a health article d_i. The term w_x are filtered by the dictionary D. Furthermore, by applying TF-IDF, the importance wight of each term appered in a set of health articles is calculated, and n terms with high importance wight are extracted. Here, we define $H_i = \{h_1, h_2, ..., h_k\}$ for the extracted health terms in d_i.

Step-2: Extraction of association rule

In a health article d_i on the Web, we regard a set of health terms H_i appeared in d_i as a transaction of d_i, so that we can, by applying the apriori algorithm, extract a set of association rules $R = \{r_1, r_2, ..., r_n\}$ from a set of the whole health articles. Here, the association rules are obtained with specified condition of mimimum confidence c and minimum support s (Fig. 2). We can choose each threshold value for mimimum confidence c and minimum support s in the ragne of values where the number of rules are not explosively increased.

Step-3: Calculation of usefulness score of health articel on the Web

Based on the association rule, a usefulness score of a health article is calculated. With the usefulness score, we can compare the degree of usefulness beten health articles. For a set of extracted association rules R, each association rule r_n is represented as a form of *lhs* (left hand side) -> *rhs* (right hand side), where lhs and rhs are a set of health term w_x. For each assorication rule r_n, a rule score S is calculated based on the confidence c, support s, lift l as the following formula (1), and asocciation rules with high rule score are extracted.

$$S = c + s + l \tag{1}$$

In the case of $lhs = \{ wl_1, wl_2, ..., wl_n \}$ and $rhs = \{ wr_1, wr_2, ..., wr_m \}$ in an association rule, if an article d_i includes health terms wl_x or wr_y, the association rule is assigned to d_i. Here, if several health terms wl_x or wr_y are appeared in the article d_i, the whole corresponding association rules are assigned to d_i. Supposing that, in a set of assoriation rules F to d_i, the number of rules in F be N_F, the rule score of r_x be $S(r_x)$, and the number of r_x assigned to d_i be $N_i(r_x)$, the usefulness score US_i of d_i is calculated as the following formula (2).

$$US_i = \sum_{x=1}^{N_F} (N_i(r_x) \times S(r_x)) \tag{2}$$

3.3   Calculation of Recommendation Score for Health Articles

The recommendation score of a health article is calculated based on the usefulness of the health article and the similarity between a set of terms that represents the user's health symptoms and the health article. The set of terms representing user's health symptoms is used as a query. For the input query, the recommendation scores for each article are calculated for the ranking and top results are suggested to a user. The recommendation score is calculated in the following steps.

Step-1: Extraction of a set of health terms from user profile

Health term t for health symptom like "stress" and "chest pain" is extracted from a user profile.

Step-2: Calculation of similarity between health terms and health articles

The similarity $Sim(t, d_i)$ between the health term t and health article d_i is calculated as follows:

1. By applying pattern matching, if the health article d_i contains the health term t, we calculate as $Sim(t, d_i) = 1$ and otherwise $Sim(t, d_i) = 0$.
2. $Sim(t, d_i)$ is calculated by the inner product or the cosine measure with the vector space model. Supposing that the number of terms consisting the vector space be n, the health term t and the d_i health article are represented in n-dimensional word vectors t and d_i, respectively. Then, the similarity $Sim(t, d_i)$ can be calculated using the inner product and the cosine measure as follows:

$$Sim(t, d_i) = \frac{t \cdot d_i}{|t||d_i|} \tag{3}$$

$$Sim(t, d_i) = t \cdot d_i \tag{4}$$

If we exlude minus value for the similarity $Sim(t, d_i)$, we set the minus values to 0, so that the range of $Sim(t, d_i)$ can be between [0, 1].

Step 3: Extraction of rule including health terms and calculation of usefulness score

In an association rule set F, which is assigned to a health article d_i, the association rule including the health term t obtained in Step-1 is extracted and a usefulness score $US(t, d_i)$ for each health term t in the health article d_i is calculated in the manner of Step-3 in Sect. 3.2.

Step 4: Calculation of recommendation score

The recommendation score RS of the health article d_i for the health term t is calculated as follows based on the similarity score and the usefulness score calculated in Steps 2 and 3.

$$
\begin{aligned}
RS &= \alpha \times \text{(Similarity score between } d_i \text{ and } t)(\text{Usefulness score of } d_i) \\
&= \alpha \times Sim(t, d_i) \times US(t, d_i)
\end{aligned} \tag{5}
$$

4 Experiment

4.1 Experimental Purpose

First, we show that the proposed method can extract association rules related to health terms by analyzing the actual health articles collected from the Web. Then, we evaluate the feasibility of the proposed method by confirming that useful health articles can be recommended by applying the usefulness score that is calculated based on the association rule as the recommendation score.

4.2 Experimental Environment

We collected 7,771 health articles in English using 22 health terms "*fatigue, stress, asthma, headache, weight loss, anorexia, cough, heartburn, fever, light-headedness, constipation, swelling, runny nose, nausea, chest pain, hay fever, palpitations, coronavirus, abdominal pain, feel listless, stiff shoulder, dizzy*", which are related to health symptoms commonly found in everyday life. Additionally, a dictionary of health terms has been created by extracting 2,073 English health terms from the website on health terms [7]. Furthermore, using this dictionary of health terminology, health terms were extracted from 7,771 health articles.

Using the proposed method, we extract the association rules for health terms from 7,771 health articles, and calculate the usefulness score. For this, the TF-IDF value is calculated for each health term, and association rule is extracted using 1,000 words with high TF-IDF value. We applied apriori algorithm extract association rules for health terms. Minimum confidence c and minimum support s are set to be $c = 0.5$ and $s = 0.05$, respectively. The maximum number of association rules is set to 10.

4.3   Experimental Method

20 health terms "*asthma, stress, cough, fatigue, nausea, constipation, weight loss, coronavirus, abdominal pain, chest pain, heartburn, headache, diabetes, cancer, exercise, stomach, infection, bowel, chronic, inflation*" are used as a query word for ranking health aritcles with the recommendation score. Recommendation accuracy is e valuated for the top 10 ranking results. Here, α is set to $\alpha = 1$ in Eq. (5) in Sect. 3.3 and only one query word is used, the similarity score is either 0 or 1. Therefore, we calculated the recommended score like recommended score $= \alpha \times$ usefulness score $\times$ similarity score $=$ usefulness score.

In addition, for calculating the accuracy of the recommendation, two points are given for the Easy-to-understand description (S), and one point is given when the cause of health sympons (C), the preventive (P) method, and the treatment method (T) are described. We defined correct health article which are regarded to be usefulness article (U) when the aritcle gets tottaly three or more points.

4.4   Experimental Result

Tables 1 and 2 show the association rules for "asthma" and "stress" as examples, respectively.

From Table 1, it can be confirmed that health keywords such as allergy could be extracted for asthma based on the association rule. Meanwhile, from Table 2, it can be confirmed that health keywords such as exercise, fatigue, blood pressure, and heart rate could be extracted for stress based on the association rule. These results show that there exists many articles containing these keywords in health articles on the Web. In addition, since association rules such as {allergy, allergic}-> {asthma}, {allergy} -> {asthma} have been extracted, and many web health articles are mentioning about the possibility that allergy (or allergic) causes asthma (asthma), it is deemed that these association rules would be useful. We can confirm that useful association rules can be extracted for other health terms as well.

Table 1. Association rule for "*asthma*" as an example

lhs	rhs	Support	Confidence	Lift	Total
Allergy, allergic	Asthma	0.053	0.654	3.952	4.660
Allergy	Asthma	0.069	0.559	3.381	4.010
Allergy, will	Asthma	0.053	0.553	3.342	3.949
Allergic	Asthma	0.072	0.541	3.268	3.882

Table 2. Association rule for *"stress"* as an example

lhs	rhs	Support	Confidence	Lift	Total
Exercise, fatigue	Will, stress	0.055	0.549	2.309	2.915
Exercise, fatigue	Stress	0.068	0.673	2.136	2.877
Blood pressure, heart rate	Stress	0.051	0.668	2.121	2.841
Blood pressure, exercise	Will, stress	0.061	0.527	2.214	2.802
Blood pressure, exercise	Stress	0.074	0.638	2.023	2.735
Hormones	Will, stress	0.051	0.505	2.122	2.678
Hormones	Stress	0.063	0.624	1.979	2.666
Exercise, trigger	Stress	0.052	0.625	1.984	2.661
Heart rate	Stress	0.068	0.605	1.919	2.593
Physical activity	Stress	0.054	0.599	1.899	2.552

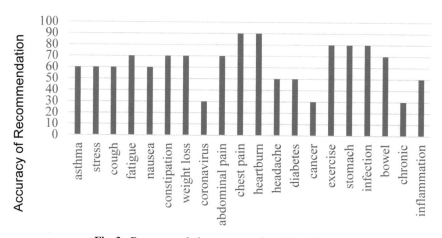

Fig. 3. Recommendation accuracy for 20 health terms

Figure 3 shows the recommendation accuracy of the top 10 for 20 health terms. Since the average of recommendation accuracy is 62.5%, and it can be confirmed that useful health articles can be recommended based on the recommendation score. However, for health terms such as coronavirus, cancer, and chronic (chronic condition), the recommendation accuracy was about 30%, and as a result the average of the recommendation accuracy was decreased.

Table 3. Calculation results of recommendation score using "asthma" as an example

Article ID	lhs	rhs	S	Rule frequency	Usefulness score
A2276	Allergy, allergic	Asthma	4.660	2	417.290
	Allergy	Asthma	4.010	2	
	Allergy, will	Asthma	3.949	2	
	Allergic	Asthma	3.882	101	
A6544	Allergy, allergic	Asthma	4.660	14	370.079
	Allergy	Asthma	4.010	14	
	Allergy, will	Asthma	3.949	4	
	Allergic	Asthma	3.882	60	
A2416	Allergy, allergic	Asthma	4.660	7	251.229
	Allergy	Asthma	4.010	7	
	Allergy, will	Asthma	3.949	5	
	Allergic	Asthma	3.882	44	
A6507	Allergy, Allergic	Asthma	4.660	13	221.534
	Allergy	Asthma	4.010	13	
	Allergy, will	Asthma	3.949	2	
	Allergic	Asthma	3.882	26	

Table 3 shows an example of a result of a recommendation score for a health term "asthma". In the result of Table 3, for example, in A2276, four association rules are assigned the usefulness score (= recommendation score) is calculated as 417.290 by considering the frequency of rules. Table 4 gives the ranking results of the for the health term "asthma". In the results, we can confirm that health articles about asthma such as A2276, A6544, A6507, A2292, A2367, A2331 are very useful in the points of view, easy desripton (S), causes of health sympton (C), preventive approach (P) and treatment (T), so that these articles can be ranked in the top 10.

From these results, we could confirm that by applying the proposed method, association rules of health terms for health articles on the Web can be extracted, and useful health articles can be recommended with the recommendation score, which is calculated based on the usefulness score of each health article.

Table 4. Recommendation results for "asthma"

Article ID	Recommendation score	Article title	S	C	P	T	U
A2276	417.290	Chronic inflammation and asthma - NCBI - NIH	X	O	O	O	O
A6544	370.079	Hay fever allergic rhinitis and your asthma - National Asthma…	O	O	O	O	O

(*continued*)

Table 4. (*continued*)

Article ID	Recommendation score	Article title	S	C	P	T	U
A2416	251.229	Allergic and Environmental Asthma: Overview, Patient History…	O	X	X	X	X
A6507	221.534	Hay fever - Better Health Channel	O	O	X	O	O
A2292	216.184	Asthma in Teens and Adults I Cigna	O	O	O	O	O
A2454	210.398	Regulatory T cells mediated immunomodulation during asthma	X	X	X	O	X
A2367	195.657	Asthma - NCBI - NIH	X	O	O	O	O
A2587	194.057	Hay Fever (Rhinitis) I Symptoms & Treatment I ACAAI Public…	X	O	X	O	X
A2331	162.656	Asthma I Allergy, Asthma & Clinical Immunology I Full Text	X	O	O	O	O
A6434	162.522	Hay fever in adolescents and adults - NCBI - NIH	X	O	X	O	X

5 Conclusion

In this study, we have proposed a method of recommending health articles based on the association rules for health terms appearing on the Web. In the experiment, we coud confirm that, by analyzing the actual health articles collected from the Web, the proposed method can calculate the recommendation score for each health article for the ranking for the input words represening user's health status as a query, and recommends useful health articles based on the recommendation score.

In our future work, we will realize a practical system of recommending health-related articles according to the daily health status of individual user with the proposed method. For this purpose, we are planning to record blood pressure, body temperature, heart rate, and on into a user health profile and implement functions for estimating user's health status based on the user health profile.

References

1. Christine, M., Chun, W.C.: A review of theoretical models of health information seeking on the web. J. Document. **68**(3), 330–352 (2012)
2. Eguchi, Y.: Health literacy as a new keyword in health education. J. Japan Diet. Assoc. **61**(10), 31–39 (2018)
3. Scott, T., Gazmararian, J.A., Williams, M.V., Baker, D.W.: Health literacy and preventive health care use among medicare enrollees in a managed care organization. Med. Care **40**(5), 395–404 (2000)
4. Baker, D.W., et al.: Functional health literacy and the risk of hospital admission among medicare managed care enrollees. Am. J. Public Health **92**(8), 1278–1283 (2000)

5. Kazuhiro, N.: Health Literacy: Ability for Deciding Health. http://www.healthliteracy.jp/. Accessed 2020
6. Ueda, H., Yanagisawa, Y., Tsukamoto, M., Nishio, S.: Applying knowledge discovery techniques to trend analysis of electronic mail. J. Inf. Proc. Soc. Japan **41**(12), 3285–3294 (2000)
7. Harvard Medical School, Medical Dictionary of Health Terms. https://www.health.harvard.edu/medical-dictionary-of-health-terms/a-through-c/. Accessed 2020

A Voronoi Edge and CCM-Based SA Approach for Mesh Router Placement Optimization in WMNs: A Comparison Study for Different Edges

Aoto Hirata[1], Tetsuya Oda[2(⊠)], Nobuki Saito[1], Yuki Nagai[2],
Tomoya Yasunaga[2], Kengo Katayama[2], and Leonard Barolli[3]

[1] Graduate School of Engineering, Okayama University of Science (OUS),
Okayama, 1-1 Ridaicho, Kita-ku, Okayama 700-0005, Japan
{t21jm02zr,t21jm01md}@ous.jp

[2] Department of Information and Computer Engineering, Okayama University
of Science (OUS), 1-1 Ridaicho, Kita-ku, Okayama 700-0005, Japan
{oda,katayama}@ice.ous.ac.jp, {t18j57ny,t18j091yt}@ous.jp

[3] Department of Information and Communication Engineering, Fukuoka Institute
of Technology, 3-30-1 Wajiro-Higashi, Higashi-Ku, Fukuoka 811-0295, Japan
barolli@fit.ac.jp

Abstract. The Wireless Mesh Networks (WMNs) enables routers to communicate with each other wirelessly in order to create a stable network over a wide area at a low cost and it has attracted much attention in recent years. There are different methods for optimizing the placement of mesh routers. In our previous work, we proposed a Coverage Construction Method (CCM), CCM-based Hill Climbing (HC) and CCM-based Simulated Annealing (SA) system for mesh router placement problem considering normal and uniform distributions of mesh clients. In this paper, we propose a Voronoi edge and CCM-based Simulated Annealing (SA) approach for mesh router placement problem. We consider a realistic scenario for mesh client placement rather than randomly generated mesh clients with normal or uniform distributions. For the simulations, we consider the evacuation areas in Okayama City, Japan, as the target to be covered by mesh routers. We also compared the proposed approach with the Delaunay edge and CCM-based SA approach. From the simulation results, we found that the proposed method was able to cover the evacuation area. Also, the Voronoi edge-based system was more stable than the Delaunay edge-based system and covered more mesh clients on average.

1 Introduction

The Wireless Mesh Networks (WMNs) [1–4] are wireless network technologies that enables routers to communicate with each other wirelessly to create a stable network over a wide area at a low cost and it has attracted much attention in recent years. The placement of the mesh routers has a significant impact

L. Barolli et al. (Eds.): AINA 2022, LNNS 451, pp. 220–231, 2022.
https://doi.org/10.1007/978-3-030-99619-2_22

on cost, communication range and operational complexity. Therefore, there are many research works to optimize the placement of these mesh routers. In our previous work [5–12], we proposed and evaluated different meta-heuristics such as Genetic Algorithms (GA) [13], Hill Climbing (HC) [14], Simulated Annealing (SA) [15], Tabu Search (TS) [16] and Particle Swarm Optimization (PSO) [17] for mesh router placement optimization. Also, we proposed a Coverage Construction Method (CCM) [18], CCM-based Hill Climbing (HC) [19] and CCM-based Simulated Annealing (SA) system. The CCM is able to rapidly create a group of mesh routers with the radio communication ranges of the mesh routers linked to each other. The CCM-based HC system covered many mesh clients generated by normal and uniform distributions. We also showed that in the two islands model, the CCM-based HC system was able to find two islands and covered many mesh clients [20]. The CCM-based HC system adapted to varying number of mesh clients, number of mesh routers and area size [21,22]. The CCM-based SA system [23] was able to cover many mesh clients in normal distribution compared with CCM. We also proposed a Delaunay edge and CCM-based SA approach [24] that focuses on a more realistic mesh clients placement.

In this paper, we propose a Voronoi edge and CCM-based SA approach. As evaluation metrics, we consider the Size of Giant Component (SGC) and the Number of Covered Mesh Clients (NCMC).

The structure of the paper is as follows. In Sect. 2, we give a short description of mesh router placement problem. In Sect. 3, we present the proposed method. In Sect. 4, we discuss the simulation results of a comparison study. Finally, in Sect. 5, we conclude the paper and give future research directions.

2   Mesh Router Placement Problem

We consider a two-dimensional continuous area to deploy a number of mesh routers and a number of mesh clients of fixed positions. The objective of the problem is to optimize a location assignment for the mesh routers to the two-dimensional continuous area that maximizes the network connectivity and mesh clients coverage. Network connectivity is measured by the SGC, while the NCMC is the number of mesh clients that is within the radio communication range of at least one mesh router. An instance of the problem consists as follows.

- An area $Width \times Height$ which is the considered area for mesh router placement. Positions of mesh routers are not pre-determined and are to be computed.
- The mesh router has radio communication range defining thus a vector of routers.
- The mesh clients are located in arbitrary points of the considered area defining a matrix of clients.

Algorithm 1. The method for randomly generating mesh routers.

Output: Placement list of mesh routers.
1: Set *Number of mesh routers.*
2: Generate mesh router [0] randomly in considered area.
3: $i \leftarrow 1$.
4: **while** $i < Number\ of\ mesh\ routers$ **do**
5:    Generate mesh router [i] randomly in considered area.
6:    **if** SGC is maximized **then**
7:       $i \leftarrow i + 1$.
8:    **else**
9:       Delete mesh router [i].
10:    **end if**
11: **end while**

---

3   Proposed Method

In this section, we describe the proposed method. In Algorithm 1, Algorithm 2 and Algorithm 3 are shown pseudo codes of CCM and CCM-based SA.

3.1   CCM for Mesh Router Placement Optimization

In our previous work, we proposed a CCM [18] for mesh router placement optimization problem. The pseudo code of randomly generating mesh routers method used in the CCM is shown in Algorithm 1 and the pseudo code of the CCM is shown in Algorithm 2. The CCM searches the solution with maximized SGC. Among the solutions generated, the mesh router placement with the highest NCMC is the final solution.

We describe the operation of the CCM in follow. First, the mesh clients are generated in the considered area. Next, randomly is determined a single point coordinate to be mesh router 1. Once again, randomly determine a single point coordinate to be mesh router 2. Each mesh router has a radio communication range. If the radio communication ranges of the two routers do not overlap, delete router 2 and randomly determine a single point coordinate and make it as mesh router 2. This process is repeated until the radio communication ranges of two mesh routers overlaps. If the radio communication ranges of the two mesh routers overlap, generate next mesh routers. If there is no overlap in radio communication range with any mesh router, the mesh router is removed and generated randomly again. If any of the other mesh routers have overlapping radio communication ranges, generate next mesh routers. Continue this process until the setting number of mesh routers.

By this procedure is created a group of mesh routers connected together without the derivation of connected component using Depth First Search (DFS) [25]. However, this method only creates a population of mesh routers at a considered area, but does not take into account the location of mesh clients. So, the procedure should be repeated for a setting number of loops. Then, determine how

Algorithm 2. Coverage construction method.

Input: Placement list of mesh clients.
Output: Placement list of best mesh routers.
1: Set *Number of loop for CCM*.
2: *i, Current NCMC, Best NCMC* ← 0.
3: *Current mesh routers* ← Alg. 1.
4: *Best mesh routers* ← *Current mesh routers*.
5: **while** *i* < *Number of loop for CCM* **do**
6:    *Current NCMC* ← *NCMC of Current mesh routers*.
7:    **if** *Current NCMC* > *Best NCMC* **then**
8:        *Best NCMC* ← *Current NCMC*.
9:        *Best mesh routers* ← *Current mesh routers*.
10:   **end if**
11:   *i* ← *i* + 1.
12:   *Current mesh routers* ← Alg. 1.
13:   *Current NCMC* ← 0.
14: **end while**

---

many mesh clients are included in the radio communication range group of the mesh router. The placement of the mesh router with the highest number of mesh clients covered during the iterative process is the solution of the CCM.

3.2 CCM-Based SA for Mesh Router Placement Optimization

In this subsection, we describe the CCM-based SA. The pseudo code of the CCM-based SA for mesh router placement problem is shown in Algorithm 3. The SA is one of the local search algorithms, which is inspired by the cooling process of metals. In SA, local solutions are derived by transitioning states and repeating the search for neighboring solutions. SA also transitions states, according to the decided state transition probability if the current solution is worse than the previous one. SA requires solution evaluation and temperature values to decide state transition probabilities. The evaluation of placement ($Eval$), the temperature (T) and the state transition probability (STP) in the proposed method are shown in Eq. (1), Eq. (2) and Eq. (3).

$$Eval \leftarrow 10 \times$$
$$(NCMC\ with\ the\ best\ results\ so\ far - NCMC\ of\ current\ solution) \quad (1)$$

$$T \leftarrow Initial\ Temperature +$$
$$(Final\ Temperature - Initial\ Temperature) \times \frac{Current\ number\ of\ loops}{Number\ of\ loops\ for\ SA} \quad (2)$$

$$STP \leftarrow e^{-\frac{Eval}{T}} \quad (3)$$

Algorithm 3. CCM-based SA.

Input: Placement list of mesh clients.

Output: Placement list of best mesh routers.

1: Set *Number of loop for SA, Initial Temperature, Final Temperature.*
2: *Current number of loop* ← 0.
3: *Current mesh routers, Best mesh routers* ← Alg. 2 (*Placementlistofmeshclients*).
4: *Current NCMC, Best NCMC* ← *NCMC of Current mesh routers.*
5: **while** *Current number of loop < Number of loop for SA* **do**
6:     Randomly choose an index of *Current mesh routers.*
7:     Randomly change coordinate of *Current mesh router [choosed index].*
8:     *Current NCMC* ← *NCMC of Current mesh router.*
9:     **if** SGC is maximized **then**
10:        r ← Randomly generate in (0.0, 100.0).
11:        $Eval$ ← $10 \times (Best\ NCMC - Current\ NCMC)$
12:        T ← *Initial Temperature* + (*Final Temperature* − *Initial Temperature*) × $\frac{Current\ number\ of\ loops}{Number\ of\ loops\ for\ SA}$
13:        **if** $e^{-\frac{Eval}{T}} \geq 1.0$ **then**
14:            *Best NCMC* ← *Current NCMC.*
15:            *Best mesh routers* ← *Current mesh routers.*
16:        **else if** $e^{-\frac{Eval}{T}} > r$ **then**
17:            *Best NCMC* ← *Current NCMC.*
18:            *Best mesh routers* ← *Current mesh routers.*
19:        **else**
20:            Restore coordinate of *Current mesh routers [choosed index].*
21:        **end if**
22:     **else**
23:        Restore coordinate of *Current mesh routers [choosed index].*
24:     **end if**
25:     *Current number of loops* ← *Current number of loops* + 1.
26: **end while**

Proposed method performs neighborhood search by changing the placement of one mesh router. The solution is basically updated when the SGC is maximized and the NCMC is larger than the previous one. In SA, the solution is also updated depending on the STP when the SGC is maximized but the NCMC is decreased. We describe the operation of the proposed method in following. First, we randomly select one of the mesh routers in the group of mesh routers as the initial solution obtained by the CCM and change the placement of the chosen mesh router randomly. Then, we decide the NCMC for all mesh routers and derive the SGC value. The SGC is derived by creating an adjacency list of mesh routers and using DFS to confirm the connected mesh routers. If the SGC is maximized and the NCMC is greater than the previous one, then the changed placement of mesh routers is the current solution. If the SGC is maximized but the NCMC is less than that of previous NCMC, then restore the placement of changed mesh router. But in this case, the changed placement of mesh routers is the current solution with a probability of STP [%]. This process is repeated until the setting number of loops.

3.3   Voronoi Edge and CCM-Based SA

In this subsection, we describe the proposed method, Voronoi edge and CCM-based SA. In the previous methods, the purpose of simulations was to cover mesh clients that were randomly generated based on normal or uniform distributions. In the previous work, we also proposed Delaunay edge and CCM-based SA as a mesh router placement optimization method focusing on more realistic scenarios. In this method, Voronoi decomposition is performed to divide the regions where mesh clients in the problem area are close to each other before performing the CCM. Each region obtained by Voronoi decomposition is called a Voronoi cell and the mesh clients of each Voronoi cell are connected by lines based on the adjacency of these Voronoi cells. This line is called Delaunay edge, and the approach using this Delaunay edge in CCM is called Delaunay edge and CCM-based SA. In the proposed method, we use Voronoi edges (edges of Voronoi cells) instead of Delaunay edges. The coordinates of the Voronoi edge are derived and listed by using the Delaunay edge and CCM-based SA. The coordinates of the listed Voronoi edges are considered as the possible regions in the CCM. By generating an initial solution with mesh routers placed in the middle of a groups of mesh clients by CCM, we can discover the distant mesh clients by the SA approach.

4   Comparison Study

4.1   Simulation Setting

In this section, we present a comparison study of Delaunay edge-based system and Voronoi edge-based system. The parameters used for simulation are shown in Table 1. We deployed mesh clients based on geographic information. The placement area is around Okayama Station in Okayama City, Okayama Prefecture, Japan. The mesh clients are the buildings used as evacuation area. We used the GIS application, QGIS, to display the geographic information. We also used the shapefiles of buildings from Open Street Map and the information of

Table 1. Parameters and values for simulations.

Width of considered area	260
Hight of considered area	180
Number of mesh routers	256
Radius of radio communication range of mesh routers	4
Number of mesh clients	3089
Number of loop for CCM	3000
Number of loop for SA	100000
Initial temperature	100
Final temperature	1

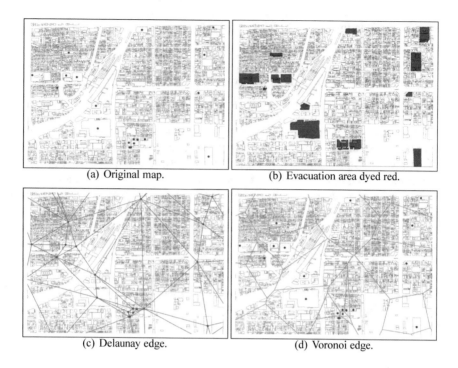

(a) Original map.   (b) Evacuation area dyed red.

(c) Delaunay edge.   (d) Voronoi edge.

Fig. 1. Visualization of evacuation area.

evacuation areas from the open data released by Okayama City [26]. The original map image is shown in Fig. 1(a). The red points in Fig. 1(a) indicate the points where evacuation areas are located. Figure 1(b) shows buildings designated as evacuation areas painted in red, and these red buildings are considered mesh clients. Figure 1(c) and Fig. 1(d) show the derived Delaunay edge and the Voronoi edge. The Voronoi edges used in the proposed system are converted into information that can be used in the proposed system by extracting color pixels from images. In Fig. 2(a), we show the converted evacuation area. Figure 2(b) and Fig. 2(c) show the converted Delaunay edge and Voronoi edge. The Voronoi edge in Fig. 2(c) becomes placement area of the CCM in proposed system. We performed the simulation 100 times for each method.

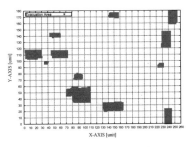

(a) Converted evacuation area.

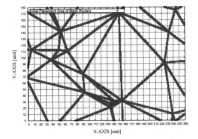

(b) Converted Delaunay edge.

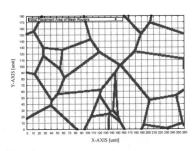

(c) Converted Voronoi edge.

Fig. 2. Converted information for the proposed system.

Table 2. Simulation results of Delaunay edge and Voronoi edge.

Method	Best SGC	Average SGC	Best NCMC	Average NCMC [%]
Delaunay edge based CCM	256	256	2118	64.788
Delaunay edge and CCM-based SA	256	256	3089	95.161
Voronoi edge based CCM	256	256	607	19.313
Voronoi edge and CCM-based SA	256	256	3089	96.268
Method	SD	Variance	Median	Mode
Delaunay edge based CCM	40.958	1677.541	1991.500	1973
Delaunay edge and CCM-based SA	67.329	7626.272	2977	2977
Voronoi edge based CCM	3.622	13.116	596.500	594
Voronoi edge and CCM-based SA	48.478	2350.123	2977	2977

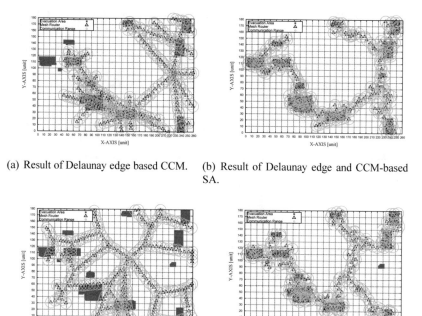

(a) Result of Delaunay edge based CCM.   (b) Result of Delaunay edge and CCM-based SA.

(c) Result of Vornoi edge based CCM.   (d) Result of Vornoi edge and CCM-based SA.

Fig. 3. Visualization results.

4.2   Simulation Results

The simulation results are shown in Table 2. We show the simulation results of the best SGC, avg. SGC, the best NCMC and avg. NCMC. We also show the simulation results of the Standard Deviation (SD), Variance, Median and Mode. In all simulations, the SGC is maximized for each method, but the method using SA covered most of the mesh clients. The visualization results of each method are shown in Fig. 3. While Fig. 4 shows the plot-box results for each method. We can see that the system using Voronoi edges has less variations than the Delaunay edge-based system. Also, the proposed system covers more mesh clients on average than the Delaunay edge-based system.

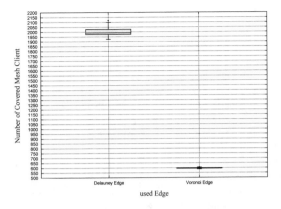

(a) Result of Delaunay edge based CCM.

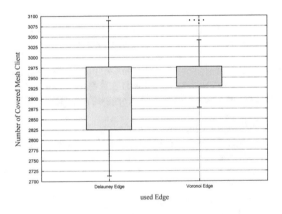

(b) Result of Delaunay edge and CCM-based SA.

Fig. 4. Five-number summary.

5   Conclusions

In this paper, we proposed a Voronoi edge and CCM-based SA approach for mesh router placement problem in WMNs. From the simulation results, we found that the proposed method covered more mesh clients compared with the system using Delaunay edges. In the future, we would like to consider other local search algorithms and Genetic Algorithms.

Acknowledgement. This work was supported by JSPS KAKENHI Grant Number JP20K19793 and Grant for Promotion of OUS Research Project (OUS-RP-20-3).

References

1. Akyildiz, I.F., et al.: Wireless mesh networks: a survey. Comput. Netw. **47**(4), 445–487 (2005)
2. Oda, T., et. al.: Implementation and experimental results of a WMN testbed in indoor environment considering LoS scenario. In: Proceedings of the IEEE 29th International Conference on Advanced Information Networking and Applications (IEEE AINA-2015), pp. 37–42 (2015)
3. Jun, J., et al.: The nominal capacity of wireless mesh networks. IEEE Wirel. Commun. **10**(5), 8–15 (2003)
4. Oyman, O., et al.: Multihop relaying for broadband wireless mesh networks: from theory to practice. IEEE Commun. Mag. **45**(11), 116–122 (2007)
5. Oda, T., et al.: Evaluation of WMN-GA for different mutation operators. Int. J. Space Based Situat. Comput. **2**(3) (2012)
6. Oda, T., et al.: WMN-GA: a simulation system for WMNs and its evaluation considering selection operators. J. Ambient Intell. Humaniz. Comput. **4**(3), 323–330 (2013)
7. Ikeda, M., et. al.: Analysis of WMN-GA simulation results: WMN performance considering stationary and mobile scenarios. In: Proceedings of The 28th IEEE International Conference on Advanced Information Networking and Applications (IEEE AINA-2014), pp. 337–342 (2014)
8. Oda, T., et al.: Analysis of mesh router placement in wireless mesh networks using Friedman test. In: Proceedings of The IEEE 28th International Conference on Advanced Information Networking and Applications (IEEE AINA-2014), pp. 289–296 (2014)
9. Oda, T., et al.: Effect of different grid shapes in wireless mesh network-genetic algorithm system. Int. J. Web Grid Serv. **10**(4), 371–395 (2014)
10. Oda, T., et al.: Analysis of mesh router placement in wireless mesh networks using Friedman test considering different meta-heuristics. Int. J. Commun. Netw. Distrib. Syst. **15**(1), 84–106 (2015)
11. Oda, T., et al.: A genetic algorithm-based system for wireless mesh networks: analysis of system data considering different routing protocols and architectures. Soft Comput. **20**(7), 2627–2640 (2016)
12. Sakamoto, S., et al.: Performance evaluation of intelligent hybrid systems for node placement in wireless mesh networks: a comparison study of WMN-PSOHC and WMN-PSOSA. In: Proceedings of The 11th International Conference on Innovative Mobile and Internet Services in Ubiquitous Computing (IMIS-2017), pp. 16–26 (2017)
13. Holland, J.H.: Genetic algorithms. Sci. Am. **267**(1), 66–73 (1992)
14. Skalak, D.B.: Prototype and feature selection by sampling and random mutation hill climbing algorithms. In: Proceedings of the 11th International Conference on Machine Learning (ICML-1994), pp. 293–301 (1994)
15. Kirkpatrick, S., et al.: Optimization by simulated annealing. Science **220**(4598), 671–680 (1983)
16. Glover, F.: Tabu search: a tutorial. Interfaces **20**(4), 74–94 (1990)
17. Kennedy, J., Eberhart, R.: Particle swarm optimization. In: Proceedings of The IEEE International Conference on Neural Networks (ICNN-1995), pp. 1942–1948 (1995)
18. Hirata, A., et al.: Approach of a solution construction method for mesh router placement optimization problem. In: Proceedings of the IEEE 9th Global Conference on Consumer Electronics (IEEE GCCE-2020), pp. 467–468 (2020)

19. Hirata, A., et al.: A coverage construction method based hill climbing approach for mesh router placement optimization. In: Proceedings of the 15th International Conference on Broadband and Wireless Computing, Communication and Applications (BWCCA-2020), pp. 355–364 (2020)
20. Hirata, A., et al.: Simulation results of CCM based HC for mesh router placement optimization considering two islands model of mesh clients distributions. In: Proceedings of The 9th International Conference on Emerging Internet, Data & Web Technologies (EIDWT-2021), pp. 180–188 (2021)
21. Hirata, A., et al.: A coverage construction and hill climbing approach for mesh router placement optimization: simulation results for different number of mesh routers and instances considering normal distribution of mesh clients. In: Proceedings of The 15th International Conference on Complex, Intelligent and Software Intensive Systems (CISIS-2021), pp. 161–171 (2021)
22. Hirata, A., et al.: A CCM-based HC system for mesh router placement optimization: a comparison study for different instances considering normal and uniform distributions of mesh clients. In: Proceedings of the 24th International Conference on Network-Based Information Systems (NBiS-2021), pp. 329–340 (2021)
23. Hirata, A., et al.: A simulation system for mesh router placement in WMNs considering coverage construction method and simulated annealing. In: Proceedings of The 16th International Conference on Broadband and Wireless Computing, Communication and Applications (BWCCA-2021), pp. 78–87 (2021)
24. Hirata, A., et al.: A delaunay edge and CCM-based SA approach for mesh router placement optimization in WMN: a case study for evacuation area in Okayama city. In: Proceedings of The 10th International Conference on Emerging Internet, Data & Web Technologies (EIDWT-2022) (2022)
25. Tarjan, R.: Depth-first search and linear graph algorithms. SIAM J. Comput. **1**(2), 146–160 (1972)
26. Integrated GIS for all of Okayama Prefecture. http://www.gis.pref.okayama.jp/pref-okayama/OpenData. Ref. 16 November 2021

Internet of Things (IoT) Enabled Smart Navigation Aid for Visually Impaired

Mriyank Roy and Purav Shah[✉]

Faculty of Science and Technology, Middlesex University, London NW4 4BT, UK
mr1330@live.mdx.ac.uk, p.shah@mdx.ac.uk

Abstract. Blindness is a disorder in which a person's ocular vision is lost. Mobility and self-reliability have been a primary concern for visually disabled and blind people. Internet of Things (IoT) enabled Smart Navigation Aid, a smart Electronic Traveling Aid (ETA), is proposed in this paper. This smart guiding ETA improves the lives of blind people since it is enabled with IoT-based sensing and is designed to help visually disabled/impaired people walk and navigate freely in both close and open areas. The proposed prototype provides highly powerful, accurate, quick responding, lightweight, low power consumption, and cost-effective solution that would enhance the lives of the visually impaired people. Within a 1m radius, ultrasonic sensors were used to locate the barrier and potholes. The location was shared with the cloud using GPS and an ESP8266 Wi-Fi module. The stick was also installed with an emergency button, which will call on the mobile when pressed. The stick can also detect wet surfaces with the help of a water sensor. The entire system was built on the Arduino UNO 3 platform. Thus, the proposed prototype is an excellent example of how IoT enabled sensing could aid in the day-to-day lives of the visually impaired people and allowing them the freedom to navigate independently.

1 Introduction

The time is not too far that the PCs would overwhelm the world by beating human processing capabilities. The Internet and computers are playing a vital role in assembling components of a rocket for space transportation. Since the mid-1990's electronic and electrical gears ("Things") has become essential in every field but it is yet to be deployed in our daily activities with full throttle. The term "Things" allude to everything around us from a little molecule to an enormous boat. During the most recent decade, the Internet of Things (IoT) has enticed intensive consideration because of a broad scope of the implementation in biomedical perception, Industrial development, monitoring environment, smart cities, agriculture, etc.

IoT is the internetworking of physical gadgets used in our regular day to day existence that utilise standard communications designs to offer new administrations/services to end-clients [1]. IoT can be summarised into an equation below: [2].

Internet of Things = Human beings + Physical Objects (controllers, sensors, storage, devices) + Internet.

L. Barolli et al. (Eds.): AINA 2022, LNNS 451, pp. 232–244, 2022.
https://doi.org/10.1007/978-3-030-99619-2_23

IoT is being used everywhere in our day-to-day life and is providing with a new vision to investigate the future. IoT is a technology which is uniting human beings by exploring emerging technologies which will help people to connect on a deeper level.

The proposed prototype includes a smart stick integrated with smart sensors for detecting obstacles around a blind person by emitting reflecting waves. These reflected signals received from the obstacles are used as inputs to micro-controller. This micro-controller then makes the person alert about the presence of some object by sounding a buzzer which will then allow the user to walk freely by detecting the obstacles. The system will have all the sensors attached to it so that the sensors can sense the environment around them and guide the blind person through all the obstacles by sounding a buzzer. This buzzer can be heard from an earphone. The smart stick is a basic and purely a mechanical gadget to make a person aware of any impediments on the path. The smart stick will also try to detect uneven surfaces while walking.

A GPS and GSM unit will be attached to the system so that it is possible to track the location of the blind person and contact with the mobile devices via messaging. Along with object detection and tracking, the system will have a rescue button. The blind person can press the rescue button whenever the person feels insecure. After pressing the button, a message will be delivered to a mobile number already stored on the system. The message will have information about the location of the blind person. The relative then can go to the location and help the blind person [3, 4]. IoT is playing a vital role in transforming the lifestyle. The smart stick for blind people will help them to be independent by trying to be their artificial eye.

The rest of the paper is organised as follows: Section 2 explores the related work on this topic, Section 3 provides an overview of the proposed system and presents results based on the operation of the smart navigation aid, and finally, Section 4 focuses on conclusions and future enhancements.

2 Related Work

Visually disabled people find it difficult to communicate with and sense their surroundings. They have very little interaction with their surroundings. Physical mobility is difficult for visually disabled people because it is challenging to discern obstacles in front of them, and they are unable to travel from one location to another. For mobility and financial support, they often have to depend on their relatives. Their mobility prevents them from engaging in social interactions and bonding with others. Different devices have been designed with shortcomings in the past due to a lack of non-visual vision. Researchers have spent decades developing an intelligent and IoT based smart stick to help visually disabled people avoid barriers and provide location information. For the past few decades, scientists have been working on new devices to provide a good and effective way for visually disabled people to sense obstacles and alert them when they are in danger [5].

The researchers developed a voice-operated outdoor navigation system in [6] for visually impaired people. The stick uses an audio output system, Global Positioning System (GPS) and ultrasonic sensors in the system. To store different locations of the blind individual, the stick is equipped with a GPS along with SD memory card. The blind person can search for any destination using the stick by using his/her voice command, and finally, the GPS will guide the person to the finalised destination. This framework will likewise give the speed and the leftover distance to arrive at the destination. Ultrasonic sensors are fixed on the stick to protect the visually impaired person from obstacles. If ever the person comes across any obstacles in the front, the ultrasonic sensor on the stick will be activated, and voice command will be sent to the person so that he/she can change direction. As the system is low-cost, it is affordable by the users. Along with that, it delivers a solution with great precision accompanied by a voice guide's assistance. To increase the operating speed of the system, it utilises ARM processors with more memory space. Since there will be no signal for the GPS indoors, the system cannot operate indoors. To use the stick efficiently, the person using it must be adequately trained.

In [7], the system uses the pulse-echo technique to help the visually impaired people detect the obstacles by providing a warning sound. United States military uses the same technique to locate submarines. They utilised ultrasound beats ranging from 21 kHz to 50 kHz, which creates reverberation pulses when it hits a solid surface. It is possible to anticipate the distance between the obstacle and the user by calculating the difference between signals transmission and receiving time. This framework is sensitive regarding recognising the deterrents. The system has a detection angle of 0 to 45° and a detection range of up to 3 m. Since the system requires receiver and transmitter circuits, it requires more power to function. So, as a result, the system must be redesigned to use less power while functioning.

The researchers in [8] propose a smart cane for assisting the mobility of blind people. ATMEL microcontroller and standard ultrasonic systems are utilised in the system. The system depends on two rechargeable battery (7.4 v) each to operate. The batteries can be recharged utilising AC adaptor or USB cable. ATMEGA328P microcontroller and ATMEL AVR microcontroller is utilised to program the control unit. When any impediments are recognised, vibration and buzzer will ring to caution the client. This system is non-complex, and it can cover a distance up to 3 m. To make it easier to carry, the stick's structure can be folded into a small piece. Since the system detects obstacles in only one direction, it cannot detect the obstacles accurately.

There has been much research carried out to assist blind people with their mobility. From the research mentioned above, Ultrasonic sensors are the best technology to detect obstacles efficiently with low power consumption. Also, to make the system user friendly, a non-complex microcontroller must be utilised. The system should also have a GPS built to track the current location of the blind adult.

Wet doors are dangerous as the person can fall and hurt himself. However, none of the research mentioned alerts the blind person about the wet door. So, the system proposed in the paper utilise a water detection sensor module to alert the blind person about the wet door. Also, the system is equipped with an emergency button which when pressed delivers the current coordinates of the location to a known person in the family of the blind adult via email already stored in the system. The coordinates of the blind person's location can be determined from the GPS system installed on the stick. The person from the family can then track the position of the blind person on the map and approach the site as soon as possible. In this way, the system will assist the blind person with their movement in any corner of the world.

K sonar [9], Palmsonar [10], iSonic cane [11], Laser cane [12], Ultra cane [13], and Virtual Eye (using Image processing) are some of the visual aid systems. These devices are not very user-friendly or simple to operate. Laser canes and Virtual Eye Aids are both expensive and inconvenient to use. The smart stick proposed in the paper enhances the lives of blind people by helping them to navigate without relying on others. Unlike in the past, blind people will now be able to live everyday lives. Aside from that, the blind person's family members would be able to locate him at any time to get his precise location from anywhere. The smart IoT blind stick is an IoT-based system that is cost effective, reliable, and easy to use.

3   Implementation of the Smart Navigation Aid

An Advanced IoT-based system with an Arduino Uno R3 microcontroller is the crucial component of the system. Several sensors and modules are connected to the Arduino Uno R3 microcontroller [14]. The following specifications are included in the suggested system:

- A low-cost navigation smart stick with a total cost of less than £ 40.
- Obstacle sensors respond quickly in close range of up to 4 m.
- Lightweight components are built into the rod, making it user-friendly and consuming less power.
- The overall circuit is easy, and the microcontroller is programmed using Java basics.
- An emergency button is provided on the stick, which, when pressed, will call a relative of the blind person notifying that the blind person is in trouble and needs assistance.
- A speaker who will alert the blind person of the type of obstacle ahead on the path of the blind person.
- A GPS module is linked to the cloud for location sharing. Every 20 ms, the location is updated on the cloud.

For the successful completion of the innovative IoT-based smart navigation aid, the proposed system is divided into two categories while implementing:

1. Obstacle Detection System (Fig. 1)
2. Location Tracking System (Fig. 2)

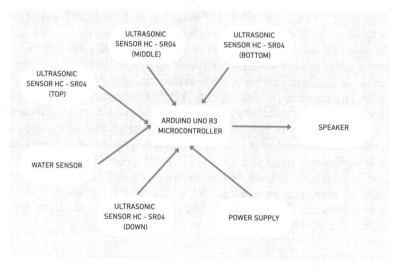

Fig. 1. Block diagram of the sub-system for obstacle detection

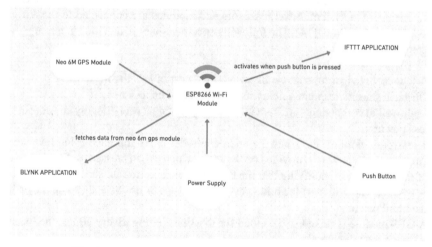

Fig. 2. Block diagram of the sub-system for location tracking

3.1 Obstacle Detection Sub-system

The obstacle detection system's primary function is detecting obstacles on the path of the blind person and alerting him/her about the type of obstacle present in front and preventing the person from a fall. The central brain of the obstacle detection system is Arduino Uno R3 Microcontroller. The hardware requirements of this sub-system include the following:

1. Arduino UNO R3 microcontroller – Main engine of the navigational-aid
2. Ultrasonic Sensor (HC-SR04) – Obstacle Detection

3. Speakers – Integrated with microcontroller to inform the blind person of obstacles
4. Water Sensor – Connected at bottom of the stick to sense water on walking surfaces

There are four ultrasonic sensors attached to the navigation-aid stick to detect various obstacles at different heights. The top sensor is used to detect obstacles that are at a height of more than 62 cm from the surface of the stick on which it is resting. The top sensor is generally used to detect obstacles above the visually impaired person's knee-length.

The down sensor is installed on the stick at a height of 33 cm from the stick's bottom. It is utilised to detect uneven surfaces. It protects the blind person from falling due to elevation or low surface.

The middle sensor is installed just above the down sensor at a height of 40 cm. It is installed just above the down sensor to detect any obstacles which come at the knee length of the blind person. As the down sensor faces downwards to detect any uneven surfaces, it cannot detect obstacles ahead. So as a result, the middle sensor will protect the down sensor from smashing against the obstacles. The bottom sensor is installed at the bottom of the stick at a height of 6 cm to detect objects on the ground. Figure 3 shows how each ultrasonic sensor interacts with the microcontroller and how the speaker activation takes place. The speakers are interfaced using the Talkie library on the Arduino UNO R3. All the programming and interfacing of the sensors for this sub-system is done using the Arduino IDE.

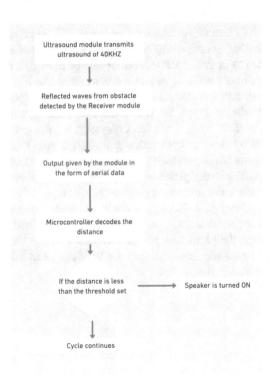

Fig. 3. Interaction of each ultrasonic sensor

3.2 Location Tracking Sub-system

The primary function of the Location Tracking System on the stick is to track the blind person everywhere. It provides live tracking of the blind person on the Blynk Application. The stick also incorporates an emergency button, which will give a call to a relative of the blind person when pressed. The central brain of the location tracking system is ESP 8266 NodeMCU WI-FI Module. The hardware requirements of this sub-system include the following:

1. ESP 8266 NodeMCU WiFi module – Complete TCP/IP stack and MCU unit.
2. GPS module – to determine exact geographical positions
3. Push buttons – IFTTT key for actions set

The software used for this sub-system is BLYNK application (IOS/Android) for controlling Arduino and Raspberry Pi with Internet connectivity. The Blynk application fetches the data from the system's GPS module and provides live graphical tracking to the user. It will assist the relative of the blind person to track the blind person when the emergency button is pressed [15].

IFTTT is a famous trigger-activity programming platform of applets that can automate more than 400 IoT and web application administrations [16]. As soon as the blind person presses the emergency button, the ESP 8266 NodeMCU Wi-Fi module receives the signal from the button and runs the action already set on the IFTTT platform using a specific key, the IFTTT platform. As soon as the IFTTT application is called using a specific key, the following action will be carried out already set against the key. The action set against the key is that it will call on the app of IFTTT using Webhooks.

The complete setup of the navigational aid is as shown in Fig. 4. It highlights all the sensors that are installed on the stick. The sensors and the wires used in the blind stick are kept open in the prototype. So, if the stick falls on the ground, then the sensors might break, and as a result, it might not work correctly. So, the sensors can be properly protected from a fall by placing them inside a protective box. Even the wires can be protected by making the wires pass through inside the rod. Finally, give the stick a finishing touch to be commercially presented in the market. To commercially present the smart stick in the market, the stick must go through six stages of a new product development process – 1. Concept, 2. Feasibility study and design planning, 3. Design and development, 4. Testing verification, 5. Validation collateral production, 6. Manufacture.

In the current stage, the smart stick has been developed up to Testing & verification stage. In future, the smart stick must go through Validation and Manufacturing stages before getting introduced in the market.

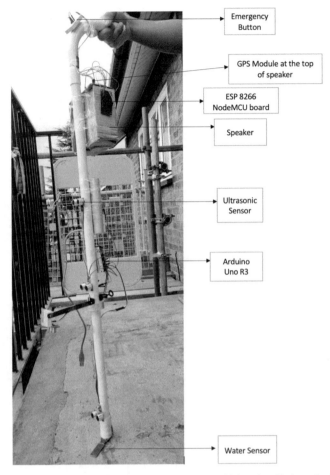

Emergency
Button

GPS Module at the top
of speaker

ESP 8266
NodeMCU board

Speaker

Ultrasonic
Sensor

Arduino
Uno R3

Water Sensor

Fig. 4. Complete IoT-based smart navigation-aid for visually impaired

3.3 Results and Evaluations

To test the effectiveness of the overall smart navigation-aid, several tests were performed.
These tests were divided into the two sub-system responses. The results for the obstacle
detection sub-system are as shown below in Table 1. Furthermore, the determination
of the speaker output is based on a pre-determined threshold, whether the object is
climbable, needs to be avoided (not climbable), or if the water sensor picks up water
presence, then wet surface detection. Figure 5 presents the water detection sensor, picking
up the wet surface. It is also observed that the down sensor is receiving the reading of
33, thus, there is a low surface ahead, which also confirms that the wet surface may be
caused by a puddle, because of an uneven surface.

Table 1. Obstacle detection sub-system evaluation

Distance between Object and Top, Middle, Bottom and Down Sensors (cm)	Threshold distance (cm)	Speaker output
9, 139,125, 33	80	Obstacle ahead not avoidable change direction
2163, 141, 3, 31	80	Obstacle ahead not avoidable change direction
2, 138, 2, 31	80	Obstacle ahead not avoidable change direction
3, 5, 122, 32	80	Obstacle ahead not avoidable change direction
126, 2, 2, 30	80	Obstacle ahead not avoidable change direction
5, 3, 1, 29	80	Obstacle ahead not avoidable change direction
136, 5, 100, 32	50	Elevated surface ahead, go slow
129, 155, 125, 8	23	Uneven surface, Go slow
146, 143, 142, 54	50	Low surface ahead, Go slow

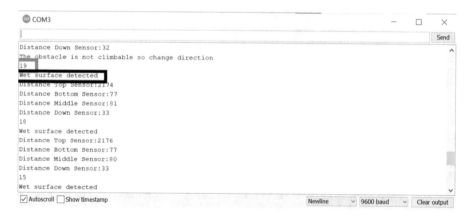

Fig. 5. Wet surface detection

For location tracking, two things were verified: One, whether the system can deter-mine the exact location of the smart stick and second, whether the outdoor connectivity is achieved or not. As soon as the NodeMCU ESP8266 board is powered, it gets connected to the network and starts fetching latitude and longitude coordinates from the Neo 6M GPS Module. The Blynk application can track the system at the current position. In Fig. 6, the arrow indicates the correct location of the smart stick.

Fig. 6. Location Detection sub-system operation

The IFFFT application operation is also verified in case the emergency button is pressed by the user. This directly makes a VoIP call and leaves a message asking for help. This is as demonstrated in Fig. 7.

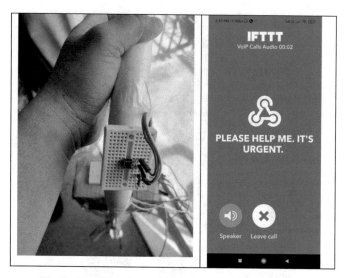

Fig. 7. Location subsystem with IFTTT VoIP messenger

4 Conclusions and Future Enhancements

It is worth noting at this stage that the study's primary goal, which was to develop and incorporate a smart walking stick for the blind, was fully met. The Smart Stick serves as a foundation for the next phase of assistive technologies to help the visually impaired navigate both indoor and outdoor environments. It is both efficient and cost-effective. It achieves successful results in detecting obstacles in the user's path within one meter. The detecting range of the stick can be changed as per user requirements. This device provides a low-cost, dependable, compact, low-power, and robust navigation solution with a noticeable fast response time. The machine is light in weight despite being hard-wired with sensors and other parts. Wireless interaction between device components can increase other facets of the system, such as increasing the range of the ultrasonic sensor and integrating technology for calculating the speed of approaching obstacles.

Visually disabled and blind communities in all developing nations were at the forefront of the mind while developing such an inspiring approach. The stick uses a speaker to alert the blind person of the various obstacles they encounter. The stick can detect uneven surfaces such as elevation and low surfaces. It can also detect wet surfaces and alert the blind person of the wet surface to protect the blind person from slippery surfaces.

With the smart stick's help created using this design, every movement of the blind person can be tracked via an app. Also, the stick provides an emergency button to the visually impaired person. If the blind person finds himself/herself in danger, he/she can press the emergency button and call a relative notifying that he/she is in danger.

4.1 Future Scope for Enhancements

The current smart stick's potential scope includes guiding the visually impaired person in independent navigation effectively while maintaining the person's safety.

- The Braille input system allows the blind person to provide the destination address for navigation in a simple manner.
- The programmable wheels will guide the stick away from potential hazards while simultaneously guiding the blind person to their destination.
- The Internet of Things (IoT) is a popular term that allows one smart stick to connect with another smart stick (PC or mobile device) nearby and use the functionality of the other stick when one stick's functionality fails.
- The stick should also connect with the tra_c lights on the road so that it will be easy for the blind person to cross the road without any assistance.
- Solar panels can be used in place of the battery to power this optimised package of hardware. A solar panel is more beneficial when it recharges itself using sunlight, a readily available green energy source.

References

1. Stallings W: Foundations of Modern Networking: SDN, NFV, QoE, IoT, and Cloud. Addison-Wesley Professional (2015)
2. Farhan, L., Shukur, S., Alissa, A., Alrweg, M., Raza, U., Kharel, R.: A survey on the challenges and opportunities of the Internet of Things (IoT). In: Eleventh International Conference on Sensing Technology (ICST), pp. 1–5, IEEE ICST (2017)
3. Mala N.S., Thushara S.S., Subbiah S.: Navigation gadget for visually impaired based on IoT. In: 2nd International Conference on Computing and Communications Technologies (ICCCT), pages 334–338, IEEE ICST (2017)
4. Nada, A., Mashelly, S., Fakhr, M., Seddik, A.F.: Effective fast response smart stick for blind people. In: Proceedings of the Second International Conference on Advances in Bioinformatics and Environmental Engineering, ICABEE (2015)
5. Gbenga, D.E., Shani, A.I., Adekunle, A.L.: Smart walking stick for visually impaired people using ultrasonic sensors and Arduino. Int. J. Eng. Technol. 9(5), 3435–3447 (2017)
6. Chaurasia, S., Kavitha, K.V.N.: An electronic walking stick for blinds. In: International Conference on Information Communication and Embedded Systems (ICICES2014), pp. 1–5 (2014)
7. Frenkel, R.S.: Coded Pulse Transmission and Correlation for Robust Ultrasound Ranging from a Long-Cane Platform. Masters Thesis, p. 104 (2008)
8. Kang, S., Ho, Y., Moon, I.: Development of an intelligent guide-stick for the blind. In: IEEE International Conference on Robotics and Automation), vol. 4, pp. 3208–3213, IEEE ICRA (2001)
9. Kim, S., Cho, K.: Usability and design guidelines of smart canes for users with visual impairments. Int. J. Des. 7(1), 99–110 (2013)
10. Fernandes, H., Costa, P., Paredes, H., Filipe, V., Barroso, J.: Integrating computer vision object recognition with location based services for the blind. In: Stephanidis, C., Antona, M. (eds.) UAHCI 2014. LNCS, vol. 8515, pp. 493–500. Springer, Cham (2014). https://doi.org/10.1007/978-3-319-07446-7_48
11. Kim, L., Park, S., Lee, S., Ha, S.: An electronic traveler aid for the blind using multiple range sensors. IEICE Electron. Exp. 6(11), 794–799 (2009)
12. Yadav, A., Bindal, L., Namhakumar, V.U., Namitha, K., Harsha, H.: Design and development of smart assistive device for visually impaired people. In: IEEE International Conference on Recent Trends in Electronics, Information & Communication Technology (RTEICT), pp. 1506–1509 (2016)

13. Hoyle, B., Waters, D.: Mobility AT: the batcane (UltraCane). In: Hersh, M.A., Johnson, M.A. (eds.) Assistive technology for visually impaired and blind people, pp. 209–229. Springer London, London (2008). https://doi.org/10.1007/978-1-84628-867-8_6

14. Saquib, Z., Murari, V., Bhargav, S.N.: Blindar: an invisible eye for the blind people making life easy for the blind with Internet of Things (IoT). In: 2nd IEEE International Conference on Recent Trends in Electronics, Information Communication Technology (RTEICT), pp. 71–75 (2017)

15. Noar, N.A.Z.M., Kamal M.M.: The development of smart flood monitoring system using ultrasonic sensor with blynk applications. In: IEEE 4th International Conference on Smart Instrumentation, Measurement and Application (ICSIMA), pp. 1–6 (2017)

16. Mi, X., Qian, F., Zhang, Y., Wang X.: An empirical characterization of IFTTT: ecosystem, usage, and performance. In: Proceedings of the 2017 Internet Measurement Conference, pp. 398–404 (2017)

Reasoning About Inter-procedural Security Requirements in IoT Applications

Mattia Paccamiccio[(⊠)] and Leonardo Mostarda

Università di Camerino, Via Andrea d'Accorso 16, 62032 Camerino, Italy
{mattia.paccamiccio,leonardo.mostarda}@unicam.it

Abstract. The importance of information security dramatically increased and will further grow due to the shape and nature of the modern computing industry. Software is published at a continuously increasing pace. The Internet of Things and security protocols are two examples of domains that pose a great security challenge, due to how diverse the needs for those software may be, and a generalisation of the capabilities regarding the toolchain necessary for testing is becoming a necessity. Oftentimes, these software are designed starting from a formal model, which can be verified with appropriate model checkers. These models, though, do not represent the actual implementation, which can deviate from the model and hence certain security properties might not be inherited from the model, or additional issues could be introduced in the implementation. In this paper we describe a proposal for a novel technique to assess software security properties from LLVM bitcode. We perform various static analyses, such as points-to analysis, call graph and control-flow graph, with the aim of deriving from them an 'accurate enough' formal model of the paths taken by the program, which are then going to be examined via consolidated techniques by matching them against a set of defined rules. The proposed workflow then requires further analysis with more precise methods if a rule is violated, in order to assess the actual feasibility of such path(s). This step is required as the analyses performed to derive the model to analyse are over-approximating the behaviour of the software.

1 Introduction

The verification of software security properties is a critical step when implementing software of any kind. More importantly, in a scenario where we want to assess the logical implementation of a software, we would like to perform, at least in a first instance, the analysis statically because of the following reasons: i) covering all possible environmental conditions can be unfeasible; ii) the code coverage achieved through concrete execution might not be optimal and the poor scaling capabilities of symbolic execution make it unfeasible for this kind of task. The assessment of the logical part of a software is often left to model-checking without considering the code. While the model and its validation can be correct, the

© The Author(s), under exclusive license to Springer Nature Switzerland AG 2022
L. Barolli et al. (Eds.): AINA 2022, LNNS 451, pp. 245–254, 2022.
https://doi.org/10.1007/978-3-030-99619-2_24

implementation can differ from the model, hence there might be the possibility for some properties of the model to be invalidated.

Other problems arising with certain implementations can be the unavailability of the source code, the toolchain employed to compile it into machine code and/or execute it in a machine with suitable computing power. Dealing with multi-threaded applications increases the complexity of such analyses, as race conditions might happen spuriously if calling order and conditions are not properly enforced. Our approach aims to tackle the detection of violations of inter-procedural requirements in a scalable, fast and static manner.

Consider the following snippet of code:

Listing 1. Example of the usage of sockets in C languase.

```c
int main(void) {
  struct sockaddr_in servername;
  int sock = socket(PF_INET, SOCK_STREAM, 0);
  if (sock < 0) { exit (EXIT_FAILURE); }
  init_sockaddr (&servername, SERVERHOST, PORT);
  if (0 > connect(sock, (struct sockaddr *)&servername,
       sizeof(servername))) { exit(EXIT_FAILURE); }
  write_to_server(sock);
  close(sock);
  exit(EXIT_SUCCESS);
}
```

Our goal is to check that whenever there is a call to write_to_server(), the methods init_sockaddr(), connect() and close() are also being called, enforcing the order of these calls. In this paper we refer to this type of requirements as rules.

These properties, which in the example above could cause a crash if the program were not to be implemented correctly, could work as well to verify, for instance, that there is no non-encrypted data being sent in certain phases of a hypothetical protocol [1]. A higher degree of precision could be obtained by enforcing a specific flow. To further generalise we want to analyse if certain rules are always followed in the implementation.

1.1 Motivation

Consider the following scenario: the company XY has in its core business certain services that are typically used via internet-enabled devices. These devices are oftentimes produced by third-parties, which we will call ABC, that take care of the development of the apps that interface with such services via APIs. For the sake of brevity we will assume the example device is a Smart Home Hub device. XY has the need to validate whether or not the software in the Smart Hub Device is performing the right operations when accessing the APIs. This to prevent having non-handled exceptions visible from the front-end application, crashes or, worse, leakage of sensitive data. XY has at its disposal the application executable but not its source files. This paper aims to cover this specific use case.

1.2 Background: Call Graph, Points-to, LLVM

A call graph is a type of control-flow graph representing calling relationships between the subroutines of a software. The property of a statically computed call graph is to be, typically, an over approximation of the real program's behaviour [2]. It is defined as a may-analysis, which means that every calling relationship happening in the program is represented in the graph and possibly some that do not. Notice that computing the exact static call graph is an undecidable problem.

Points-to analysis is a static analysis technique that computes a model of the memory that is used to retrieve what value or set of values a pointer variable will have when de-referenced. Computing a precise call graph for languages that allow for dynamic dispatching requires a precise points-to analysis and vice-versa. Another way to put it is points-to analysis precision is boosted by a precise call graph computation and vice-versa, because they reduce the sets of possible call relationships and points-to relationships by discarding the impossible ones.

The construction of both call-relationships and points-to relationships have various degrees of sensitivity, which can be used to enhance the precision of the analysis. It can be based on: context (call-site, object), flow [2].

LLVM [3] is a widely used compiler and tool-chain infrastructure. Its intermediate representation, called LLVM IR, acts as intermediate layer between high-level programming languages and machine code, so that the compiler's duty is to translate the language to LLVM, which is then assembled into the targeted architecture by LLVM itself. The purpose behind this is to simplify the source to binary translation process, by offloading architecture-specific logic from the compiler, task which is instead performed by the LLVM assembler.

2 Paper Contribution

This paper presents a novel approach for the verification of software security properties, specifically targeting IoT software but potentially extensible to other software domains, such as security protocols. Our goal consists in obtaining a inter-procedural control-flow graph, apply transformations to it and using efficient techniques to performs analyses against a set of defined rules. To reach this goal we need to address the following basic questions: (i) would the approach taken be precise enough for reconstructing the behaviour of the software?; (ii) will using this method provide an advantage in terms of performance compared to other methods of analysis?; (iii) what possible limitations can arise compared to other methodologies?; (iv) how can those be addressed or mitigated?. In order to answer these questions we reviewed the literature on the subject and performed several tests. Such experiments targeted known edge-cases in the analyses we performed. More specifically we assessed the precision for static call-graph generation and points-to and what could be done, if anything, to improve the performance of the analysis. Our experiments show that our set of approaches could allow for fast and sound analyses, appropriate with the issue being tackled, as also confirmed by the literature.

3 Related Work

In this section we review the state of the art in analyzing and reconstructing the behaviour of a software and give a brief overview on the techniques used, their strengths and weaknesses.

3.1 Concrete Execution

Approaches based on concrete execution consist in executing the software in instrumented environments. The way they typically work is by the use of test cases, aiming for maximum code coverage. As mentioned this approach is accurate when it comes to vulnerability discovery and the reconstruction of a program's behaviour but is not best suited for our scope of analysis. It lacks in terms of performance when computing call relationships and, most importantly, the coverage offered is limited to the paths that are actually executed.

afl++[1] (american fuzzy lop) is a state of the art fuzzer for concrete executions. It works like a traditional fuzzer: a binary is instrumented and random mutated inputs are fed to the program to explore its states. It also supports LLVM bitcode fuzzing.

3.2 Dynamic Symbolic Execution

Approaches based on symbolic execution consist in statically exploring a binary, representing, ideally, all possible combination of paths it can take as states [4]. Its scalability is poor due to the state-explosion problem which is an open research problem. Methods to mitigate the issue have been proposed and implemented, by selecting promising paths [5], by the means of executing the program backwards [6,7], and merging paths on loops [8].

Valgrind [9] is an instrumentation framework for building dynamic analysis tools. There are Valgrind tools that can automatically detect many memory management and threading bugs, and profile programs in detail.

klee [10] is a symbolic execution tool capable of automatically generating tests that achieve high coverage on a diverse set of complex and environmentally-intensive programs. There are two main components: i) the symbolic virtual machine engine built on top of LLVM, ii) a POSIX/Linux emulation layer, which also allows to make parts of the operating system environment symbolic.

3.3 Hybrid Approaches

The capabilities of symbolic execution and dynamic execution can be combined in order to mitigate each techniques' limitations. Dynamic execution's main weakness is the lack of semantic insights inside the program's reasoning and is prone to getting stuck. This can be mitigated with symbolic execution, which offloads dynamic execution by solving complex constraints, while the dynamic

---

[1] https://aflplus.plus, https://github.com/AFLplusplus/AFLplusplus.

execution engine is fast at exploring the'trivial' parts of a program and does not suffer from state explosion.

angr [11] is a programmable binary analysis toolkit that performs dynamic symbolic execution and various static analyses on binaries. It allows for `concolic` execution: a virtual machine emulates the architecture of the loaded binary and there is a concrete execution driven by the symbolic exploration engine. This allows to retrieve concrete data from symbolic constraints, allowing for SMT solver offloading and state thinning.

driller [5], based on angr, is a successful example of a symbolic execution engine combined with the traditional fuzzer, afl, with the benefits mentioned above.

3.4 Static Analysis

Approaches based on static analysis work by reasoning about the software without executing it. Applications of the usage of static analysis techniques are: i) disassemblers, ii) linters, iii) data-flow analyses, iv) model checking.

Ddisasm [12] is a fast and accurate disassembler. It implements multiple static analyses and heuristics in a combined Datalog implementation. Ddisasm is, at the moment, the best-performing disassembler aimed at reassembling binaries.

cclyzer [13] is a tool for statically analyzing LLVM bitcode. It is built on Datalog and works by querying a parsed version of the LLVM bitcode. It then performs points-to analysis and call-graph construction on the facts generated. It is capable of yielding highly precise analyses. It supports call-site sensitivity.

RetDec [14] is a retargetable decompiler based on LLVM. Its main feature is the conversion of machine code to LLVM bitcode. It supports reconstruction of functions, types, and high-level constructs. It is capable of generating call graphs and control-flow graphs.

McSema[2] offers very similar functionalities and can be applied to the same scope as RetDec. Its main difference from it is to use external software to deal with control-flow graph recovery (a step of paramount importance for disassembling).

4   Proposed Approach

Our approach can be summarised by the diagram in Fig. 1 and will be further explained in the following sections.

---

[2] https://github.com/lifting-bits/mcsema.

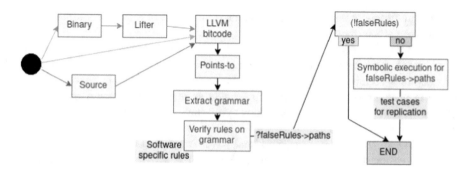

Fig. 1. Our approach.

4.1   Call Relationships

We chose to base our workflow on LLVM, which allows to be quasi-agnostic to binary, source or LLVM bitcode as a starting point, as source can be compiled to LLVM and binary can be lifted to LLVM bitcode. The LLVM bitcode will then be used to perform points-to analysis with online call-graph construction, so to get a accurate inter-procedural and infra-procedural view of the software. We would obtain a model which would then be validated against our set of rules.

This could be formalised as follows:

$$\forall f, \forall c(f) \rightarrow \exists \varphi \in \bar{f}(c(f)) \cup \bar{f} : \varphi \uparrow \tag{1}$$

The previous formula is read as follows: for each function f and $c(f)$ (c calls function f), there exists a set φ that is a subset of $\bar{f}(c(f))$ (set of functions that call functions calling f) and $\bar{f}$ (that do not call f and c) so that φ is converging (up arrow). The analysis, as introduced here, only reasons about call-relationships with no considerations regarding flow. We call this property converge. Our goal for the scope of this paper has been to assess whether or not the model of tested software can be tested against this rules and similar ones.

Consider the following example rule that encodes the fact that the call order is connect, followed by write_to_server, followed by close:

```
rule { properties[] = ["converge","order"],
       f[] = ["connect", "write_to_server", "close"]; }
```

What we want to verify is that for each of the functions specified in f[] the property is ensured. The property converge is explained in the Eq. 1. The property order means: the functions in f[] are executed in the specified order.

4.2   Program Behaviour Extraction

For complexity reasons we deem necessary to further simplify the model we analyse, instead of working directly with control-flow graphs. As mentioned in [15] we can transform control-flow graphs into context-free grammars. This allows for very efficient exploration and further simplification of the model without compromising soundness and accuracy for the intended scope.

4.3   Validation and Refinement

Once the behaviour has been extracted and matched against our set of rules, we obtain the violations, if present. Since we are working in an over-approximated realm we propose to apply selective symbolic execution to the paths that failed the validation in order to retrieve if: i) if such paths can exist; ii) test cases to replicate the failure; iii) information to help debugging by locating the failure.

5   Experiments

To evaluate if our methodology is feasible we tested our workflow on edge-cases that make points-to analysis necessary to obtain an accurate call-graph. These tests included: i) utilisation of pointers to data, ii) utilisation of calls and branched calls to function pointers. The following snippets of code in C (compiled to LLVM) are the examples we performed validations on:

Listing 2. "branched_funptr" snippet.

```
int main()
{   int in = 3;
    void (*one)();
    if (in > 5) {
      one = &first;
    } else {
      one = &second;
    }
    (*one)();
    return 0; }
```

Listing 3. "branched_funptr_order" snippet.

```
int main()
{   int in = 3;
    void (*one)(), void (*two)();
    if (in > 5) {
      one = &first;
      two = &second;
    } else {
      one = &second;
      two = &first;
    }
    (*one)();
    (*two)();
    return 0; }
```

Listing 4. Methods "first" and "second".

```
void first(){ puts("First"); } void second(){ puts("Second"); }
```

For the sake of brevity we included only the snippets involving branched function pointers. At the time of writing we did not account for control-flow in our experiments but just on pure call relationships, as control-flow recovery can be obtained once the inter-procedural relationships are accurate. Additional analyses on GNU binaries have been performed to test how fast the analysis would have been on more sizeable software, considering performances only and not accuracy. The validation of the rule formalised in Sect. 4.1 is left behind because the performance load is negligible compared to the one given by computing the call-graph and points-to relationships.

5.1   Experimental Results

We obtained positive results using Cclyzer [13] when performing points-to and call-graph generation on LLVM bitcode. We also noted how the analysis becomes slower when working with higher call-site sensitivities. We also tested RetDec [14] and McSema[3] as lifters, with the latter being the most accurate one. We did not perform more detailed experiments on this aspect at the time of writing because we assumed that we can have correct LLVM bitcode. Table 1 summarises the experimental results obtained with CClyzer. Such results show performance (time taken) only, based on sensitivity and complexity (file size).

Table 1. CClyzer performance tests for points-to analysis and call-graph construction. "insensitive" analyses do not take into account call-site, "call-site-2" account for call-site up to depth 2. Among the snippets the words: fptr, br, order mean respectively: call to function pointer, branched and change of order, as shown in Sect. 5.

Target	Sensitivity	Size (KB)	Time (s)
fptr (snippets)	call-site-2	2.5	6.39
br_fptr (snippets)	call-site-2	2.8	6.12
fptr_order (snippets)	call-site-2	2.6	6.33
br_fptr_order (snippets)	call-site-2	2.8	6.24
dd (coreutils)	insensitive	273	14.27
dd (coreutils)	call-site-2	273	16.23
ls (coreutils)	insensitive	612.3	22.57
ls (coreutils)	call-site-2	612.3	27.38
cp (coreutils)	insensitive	733	28.50
cp (coreutils)	call-site-2	733	31.49

---

[3] https://github.com/lifting-bits/mcsema.

6 Conclusion and Future Work

The first results of our methodology are positive and allow for basic validations on our test samples. The test samples were designed to target applications where points-to analysis is made necessary and met our expectations, both in precision and performance.

6.1 Future Developments

Future work includes adding parallel and flow reasoning to the framework, thus extending the application to detection of call-order violations on both single and multi-threaded application, enabling validation of operational order. This task is made easier by LLVM being a Single Static Assignment, hence having intrinsic flow-sensitivity [2]. To implement this step we propose to derive context-free grammars from control-flow graphs, as discussed in Sect. 4.2 and demonstrated in [15], allowing for both call-order enforcement and verification of atomicity properties in parallel software. To narrow down unfeasible paths, we propose to use symbolic execution to generate use-cases targeting the path(s) violating the rule(s) as it is more precise than the previously performed analyses.

References

1. Sekaran, R., Patan, R., Raveendran, A., Al-Turjman, F., Ramachandran, M., Mostarda, L.: Survival study on blockchain based 6G-enabled mobile edge computation for IoT automation, IEEE Access **8**, 143453–143463 (2020)
2. Yannis, S., George, B.: Pointer analysis. Found. Trends Program. Lang. **2**(1), 1–69 (2015)
3. Lattner, C.: LLVM: an infrastructure for multi-stage optimization, Master's thesis, Computer Science Dept., University of Illinois at Urbana-Champaign, Urbana, IL, December 2002
4. Vannucchi, C., et al.: Symbolic verification of event-condition-action rules in intelligent environments. J. Reliab. Intell. Environ. **3**(2), 117–130 (2017)
5. Stephens, N., et al.: Driller: augmenting fuzzing through selective symbolic execution (2016)
6. Ma, K.-K., Yit Phang, K., Foster, J.S., Hicks, M.: Directed symbolic execution. In: Yahav, E. (ed.) SAS 2011. LNCS, vol. 6887, pp. 95–111. Springer, Heidelberg (2011). https://doi.org/10.1007/978-3-642-23702-7_11
7. Basile, C., Canavese, D., d'Annoville, J., Sutter, B.D., Valenza, F.: Automatic discovery of software attacks via backward reasoning. In: Falcarin, P., Wyseur, B. (eds.) 1st IEEE/ACM International Workshop on Software Protection, SPRO 2015, pp. 52–58. IEEE Computer Society (2015)
8. Avgerinos, T., Rebert, A., Cha, S.K., Brumley, D.: Enhancing symbolic execution with veritesting. In: Jalote, P., Briand, L.C., van der Hoek, A. (eds.) 36th International Conference on Software Engineering, ICSE 2014, pp. 1083–1094. ACM (2014)
9. Nethercote, N., Seward, J.: Valgrind: a framework for heavyweight dynamic binary instrumentation. In: Ferrante, J., McKinley, K.S. (eds.) Proceedings of the ACM SIGPLAN 2007 Conference on Programming Language Design and Implementation, pp. 89–100. ACM (2007)

10. Cadar, C., Dunbar, D., Engler, D.R.: KLEE: unassisted and automatic generation of high-coverage tests for complex systems programs. In: Draves, R., van Renesse, R. (eds.) 8th USENIX Symposium on Operating Systems Design and Implementation, OSDI 2008, pp. 209–224. USENIX Association (2008)

11. Shoshitaishvili, Y., et al.: SOK: (state of) the art of war: offensive techniques in binary analysis (2016)

12. Flores-Montoya, A., Schulte, E.M.: Datalog disassembly. CoRR abs/1906.03969 (2019)

13. Balatsouras, G., Smaragdakis, Y.: Structure-sensitive points-to analysis for C and C++. In: Rival, X. (eds.) SAS 2016. LNCS, vol. 9837, pp. 84–104. Springer, Heidelberg (2016). https://doi.org/10.1007/978-3-662-53413-7_5

14. Křoustek, J., Matula, P., Zemek, P.: RetDec: an open-source machine-code decompiler, [talk], presented at Botconf 2017, Montpellier, FR, December 2017

15. Sousa, D.G., Dias, R.J., Ferreira, C., Lourenço, J.: Preventing atomicity violations with contracts. CoRR abs/1505.02951 (2015)

Blockchain and IoT Integration for Pollutant Emission Control

Stefano Bistarelli[1], Marco Marcozzi[2], Gianmarco Mazzante[2],
Leonardo Mostarda[1(✉)], Alfredo Navarra[1], and Davide Sestili[2]

[1] Mathematics and Computer Science Department,
University of Perugia, Perugia, Italy
{stefano.bistarelli,alfredo.navarra}@unipg.it,
leonardo.mostarda@unicam.it
[2] Computer Science Division, University of Camerino, Camerino, Italy
{marco.marcozzi,gianmarco.mazzante,davide.sestili}@unicam.it

Abstract. A recent and prominent trend is the integration of
Blockchain and the Internet of Things (IoT). Blockchain can enable completely decentralised IoT applications where autonomous IoT devices
securely interact without the need of a centralised trusted party. The
perfect match of these two technologies still needs many challenges to
be faced. IoT devices can produce a huge amount of data which can
pose blockchain scalability problems in terms of fees and memory. IoT
transactions should occur in real-time, but in reality it seldom happens
due to transactions validation and involved consensus techniques. In this
paper we introduce a novel industrial IoT case study scenario, where
Blockchain technology has been used to check the compliance of industrial incinerators with the Italian emission regulations of pollutant. We
have explored various blockchain based alternatives that are on-chain,
off-chain and permissioned based solutions. These have been compared
in order to find a viable one.

1 Introduction

The Internet-of-Things (IoT) is one of the modern technological revolutions that
enables communication amongst a plethora of different devices [2,5]. IoT devices
range from tiny sensing ones to more complex objects such as smart buildings, autonomous connected cars, smartphones, laptops and tablets. These can
directly communicate in order to allow the implementation of innovative applications such as intelligent health care, smart supply chain, smart agriculture,
smart cities and smart industry. No doubt, blockchain will play a prominent role
for building decentralised IoT applications. The trusted, decentralised, and self-
regulating nature of blockchain technology can be the key driving force behind
a secure decentralisation of the IoT. Blockchain smart contracts (SC) can allow
secure IoT device interaction without the need of a centralised trusted party.
While this integration seems quite promising there are still many challenges to
be solved [9,12]. SCs are run by miners which can result in slow execution time
(e.g., 10 min to mine a Bitcoin block while 40 s to mine an Ethereum one) and

L. Barolli et al. (Eds.): AINA 2022, LNNS 451, pp. 255–264, 2022.
https://doi.org/10.1007/978-3-030-99619-2_25

expensive fees. These issues are exacerbated in Internet-of-Things (IoT) systems, where a large amount of devices can produce a high volume of data which usually requires real-time elaboration. The use of the main chain can require high monetary cost for storing and elaborating data and the time required to mine and validate blocks can be too high [10]. To date there are several research directions that try to tackle the integration of IoT and Blockchain. Off-chain approaches [3,4,7,8,11] such as state channel and side channels seem to be viable but the tools are still immature. The objective of this paper is to introduce a novel industrial IoT case study where multiple untrusted parties are involved. In detail, the case study addresses the problem of regulating emissions of industrial incinerators. Incinerator companies are required to keep their emissions below a certain pollution threshold, otherwise they will be fined by the health authority. We have explored various blockchain based alternatives that are on-chain, off-chain and permissioned based solutions. These have been analysed in order to find a viable one. The case study demonstrates that blockchain technology and IoT can be integrated to operate pollutant emission control systems. In particular, for the permissioned based solution we have performed a benchmark to show that the amount of transactions can be handled by a permissioned blockchain. For our use case the benchmark requires a peculiar setting on the IoT transactions (high send rate of transactions containing few bytes of data). The contribution of this paper can be summarised as follows: (i) a novel blockchain-based case study for IoT pollution control; (ii) a benchmark of a permissioned blockchain with a high volume of transactions, each of them with few bytes.

2   Pollutant Emission Control Case Study

Incinerating waste to energy power provides a significant share of the EU total energy supply. Incinerators use household garbage to generate power. The burning of the waste heats water whose steam drives a turbine that generates electricity. Left unsupervised, this process would lead to a massive air pollution. Thus, Italian and EU law and regulations set a threshold on the amount of pollutant emission. A fine, that is proportional to the time the threshold is exceeded, must be payed by the incinerator owner.

The entities that are involved in a waste to energy pollutant emission control are a sensor provider, the incinerator owners, the health authority and the industries. A sensor provider installs at the incinerator sites various sensors. These can measure different parameters such as nitrogen oxides, carbon monoxide (CO) and oxygen. The CO emission is regulated in the *Italian and European emission regulations*. This establishes that the concentration of *carbon monoxide* emissions must not exceed the following **law limit values**: $150 \, \text{mg/Nm}^3$, $100 \, \text{mg/Nm}^3$ and $50 \, \text{mg/Nm}^3$ as an average value over 10 min, 30 min and 1 day. Values are calculated by using a normalisation formula that is described by the Eq. 1, where Es is the concentration of emission calculated to the reference oxygen content; Em is the concentration of measured emission; Os and Om are the reference (11%) and the measured, resp., oxygen content.

$$Es = \frac{21 - Os}{21 - Om} \cdot Em \qquad (1) \qquad\qquad E_{min} = \frac{\sum_{i=1}^{n} Es_i}{n} \qquad (2)$$

The solution adopted by the company is to read the emission every 5 s (this is referred to as instantaneous emission). We denote with Es_i the instantaneous emission calculated at time i. The instantaneous emission are averaged in order to calculate the emission per minute. This is defined by the Eq. 2 where E_{min} is the minute average emission, Es_i is a reading that is performed inside the minute min and n is the number of readings per minute that is 12 in our case (every 5 s).

The owner of the incinerator has implemented an **emission control policy** that tries to avoid exceeding the law limit values. This policy monitors the minute average emission E_{min}. When this exceeds a threshold δ (this is set to be less than the average daily value) the incinerator burners reduce their activity. This avoids exceeding the law limit values while maximising the energy production. A health authority body uses the reading values in order to check when the law limit values are exceeded. In this case a fine that is proportional to the exceeding time must be paid by the owner of the incinerator. The emission data can be analysed by the energy consumers (i.e., industries). This can be used to certify the green behaviour of incinerator companies. The aforementioned *Internet of Things* infrastructure is a solution for analysing the emissions and for reacting promptly to critical states. However the retrieved - and analysed - reading data cannot be managed by a single party because of trust issues. The health authority does not usually have the infrastructure to hold data and would not trust the incinerator company and the industries to hold the reading data. A dishonest incinerator company could counterfeit data to avoid fines. At the same time, the incinerator company would not trust the health authority since it can place fines. In the first version of the system the health authority, the incinerator companies and the industries trust a third party that is the company installing the sensors. This would calculate the fine that the incinerator companies must pay to the health authority. The sensor company would also expose the reading data to the incinerator companies, the industries and the health authority in order to perform their activities. This solution was immediately found inappropriate by the incinerator companies and health authority that would rather prefer the use of the blockchain in order to ensure data immutability and avoid the need of a third trusted party. We have considered three blockchain based solutions for the implementation of the pollutant emission control case study; (i) a public blockchain; (ii) an off-chain solution and (iii) a permissioned blockchain;

The main problem of the public blockchain solution is that an high volume of data would be stored in the blockchain. In our case study an incinerator generates about 500 data transactions every 5 s with a total amount of 4 MB. This would mean 69.12 GB per day and about 25 TB per year. Data must be kept at least for 5 years. This would make 125 TB for a single incinerator system. This storage of data would cost millions of dollars in an Ethereum public blockchain. A smart contract that calculates the fine should also run. This would transfer the money from the wallet of the incinerator companies to the health authority one. Its execution would cost money as well. Moreover the **emission control policy** should be implemented: the activity of incinerator burners must be managed in order to promptly reduce the pollution emission before exceeding the limit and it is not feasible with public blockchains due to their latency issues.

We have also explored the feasibility of off-chain solutions [3,7,8,11]. Lightning [11] and, more in general, payment channels are an effective solution to improve scalability in terms of transaction throughput and fees. Their technology is mature and several implementations are available, but smart contract execution is not supported. State channels [3][1] generalise the concept of payment ones and allow the execution of smart contracts. Iota 2.0[2] and Ethereum 2.0 also promise scalability by means of side chain and sharding technology, respectively. To date off-chain technology tools are still immature thus we decided to implement the system by proposing a permissioned blockchain solutions.

3   Evaluation of a Hyperledger Besu Based Solution

Hyperledger Besu[3] is an open-source Ethereum client allowing enterprises to design blockchain solutions in both public and private networks, enhancing the development of DApps (Distributed applications) with high-performances in a secure environment. Besu implements different consensus protocols, such as: Ethash (Proof of Work), IBFT 2.0 (Proof of Authority), and Clique (Proof of Authority). Our interest is to test the overall performances of a Besu-based network in order to see whether or not satisfies the performance requirements of our case study scenario. We first provide some background on IBFT 2.0 and Clique protocols afterwards we test the performance of both protocols.

Clique[4] is a Proof of Authority (PoA) algorithm for achieving consensus in a Ethereum-like network. Clique works in epochs, defined as a fixed amount of committed blocks in the blockchain. At the beginning of each epoch, there is a special block with a list of authorities' IDs, and this block can be used as a snapshot of the blockchain for new authorities having to synchronise. Votes to add or remove validators can be cast by already accepted validators within an epoch. A vote passes if more than 50% of validators voted the same. The protocol selects in a round-robin fashion the authorities (peers) that are allowed to submit blocks to the blockchain, specifying the role of a leader. If the leader doesn't propose a block in time, other validators can propose it, but only those that didn't sign any of the previous $N/2$ blocks, where N is the number of authorities on the network. The fact that, in general, more than one peer is allowed to propose blocks may lead to forks (i.e. distinct conflicting chain of blocks) in the blockchain. Clique is using a modified version of the GHOST protocol (also used in Bitcoin and Ethereum) for solving forks: blocks with the heaviest score are retained as the main chain. Blocks proposed by leaders have

---

[1] https://magmo.com/force-move-games.pdf.

[2] Popov, S.: The tangle. White paper 1(3) (2018).

[3] https://besu.hyperledger.org/en/stable/.

[4] https://eips.ethereum.org/EIPS/eip-225.

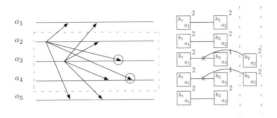

Fig. 1. Clique example

heavier score, such that, in case of a fork, blocks proposed by leaders, if any, will be considered. Figure 1 shows a clique example taken from [6]. In red dashed box there are authorities a_2, a_3 and a_4 allowed to propose blocks, where a_2 is the leader. Red circles and blocks on the right side explain how a fork is solved: a_3 and a_4 would commit the block proposed by a_3, but since a_2 is the leader, both eventually replace the block proposed by a_3 with the one proposed by a_2, since it has a heavier score.

Istanbul Byzantine Fault Tolerance (IBFT)[5] is a PoA protocol with a Byzantine Fault Tolerant (BFT) scheme to achieve consensus. Validators are nodes of the blockchain involved in the consensus protocol and they are in number of $N = 3F + 1$, where F is the number of Byzantine nodes that can be tolerated in the system. A Byzantine node is a node that either is not available or it may act maliciously against the system itself. As shown in Fig. 2, there are four distinct states, in which a validator may be: proposal, prepare, commit, and round change. At the beginning of a round, the elected proposer enters the state *proposal*, thus proposer submits a block of transactions to validators. On the arrival of the message, validators enter the *prepare* phase, in which each of them checks the validity of the incoming set of transactions, then validators send a message to other nodes with the round identifier and the digest of the block. When $2F + 1$ prepare messages have been received, validators enter the *commit* round. Validators broadcast a signed message with round identifier, the digest of the block, and a "commitSeal", which is a cryptographic signature. *Round change* is a special turn, triggered both when a certain timeout is reached or when a block is successfully committed, in which validators elect the proposer for the next round. The freshly elected proposer can submit to validators a block of transactions, starting a new round.

When comparing the performances of Hyperledger Besu networks implemented with different consensus protocols (IBFT 2.0 and Clique), it is fundamental to find the key features to be measured. For this purpose, the Hyperledger's Performance and Scale Working Group (PSWG) published a white paper analysing the metrics needed in both performance evaluation and benchmark[6]. There are further papers on the performance of permissioned blockchains, e.g.

---

[5] https://github.com/ethereum/EIPs/issues/650.
[6] https://www.hyperledger.org/learn/publications/blockchain-performance-metrics.

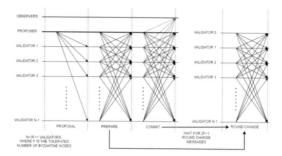

Fig. 2. IBFT 2.0. A line is a potential proposal, the red line, i.e., "observers", keep a blockchain copy without being part of the consensus.

[1], and they all rely, with sightly different definitions, on the same metrics. In our analysis, we found that all those different metrics can be grouped in three main classes of indicators: efficiency, security, and scalability. Efficiency includes metrics related to time consumption; security is related to liveness and finality properties of the blockchain; finally, scalability reports the variation of all the above parameters respect to the number of nodes in the network (hence also the number of transaction handled by the network itself). To obtain such data, we chose to use Caliper[7], a software developed by the Hyperledger community itself to benchmark blockchain solutions. The following indicators can be retrieved from Hyperledger Caliper:

- **Transaction throughput (TPS)** indicates the total committed transaction in a unitary time period, i.e. $TPS = \frac{\#\, transactions(t_1) - \#\, transactions(t_0)}{t_1 - t_0}$ where $t_1 \geq t_0$
- **Transaction latency (latency)** measures the amount of time needed between the submission of the transaction and the point when the transaction is committed; this time interval includes the consensus process and the transaction propagation time.

Where TPS and latency assess the efficiency of the network and the presence of forks in the resulting blockchain determines the finality. To retrieve, instead, data about scalability, i.e. all the above metrics measured for different number of nodes, we need to repeat our experiment with a different number of validators. Experimentally, we run benchmarks to compare performances in terms of latency and throughput on different network configurations. The metrics collected are transaction latency and throughput. These metrics have been collected on blockchains with different parameters. The difference between the blockchains lie in the following parameters: (i) BPS (Block Period Seconds) that is the number of seconds after which a node can propose a new block to be added to the blockchain; (ii) validator number that is the number of nodes playing a role in the validation of new blocks; (iii) Consensus protocol that is either Clique

---

[7] https://hyperledger.github.io/caliper.

or IBFT 2.0. Each blockchain configurations has been deployed and tested on a single local machine having the following characteristics: Operating System: Ubuntu 20.04.3 LTS, 2 CPU: Intel(R) Xeon(R) Gold 6256 CPU @ 3.60 GHz, RAM: 128 GB. The version of the Besu client used for running the nodes is the 21.7.0-RC1. The transactions sent to the nodes are smart contract method calls. The smart contract exposes methods to store readings of sensor data and to query them. The first method changes the state of the smart contract thus it is necessary to commit a transaction to the blockchain to call it. The query method is a *view* one, which means that it cannot change the state of the blockchain and does not require a transaction in order to be called, therefore the latency of their responses is not affected by the consensus protocol used or the block period of the blockchain. For each benchmark a set of store method calls have been sent to a validator node at different rates. *Validator adding time* has also been analysed. With *validator adding time* we refer to the time required for a validator (IBFT 2.0) or signer (Clique) to be added to the blockchain.

4   Results

Figures 3a and 3b show the average transaction latency for an increasing number of transaction per seconds and a increasing number of validators. We have set the block period time (BPS) to 1 s. Figure 3a shows that the number of validators seem to affect consistently transaction latency in IBFT 2.0 blockchains while it does have a milder effect in Clique blockchains (Fig. 3b). This is probably due to the fact that each block proposed in an IBFT 2.0 blockchain has to be signed by a super-majority (more than 66%) of validators before being added to the blockchain. The situation is different in a Clique blockchain. In a Clique blockchain a block is only signed by the proposer and do not require other signers to sign it. This difference allows Clique blockchains to maintain a stable block period even when there are several signers on the network. This advantage comes at the cost of possible forks. As seen in Fig. 3a, 3b, 4a and 4b, the transaction send rate seems to play a role in the performance of the blockchain. The higher the transaction rate, the higher the transaction latency, even though the increase in latency is minimal. This statement holds true both in IBFT 2.0 and

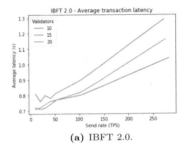
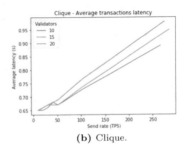

(a) IBFT 2.0.   (b) Clique.

Fig. 3. Average transaction latency using Clique and IBFT 2.0 with a BPS of 1 s.

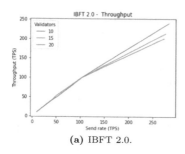

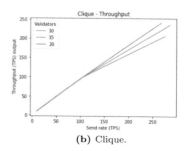

(a) IBFT 2.0.   (b) Clique.

Fig. 4. Average throughput using IBFT 2.0 and Clique with a BPS of 1 s.

Clique blockchains. The throughput of the network increases as the send rates increases. In our setup we were limited to sending a maximum of around 300 transactions per second, which is not enough to reach saturation throughput in the blockchains tested. Gas limit specifies the maximum size of the blocks. In the blockchain tested the gas limit has been set to the maximum allowed by Besu. This enabled our blockchains to create blocks as big as they could. In this way the gas limit didn't provide to be a bottleneck in our benchmark runs.

Proposals to add one validator have been sent to blockchains with different configurations. Eight different blockchain configurations have been benchmarked; four use the IBFT 2.0 consensus protocol while the others Clique. The blockchains for each consensus protocol vary in their BPS. The BPS used in our benchmarks are 1, 10, 30 and 60. Every blockchain tested is made up of 20 nodes. In our benchmark runs proposals to add a validator have been sent at the same time to every node of the chain. The *validator adding time* has been measured as the amount of time taken by the blockchain to add the proposed validator since it was first proposed by the nodes. The impact of the number of already accepted validators on *validator adding time* has also been measured by varying the number of validators in our chain at each benchmark run. The results obtained shown in Fig. 5b and Fig. 5a show that every parameter taken into consideration (BPS, consensus protocol, number of validators already in the chain) affect the *validator adding time*. Results show that the amount of time taken to add a validator in an IBFT 2.0 blockchain is always higher than in a Clique blockchain. The difference between the results obtained testing the two consensus protocol is always almost equal to the BPS used in those configurations, suggesting that IBFT 2.0 requires the generation of a block more than Clique to add a validator. The number of pre-approved validators already on the chain also plays a role. The more the validators the higher the time required to the network to add a validator regardless of the consensus protocol used. This is due to the fact that it requires a majority of validators to add successfully a new validator and only one vote at a time is stored in each block. Votes are stored in the block header and one vote can be added to the block only by the proposer of that block. BPS also affects *validator adding time* since votes are stored in blocks. *Validator adding time* increases linearly as the BPS increases.

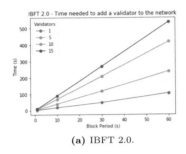

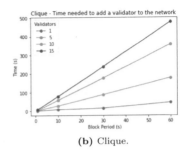

(a) IBFT 2.0.   (b) Clique.

Fig. 5. Experiment testing the time needed to add a new validator.

5   Discussion and Conclusions

This paper demonstrates that blockchain technology and IoT can be integrated to operate on a pollutant emission control system. In particular, our Besu solution shows that the amount of transactions can be handled by a permissioned blockchain. Our benchmark required a peculiar setting on the IoT transactions (high send rate of transactions containing few bytes of data). Both IBFT and Clique consensus protocol can handle the transactions coming from up to three incinerators (i.e., 300 per seconds) with one block per second setting. The delay for transaction validation is 1.3 s with IBFT and 1 s for Clique when 20 validators are considered. This low validation time allows the implementation of the **emission control policy**. This must run every minute and promptly reduce the activity of incinerator burners when the pollution emission limit is exceeded. The system can support 20 validators that is health authority, various incinerators and several industries can participate to the blockchain with a validator. As future work we plan to integrate the reaction logic (i.e., the reduction of the burner power) into the smart contract.

References

1. Angelis, S.D.: Assessing security and performances of consensus algorithms for permissioned blockchains. Master's thesis, "Sapienza" University of Rome (2018)
2. Balaji, S., Nathani, K., Santhakumar, R.: IoT technology, applications and challenges: a contemporary survey. Wireless Pers. Commun. **108**(1), 363–388 (2019)
3. Cacciagrano, D., Corradini, F., Mazzante, G., Mostarda, L., Sestili, D.: Off-chain execution of IoT smart contracts. In: Barolli, L., Woungang, I., Enokido, T. (eds.) AINA 2021. LNNS, vol. 226, pp. 608–619. Springer, Cham (2021). https://doi.org/10.1007/978-3-030-75075-6_50
4. Cacciagrano, D., Corradini, F., Mostarda, L.: Blockchain and IoT integration for society 5.0. In: Gerber, A., Hinkelmann, K. (eds.) Society 5.0 2021. CCIS, vol. 1477, pp. 1–12. Springer, Cham (2021). https://doi.org/10.1007/978-3-030-86761-4_1

5. Cacciagrano, D., Culmone, R., Micheletti, M., Mostarda, L.: Energy-efficient clustering for wireless sensor devices in Internet of Things. In: Al-Turjman, F. (ed.) Performability in Internet of Things. EICC, pp. 59–80. Springer, Cham (2019). https://doi.org/10.1007/978-3-319-93557-7_5

6. De Angelis, S., Aniello, L., Baldoni, R., Lombardi, F., Margheri, A., Sassone, V.: PBFT vs proof-of-authority: applying the cap theorem to permissioned blockchain (2018)

7. Hafid, A., Hafid, A.S., Samih, M.: Scaling blockchains: a comprehensive survey. IEEE Access **8**, 125,244–125,262 (2020)

8. Hu, B., et al.: A comprehensive survey on smart contract construction and execution: paradigms, tools, and systems. Patterns **2**(2), 100,179 (2021)

9. Panarello, A., Tapas, N., Merlino, G., Longo, F., Puliafito, A.: Blockchain and IoT integration: a systematic survey. Sensors **18**(8), 2575 (2018)

10. Peker, Y.K., Rodriguez, X., Ericsson, J., Lee, S.J., Perez, A.J.: A cost analysis of internet of things sensor data storage on blockchain via smart contracts. Electronics **9**(2), 244 (2020)

11. Poon, J., Dryja, T.: The bitcoin lightning network: scalable off-chain instant payments (2016)

12. Sekaran, R., Patan, R., Raveendran, A., Al-Turjman, F., Ramachandran, M., Mostarda, L.: Survival study on blockchain based 6G-enabled mobile edge computation for IoT automation. IEEE Access **8**, 143,453–143,463 (2020)

Robot Based Computing System: An Educational Experience

Diletta Cacciagrano[1], Rosario Culmone[1], Leonardo Mostarda[1], Alfredo Navarra[2], and Emanuele Scala[1(✉)]

[1] Computer Science Division, University of Camerino, Camerino, Italy
{diletta.cacciagrano,rosario.culmone,leonardo.mostarda,
emanuele.scala}@unicam.it
[2] Mathematics and Computer Science Department,
University of Perugia, Perugia, Italy
alfredo.navarra@unipg.it

Abstract. Robot based computing systems have been widely investigated in the last years. One of the main issues is to solve global tasks by means of local and simple computations. Robots might be cooperative or competitive, still the algorithm designer has to detect a way to accomplish the desired task. In this paper, we propose a platform made up of small and self-propelled robots with very limited capabilities in terms of computing resources, storage and sensing. In particular, we consider cheap robots moving within a confined area. The area is suitably coloured so as to be able for a robot endowed with a light sensor to reasonably detect its position. Moreover, robots can communicate with each other by exchanging short messages. Based only on those weak capabilities, we show how it is possible to realise interesting basic tasks. Apart for the relevance in educational contexts, our platform also represents an interesting case study for the main question posed in the literature about the minimal settings under which interesting tasks can be distributively solved.

1 Introduction

Robotic systems have been widely studied in the last decades. Robotics finds applications in many field of research and it involves both computer scientists and engineers. In fact, facing robotic issues concerns the design, the construction, the operation, and the use of robots. In particular, two main field of research have been investigated so far: *swarm robotics* (e.g., see [2,20]) and *modular robotics* (e.g., see [3]).

The former is mainly about robotic systems in which interconnected entities can somehow recover from failures or rearranging themselves in order to accomplish the required task (e.g., see [23]). Their main objective is to achieve systems that are more versatile, affordable, and robust than their standard counterparts, at the cost of a probable reduced efficacy for specific tasks, see, e.g. [1,22].

Work supported by the Italian National Group for Scientific Computation GNCS-INdAM.

The latter mainly differ from modular robotics as individual robots in the system do not need to be connected with each other all the time, but they are usually fully autonomy mobile units (see, e.g., Kilobot [19]). The interaction among robots specified by means of robots' capabilities should lead to a desired collective behaviour. The approach has been mainly investigated in the field of artificial swarm intelligence, but it finds applications also in biological studies of insects, ants and other fields in nature, where swarm behaviour occurs. The research concerning swarm robotics is mainly focused on theoretical aspects, by considering robot systems in the abstract, where capabilities of the robots as well as the complexity of the environment are reduced to their minimum. Basic models widely investigated in this context are the Amoebot model [12,13], and the more recent Silbot [8,9], Moblot [4], and Pairbot [15] models. One of the main issue faced when dealing with such models is that they help, in general, in rigorously analyse the designed algorithms, hence providing new theoretical insights that subsequently also extend the practical aspect of the studied systems.

One of the models well investigated in swarm robotics is certainly $\mathcal{OBLOT}$ (see, e.g., [5–7,10]). Such a model can be considered as a sort of framework within which many different settings can be manipulated, each implied by specific choices among a range of possibilities, with respect to fundamental components like time synchronisation as well as other important elements, such as memory, orientation, mobility and communication. Settings are often maintained at their minimum.

From a practical view point, the technology required to implement algorithms designed within $\mathcal{OBLOT}$ does not rely on special sensors nor actuators. Hence, cheap hardware might be used and experimented. An example of real robots working this way can be found in [11,18,21]. In general, robots' capabilities are maintained as weak as possible so as to understand what is the limit for the feasibility of the problems. Moreover, the less assumptions are made, the more a resolution algorithm is robust with respect to possible disruptions.

In this paper, we propose a new practical swarm robotic system. Our study started from educational purposes but we believe it can find many interesting applications, suggesting non-trivial solutions in case of reduced capabilities allowed to the robots.

2 System Model and Communication Protocol

We envision and implement a robotic system where a set of identical robots is equipped with the following minimum components: one sensor, one actuator and one communication subsystem.

All robots operate in a finite environment, acquiring data from it through the sensor (position) and interacting with it through the actuator (motion). In what follows we frequently refer to robots as *ANTs*. An optional *monitor* (laptop) module is used to start, stop and monitoring the robots; it is connected to the same communication subsystem of ANT_s, but it does not interact with the environment nor participate to the robots' behaviour. The robot model is depicted in Fig. 1.

Other important constraints for robots regard the limited computational and memorisation capabilities. Those lead to keep main attention on efficient algorithms that cannot count on complex sensors and actuators.

Further, this choice is aimed at experimenting with real-time systems for resource constrained devices. To this ending, it is important to accurately define the memory settings for the data structures in each single task, as well as the composition of tasks at a higher level.

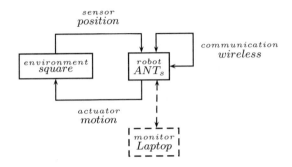

Fig. 1. System model.

The communication subsystem we consider implements a mutual-exclusive ring protocol where a set of z robots, denoted with $R = \{r_0, \ldots, r_z\}$, provide to each other their position in the environment. What a robot sends on its turn t, with $t \in [0, z]$, is a broadcast message $b_t = r_t || r_{t+1} || currPos$ where r_t is the identifier of itself, r_{t+1} is the identifier of the robot that will broadcast the message b_{t+1} on the next turn and $currPos$ is the current position of r_t on the environment. An additional payload can be provided with b_t, consisting of information for verifying the correct reception of the message and a bit to control the start and stop of all robots by the monitor.

Each robot is configured to be in two states, the listening state S_L and the broadcasting state S_B. When it is not his turn to communicate, a robot is always in S_L, parsing the message received form the only robot r_t of the ring enabled to broadcast. Once received, a robot in S_L obtains from the message b_t the current position sent by r_t, storing it locally, and checks if r_{t+1} matches its identifier. If so, the robot switches to state S_B. At each turn t, the r_t robot in S_B assembles and broadcasts the message b_t by deriving the identifier of the next ring member and the current position. In order to detect faulty ring members, two local timeouts are set in each robot, a global timeout T_g and a response timeout T_r. The first is of the order of seconds, the time required for a complete ring round trip. It is useful both to start the protocol, ensuring that the first robot started enters in S_B, and to prevent the protocol from ending. The second is used in the S_B state so that the broadcasting robot r_t verifies that the next member of the ring is active. If not, the robot r_t will continue this acknowledgement i times with the successors of r_{t+1} until a robot r_{t+1+i} in the ring responds with a broadcast message.[1] All the robots $r_{t+1}...r_{t+i}$ are declared inactive for the turn t and will be allowed to re-enter in the successive turns. Our ring protocol is implemented by a single infinite loop algorithm installed in all robots, see Algorithm 1. We call this program *RACom* together with two subroutines *listen* and *broadcast*.

2.1   Implementation Details

Our implementation of the model is made up of small ANTs, a platform on which they operate and an optional monitoring computer. Each ANT (see Fig. 2a) consists of:

- One load-bearing aluminium structure on which it is positioned a Arduino UNO (ATmega328P at 16 MHz, 2 KB SRAM, 32 KB);
- Two stepper motor SY35ST26-0284A with 200 step and Adafruit Motor Shield v2.3;
- One RGB light sensor TCS34725;
- One serial wireless module HC-12 433 MHz SI4438;
- Two 9 V rechargeable battery 900 mAh.

The platform on which the ANTs move is a two meters square. The surface is made of coloured photographic paper (see Fig. 2b). Colours are distributed so as a robot can deduce its position by sensing the corresponding colour. The system can be monitored using a notebook with a specific ANT (Arduino and communication module only) connected via USB. The software has been designed in C with FreeRTOS [14]. Four subroutines have been realised:

- *Communication routine.* It deals with the communications among ANTs. The driver of the HC-12 module performs a timed reading of the buffer and transfers the data to the Communication routine which interprets them and sends them to the other ANTs involved in the communication;
- *Position routine.* It deals with the identification of the positioning of the ANT. The TCS34725 module drive takes timed readings of the RGB values of the surface under the ANT. The values are transferred to the *Position routine* which interprets them by two functions:

    $map(r, g, b) \rightarrow$ *sites*: Given a tuple of red, green, blue values, *map* produces the set *sites* composed by all the coordinates of the plane that are consistent with these values. Note that although the plane has been generated in such a way that the uniqueness of colours for each point is guaranteed, the low sensitivity of the sensors or interference can produce incorrect reading values;

    $trace(traces, sites) \rightarrow (x, y)$: Given two sets of points, *traces* and *sites*, the trace function identifies the point where the ANT is located, selecting among the sites those that are consistent with the path taken.

---

[1] Sums in the subscript of the robots' identifiers must be considered modulo z.

Algorithm 1. RACom

1: **while** true **do**
2:     **if** init = false **then**
3:         Let z be the number of robots
4:         Let $myID$ be the robot's identifier
5:         Start global timeout T_g
6:         Set state = S_L
7:         Set init = true
8:     **end if**
9:     **if** T_g not expired **then**
10:         listen()
11:     **end if**
12:     **if** T_g expired **then**
13:         broadcast()
14:     **end if**
15: **end while**
16:
17: **function** BROADCAST :
18:     Set $i = 1$, $t = myID$, state = S_B
19:     **while** state = S_B **do**
20:         Set $t = (t + i) \bmod z$
21:         Set $currPos$
22:         Set $b_{myID} = myID||r_t||currPos$
23:         Broadcast message b_{myID}
24:         Start response timeout T_r
25:         **while** T_r not expired **do**
26:             listen()
27:         **end while**
28:         $i = i + 1$
29:     **end while**
30: **end function**
31:
32: **function** LISTEN :
33:     Read b_t from serial
34:     **if** b_t not empty **then**
35:         Parse r_{t+1} from b_t
36:         **if** $r_{t+1} = myID$ **then**
37:             broadcast()
38:         **end if**
39:         **if** $r_{t+1} \neq myID$ **then**
40:             Set state = S_L
41:         **end if**
42:         Restart T_g
43:     **end if**
44: **end function**

(a) One of our own built ANTs.   (b) The coloured surface among which ANTS move.

Fig. 2. Our implementation of the ANTs system.

- *Motion routine.* It deals with the motion of the ANT. The motion is made by sending commands to *Adafruit Motor Shield*. A control has been created that allows to issue commands for the advancement in a straight line, the curves with a given arc and control of the accelerations in order to have a fluid motion without jerks. This is also useful for preventing the wheels from losing grip. This feature makes it possible to monitor, by means of a feedback system, the real position of the ANT with respect to that reached after a movement.
- *Brain routine.* It deals the autonomous control of all the system. This routine allows the coordination of the system by interacting with all the other routines and realising its objectives. A mini control language has been created that allows to load the objectives with a laptop connected to the ANTs network. Once the relative objectives have been loaded on the ANTS, the function of the laptop is for supervision only (position and states of the ANTs) or it can be disconnected from the network.

Timed callbacks have been created in order to manage the hardware modules TCS34725 (colour sensor) and SI4438 (wireless communication). The routines and callbacks are connected via queues for correct interactions. The general architecture is shown in Fig. 3.

2.2 Patrolling

An example of task for which we applied the ANTs system previously described concerns the *Patrolling*. The requirement is to continuously monitor/visit the area of interest (our coloured square surface) without incurring in colliding ANTs. Moreover, ANTs are subject to disruptions, that may interrupt their functioning. Hence, we aim to design a fault-tolerant distributed algorithm to solve Patrolling. Due to the reduced capabilities of the ANTs, a simple solution in terms of running complexity is also required.

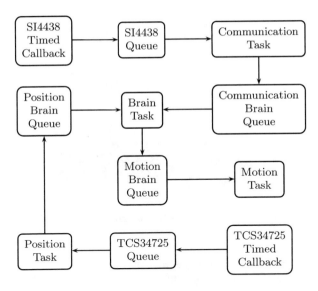

Fig. 3. Software architecture.

In our system each ANT broadcasts its position according to the described mutual-exclusive ring protocol. In so doing, all ANTs have a map with the coordinates of all ANTs continuously updated. Referring to this map and its current position, an ANT calculates two movements, "going straight" or "dodging". The idea is that, based on two consecutive positions of each other ANT, an ANT can understand whether along its (straight) way a collision may occur, in which case it decides to slightly deviate in order to avoid the positions occupied by the other ANTs. Deviations are also required if the border of the area to monitor is reached. If an ANT breaks, clearly it cannot communicate its position during its turn. Hence, all other ANTs can deduce that the ANT is broken by exploiting the associated timeouts not met by the broken ANT. In so doing, the failure will not alter the correct functioning of the ANTs. An ANT_i can avoid the border or any other ANT_j, with $i \neq j$, along its way, by calculating the dodging movement. ANT_i can in fact deduce the direction (in terms of angle α_{ij} wrt the coordinate system given by the area of interest) and hence the vector between the two coordinates (x_i, y_i) and (x_j, y_j). If α_{ij} is large enough and the obstacle is close enough (with respect to a predefined safety measures), ANT_i will turn according to the sign of the angle, otherwise it continues to go straightforward. A similar approach is used in [16]. Our solution differs in the use of less expensive hardware and that the calculation is done on the basis of the positions of the ANTs.

We build a simulator for our solution in ARGoS [17]. The goal is to trace the coverage of the area of interest where ANTs perform the patrolling. The environment is configured to be a square of $(-2, 2)$ length per side in the ARGoS unit. In there, 10 ANTs are randomly placed with uniform distribution at the

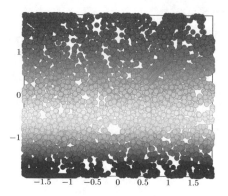

Fig. 4. ANTs patrol coverage.

start of the simulation. We run 80 s simulation and we detect 5 coordinates per second from each ANT, collecting 4000 coordinates in total. The speed of the ANTs is set in a reasonable way to avoid errors in the simulation. In Fig. 4 we show the coverage obtained, it is possible to notice a coverage of more than the 90% of the area.

3 Conclusions and Future Work

We have proposed a new platform for the implementation of modular robotics solutions where robots' capabilities have been reduced to their minimum. Although we started from educational intents, we believe the composed system has some potentials that can be exploited for designing non-trivial solutions in environments where robots cannot count on sophisticated hardware. We have reported our experience by means of a basic task, namely the Patrolling. It is worth noting that in the proposed solution each ANT moves in a deterministic way. Instead, we could modify this behaviour by considering that each ANT calculates future positions in a random way such that, by knowing the future locations of the other ANTs, each ANT can decide in advance to travel a safe path. We conjecture that such an algorithm might be a good solution to the Patrolling task and it might be even faster in terms of number of iterations. We plan to verify the conjecture by making comparisons with our current system via the simulator, possibly defining also different movement policies.

Another direction of research that deserves investigation is about the possible tasks that can be approached within the simple proposed system. For instance, an interesting task where a competitive rather than collaborative behaviour must be designed has been faced while investigating the so-called *Puss in the corner* puzzle. Six ANTs are placed at the corner of a regular hexagon and one is placed in the centre. From this initial configuration, a configuration must be reached where each ANT does not occupy its initial position. In this case, the choice of each ANT competes with all the other ANTs to achieve its goal.

References

1. Ahmadzadeh, H., Masehian, E., Asadpour, M.: Modular robotic systems: characteristics and applications. J. Intell. Robotic Syst. **81**(3–4), 317–357 (2016). https://doi.org/10.1007/s10846-015-0237-8

2. Amir, M., Bruckstein, A.M.: Minimizing travel in the uniform dispersal problem for robotic sensors. In: Proceedings of 18th International Conference on Autonomous Agents and MultiAgent Systems (AAMAS), pp. 113–121. International Foundation for Autonomous Agents and Multiagent Systems (2019)

3. Christensen, A.L.: Self-reconfigurable robots - an introduction. Artif. Life **18**(2), 237–240 (2012). https://doi.org/10.1162/artl_r_00061

4. Cicerone, S., Di Fonso, A., Di Stefano, G., Navarra, A.: MOBLOT: molecular oblivious robots. In: Dignum, F., Lomuscio, A., Endriss, U., Nowé, A. (eds.) AAMAS 2021: 20th International Conference on Autonomous Agents and Multiagent Systems, Virtual Event, United Kingdom, 3–7 May 2021, pp. 350–358. ACM (2021)

5. Cicerone, S., Di Stefano, G., Navarra, A.: Asynchronous arbitrary pattern formation: the effects of a rigorous approach. Distrib. Comput. **32**(2), 91–132 (2019)

6. Cicerone, S., Di Stefano, G., Navarra, A.: Solving the pattern formation by mobile robots with chirality. IEEE Access **9**, 88177–88204 (2021). https://doi.org/10.1109/ACCESS.2021.3089081

7. Cicerone, S., Di Stefano, G., Navarra, A.: A structured methodology for designing distributed algorithms for mobile entities. Inf. Sci. **574**, 111–132 (2021). https://doi.org/10.1016/j.ins.2021.05.043

8. D'Angelo, G., D'Emidio, M., Das, S., Navarra, A., Prencipe, G.: Asynchronous silent programmable matter achieves leader election and compaction. IEEE Access **8**, 207,619–207,634 (2020)

9. D'Angelo, G., D'Emidio, M., Das, S., Navarra, A., Prencipe, G.: Leader election and compaction for asynchronous silent programmable matter. In: Proceedings of 19th International Conference on Autonomous Agents and Multiagent Systems (AAMAS), pp. 276–284. International Foundation for Autonomous Agents and Multiagent Systems (2020)

10. D'Angelo, G., Di Stefano, G., Navarra, A.: Gathering on rings under the look-compute-move model. Distrib. Comput. **27**(4), 255–285 (2014)

11. Das, S., Focardi, R., Luccio, F.L., Markou, E., Squarcina, M.: Gathering of robots in a ring with mobile faults. Theor. Comput. Sci. **764**, 42–60 (2019). https://doi.org/10.1016/j.tcs.2018.05.002

12. Daymude, J.J., Hinnenthal, K., Richa, A.W., Scheideler, C.: Computing by programmable particles. In: Flocchini, P., Prencipe, G., Santoro, N. (eds.) Distributed Computing by Mobile Entities, Current Research in Moving and Computing. LNCS, vol. 11340, pp. 615–681. Springer, Cham (2019). https://doi.org/10.1007/978-3-030-11072-7_22

13. Derakhshandeh, Z., Gmyr, R., Strothmann, T., Bazzi, R., Richa, A.W., Scheideler, C.: Leader election and shape formation with self-organizing programmable matter. In: Phillips, A., Yin, P. (eds.) DNA 2015. LNCS, vol. 9211, pp. 117–132. Springer, Cham (2015). https://doi.org/10.1007/978-3-319-21999-8_8

14. Fei, G., Long, P., Luc, P., Martin, T.: Open source FreeRTOS as a case study in real-time operating system evolution. J. Syst. Softw. **118**, 19–35 (2016)

15. Kim, Y., Katayama, Y., Wada, K.: Pairbot: a novel model for autonomous mobile robot systems consisting of paired robots (2020)

16. Piardi, L., Lima, J., Oliveira, A.S.: Multi-mobile robot and avoidance obstacle to spatial mapping in indoor environment. In: 11th International Conference on Simulation and Modeling Methodologies, Technologies and Applications, SIMULTECH 2021, pp. 21–29 (2021)

17. Pinciroli, C., et al.: ARGoS: a modular, parallel, multi-engine simulator for multi-robot systems. Swarm Intell. **6**(4), 271–295 (2012)

18. Pásztor, A.: Gathering simulation of real robot swarm. Tehnicki vjesnik **21**(5), 1073–1080 (2014)

19. Rubenstein, M., Ahler, C., Nagpal, R.: Kilobot: a low cost scalable robot system for collective behaviors. In: IEEE International Conference on Robotics and Automation (ICRA), pp. 3293–3298. IEEE (2012). https://doi.org/10.1109/ICRA.2012.6224638

20. Şahin, E.: Swarm robotics: from sources of inspiration to domains of application. In: Şahin, E., Spears, W.M. (eds.) SR 2004. LNCS, vol. 3342, pp. 10–20. Springer, Heidelberg (2005). https://doi.org/10.1007/978-3-540-30552-1_2

21. Salman, M., Garzón-Ramos, D., Hasselmann, K., Birattari, M.: Phormica: photochromic pheromone release and detection system for stigmergic coordination in robot swarms. Front. Robot. AI **7**, 591,402 (2020). https://doi.org/10.3389/frobt.2020.591402

22. Tucci, T., Piranda, B., Bourgeois, J.: A distributed self-assembly planning algorithm for modular robots. In: André, E., Koenig, S., Dastani, M., Sukthankar, G. (eds.) Proceedings of the 17th International Conference on Autonomous Agents and MultiAgent Systems (AAMAS), pp. 550–558. International Foundation for Autonomous Agents and Multiagent Systems Richland, SC, USA/ACM (2018)

23. Yim, M., et al.: Modular self-reconfigurable robot systems [grand challenges of robotics]. IEEE Robot. Autom. Mag. **14**(1), 43–52 (2007)

ARM vs FPGA: Comparative Analysis of Sorting Algorithms

Yomna Ben Jmaa[1]([✉]), David Duvivier[2], and Mohamed Abid[3]

[1] REDCAD Laboratory, University of Sfax, Sfax, Tunisia
yomna.benjmaa@redcad.org
[2] LAMIH UMR CNRS 8201, Polytechnic University of Hauts-de-France,
Valenciennes, France
David.Duvivier@uphf.fr
[3] CES Laboratory, University of Sfax, Sfax, Tunisia
mohamed.abid_ces@yahoo.fr

Abstract. Sorting is a crucial operation for many real-time computing applications in several fields such as avionics and real time decision support systems. Meanwhile, because of the limited scaling of CPUs, FPGAs has appeared as an interesting alternative to accelerate software applications owing to their high performance and energy efficiency. In this paper, we aim to compare efficient hardware and software implementations of different sorting algorithms (Quicksort, Heapsort, Shellsort, Mergesort and Timsort) from high-level descriptions using different numbers of elements ranging from 8 to 4096 on the zynq-7000 platform and ARM Cortex A9 processor. In this paper, we need to sort 4096 elements at most because these elements are provided to a real-time decision support system as solutions to be sorted. As experimental results, we compare the performance of different algorithms in terms of average and standard-deviation of computational times. On the ARM Cortex A9 processor, we showed that Shellsort is 14–72% faster than others sorting algorithms if $n > 64$, otherwise Timsort algorithm is the best. Furthermore, we propose the "IShellsort" composed of Insertionsort and Shellsort which appears to provide the best computational times for all n on the ARM processor. On FPGA, Timsort algorithm is 1.16x–1.21x, 1.08x–1.23 and 1.25x–1.61x faster than Mergesort, Heapsort and Shellsort respectively running on FPGA if $n > 64$, otherwise Heapsort is the fastest. Also, we notice that the FPGA platform gives a better performance compared to the ARM Cortex A9 processor in terms of average and worst-case temporal stability considering the frequency 667 MHz for ARM and 100 MHz for FPGA.

1 Introduction

Intelligent Transportation Systems (ITS) [8] provide innovative services for different modes of transport and traffic management. They must offer a high performance computing, adaptability to the environment, reliability, etc. In addition, Intelligent Transportation Applications (ITA) need to compare and sort

L. Barolli et al. (Eds.): AINA 2022, LNNS 451, pp. 275–287, 2022.
https://doi.org/10.1007/978-3-030-99619-2_27

many solutions on various platforms to make decisions. Among these platforms, Field Programmable Gate Arrays (FPGAs) [9] provide very low computational times and energy consumption. The increase in the complexity of the applications has led to high-level design methodologies [10]. The new generation of High-Level Synthesis (HLS) allows increasing design productivity and detaching the algorithm from architecture [11]. In this case, the vivado HLS [12] tool is used to generate hardware accelerators from C/C++ language. For their part, recent CPUs have a potential for high performance thanks to multi-cores, with improved SIMD instructions.

Table 1. Complexity of sorting algorithms

	Best time complexity	Average time complexity	Worst time complexity
Shellsort	$O(n \log(n))$	$O(n^{\frac{4}{3}})$	$O(n^{\frac{3}{2}})$
Quicksort	$O(n \log(n))$	$O(n \log(n))$	$O(n^2)$
Heapsort	$O(n \log(n))$	$O(n \log(n))$	$O(n \log(n))$
Mergesort	$O(n \log(n))$	$O(n \log(n))$	$O(n \log(n))$
Timsort	$O(n)$	$O(n \log(n))$	$O(n \log(n))$
Insertionsort	$O(n)$	$O(n^2)$	$O(n^2)$

We focus on sorting a linear structure (table, list, etc.) of elements. Considering these target computational platforms, sorting algorithms are compared to propose an efficient sorting algorithm for an ITA dedicated to an avionic decision support system [14,18,20]. Hence, Table 1 shows that Insertionsort has a high "theoretical time complexity". Consequently, it is highly preferable to focus on sorting algorithms with $O(n \log(n))$ worst time complexity [7,13] such as Mergesort, Timsort and Heapsort. It is also worth noting that the complexity of Shellsort, which is an enhancement of Insertionsort, depends on gaps sequence [19]. On another way, Timsort is an example of a hybrid sort combining Insertionsort and Mergesort. The objective of this work is to compare optimized hardware and software implementations of Heapsort, Shellsort, Quicksort, Timsort and Mergesort on FPGA and ARM processor in terms of average and standard-deviation of computational times using a limited number of elements ranging from 8 to 4096 elements. Indeed, contrary to most of related works in the literature, there are 4096 elements at most because they are provided to a real-time decision support system as solutions to be sorted. To be more precise, each solution (i.e. a short-term path to follow) is identified by a unique integer, so-called index, and evaluated on the basis of various performance criteria (such as distance...). Consequently, index sorting algorithms are used in the real-life application. However, sorted elements are integers in this paper to simplify the problem. The only difference lies in the way the solutions are "evaluated": index's value in this paper, multicriteria evaluation of each indexed solution thanks to

PROMETHEE [16,17] in the real-life application. This justifies that sorted elements are permutations of integers encoded in 4 bytes. This also explains that the space complexity (in terms of memory consumption) is not considered in this paper because of the small size and number of elements, but also almost all algorithms are able to sort "in-place" using a minimum amount of space. Indeed, these limitations (i.e. maximum number of elements to sort, elements constitute permutations of integers, the target hardware platforms, etc.) have to be compatible with the field of real-time decision support as solutions to choose the best trajectory. Finally, thanks to the selected index sorting algorithm, the target decision making system allows to avoid obstacles and reduce the number of accidents for avionics systems as mentioned in [14,20]. Consequently, the purpose of this study is to select the fastest algorithm while considering its temporal stability (i.e. relative standard deviation of computational times and the number of outliers) and the specific features of the target application.

The paper is structured as follows: Sect. 2 presents a state of the art on several sorting algorithms using different platforms. Section 3 shows experimental results. Section 4 gathers our conclusions and some of our future works.

2   Related Works

Sorting is present in our daily life [1], in decision-making systems, avionics systems, etc. For this reason, researchers are always attempting to increase the performance of sorting algorithms [2]. The authors in [6] present an efficient implementation of Mergesort to sort up to 256M elements on a single Intel Q9550 quadcore processor. This work exploits modern processors and shows that the SIMD implementation with 128-bit SSE is 3.3X faster than the scalar version and the excellent scalability of this implementation with SIMD width scaling up to 16X wider than the current SSE width of 128-bits, and CMP core-count scaling well beyond 32 cores. The authors in [3] propose several approaches for data sorting acceleration of large data streams where sorting and/or merging are executed on CPU and/or FPGA. The algorithms in [4] are based on openMP and CUDA: the quick-merge parallel algorithm (a Quicksort algorithm sorts each sub-set of data, which are merged using the Mergesort algorithm) and the hybrid algorithm (parallel bitonic algorithm on GPU and sequence Mergesort on CPU). This work shows that running the sorting algorithms on a multicore processor is more efficient for the maximum of elements considered in [4]. Moreover, the computational times of these algorithms on GPU is faster for a small number of elements. In addition, authors show that the hybrid algorithm is a little slower than the quick-merge parallel algorithm on CPU. Thereafter, the approach presented in [5] provides a high-level tool for the rapid prototyping of parallel Divide and Conquer algorithms on multicore platforms. The objective is to propose an implementation of a model for the multicore architecture using OpenMP, Intel Threading Building Block framework and Fastflow parallel programming. In conclusion, this work shows that the prototype parallel algorithms consumes less time and requires minimal programming effort compared to manual parallelization. The authors in [1] accelerate sorting by leveraging models that approximate the empirical CDF of the input. The results show that this approach provides 3.38 performance improvement over STL sort, it is also 1.49 faster than Radixsort and 5.54 faster than Timsort. In addition, The authors in [21] propose a new sorting algorithm and show that this algorithm requires approximately 4–6 us to sort 1024 elements with a clock cycle of 0.5 GHz and consumes 1.6 mw.

In the following sections, we compare software and hardware implementations of sorting algorithms on ARM and FPGA in terms of average and worst-case temporal stability.

3   Experimental Results

The fixed frequency is 667 MHz for the software version on ARM Cortex A9 and 100 MHz for the hardware version on FPGA. Our hardware implementation uses a Zedboard platform and is synthesized via the vivado suite 2015.4. First of all, the average and standard deviation of computational times of the sorting algorithms are compared. Then, the relative standard deviations of computational times and the number of outliers are also calculated in order to evaluate their temporal stability.

3.1   Average and Standard-Deviation of Computational Times

In order to evaluate the "average temporal stability", in this subsection the average and standard-deviation of computational times of sorting algorithms are studied using 8 to 4096 elements, via 47 permutations of integers generated using Lehmer's method [15] and R = 1000 replications.

3.1.1   Computational Times on ARM Cortex A9 Processor

Table 2 reports the average and standard-deviation of computational times using different numbers of elements on ARM. Figure 1 and Table 2 show that Mergesort and Timsort give quite similar computational times if the size of the array is large. Also, Quicksort has lengthy computational times compared to others. In conclusion, Shellsort is 14% faster up to 72% when running on ARM if $n >$ 64, otherwise Timsort is the best. Table 2 illustrates that Shellsort is the best algorithm in terms of average and standard-deviation if $n > 64$. For example, if n = 2048, the standard-deviation is equal to 62.61 for Heapsort, 85.94 for Quicksort, 71.61 for Mergesort, 50.12 for Timsort and 49.94 for Shellsort.

Figure 2 shows the average and standard deviation of computational times on a vertical logarithmic scale when the number of elements are powers of two in order to highlight two specific "events" located by red circles: The first one on the right, for n = 64, illustrates the "switch" from Insertionsort to Mergesort according to the OP parameter. On the left, the second one, for n = 32, shows that (even when considering standard deviation of computational times) an hybrid sort composed of Insertionsort and Shellsort with OP = 32 provides us

Table 2. Average and standard-deviation of computational times on ARM Cortex A9

	Heapsort (us)	Shellsort (us)	Quicksort (us)	Mergesort (us)	Timsort (us)	IShellsort (us)
8	2.72 (0.37)	2.13 (0.05)	3.87 (0.06)	3.29 (0.06)	1.18 (0.05)	1.18 (0.05)
16	6.71 (0.63)	5.63 (0.09)	10.41 (0.31)	7.79 (0.09)	4.09 (0.08)	4.09 (0.08)
32	17.36 (0.67)	13.43 (0.18)	31.94 (0.20)	17.94 (1.65)	14.37 (0.62)	13.43 (0.18)
64	44.4 (2.19)	31.55 (0.18)	105.76 (0.20)	40.61 (2.55)	46.32 (3.28)	31.55 (0.18)
128	107.96 (4.28)	76.59 (0.72)	164.24 (0.19)	93.5 (4.07)	103.04 (7.48)	76.59 (0.72)
256	255.76 (6.40)	168.58 (4.11)	266.49 (5.2)	207.84 (8.1)	220.07 (14.39)	168.58 (4.11)
512	583.33 (14.18)	437.38 (11.08)	600.2 (12.55)	536.57 (36.26)	553.82 (21.7)	437.38 (11.08)
1024	1323.75 (35.88)	1029.52 (28.86)	1414.4 (40.21)	1198.48 (32.83)	1205.97 (30.69)	1029.52 (28.86)
2048	2949.01 (62.61)	2116.45 (49.94)	3499.45 (85.94)	2629.1 (71.61)	2628.01 (50.12)	2116.45 (49.94)
4096	6451.2 (79.51)	3291.38 (59.6)	11843.88 (308.51)	5684.51 (60.08)	5741.88 (67.45)	3291.38 (59.6)

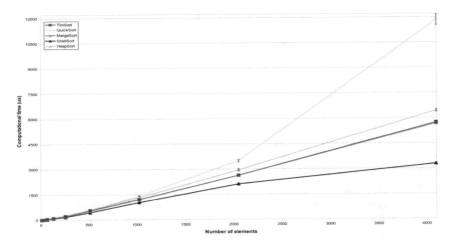

Fig. 1. Computational times of sorting algorithms on ARM Cortex A9

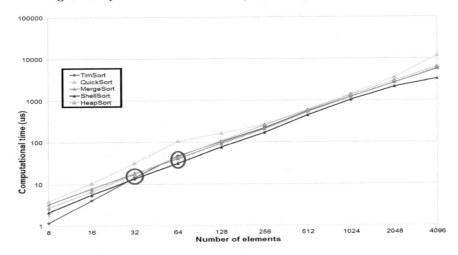

Fig. 2. Computational times on a vertical logarithmic scale when the number of elements are powers of two, on ARM

Algorithm 1. IShellsort

---

procedure ISHELLSORT(Table T)
    $n \leftarrow size(T)$
    if $n \leq 32$ **then**
        for $i \leftarrow 1$ **to** n **do**
            $j \leftarrow i$
            $tmp \leftarrow T[i]$
            while $j > 0$ and $T[j - 1] > tmp$ **do**
                $T[j] \leftarrow T[j - 1]$
                $j \leftarrow j - 1$
            $T[j] \leftarrow tmp$
    else
        for k in [1750, 701, 301, 132, 57, 23, 10, 4, 1] **do**
            for $i \leftarrow k$ **to** n **do**
                $j \leftarrow i$
                 $tmp \leftarrow T[i]$
                while $(j >= k)$ and $(T[j - k] > tmp)$ **do**
                    $T[j] \leftarrow T[j - k]$
                    $j \leftarrow j - k$
                $T[j] \leftarrow tmp$

---

with better average temporal performances than TimSort. Therefore, from these results, we can propose a new algorithm based on Insertionsort and Shellsort, called IShellsort and presented in a simplified version in Algorithm 1. To improve its efficiency, we use Ciura's gaps [19] in the "for k loop" and additional tests have been added to avoid useless assignments. Moreover, the "while loop" in the fist part of the algorithm (i.e. when $n \leq 32$) has been replaced by an efficient built-in memory move procedure. IShellsort takes advantage of basic but quick Insertionsort when sorting less than 32 elements and switches to Shellsort for more elements, taking advantage of a more sophisticated (although slightly more costly) sorting algorithm. As shown in Table 2, in any case, IShellsort provides us with best computational times. For $n \geq 32$, it corresponds also to the smallest values of standard deviation of computational times.

3.1.2   Computational Times on FPGA

This part presents the optimized hardware implementations using HLS directives (Loop unrolling, Loop pipelining, Input/output Interface) so as to enhance the performance of sorting algorithms. Table 3 gathers the average and standard deviation of computational times, complemented by Fig. 3. It can be deduced that Heapsort is 1.01x–1.12x faster than other algorithms if $n \leq 64$.

Table 3 and Fig. 3 also show that when N = 4096, the average computational times are 3756 us, 4740.7 us, 4848.5 us, 9756.4 us and 38599 us for Timsort, Mergesort, Heapsort, Shellsort and Quicksort respectively. Hence, Timsort is 1.16x–1.21x, 1.08x–1.23 and 1.25x–1.61x faster than Mergesort, Heapsort and Shellsort respectively if $n > 64$, otherwise Heapsort is the fastest. Moreover, the standard deviation is almost constant for all n. This makes sense due to

Table 3. Average and standard-deviation of computational times on FPGA

	Heapsort (us)	Shellsort (us)	Quicksort (us)	Mergesort (us)	Timsort (us)	IShellsort (us)
8	17.8 (3.49)	17.88 (3.48)	18.68 (3.51)	18.93 (3.50)	18.12 (3.47)	18.12 (3.47)
16	21.85 (3.48)	22.25 (3.48)	23.22 (3.49)	23.6 (3.49)	22.07 (3.47)	22.07 (3.47)
32	31.1 (3.48)	33.25 (3.48)	34.3 (3.48)	34.48 (3.48)	31.18 (3.48)	33.25 (3.48)
64	50.78 (3.48)	60.24 (3.48)	62.8 (3.48)	58.465 (3.48)	50.89 (3.48)	60.24 (3.48)
128	102.665 (3.48)	126.3 (3.48)	137.9 (3.48)	111.88 (3.48)	94.5 (3.48)	126.3 (3.48)
256	228.57 (3.48)	293.55 (3.48)	351.2 (3.48)	231.3 (3.48)	189 (3.48)	293.55 (3.48)
512	466.05 (3.67)	668.2 (3.48)	1001.3 (3.48)	482.5 (3.70)	393.17 (3.48)	668.2 (3.48)
1024	1015.5 (4.08)	1625.1 (3.56)	3121 (3.49)	1031.5 (3.48)	832.85 (3.47)	1625.1 (3.56)
2048	2227.1 (3.49)	4027.67 (3.48)	10660 (3.48)	2211.5 (3.49)	1769.5 (3.47)	4027.67 (3.48)
4096	4848.5 (3.48)	9756.4 (3.48)	38599 (3.49)	4734.9 (3.48)	3756 (3.47)	9756.4 (3.48)

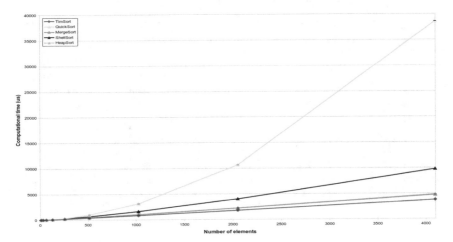

Fig. 3. Computational times of sorting algorithms on FPGA

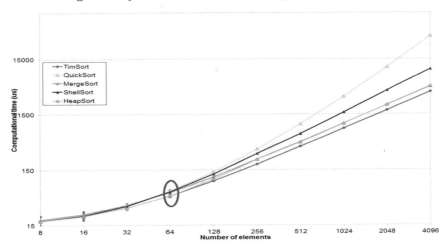

Fig. 4. Computational times on a vertical logarithmic scale when the number of elements are powers of two, on FPGA

the internal functioning of the FPGA, based on synchronous design, therefore the average temporal variation does not depend on the number of elements (considering 8 to 4096 elements).

Figure 4 shows the average and standard deviation of computational times on a vertical logarithmic scale, when the number of elements are powers of two, to highlight the "switch" from Insertionsort to Mergesort according to the OP parameter (i.e. when n = 64), located by a red circle. Contrary to the sorting algorithms on ARM, it is almost impossible to clearly distinguish between the sorting algorithms to try to select "the best one" for $n \leq 128$.

Table 4. Utilization of resources of the sorting algorithms

Size	Heapsort			Shellsort			Quicksort			Mergesort			Timsort		
	LUT	FF	BRAM	LUT	FF	BRAM	LUT	FF	BRAM	LUT	FF	BRAM	LUT	FF	BRAM
8	524	444	0	356	224	0	733	541	2	955	877	0	5299	2480	35
16	526	459	1	355	232	0	750	552	2	985	883	0	5326	2498	35
32	541	474	1	314	208	0.5	776	564	2	856	809	1.5	5293	2482	35
64	556	489	1	319	216	0.5	814	608	2	871	815	1.5	5320	2499	35.5
128	583	504	1	337	224	0.5	820	588	2	885	821	1.5	5487	2516	35.5
256	606	519	1	345	232	0.5	835	600	2	893	827	1.5	5468	2533	35
512	611	534	1	359	240	0.5	841	612	2	925	849	1.5	5494	2552	35.5
1024	651	551	1	365	248	1	860	624	3	936	855	3	5545	2569	36
2048	669	566	2	367	258	2	909	638	6	942	861	6	5668	2586	38
4096	689	581	4	384	266	4	976	682	15.5	954	867	12	5841	2602	44

Considering resources utilization for the sorting algorithms, Table 4 shows that Timsort consumes more resources than the other algorithms. It consumes 44% more for slice LUT in hardware implementation and around 9% for slice register.

3.1.3 Comparison of the Computational Platforms

Despite a lower frequency, computational times on FPGA are reduced compared to those on ARM because Xilink's tool is able to extract the parallelism of the algorithms by means of the optimizations used (HLS directives). For example, for Timsort when n = 4096, the computational times on FPGA (100 MHz) is 3756 us and 5741.88 us on ARM (667 MHz). However, a decrease in terms of computational time leads to an increase in resource utilization on FPGA.

3.2 Temporal Stability of Sorting Algorithms

In this subsection, the average and worst-case temporal stability are compared on ARM and FPGA. The relative standard deviation (RSD) is computed as the division of the standard-deviation of computational times by the average to obtain a relative measure of the dispersion of the computational times around the mean. This ratio is used to compare the degree of variation from one sample to another, even if the means are different: a RSD less than 1% is considered excellent for the average to be representative, provided that the size of the population is sufficient (i.e. *empirically* ≥ 30). It is an important measure of average stability when considering a given amount of time to perform a sort. Thus, it can be admitted that the stability of sorting algorithms is significant if the RSD is less than 1%. In addition, the worst-case temporal stability is given by the "upper outliers". The total number of outliers is the sum of the upper outliers which are located above the value Q3 + 1.5 * IQR with the interquartile range IQR = Q3 − Q1 and the lower outliers which are below Q1 − 1.5 * IQR. In addition, the percentage of outliers is calculated as follows:

$$Percentage\_outliers = 100 * \frac{Nb\_outliers}{Nb\_R * Nb\_P}$$

where Nb_R is the number of replications (1000 replications) and Nb_P is the number of permutations (47 permutations).

3.2.1   RSD and Outliers on ARM Cortex A9

Table 5. Relative standard deviation of sorting algorithms on ARM Cortex A9

	Heapsort (%)	Shellsort (%)	Mergesort (%)	Timsort (%)	IShellsort (%)
8	13.6	2.34	1.82	4.23	4.23
16	9.38	1.6	1.15	1.95	1.95
32	3.86	1.34	9.16	4.31	4.31
64	4.93	0.57	6.26	7.06	0.57
128	3.96	0.94	4.35	7.26	0.94
256	2.5	2.43	3.89	6.53	2.43
512	2.43	2.53	6.76	3.91	2.53
1024	2.71	2.8	2.73	2.54	2.8
2048	2.2	2.35	2.72	1.9	2.35
4096	1.23	1.81	1.05	1.17	1.81

Table 6. Percentage of outliers for Timsort on ARM

Size	% of total outliers	% of upper outliers
128	6.33	3.23
256	4.3	2.44
512	3.17	1.58
1024	7.22	3.2
2048	5.6	5
4096	2.66	1.98

On ARM, Table 5 shows that the RSD are almost always greater than 1%. Consequently, there is a significant variation of the average temporal stability. However, the RSD of the IShellsort is always lesser than 4.32%, which might be sufficient for some applications, even if the RSD is greater than 1%. In order to save space in this paper, we propose to focus on the Timsort on ARM. For this, Table 6 shows the percentage of outliers for the Timsort. The percentage of upper outliers is lower than 3.24% which is negligible this shows a fairly good worst-case temporal stability. These outliers are explained by noises generated by the operating system.

3.2.2 RSD and Outliers on FPGA

Table 7. Relative standard deviation of sorting algorithms on FPGA

	Heapsort (%)	Shellsort (%)	Mergesort (%)	Timsort (%)	IShellsort (%)
8	19.58	19.43	18.44	19.15	19.15
16	15.91	15.61	14.75	15.72	15.72
32	11.1	10.46	10.07	11.13	11.13
64	6.84	5.76	5.94	6.82	5.76
128	3.38	2.75	3.1	3.67	2.75
256	1.51	1.18	1.5	1.83	1.18
512	0.78	0.52	0.76	0.88	0.52
1024	0.4	0.22	0.34	0.42	0.22
2048	0.16	0.09	0.16	0.2	0.09
4096	0.07	0.04	0.07	0.09	0.04

Table 8. Percentage of outliers for Shellsort on FPGA

Size	% of total outliers	% of upper outliers
128	0.3	0.13
256	0.3	0.13
512	2.41	0.7
1024	3.74	0.2
2048	1.98	1.16
4096	2.2	0.8

Table 7 shows that for $n \geq 64$, the RSD on FPGA and on ARM are comparable whereas the RSD is less than 1% if $n > 256$. Consequently, the variations of the average temporal stability of sorting algorithms are not significant for $n > 256$. Indeed, this is confirmed by the standard deviation, which is almost constant in Table 3 for all n. When focusing on the Shellsort – in order to save space in this paper – Table 8 shows that the percentage of total outliers is negligible since the percentage of upper outliers is below 1%, except for n=2048, but the total number of outliers remains below 2%. In conclusion, the number of outliers gives a fairly accurate additional measure of the worst-case temporal stability of the Shellsort on ARM even if some outliers are due to intrinsic noises generated by input/output operations and the Operating System while measuring computational times.

3.2.3   Comparison of the Computational Platforms

In this part, we notice that the temporal stability of the hardware implementation is better than that on ARM. Therefore, FPGA gives a better performance compared to ARM in terms of average and worst-case temporal stability. It is also worth noting that, when comparing software to hardware implementations, the communications-to-computations ratio should also be considered if all elements to be sorted are not directly available on the FPGA platform.

4   Conclusion

In this paper, we presented an optimized hardware implementation of sorting algorithms to improve the performance in terms of average and standard-deviation of computational times using 8 to 4096 elements, 47 permutations and 1000 replications. We used a High-Level Synthesis tool to generate the RTL design from behavioral description. On the basis of these results, we can conclude that Shellsort is 14–72% faster than other sorting algorithms running on the ARM Cortex A9 processor if $n > 64$, otherwise Timsort is the best. Furthermore, we propose the "IShellsort" composed of Insertionsort and Shellsort which appears to provide the best computational times on the ARM processor. In addition, Timsort is 1.16x–1.21x, 1.08x–1.22 and 1.25x–1.61x faster than Mergesort, Heapsort and Shellsort respectively running on FPGA if $n > 64$ otherwise Heapsort is the fastest. Also, we notice that the computational times of the hardware implementation is reduced compared to running on ARM. In conclusion, the FPGA platform gives better performance compared to the ARM Cortex A9 processor in terms of average and worst-case temporal stability.

As future work, we plan to use the hardware Timsort algorithm and the software IShellsort in our targeted avionics decision support system [14,18,20]. We have also performed preliminary researches and tests on a modified version of a "Bitonic Mergesort" to implement this sorting network on a FPGA while taking into account the specificities of the target application. For the moment, several improvements are still needed to compare this sorting network with other sorting algorithms. The next step is to include the "best sorting algorithm" in hardware path planning and PROMETHEE algorithms. The present work is also linked to other researches dedicated to the optimization of matching and scheduling on heterogeneous CPU/FPGA architectures [18] where efficient sorting algorithms are also required.

References

1. Kristo, A., Vaidya, K., Çetintemel, U.: The case for a learned sorting algorithm. In: Proceedings of the 2020 ACM SIGMOD International Conference on Management of Data, pp. 1001–1016 (2020)
2. Usmani, A.R.: A novel time and space complexity efficient variant of counting-sort algorithm. In: 2019 IEEE International Conference on Innovative Computing (ICIC), pp. 1–6 (2019)
3. Liu, B.: A data sorting hardware accelerator on FPGA. Ph.D. thesis, Kth Royal Institute of Technology (2020)
4. Zurek, D., Pietro'n, M., Wielgosz, M., Wiatr, K.: The comparison of parallel sorting algorithms implemented on different hardware platforms. Comput. Sci. **14**, 679–691 (2013)
5. Danelutto, M., De Matteis, T., Mencagli, G., Torquati, M.: A divide-and-conquer parallel pattern implementation for multicores. In: Proceedings of the 3rd International Workshop on Software Engineering for Parallel Systems, pp. 10–19 (2016)
6. Chhugani, J., et al.: Efficient implementation of sorting on multicore SIMD CPU architecture. In: Proceedings of the VLDB Endowment, pp. 1313–1324 (2008)
7. Ben Jmaa, Y., Ben Atitallah, R., Duvivier, D., Ben Jemaa, M.: A comparative study of sorting algorithms with FPGA acceleration by high level synthesis. Computacion y Sistemas, pp. 213–230 (2019)
8. Arena, F., Pau, G., Severino, A.: A review on IEEE 802.11 p for intelligent transportation systems. J. Sens. Actuator Netw. **9**, 22–33 (2020)
9. Grozea, C., Bankovic, Z., Laskov, P.: FPGA vs. multi-core CPUs vs. GPUs: hands-on experience with a sorting application. In: Keller, R., Kramer, D., Weiss, J.-P. (eds.) Facing the Multicore-Challenge. LNCS, vol. 6310, pp. 105–117. Springer, Heidelberg (2010). https://doi.org/10.1007/978-3-642-16233-6_12
10. Cong, J., Liu, B., Neuendorffer, S., Noguera, J., Vissers, K., Zhang, Z.: High-level synthesis for FPGAs: from prototyping to deployment. IEEE Trans. Comput.-Aided Des. Integr. Circ. Syst. **30**, 473–491 (2011)
11. Coussy, P., Gajski, D.D., Meredith, M., Takach, A.: An introduction to high-level synthesis. IEEE Des. Test Comput. **26**, 8–17 (2009)
12. Srivastava, A., Chen, R., Prasanna, V.K., Chelmis, C.: A hybrid design for high performance largescale sorting on FPGA. In: 2015 IEEE International Conference on ReConFigurable Computing and FPGAs (ReConFig), pp. 1–6 (2015)
13. Ben Jmaa, Y., Ali, K.M., Duvivier, D., Ben Jemaa, M., Ben Atitallah, R.: An efficient hardware implementation of Timsort and Mergesort algorithms using high level synthesis. In: 2017 IEEE International Conference on High Performance Computing & Simulation (HPCS), pp. 580–587 (2017)
14. Nikolajevic, K.: Dynamic autonomous decision-support function for piloting a helicopter in emergency situations. Ph.D. thesis, UPHF Valenciennes (2016)
15. Diallo, A., Zopf, M., Furnkranz, J.: Permutation learning via Lehmer codes. In: 24th European Conference on Artificial Intelligence, pp. 1095–1102 (2020)
16. Brans, J., Mareschal, B.: PROMCALC & GAIA: a new decision support system for multicriteria decision aid. Decis. Support Syst. (DSS) **12**, 297–310 (1994)
17. Brans, J., Vincke, P., Mareschal, B.: How to select and how to rank projects: the Promethee method. Eur. J. Oper. Res. **24**, 228–238 (1986)
18. Souissi, O., Ben Atitallah, R., Duvivier, D., Artiba, A.: Optimization of matching and scheduling on heterogeneous CPU/FPGA architectures. In: 7th IFAC Conference on Manufacturing Modelling, Management, and Control, Saint Petersburg (2013)

19. Ciura, M.: Best increments for the average case of ShellSort. In: Freivalds, R. (ed.) FCT 2001. LNCS, vol. 2138, pp. 106–117. Springer, Heidelberg (2001). https://doi.org/10.1007/3-540-44669-9_12

20. Ollivier-Legeay, H., Cadi, A.A.E., Belanger, N., Duvivier, D.: A 4D augmented flight management system based on flight planning and trajectory generation merging. In: Mohammad, A., Dong, X., Russo, M. (eds.) TAROS 2020. LNCS (LNAI), vol. 12228, pp. 184–195. Springer, Cham (2020). https://doi.org/10.1007/978-3-030-63486-5_21

21. Abdel-Hafeez, S., Gordon-Ross, A.: An efficient O (N) comparison-free sorting algorithm. IEEE Trans. Very Large Scale Integr. Syst. **25**, 1930–1942 (2017)

A Review on Recent NDN FIB Implementations for High-Speed Switches

Eduardo Castilho Rosa[1(✉)] and Flávio de Oliveira Silva[2]

[1] Goiano Federal Institute, Catalão, GO, Brazil
eduardo.rosa@ifgoiano.edu.br
[2] Federal University of Uberlândia, Uberlândia, MG, Brazil
flavio@ufu.br

Abstract. Forwarding Information Base (FIB) plays an essential role in Named-Data Networking (NDN) since it allows contents identified by unique hierarchical names to be reachable anywhere. Over the last few years, the advances in programmable switches have become possible to implement data structures for FIB in hardware to run at line rate. However, such implementations are not trivial in these devices, taking into account its architectural constraints and some NDN features like the complexity of dealing with variable-length names and the FIB size being orders of magnitude larger than the current IP routing tables. Despite all the benefits that high-speed switches may bring to NDN as a whole, the literature has been missing a survey that covers the data structures for FIB designed specifically to run in physical switches. To this end, we present a review on recent FIB implementations for both fixed-function and programmable high-speed switches. Our main contribution includes a fair and new comparative analysis among different approaches to implement the FIB highlighting its features and limitations. We also provide new insights and future research directions in this field.

1 Introduction

Information-Centric Networking (ICN) [1] is a clean-slate approach aiming to address the limitations of the current Internet. By using concepts like in-network caching, name-based forwarding, and data-centric security, ICN can better support emerging applications in Vehicular Ad Hoc Network (VANET), the Internet of Things (IoT), and 5G/6G.

Among several ICN-based architectures in literature, Named-Data Networking (NDN) [2] is currently the most mature. In contrast to IP, NDN forward packets by using hierarchical and variable-length names. The communication is based on the exchange of two types of packets: 1) Interest Packet (Ipckt) and 2) Data Packet (Dpckt). The former contains a unique name the consumer uses to retrieve the data. A Dpckt stored at any intermediate node is sent back to the consumer following the reverse path. The routing is performed through a data structure called Forwarding Information Base (FIB). FIB plays an essential role in NDN because it guarantees the reachability of any data content in the NDN

L. Barolli et al. (Eds.): AINA 2022, LNNS 451, pp. 288–300, 2022.
https://doi.org/10.1007/978-3-030-99619-2_28

domain. Such importance of FIB has been motivating the research community to adapt existing data structures like Trie, Bloom-Filter (BF), and Hash Tables (HT), to store name prefixes taking into account parameters such as lookup speed and memory footprint.

There are many FIB implementations in software such as NFD [3], NDN-DPDK [4], and YaNFD [5]. However, to scale up the NDN, it is necessary to move forward and implement its main data structures in hardware to enable forwarding packets at high speed. Recent developments in hardware and programmable switches make this possible, even though in most cases, native features in NDN such as caching and name-based forwarding do not fit naturally with the constraints of such devices like the limited on-chip SRAM/TCAM memory. Many solutions have been proposed recently to address these issues that use different techniques in several target architectures. However, although some surveys in the literature cover how NDN forwarding strategies are implemented in software [6–8], we still lack a comprehensive survey on specific FIB implementations for physical switches.

This work provides a review and a comparative analysis of recent FIB implementation for high-speed switches. Our goal is to answer the following questions: *What do we know about how to implement the FIB in both programmable and fixed-function switches? What is the most used method to do so?*, and *What are the gaps in this field?*. To answer these questions, the papers we reviewed are classified according to what switch technology the FIB is designed for and what method they used. We do not intend to provide a full systematic literature review. Instead, the idea is to present an overview of the most used methods to implement the FIB that is deployable in hardware and provide a fair comparison among relevant aspects of each paper, highlighting its main features and limitations.

The remain of this paper is organized as follows: In Sect. 2 we present a brief overview of the NDN architecture. In Sect. 3 we present 15 papers that focus on FIB implementations for both programmable and fixed-function switches. Section 4 brings a discussion and future research directions. Finally, we conclude the paper in Sect. 5.

2   NDN Overview

The NDN project was funded by the National Science Foundation (NSF) as part of the Future Internet Architecture Program. The central concept behind NDN is to replace the thin waist of today's Internet with a content-oriented network layer. In contrast to IP, the core communication of the NDN layer-3 protocol is based on the asynchronous exchange of Ipckts and Dpckts. The NDN Routers deliver contents based on hierarchical variable-length names carrying on both Ipckt and Dpckt. To do so, NDN Routers implements three data structures: Content Store (CS), Pending Interest Table (PIT), and Forwarding Information Base (FIB).

Overall, CS plays an essential role in providing a Quality-of-Service (QoS) to applications because it can significantly alleviate the traffic in the core network, improving both latency and throughput. The CS is responsible for storing

Dpckts in buffers to support in-network caching, one of the main features of an ICN-based architecture. In-network caching is used to ensure the independent-location property in NDN, seeing as not only the producer can serve the consumer with data but also any intermediate node storing a copy of it. The NDN specification foresees the need for CS to support a cache replacement policy to avoid overflowing the buffer capacity with too many Dpckts. The replacement policy uses the metadata information in Dpckt that includes the freshness time.

From a perspective of an incoming Ipckt, when the CS cannot find a Dpckt that matches exactly with the name in the Ipckt, the PIT table needs to be checked. The PIT stores all pending Ipckt that have been forwarded but not been satisfied yet, to avoid sending unnecessary information to the network. In other words, only the first Ipckt needs to be sent out to the network, and all the subsequent requests for the same data are aggregated in PIT together with the incoming port. Thus, similarly to CS, PIT can alleviate the traffic in the network.

FIB is the most important data structure in NDN because it is responsible for interconnecting consumers and producers worldwide. When a given Ipckt arrives at the NDN router for the first time, and both CS and PIT can not satisfy that request, FIB comes into play and forwards such Ipckt to one or more interfaces configured by a routing strategy. The FIB stores name prefixes and performs a Longest Name Prefix Matching (LNPM) whose definition is the same as the longest prefix matching of IP, except in FIB, the entries are variable length. Even though FIB can be implemented as HT [9] and BF [10], trie-based data structures are primarily used to reduce memory consumption in software, as we can see in [11–13]. Figure 1 shows how CS, PIT, and FIB are connected.

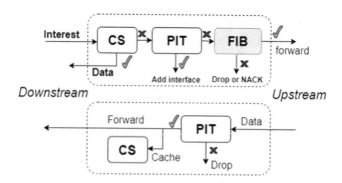

Fig. 1. Operational flow in the NDN forwarding plane and its main data structures.

3   FIB Implementations in Hardware

The FIB implementations we present in this review are classified into two categories: 1) FIB designed for Fixed-Function Switches, and 2) FIB designed for

Programmable Switches. In each category, we classify the papers according to their methods (HT, Trie, and BF) and present the publications in chronological order. Both fixed and programmable switches mean that the proposed mechanisms can be deployed either in traditional routers or in real programmable devices, not necessarily meaning they were implemented in such real devices. As we will see, some of them were implemented in software like BMv2 [14], others in FPGAs, and some of them the authors provide simulation analyses.

3.1 FIB for Fixed-Function Switches

3.1.1 Hash Table

The first content router that supports name-based forwarding at high speed is Caesar [15]. Caesar distributes the FIB across multiple line cards and performs the LNPM accordingly. To optimize memory allocation, rather than duplicating the same forwarding table with S entries at each line card, each line card stores a different subset of entries S' such that, summing up all of them, we get S. The downside of this approach is a possible increase in switching operations and complexity that may harm the performance. The authors suggest using CRC-64 as a hash function to store the prefixes into the FIB. A numerical evaluation of Caesar is performed by using the Xilinx Virtex-6 FPGA family as reference design. The main metric evaluated is the number of prefixes supported that depends on the number of line cards. A Caesar router with 600 line cards handles up to 600 million content prefixes.

The authors in [16] proposed a framework focusing on LNPM in NDN. Such a framework consists of two key components: a name prefix transformation and FIB instrumentation. When a router receives a name prefix announcement, it first applies the name reduction algorithm to transform the hierarchically structured name prefix into a compressed key. This transformation is achieved through hashing. For instance, if they have the NDN name /ufu/facom/mehar, they generate three keys by hashing /ufu, /ufu/facom, and /ufu/facom/mehar. As we can see, the redundant information (/ufu, /ufu/facom) is incorporated into the keys, which harms memory consumption. On the other hand, such an approach can easily be adapted to run in programmable switches. The evaluation was performed in both CPU and GPU and it used three name datasets. To measure the performance of the LNPM algorithm, the throughput was the only metric evaluated.

SACS [17] is another method that focuses on LNPM, like [16]. It consists of a shape and content search framework for TCAM and SRAM. In SACS, a TCAM-based shape search module is first used to determine a subset of possible matching name prefixes, and then an SRAM-based content search module is used on the subset to find the longest matching prefix. The shape of a given prefix is a sequence of its lengths. For instance, the shape for the prefix /a/ab/abc is /1/2/3. These shapes are stored in TCAMs, pointing to hash tables located in SRAM. Compared to traditional hash-based methods, such an approach improves latency because many memory accesses are eliminated since the shape guides the search to a small subset of keys. However, SACS requires more memory than NFD [3], for instance, because it needs to maintain dual fingerprints and cells in a slot of hash tables. The experiments were conducted on a set of public name datasets and a public TCAM modeling tool [18] were used to test the performance of TCAM-based shape search. For the SRAM-based content search, the experiments were conducted on a server with Intel Xeon CPU E5-2640x2 and 94 GB of main memory. The main metrics evaluated were throughput and memory consumption.

3.1.2   Trie

Name Component Encoding (NCE) [19] is a mechanism to reduce memory consumption by assigning a unique code to a given name component in such a way that LNPM can be applied with the same semantics. For instance, the NDN names /ufu/facom/mehar and /facom/lsi can be converted to /1/2/3 and /2/4, respectively, by mapping a name component into an integer. The encoded name is then stored in the FIB implemented as a trie-based data structure. The LNMP is performed starting from the root to the full NDN name. The main limitation of NCE is the frequent access to a slow memory and the extra time to perform the encoding. Theoretical analysis and experiments conducted by using real dataset on a PC with an Intel Core 2 Duo CPU of 2.8 GHz and DDR2 SDRAM of 8 GB demonstrate that NCE can compress a FIB containing 3M entries to about 272 MB, 32.45% less than the baseline. Packet delay and average packet lookup time are also evaluated. For the same dataset, the former correspond to 1.9 us (7.7 faster than baseline) and the ladder 2,975.26 cycles. The benefits in terms of memory reduction and latency in NCE comes at the expense of a building time of 34 s to encode the prefixes.

The second trie-based method proposed to fixed-function switches is [20]. Its main design goal is to have a compact data structure so that FIB with a few million entries can still fit in SRAM for fast lookup. Thus, the authors propose to use dual binary Patricia trie to minimize the redundant information stored. The binary representation provides more opportunities to compress shared parts between different prefixes. On the other hand, binary tries tend to increase the depth of the tree, impacting the lookup speed. Moreover, they introduce the idea of speculative forwarding, which uses the Longest-Prefix Classification (LPC) instead of the Longest Prefix Match (LPM). Unlike the LPM, LPC lookup guarantees that if there is a match, the packet will be forwarded to the same next hop that LPM would use but, if there is no match, the packet is still forwarded. However, for speculative forwarding, the imprecise forwarding of the LPC may cause forwarding loops. The evaluation is performed analytically and the metric used is the memory footprint. For a dataset containing about 3.7M name prefixes, the memory footprint is around 31 MiB, 50% less compared to tokenized Patricia trie.

3.1.3   Bloom-Filter

An Adaptive Prefix Bloom Filter (NLAPB) is presented in [21]. The key idea of NLAPB is to split NDN prefixes into two segments and conduct the lookup operation with a combination of CBF (Counting Bloom Filter) and trie. The first segment contains prefixes with m name components (B-prefix) stored in BF, whereas the second (T-suffix) contains variable-length prefixes stored in trie. NLAPB is implemented in a commodity router equipped with eight interfaces. The experiments illustrate that NLAPB achieves a fairly guaranteed scalability in terms of memory consumption when huge namespaces are considered. The limitation of NLAPB includes the possibility of false positives, which can impair the forwarding. Memory cost, lookup processing rate and false positive rate are the main metrics measured.

MaFIB [22] is another BF-based data structure for FIB. It uses a data structure called Mapping Bloom Filter (MBF), proposed in [23]. The MBF consists of a regular BF and a Mapping Array (MA), which are bit arrays in fast on-chip memory (SRAM). With the BF, the elements are verified whether they are in the MBF or not. The value in MA is utilized as the offset address to access the output face(s) stored in slow off-chip memory (DRAM). In comparison with [19] and other methods, MaFIB can provide a better compression ratio. However, its main limitations include frequent high access to DRAM to extract the output faces(s), and false positives may occur, impairing the accuracy of the forwarding. The performance of MaFIB is compared with the FIB based on NCE [19] in terms of the on-chip memory consumption, false positive probability, and building time.

B-MaFIB [24] is an improvement of MaFIB. It has the same features as MaFIB, but the authors changed MBF by using a bitmap-mapping bloom filter (B-MBF) index. The idea of B-MBF is to allow dynamic memory allocation to reduce memory consumption. However, B-MaFIB still needs to access slow DRAM to read the packet information, such as output port(s) and stale time. Besides the evaluated metrics in Ma-FIB [22], B-MaFIB also includes throughput.

3.2 FIB for Programmable Switches

3.2.1 Hash Table

NDN.p4 [25] is the first attempt to implement the NDN in the P4 language. They considered the original TLV name encoding to parse the NDN name content, although they concluded the P4 language restricts many operations on this kind of field. The FIB in NDN.p4 is implemented as a single P4 table with fixed-size entries. The number of table entries for a single FIB entry is proportional to the maximum number of components in processed packets. For instance, to store a prefix /ufu/facom/mehar, they calculate the hash of each part of the name, which is h(/ufu), h(/ufu/facom), and h(/ufu/facom/mehar), similarly to [16], and store all of them in the FIB. The LNPM is performed very quickly, in only one cycle, by using a ternary-matching operation. However, in a realistic scenario, this process can lead to inefficient memory use as the need to store m entries in the FIB for only one name prefix, where m is the maximum number of components. Also, the collision problem can increase memory consumption significantly.

To improve the NDN.p4, the authors in [26] designed an NDN router to address the scalability issues of the proposed FIB in NDN.p4 and to extend the NDN functionality, including the CS and multicast-capability. Similar to

NDN.p4, the FIB in [26] is implemented using only one P4 table. To deal with the problem that names have variable length, they proposed a data structure called *hashtray*. The *hashtray* is constructed from a name of *max* components, and it is divided into *max* blocks, with each block i containing the result of the hash of component i. In contrast to NDN.p4, such a technique requires a single table entry per FIB route, significantly improving memory usage. However, using a 16-bit hash function to construct the *hashtray* can cause lots of collisions. Furthermore, to store the *hashtrays* in the FIB properly, it is necessary padding it from the index i up to the index *max*, where i is the current number of name components in the name, and *max* is the maximum name components supported by the architecture.

A similar approach based on hashtrays is presented in [27]. However, they focus on the NDN routing mechanism primarily instead of forwarding. In addition to the limitations of a hashtray-based method [26,27] stores the only single FIB table at the egress pipeline. As SRAM/TCAM are distributed equally across both ingress and egress pipelines, such an approach wastes half of the on-chip memory available in the switch.

The first attempt to implement the NDN router in a real programmable ASIC is [28]. The key idea is that Dpckts are forwarded by a switch ASIC alone, whereas an NDN engine forwards Ipckts at a server. The main limitation of such an approach is that FIB implemented in DRAMs can severely impact the latency. The data structure used to implement the FIB is HT. However, the authors do not provide any details about it. In terms of memory usage, the PIT is distributed across many different stages of pipeline to optimize TCAM and SRAM resources. However, there are a couple of open research issues such as re-designing the integrity maintenance method of PIT entries between pipelines, using the server as testers is not a good solution since the number of ports on a router is more than 100, and the degradation from the ideal throughput due to smaller Dpckt sizes.

3.2.2   Trie

On-Chip-FIB [29] is a FIB implemented as a Binary Search Tree (BST) in FPGA. Although FPGA is a P4-capable target [30], the OnChip-FIB was not designed to be P4 compatible. The content name is represented as a collection of strides of the same size. For instance, the name /ufu/facom is represented as '/ufu', '/fac' and 'om', using stride size four. Those strides are inserted in the BST, and the LNPM is performed by traversing the BST starting from the root node down to a leaf node. In terms of memory consumption, the complexity of OnChip-FIB is $O(n)$, where n is the number of names in the dataset. However, in the worst case, OnChip-FIB is $O(nk)$, where k is the number of strides. When it comes to lookup speed, the time complexity of OnChip-FIB is $O(n)$, where n is the name size.

A second trie-based implementation of the FIB is presented in ENDN [31]. In summary, ENDN extends the previous NDN data plane with a new P4 target architecture. The authors claim this novel architecture allows multiple isolated

P4 forwarding functions to be defined and executed. They also extend the P4 language with several extern functions to enable the processing of strings. The FIB in ENDN (EFIB) is implemented as a data structure called FCTree [32]. FCTree uses trie and can compress name prefixes storing common sub-prefixes only one time. However, as FCTree is proposed to run in software, the authors do not explain how FCTtree is implemented in P4. ENDN uses a combination of trie and hashing because the authors add a function that takes a name component position in parameter and returns the hash of the component, which is used to query match tables and thus perform a specific P4 action based on the value of a content name component. The time complexity of LNPM, insertion, and deletion is $O(n)$, which is the main limitation of FCTree. Also, when it comes to implementing FCTree in P4, which the authors did not address, is how to use a self-balancing binary search tree taking into account the constraints that usually P4 targets impose.

3.2.3   Bloom-Filter

Finally, in [33] is proposed an FPGA-based FIB implementation for programmable switches. Just like [29], the mechanism in [33] is not P4-compliant. However, in contrast to [29], in which all FIB entries are stored in the same data structure (BST), in [33] the name prefixes in the FIB are divided into two distinct groups: 1) Group 1: Name prefixes with up to 4 components and 2) Group 2: Name prefixes with five or more components. The Group 1 name prefixes are stored in an external lookup table, i.e., in software, and the Group 2 name prefixes are stored in on-chip memory. Such division makes sense considering the limited amount of on-chip memory available in today's FPGA boards. The main data structure used is BF [34] combined with a hash function called H_3 to store the prefixes into the FIB on an FPGA. The main limitation of this approach is the high latency caused by the external name lookup. A proof of concept system is implemented on a virtex-7 FPGA where both memory consumption and packet processing rate are the metrics measured.

4   Discussion and Future Directions

Table 1 shows a preference for implementing the FIB by using HT weather in fixed-function or programmable switches. The main reason is probably the simplicity of calculating hashing in such devices since there are several dedicated chips to do so at line rate. Also, hashing optimizes the latency, which fits well with high-speed switches. However, when it comes to FIB in NDN, hashing increases the memory footprint due to the redundancies in the name structure and the space necessary to deal with collisions [35].

On the other hand, BF is also a compelling data structure but is less used. False positives and large memory to deal with it could explain that. Considering that the most memory-efficient methods reported in this review are MaFIB [22] and B-MaFIB [24], both using BF as its primary data structure, such techniques should be better explored in future research.

Table 1. High-level comparison among FIB implementations for high-speed switches.

Cat.	M	Ref.	Year	Key features	Limitations	Target
F	HT	[15]	2014	• Distributes FIB across multiple line cards • FIB entries are stored by us-ing CRC-64 hash	• High #of switching operations, which impacts the latency	FPGA
		[16]	2014	• Focus on LNPM • Transform the name into a compressed key	• Redundant information increases memory consumption	GPU
		[17]	2018	• Stores shape of FIB entry • Content search framework • Low #of memory access	• Need to maintain dual fingerprint which increases memory usage	TCAM
	Trie	[19]	2012	• Compress name prefixes • Component trie • Use State Transition Arrays	• Frequent access to slow memory • Encoding process increases the latency	CPU
		[20]	2015	• Speculative forwarding • Longest-Prefix Classification	• Forwarding loop may occur • Higher depth of the tree($>$latency)	SRAM
	BF	[21]	2014	• Split NDN prefixes into two segments • Lookup based on popularity	• False positives may occur • Dynamic memory allocation is hard to achieve in hardware	Router
		[22]	2017	• Low on-chip memory cost • Support update operations • Stores large name datasets • Low building time	• Frequent access to DRAM • High memory usage due to CBFs in off-chip memory • Lots of hash functions executed in the dataplane	SRAM DRAM
		[24]	2018	• Idem [22] • Leverage only one hash function in the data plane • Dynamic memory allocation • Use of Bitmap as BF	• Frequent access to DRAM to access output faces increase latency	SRAM DRAM
P	HT	[25]	2016	• FIB implemented as a single P4 table • #table entries are proportional to #name components • LNPM executed at line rate	• Waste of memory because many tables entries are inserted for each NDN name prefix • Not scale to millions of prefixes	BMv2
		[26]	2018	• FIB implemented as a single P4 table • FIB can scale better • Use of hashtrays in TCAMs	• Use of 16-bit hash functions can cause lots of collisions, requiring more scarce TCAMs	BMv2
		[27]	2021	• Idem [26] • Focus on routing	• Idem [26] • A large P4 table is used at egress wasting TCAM/SRAM at ingress	BMv2
		[28]	2021	• FIB is stored in DRAM • Optimize memory usage by splitting the PIT into stages	• LNPM in DRAM adds to latency	ASIC
	Trie	[29]	2019	• Content name is represented as strides • Exploits the massive parallel processing of FPGAs	• Worst-case memory complexity is $O(nk)$ • Transverse the trie is time consuming	FPGA
		[31]	2020	• Extern to process strings • Support of wildcards searches • Compress prefixes whose share common sub-prefixes	• NLPM, insertion and deletion is $O(n)$ impacting the latency • Supporting self-balance trees is costly in P4 targets	BMv2
	BF	[33]	2019	• Prefixes are splitted into 2 groups	• External lookup increases latency	FPGA

(Cat = Category;   F = Fixed-Function Switches;   P = Programmable Switches; M = Method; HT = Hash Table; BF = Bloom-Filter)

Trie-based methods are widely adopted in software implementations of the FIB [6] but, as Table 1 shows, when it comes to FIB implemented in hardware, there are just a few in literature. All of them were implemented in FPGAs and software switches like BMv2, which are more flexible but less restrictive than ASICs. The main reason for that resides in the limitations in hardware such as no-loops, no pointers, and no dynamic memory allocation that all trie-based methods rely upon.

Since HT, tries, and BF are the methods used in all the papers we reviewed and also since such methods are widely used in software as well [6], one possible research direction is to come up with solutions to implement such data structures in hardware overcoming its limitations and constraints. Given the fact that some programmable ASICs like Tofino have been open-source recently [36] and the open P4 specification allows the definition of new target architectures, researchers can even develop new hardware design in the future to better support the implementation of the data structures aforementioned.

Finally, from all papers we reviewed, we can also observe that [28] was the only one that provided a prototype of an NDN router in an actual programmable switch (Tofino ASIC). However, only the PIT data structure is fully implemented in the hardware. Both CS and FIB are stored externally in DRAMs, which may impact the performance negatively. Given the crucial role that FIB plays in NDN, this gap needs to be filled. Therefore, efficient implementations of FIB and CS in the ASIC besides HT, tries, and BF is a worthy topic of research with excellent opportunities to innovate, mainly because P4 language is a trend now, and NDN has been gaining momentum over the last few years due to its standardization process at NIST [37]. Since on-chip memory in today's programmable ASICs is minimal but, at the same time, the scalability of NDN depends on processing packets at Tbps, the study of new compression techniques that takes into account both the flexibility of current programmable pipelines and the hardware constraints is a good direction towards filling this gap. Such an approach goes beyond the scope of NDN and can benefit all fields in computing that use cache-based systems implemented in hardware that makes use of string processing. Examples of such use-cases may include Directory Caches in Operating Systems (DCache), Content Delivery Networks (CDN), Domain Name System (DNS), and many others.

5 Conclusion

Forwarding Information Base (FIB) is one crucial data structure in Named-Data Networking (NDN) that aims to provide connectivity between consumers and producers. This work reviewed some recent FIB implementations for both fixed and programmable switches. Our observation is that implementing FIB data structure in hardware is quite challenging due to matching NDN rules to switch constraints and limitations. A total of 15 reviewed papers have shown that Hash Table is the most used method to implement the FIB in hardware, followed by Trie and Bloom-Filter. Such data structures are compelling, but, at the same time, they were not designed for the specific purpose of FIB.

Therefore, we argue that the research community needs to fill this gap and come up with new data structures for FIB that are aware of the architecture constraints while providing a good balance between memory footprint and lookup speed.

References

1. Ahlgren, B., Dannewitz, C., Imbrenda, C., Kutscher, D., Ohlman, B.: A survey of information-centric networking. IEEE Commun. Mag. **50**(7), 26–36 (2012)
2. Zhang, L., Estrin, D., Burke, J.: Named data networking (NDN) project. University of California, Los Angeles, CA, NDN Project Technical Report NDN-0001 (October 2010)
3. Afanasyev, A., Shi, J.: NFD Overview - Named Data Networking Forwarding Daemon (NFD) 0.6.6-24-g1402fa1 documentation (2019). http://named-data.net/doc/NFD/current/overview.html
4. Shi, J., Pesavento, D., Benmohamed, L.: NDN-DPDK: NDN forwarding at 100 gbps on commodity hardware. In: Proceedings of the 7th ACM Conference on Information-Centric Networking, ICN 2020, pp. 30–40. Association for Computing Machinery, New York (2020). https://doi.org/10.1145/3405656.3418715
5. Newberry, E., Ma, X., Zhang, L.: YaNFD: yet another named data networking forwarding daemon, pp. 30-41. Association for Computing Machinery, New York (2021). https://doi.org/10.1145/3460417.3482969
6. Li, Z., Xu, Y., Zhang, B., Yan, L., Liu, K.: Packet forwarding in named data networking requirements and survey of solutions. IEEE Commun. Surv. Tut. **21**(2), 1950–1987 (2019)
7. Tariq, A., Rehman, R.A., Kim, B.-S.: Forwarding strategies in NDN-based wireless networks: a survey. IEEE Commun. Surv. Tut. **22**(1), 68–95 (2020)
8. Majed, A., Wang, X., Yi, B.: Name lookup in named data networking: a review. Information **10**(3) (2019). https://www.mdpi.com/2078-2489/10/3/85
9. Shubbar, R., Ahmadi, M.: Efficient name matching based on a fast two-dimensional filter in named data networking. Int. J. Parallel Emergent Distrib. Syst. **34**(2), 203–221 (2019)
10. Muñoz, C., Wang, L., Solana, E., Crowcroft, J.: I(fib)f: iterated bloom filters for routing in named data networks. In: 2017 International Conference on Networked Systems (NetSys), pp. 1–8 (2017)
11. Saxena, D., Raychoudhury, V.: N-FIB: scalable, memory efficient name-based forwarding. J. Netw. Comput. Appl. **76**, 101–109 (2016)
12. Ghasemi, C., Yousefi, H., Shin, K.G., Zhang, B.: A fast and memory-efficient trie structure for name-based packet forwarding. In: 2018 IEEE 26th International Conference on Network Protocols (ICNP), pp. 302–312 (2018)
13. Seo, J., Lim, H.: Bitmap-based priority-NPT for packet forwarding at named data network. Comput. Commun. **130**, 101–112 (2018). https://www.sciencedirect.com/science/article/pii/S0140366417303298
14. Bas, A., Fingerhut, A., Sivaraman, A.: The behavioral model (May 2021). https://github.com/p4lang/behavioral-model. Accessed 26 Jan 2015
15. Perino, D., Varvello, M., Linguaglossa, L., Laufer, R., Boislaigue, R.: Caesar: a content router for high-speed forwarding on content names. In: 2014 ACM/IEEE Symposium on Architectures for Networking and Communications Systems (ANCS), pp. 137–147 (2014)

16. Li, F., Chen, F., Wu, J., Xie, H.: Longest prefix lookup in named data networking: how fast can it be? In: 2014 9th IEEE International Conference on Networking, Architecture, and Storage, pp. 186–190 (2014)
17. Huang, K., Wang, Z.: A hybrid approach to scalable name prefix lookup. In: 2018 IEEE/ACM 26th International Symposium on Quality of Service (IWQoS), pp. 1–10 (2018)
18. Agrawal, B., Sherwood, T.: Ternary cam power and delay model: extensions and uses. IEEE Trans. Very Large Scale Integr. (VLSI) Syst. **16**(5), 554–564 (2008)
19. Wang, Y., et al.: Scalable name lookup in NDN using effective name component encoding. In: 2012 IEEE 32nd International Conference on Distributed Computing Systems, pp. 688–697 (2012)
20. Song, T., Yuan, H., Crowley, P., Zhang, B.: Scalable name-based packet forwarding: from millions to billions. In: Proceedings of the 2nd ACM Conference on Information-Centric Networking, ACM-ICN 2015, pp. 19–28. Association for Computing Machinery, New York (2015). https://doi.org/10.1145/2810156.2810166
21. Quan, W., Xu, C., Guan, J., Zhang, H., Grieco, L.A.: Scalable name lookup with adaptive prefix bloom filter for named data networking. IEEE Commun. Lett. **18**(1), 102–105 (2014)
22. Li, Z., Liu, K., Liu, D., Shi, H., Chen, Y.: Hybrid wireless networks with FIB-based named data networking. EURASIP J. Wirel. Commun. Netw. **2017**(1), 54 (2017). https://doi.org/10.1186/s13638-017-0836-0
23. Li, Z., Liu, K., Zhao, Y., Ma, Y.: MaPIT: an enhanced pending interest table for NDN with mapping bloom filter. IEEE Commun. Lett. **18**(11), 1915–1918 (2014)
24. Li, Z., Xu, Y., Liu, K., Wang, X., Liu, D.: 5G with B-MaFIB based named data networking. IEEE Access **6**, 30501–30507 (2018)
25. Signorello, S., State, R., François, J., Festor, O.: NDN.p4: programming information-centric data-planes. In: 2016 IEEE NetSoft Conference and Workshops (NetSoft), pp. 384–389 (2016)
26. Miguel, R., Signorello, S., Ramos, F.M.V.: Named data networking with programmable switches. In: 2018 IEEE 26th International Conference on Network Protocols (ICNP), pp. 400–405 (2018)
27. Guo, X., Liu, N., Hou, X., Gao, S., Zhou, H.: An efficient NDN routing mechanism design in p4 environment. In: 2021 2nd Information Communication Technologies Conference (ICTC), pp. 28–33 (2021)
28. Takemasa, J., Koizumi, Y., Hasegawa, T.: Vision: toward 10 Tbps NDN forwarding with billion prefixes by programmable switches, pp. 13–19. Association for Computing Machinery, New York (2021). https://doi.org/10.1145/3460417.3482973
29. Saxena, D., Mahar, S., Raychoudhury, V., Cao, J.: Scalable, high-speed on-chip-based NDN name forwarding using FPGA. In: Proceedings of the 20th International Conference on Distributed Computing and Networking, ICDCN 2019, pp. 81–89. Association for Computing Machinery, New York (2019). https://doi-org.ez34.periodicos.capes.gov.br/10.1145/3288599.3288613
30. Wang, H., et al.: P4FPGA: a rapid prototyping framework for P4. In: Proceedings of the Symposium on SDN Research, Santa Clara CA USA, April 2017, pp. 122–135. ACM. https://dl.acm.org/doi/10.1145/3050220.3050234
31. Karrakchou, O., Samaan, N., Karmouch, A.: ENDN: an enhanced NDN architecture with a P4-programmabie data plane. In: Proceedings of the 7th ACM Conference on Information-Centric Networking, ICN 2020, pp. 1–11. Association for Computing Machinery, New York (2020). https://doi.org/10.1145/3405656.3418720

32. Karrakchou, O., Samaan, N., Karmouch, A.: FCTrees: a front-coded family of compressed tree-based FIB structures for NDN routers. IEEE Trans. Netw. Serv. Manage. **17**(2), 1167–1180 (2020)

33. Yu, W., Pao, D.: Hardware accelerator for FIB lookup in named data networking. Microprocess. Microsyst. **71**, 102877 (2019)

34. Luo, L., Guo, D., Ma, R.T.B., Rottenstreich, O., Luo, X.: Optimizing bloom filter: challenges, solutions, and comparisons. IEEE Commun. Surv. Tut. **21**(2), 1912–1949 (2019)

35. Rosa, E.C., Silva, F.O.: A hash-free method for FIB and LNPM in ICN programmable data planes. In: 2022 International Conference on Information Networking (ICOIN), pp. 186–191 (2022)

36. Open Tofino (December 2021). https://github.com/barefootnetworks/Open-Tofino. Accessed 14 Oct 2020

37. NDN Community Meeting (June 2020). https://www.nist.gov/news-events/events/2020/09/ndn-community-meeting. Accessed 23 Sept 2020

Formal Specification of a Team Formation Protocol

Rajdeep Niyogi[✉]

Indian Institute of Technology Roorkee, Roorkee, India
rajdeep.niyogi@cs.iitr.ac.in

Abstract. A high level specification provides the behavioral aspects of a protocol, i.e., the functional or logical properties. Such a specification should say *what* a protocol is allowed to do and not *how* it is achieved or implemented. State transition systems are mostly used to specify behavior of the agents (or robots) and temporal logic formulas are used to specify desirable properties of a system. TLA$^+$ is a formal specification language designed to provide high level specifications of concurrent and distributed systems. We provide a formal specification of a team formation protocol using TLA$^+$. TLC model checker is used to verify that the TLA$^+$ specification satisfies some desirable properties of the protocol.

1 Introduction

A team formation protocol is a communication protocol where multiple agents interact to form a team. A team formation protocol is an instance of a concurrent program. According to Lamport "Concurrent systems are not easy to specify. Even a simple system can be subtle, and it is often hard to find the appropriate abstractions that make it understandable. Specifying a complex system is a formidable engineering task [1]." A method is developed that "provides a conceptual and logical foundation for writing formal specifications. The method determines *what* a specification must say [1]." The main idea of the method is the following model.

The Model [2]: The execution of a concurrent program can be represented as a sequence of state transitions of the form $s \xrightarrow{a} s'$ where the action a took the program from state s to state s'. Thus we have a set S of states, a set A of actions, and a set Σ of program executions of the form $s_0 \xrightarrow{a_1} s_1 \xrightarrow{a_2} s_2 \cdots$, where each state denotes a global state of the system.

This model is the basis for the formal specification language TLA$^+$ [17]. We developed a Team Formation Protocol (TFP) [24–26] that is similar to auction-based protocols for team formation [3,4] but unlike these protocols TFP is state based. This means that the robot's internal state determines whether it will respond to a request message seeking possible team members for a task execution. In an auction-based protocol [3,4] such a notion of a state of an agent is

L. Barolli et al. (Eds.): AINA 2022, LNNS 451, pp. 301–313, 2022.
https://doi.org/10.1007/978-3-030-99619-2_29

irrelevant. TFP is expressed using communicating automata [24] and distributed algorithms [25, 26]. In this paper we consider the problem of formal specification of TFP using TLA$^+$. In the remainder of this paper, team formation protocol would refer to [24–26].

A high level specification [5, 17] allows reasoning of a software above code level and of a hardware above circuit level. The specification provides the behavioral aspects of a program/protocol, i.e., the functional or logical properties. In essence, such a specification should say *what* a program is allowed to do and not *how* it is achieved or implemented. In this paper we make an attempt to specify the logical properties of the executions of the team formation protocol that result in team formation.

Formal specification languages, like Alloy [11], TLA$^+$ [17], VDM [12], Z [21], have been used to specify software requirements. TLA$^+$ [17], a formal specification language developed by Lamport, provides a notation to specify a system as a set of actions. It was designed to provide high level specifications of concurrent and distributed systems. The elegance of TLA$^+$ lies in the fact that a system can be described by a single formula, that is a combination of first order predicate calculus and a temporal logic with only the "always" operator. TLA$^+$ comes with a model checker TLC that can verify any TLA$^+$ specification. TLA$^+$ is also widely used in industry [5, 20].

We need a formal specification language for our work that enables modelling distributed systems, has good tool support, and is simple to use. TLA$^+$ fits our requirements perfectly well. So, in this paper we chose TLA$^+$ for specifying the requirements and desirable properties (invariants) of executions of TFP that result in team formation.

The remaining part of the paper is organized as follows. Related work is discussed in Sect. 2. The team formation protocol is given in Sect. 3. A brief discussion of TLA$^+$ is given in Sect. 4. Modeling of the protocol in TLA$^+$ is given in Sect. 5. Conclusions are given in Sect. 6.

2 State of the Art and Related Work

According to the recent survey [19] on formal specification and verification of autonomous robotic systems, state transition systems are mostly used to specify the behavior of agents (or robots) and some temporal logic formulas are used to specify desirable properties. We discuss some works that relate to robot teams.

In [6] a formal specification language STOKLAIM [9] (stochastic version of KLAIM [8]) is used to specify the behaviors of robots that collectively transport an object. The specification provides low-level details of the robots, obstacles, and the goal. Behaviors are expressed as processes in KLAIM. Examples of some processes considered in [6] are: a process that provides each robot information about the neighboring obstacles, and a process that periodically updates the robots' positions according to their directions. The properties of interest are specified using the stochastic logic MoSL [9]. The model checking tool SAM [9, 18] is used for quantitative analysis of the MoSL formulas for the STOKLAIM specification. Team formation is not considered in [6].

In the domestic assistant robot model [22,23], SPIN model checker [10] has been used for verification of each individual safety property of the robots. SPIN allows verification of qualitative properties, ones without probability and explicit time. PRISM (probabilistic symbolic model checker) model checker [15] has been used in [7,13,14] for some swarm robotics applications. PRISM allows verification of quantitative properties. Discrete-time Markov chains (DTMCs) are used in [7] to model the aggregation algorithm and the properties of swarm behavior are specified using probabilistic computation tree logic* (PCTL*). DTMCs are used in [13,14] to model the foraging scenario and safety properties are specified using PCTL. A communicating automata based model is used to capture the behavior of robots [24].

Based on the survey given in [19] and to the best of our knowledge, there are no published works that deal with formal specification of a team formation protocol using formal notations like Alloy, VDM, TLA$^+$. In this paper we give a first formal specification of the team formation protocol in TLA$^+$ and verify some properties of the protocol using the TLC model checker.

3   Team Formation Protocol

We first describe the team formation process. Consider a closed environment that can be represented as a grid of size $m \times m$ units. There are some objects placed in some cells of the grid. There are some robots that are moving in the grid. The task of the robots is to clear the grid of these objects collaboratively. The robots can communicate only via exchanging messages.

We now discuss the team formation protocol for exchanging messages among the robots, where communication is assumed to be lossless and message delay is arbitrary but finite. We call robot A initiator (one that initiates the team formation process) and the other robots that participate in the team formation are called non-initiators. Assume that there are available three other robots B,C,D within the vicinity of A. Each robot has at least one of the three capabilities that are required for executing a task. The robot A *broadcasts* a request message that contains the details of the task (name, location, capabilities). As the robots B,C,D are in the vicinity of A, so all of them will receive the request message since there is no loss of messages. Now B,C,D send willing messages to A that contains their current location, cost, capabilities. When A *receives* these willing messages it selects the best possible team. Assume that A *selects* B and C and *rejects* D. Now A *notifies* its decision to the non-initiators by sending confirm message to B,C and notRequired message to D. The actions italicized above (i.e., *broadcast, receive, select, notify*) are the ones abstracted in the TLA$^+$ specification given in Sect. 5. Note that in this protocol no message is sent more than once by a robot.

At different points of the execution of the protocol the robots are assumed to be in different states. For instance, the initiator is in ready state when it makes the broadcast. All the non-initiators are in idle state before receiving the request message. Upon receipt of the request message, their states change to promise. After receiving the notifications B,C change their state to busy, whereas D changes its state to idle. At the same time, A makes its state busy.

The team formation process does not guarantee that a team would always be formed. We now illustrate such a situation. The initiator after broadcasting the request message waits for some unit of time Δ. It records all the willing messages received within Δ and ignores any willing message that arrives after Δ. If sufficient number of willing messages are not received by the initiator within Δ, then team formation would be unsuccessful.

4   TLA$^+$: A Brief Introduction

The following description of TLA$^+$ is adapted from [16,17], where the keywords of TLA$^+$ are written in uppercase sans serif font. We describe briefly the notions that would be useful for understanding the model given in Sect. 5. We refer to [16,17] for a comprehensive account of TLA$^+$.

A state (global state) is an assignment of values to every variable. A behavior is any infinite sequence of states. A system is specified by specifying a set of possible behaviors–the ones representing a correct execution of the system.

An initial predicate $Init$ specifies the possible initial values of variables. The next-state relation $Next$ specifies the relation between their values in the current state and their possible values in the next state. A pair of successive states $\langle currentstate, nextstate \rangle$ is called a step. Thus $Next$ specifies how the values of the variables change in any step. If v denotes a value in the current state, then v' (v prime) denotes a value in the next state. $Next$ is a formula that contains primed and unprimed variables. It also contains the keyword UNCHANGED to specify the variables whose values remain unchanged. So UNCHANGED v is equivalent to writing $v' = v$. Such a formula is called an action that is true or false of a step.

A TLA$^+$ specification is a single formula. This formula must assert about a behavior that (i) its initial state satisfies $Init$, and (ii) each of its steps satisfies $Next$. We express (i) as the formula $Init$, which we interpret as a statement about behaviors to mean that the initial state satisfies $Init$. To express (ii), we use the temporal-logic operator $\Box$ (always), that is the *only* temporal operator in TLA$^+$. The temporal formula $\Box P$ asserts that formula P is always true. In particular, $\Box Next$ is the assertion that $Next$ is true for every step in the behavior. So, $Init \wedge \Box Next$ is true of a behavior if and only if the initial state satisfies $Init$ and every step satisfies $Next$.

A step that leaves one or more variables unchanged is called a stuttering step. In TLA$^+$, $[Next]_v$ denotes $Next \vee (v' = v)$. In TLA$^+$ a "safety specification" has the form $Init \wedge \Box [Next]_{vars}$, where $vars$ is a tuple of specification variables.

A state function is an ordinary expression (one with no prime or $\Box$) that can contain variables and constants. A state predicate is a Boolean-valued state function. An invariant Inv of a specification $Spec$ is a state predicate such that $Spec \implies \Box Inv$ is a theorem. A temporal formula satisfied by every behavior is called a theorem. This formula says that any behavior that satisfies $Spec$ also satisfies $\Box Inv$. TLA$^+$ is an untyped language. In TLA$^+$ every value is a set. An expression like $v \in Nat$ associates to the variable v a type which in this case is the set of natural numbers Nat. Formally, a variable v has type T in $Spec$ iff $v \in T$ is an invariant of $Spec$.

5   Specification of the Team Formation Protocol in TLA$^+$

In this section we give a detailed step-by-step modeling of the team formation protocol for one initiator. We present the reasoning that leads to the specification.

We distinguish the robots based on their roles in the team formation process. There is an initiator and a set of non-initiators. An initiator is a robot that initiates the team formation process. A non-initiator is a robot whose willing message, in response to the initiator's request message, is received by the initiator within a predefined time period.

In order to specify the behavioral aspects of these two types of robots we need to make suitable abstractions. So we abstract the initiator by giving the attributes: *state* (to denote the current state), *member* (to denote the other team members i.e., the selected non-initiators), and *nonmember* (to denote the non-members, i.e., the rejected non-initiators). For each non-initiator, the attributes are: *state* (to denote the current state), *sent, recd* (to denote the last message sent and received respectively).

Let *NonInitiator* denote the set of non-initiators. This is a constant parameter, one whose value remains same at all the states. We have a variable *initiator*, a record, to capture the attributes (fields of the record) of the initiator. We have a variable *Data*, an array, where *Data[p]* is a record for a non-initiator p, to capture the attributes of the non-initiator.

The parameters are declared as:

CONSTANT   *NonInitiator*
VARIABLES   *initiator, Data*

The sets of records of initiator and non-initiators are defined as:

$initiatorRec \triangleq$
  $[state : \{$"ready", "busy"$\}$,
  $member :$ SUBSET $NonInitiator$,
  $nonmember :$ SUBSET $NonInitiator]$

$nonInitRec \triangleq$
  $[state : \{$"idle", "promise", "busy"$\}$,
  $sent : \{$"willing", "nil"$\}$,
  $recd : \{$"request", "confirm", "notRequired", "nil"$\}]$

The definition, denoted by the symbol $\triangleq$, of *initiatorRec* says that *initiatorRec* is a set of records. If r is a record from *initiatorRec*, then $r.state$ is its *state* field, that is an element of the set $\{$"ready", "busy"$\}$. The fields $r.member$ and *r.nonmember* are elements of the set of all subsets of *NonInitiator*. The operator SUBSET constructs the set of all subsets of a set. "nil" denotes nothing that is used to capture the situation in the initial state. The definition of *nonInitRec* is analogous.

Types of Variables
The types of *initiator* and *Data* are given by the expressions:

$initiator \in initiatorRec$ and
$Data \in [NonInitiator \rightarrow nonInitRec]$

Recall that a variable v has type T iff $v \in T$ is an invariant of *Spec*. The variables *initiator* and *Data* have types *initiatorRec* and $[NonInitiator \rightarrow nonInitRec]$ respectively, since both the expressions are type invariants of *Spec*. A detailed discussion of the meaning of these expressions is given in the Subsect. 5.2 where we describe the type invariant $TypeOK$.

Initial State
The variable *initiator* is a record with fields *state*, *member*, and *nonmember*. In the initial state, the fields of *initiator* take on the values: *initiator.state* = "ready", *initiator.member* = *initiator.nonmember* = {}. The value of *initiator* in the initial state is written as:

$initiator = [state \mapsto$ "ready", $member \mapsto$ {}, $nonmember \mapsto$ {}]

Similarly, the value of *Data* in the initial state is written as:

$Data = [p \in NonInitiator \mapsto$
$[state \mapsto$ "idle", $sent \mapsto$ "nil", $recd \mapsto$ "nil"]]

So the record corresponding to a non-initiator p is written as:

$Data[p] = [state \mapsto$ "idle", $sent \mapsto$ "nil", $recd \mapsto$ "nil"]

In the following some expressions would be used multiple times. For the sake of clarity, we define the following.

$InitialInitiator \triangleq$
$[state \mapsto$ "ready", $member \mapsto$ {}, $nonmember \mapsto$ {}]
$InitialNoninit \triangleq [state \mapsto$ "idle", $sent \mapsto$ "nil", $recd \mapsto$ "nil"]
$AfterReceiveNoninit \triangleq [state \mapsto$ "promise", $sent \mapsto$ "willing",
$recd \mapsto$ "request"]

Init, the predicate describing the initial state, is defined as the conjunction of the values of the two variables *initiator* and *Data*.

$Init \triangleq \wedge initiator = InitialInitiator$
$\wedge Data = [p \in NonInitiator \mapsto InitialNoninit]$

5.1   Actions

1. *BroadcastRequest*

The action is enabled at the initial state (the first two conjuncts of the definition). The effect of the action is a state where the attributes *state* and *recd* of the non-initiators change and none of the attributes of the initiator change (the last two conjuncts of the definition). The action is formally defined as:

$$
\begin{aligned}
&BroadcastRequest \;\overset{\Delta}{=}\\
&\wedge\; initiator = InitialInitiator\\
&\wedge\; Data = [p \in NonInitiator \mapsto InitialNoninit]\\
&\wedge\; Data' = [p \in NonInitiator \mapsto\\
&\quad [Data[p] \text{ EXCEPT } !.state = \text{``promise''}, \ !.recd = \text{``request''}]]\\
&\wedge\; \text{UNCHANGED } initiator
\end{aligned}
$$

$[Data[p]$ EXCEPT $!.state =$ "promise", $!.recd =$ "request"]] is the new record obtained by changing only the fields *state* and *recd* of the record in the previous state; the value of the other fields remain same.

2. *ReceiveWilling*

The action is enabled at a state where the fields of a non-initiator are *state* = "promise", *recd* = "request" and the initiator is in the initial state (the first two conjuncts of the definition). The effect of the action is to change the *sent* field of a non-initiator to "willing" and the value of the initiator remains unchanged (the last two conjuncts of the definition). The action is formally defined as:

$$
\begin{aligned}
&ReceiveWilling \;\overset{\Delta}{=}\\
&\wedge\; initiator = InitialInitiator\\
&\wedge\; Data = [p \in NonInitiator \mapsto\\
&\quad [state \mapsto \text{``promise''}, \ sent \mapsto \text{``nil''}, \ recd \mapsto \text{``request''}]]\\
&\wedge\; Data' = [p \in NonInitiator \mapsto\\
&\quad [Data[p] \text{ EXCEPT } !.sent = \text{``willing''}]]\\
&\wedge\; \text{UNCHANGED } initiator
\end{aligned}
$$

3. *SelectTeam*

The action is enabled in a state where the value of *initiator* is that in the initial state, and the value of *Data* corresponds to the record after obtaining the "willing" messages (the first two conjuncts of the definition). The initiator selects the team members (i.e., the fields *member* and *nonmember* are updated in the third conjunct of the definition). We want to specify that a team is chosen, not how it is chosen. This is abstracted by taking a nonempty subset of *NonInitiator* (the third conjunct of the definition). The formula

$\exists q \in S : F$ asserts that there exists a q in the set S satisfying the formula F; q is a local variable whose scope is the formula F. The value of $Data$ is unchanged (the fourth conjunct of the definition). The action is formally defined as:

$$
\begin{aligned}
SelectTeam \;\triangleq\; \\
&\wedge\; initiator = InitialInitiator \\
&\wedge\; Data = [p \in NonInitiator \mapsto AfterReceiveNoninit] \\
&\wedge\; \exists v \in \text{SUBSET } NonInitiator : \wedge\; C1 \\
&\hspace{6.2cm} \wedge\; C2 \\
&\wedge\; \text{UNCHANGED } Data
\end{aligned}
$$

$\qquad C1 = Cardinality(v) > 0$

$\qquad C2 = initiator' = [initiator \text{ EXCEPT } !.member = v, \; !.nonmember = NonInitiator \backslash v]$

In the above definition, v is a local variable used to obtain a nonempty subset of $NonInitiator$. Then the value of the field $member$ is set equal to v and the value of the field $nonmember$ is set equal to the remaining elements of $NonInitiator$ ($\backslash$ is the set difference operator). Note that $C1$ can be designed according to the requirement. $C1 = Cardinality(v) > 0 \wedge v \neq NonInitiator$ says that not all the non-initiators are selected. If $Cardinality(v) = k$ where k is declared as a constant parameter to denote the number of other members required by the initiator, then we obtain the situation where the team size is known in advance. For an auction protocol set $k = 1$.

The previous actions exhibit interaction between different components of a system, where the initiator sends/receives messages to/from non-initiators. Unlike these actions, $SelectTeam$ is an internal action that is not in response to any message.

4. *Notify*

This action is enabled at a state where $initiator.state =$ "ready" (first conjunct of the definition), $initiator.member \neq \{\}$ (second conjunct of the definition), and the value of $Data$ corresponds to the record after obtaining the "willing" messages (third conjunct of the definition). All elements of $initiator.member$ should have the values of the fields $state$ equal to "busy", $recd$ equal to "confirm", and all elements of $initiator.nonmember$ should have the values of the fields $state$ equal to "idle", $recd$ equal to "notRequired", $sent$ equal to "nil" (fourth conjunct of the definition). This is done using the IF/THEN/ELSE construct. The initiator also changes its state to "busy" (last conjunct of the definition). The action $Notify$ is formally defined as:

$Notify \triangleq$
$\wedge\ initiator.state = $ "ready"
$\wedge\ initiator.member \neq \{\}$
$\wedge\ Data = [p \in NonInitiator \mapsto AfterReceiveNoninit]$
$\wedge\ Data' = [p \in NonInitiator \mapsto$
IF $p \in initiator.member$ THEN
$[Data[p]$ EXCEPT $!.state = $ "busy", $!.recd = $ "confirm"]
ELSE $[state \mapsto$ "idle", $sent \mapsto$ "nil", $recd \mapsto$ "notRequired"]]
$\wedge\ initiator' = [initiator$ EXCEPT $!.state = $ "busy"]

From the above definitions, exactly one action is enabled at a state, and all the enabled actions are distinct. Since the number of actions is finite, so a finite sequence of states is obtained. In TLA$^+$, proper execution refers to an infinite sequence of states. So the final state of the sequence is repeated infinitely many times by a stuttering step, *StutStep*, that keeps the values of the variables unchanged.

$StutStep \triangleq \wedge\ initiator.member \neq \{\}$
$\qquad\qquad \wedge$ UNCHANGED $\langle Data, initiator \rangle$

So the sequence of states and actions of the protocol would be:
$$s_0 \xrightarrow{BroadcastRequest} s_1 \xrightarrow{ReceiveWilling} s_2 \xrightarrow{SelectTeam} s_3 \xrightarrow{Notify} s_4 \xrightarrow{StutStep} s_4, \ldots$$
There are 5 distinct states $s_0, \ldots, s_4$. The final state s_4 repeats by virtue of *StutStep*. Thus the next step relation *Next* is defined as follows.
$Next \triangleq BroadcastRequest \vee ReceiveWilling \vee SelectTeam \vee Notify \vee StutStep$
Thus the specification is defined as: $Spec \triangleq Init \wedge \square[Next]_{vars}$ where $vars \triangleq \langle initiator, Data \rangle$. That the specification *Spec* is complete (meaning that it specifies what should occur for any possible action), has been verified by TLC.

5.2 Predicates

We define the state predicate *TypeOK*, that describes the types of the variables, as:

$TypeOK \triangleq \wedge\ initiator \in initiatorRec$
$\qquad\qquad \wedge\ Data \in [NonInitiator \to nonInitRec]$

The type invariant *TypeOK* asserts that (i) the value of *initiator* is an element of the set of records *initiatorRec*, and (ii) the value of *Data* is an element of the set of functions f with $f[x] \in nonInitRec$ for $x \in NonInitiator$ (i.e., the value of $Data[x]$ is an element of the set of records *nonInitRec*).

Correctness of the protocol is expressed by the invariance of a state predicate *Consistency* that is a conjunction of two conditions. The first asserts that the protocol is not in an inconsistent final state where either the initiator is in busy state and the team members are in some other state or the team members are in busy state and the initiator is in some other state. The second asserts that the sum of the cardinality of the set of team members and non-members is less than or equal to the cardinality of the set of non-initiators.

$$
\begin{aligned}
&Consistency \;\overset{\Delta}{=} \\
&\wedge\; (initiator.state = \text{``busy''}) \equiv C3 \wedge C4 \\
&\wedge\; Cardinality(initiator.member)+ \\
&\;Cardinality(initiator.nonmember) \le Cardinality(NonInitiator)
\end{aligned}
$$

$C3 = Cardinality(initiator.member) > 0$
$C4 = \forall p \in initiator.member : Data[p].state = \text{``busy''}$

We have a theorem asserting the invariance of $TypeOK$ and $Consistency$. The theorem is verified by TLC model checker.

THEOREM $Spec \implies \Box(TypeOK \wedge Consistency)$
where $Spec \overset{\Delta}{=} Init \wedge \Box[Next]_{vars}$

5.3   Verification Using TLC Model Checker

Microsoft Visual Studio Code with TLC 2 Version 2.16 is used for writing the TLA^+ specifications and verification of the codes on a computer having the configuration: Intel(R) Core(TM) i7-10510U CPU @ 1.80 GHz 2.30 GHz, 16.0 GB (15.7 GB usable) RAM, 64-bit operating system, x64-based processor, Windows 10 Pro, version 21H1.

We checked with TLC different models by varying the cardinality of $NonInitiator$. Table 1 gives the summary of the output obtained for different models. When we checked the model ($|NonInitiator| = 7$), the output obtained is shown in Fig. 1, where Success means no error has been found, Diameter is the depth of the complete state graph search, Found indicates the number of states generated, Distinct indicates the number of distinct states, and the entry for Queue indicates 0 states left on queue. The Coverage shows the states generated for each action and $Init$. If we remove $StutStep$, TLC reports an error "deadlock reached".

Table 1. Summary of output of TLC for different models

$|NonInitiator|$	Found	Distinct	Finished in (ms)
5	127	65	520
7	511	257	551
9	2047	1025	662

```
Checking TeamFormation.tla / TeamFormation.cfg
Success Fingerprint collision probability: 3.5E-15
Start: 19:29:13 (Jan 21), end: 19:29:14 (Jan 21)
```

States

Time	Diameter	Found	Distinct	Queue
00:00:00	5	511	257	0

Coverage

Module	Action	Total	Distinct
TeamFormation	Init	1	1
TeamFormation	BroadcastRequest	1	1
TeamFormation	ReceiveWilling	1	1
TeamFormation	SelectTeam	127	127
TeamFormation	Notify	127	127
TeamFormation	StutStep	254	0

Fig. 1. Output of TLC for the model ($|NonInitiator| = 7$)

6 Conclusions

In this paper we made a first attempt to give a formal, complete specification of the executions of a team formation protocol, that result in team formation, in TLA$^+$. The next step in understanding a protocol is to come up with suitable invariant conditions. We suggested an invariant condition to express the correctness of the protocol, which was verified using TLC model checker. The TLA$^+$ specification, given as a single formula, provides a better understanding of the logical properties of the executions of the protocol. As part of future work, we would like to explore the scope of TLA$^+$ for specifying other agent interaction protocols in multi-agent systems. We hope this work serves as a motivating case study for research in formal specification of behaviors of autonomous robotic systems.

Acknowledgements. The author was in part supported by a research grant from Google.

References

1. Lamport, L.: A simple approach to specifying concurrent systems. Commun. ACM **32**(1), 32–45 (1989)
2. Lamport, L.: Specifying concurrent program modules. ACM Trans. Program. Lang. Syst. **5**(2), 190–222 (1983)
3. Gerkey, B.P., Mataric, M.J.: Sold!: auction methods for multirobot coordination. IEEE Trans. Robot. Autom. **18**(5), 758–768 (2002)
4. Kong, Y., Zhang, M., Ye, D.: An auction-based approach for group task allocation in an open network environment. Comput. J. **59**(3), 403–422 (2015)

5. Batson, B., Lamport, L.: High-level specifications: lessons from industry. In: de Boer, F.S., Bonsangue, M.M., Graf, S., de Roever, W.-P. (eds.) FMCO 2002. LNCS, vol. 2852, pp. 242–261. Springer, Heidelberg (2003). https://doi.org/10.1007/978-3-540-39656-7_10

6. Gjondrekaj, E., et al.: Towards a formal verification methodology for collective robotic systems. In: Aoki, T., Taguchi, K. (eds.) ICFEM 2012. LNCS, vol. 7635, pp. 54–70. Springer, Heidelberg (2012). https://doi.org/10.1007/978-3-642-34281-3_7

7. Brambilla, M., Pinciroli, C., Birattari, M., Dorigo, M.: Property-driven design for swarm robotics. In: International Conference on Autonomous Agents and Multiagent Systems, pp. 139–146 (2012)

8. De Nicola, R., Ferrari, G., Pugliese, R.: KLAIM: a kernel language for agents interaction and mobility. IEEE Trans. Softw. Eng. 24(5), 315–330 (1998)

9. De Nicola, R., Katoen, J., Latella, D., Loreti, M., Massink, M.: Model checking mobile stochastic logic. Theor. Comput. Sci. 382(1), 42–70 (2007)

10. Holzmann, G.: Spin Model Checker, The Primer and Reference manual. Addison Wesley Professional (2003)

11. Jackson, D.: Software Abstractions: Logic, Language, and Analysis. MIT Press (2006)

12. Jones, C.B.: Systematic Software Development using VDM. Prentice Hall (1990)

13. Konur, S., Dixon, C., Fisher, M.: Analysing robot swarm behavior via probabilistic model checking. Robot. Auton. Syst. 60(2), 199–213 (2012)

14. Konur, S., Dixon, C., Fisher, M.: Formal verification of probabilistic swarm behaviours. In: Dorigo, M., et al. (eds.) ANTS 2010. LNCS, vol. 6234, pp. 440–447. Springer, Heidelberg (2010). https://doi.org/10.1007/978-3-642-15461-4_42

15. Kwiatkowska, M., Norman, G., Parker, D.: PRISM: probabilistic symbolic model checker. In: International Conference on Modelling Techniques and Tools for Computer Performance Evaluation, pp. 200–204 (2002)

16. Lamport, L.: The TLA home page. http://research.microsoft.com/en-us/um/people/lamport/tla/tla.html

17. Lamport, L.: Specifying Systems. The TLA$^+$ Language and Tools for Hardware and Software Engineers. Addison-Wesley (2002)

18. Loreti, M.: SAM: Stochastic Analyser for Mobility. http://rap.dsi.unifi.it/SAM

19. Luckcuck, M., Farrell, M., Dennis, L.A., Dixon, C., Fisher, M.: Formal specification and verification of autonomous robotic systems: a survey. ACM Comput. Surv. 52(5), 1–41 (2020)

20. Newcombe, C., Rath, T., Zhang, F., Munteanu, B., Brooker, M., Deardeuff, M.: How Amazon web services uses formal methods. Commun. ACM 58(4), 66–73 (2015)

21. Spivey, M.: The Z Notation. Prentice Hall International (1992)

22. Webster, M.: Toward reliable autonomous robotic assistants through formal verification: a case study. Trans. Hum. Mach. Syst. 46(2), 186–196 (2016)

23. Webster, M., et al.: Formal verification of an autonomous personal robotic assistant. In: AAAI FVHMS, pp. 74–79 (2014)

24. Nath, A., Arun, A.R., Niyogi, R.: An approach for task execution in dynamic multirobot environment. In: Mitrovic, T., Xue, B., Li, X. (eds.) AI 2018. LNCS (LNAI), vol. 11320, pp. 71–76. Springer, Cham (2018). https://doi.org/10.1007/978-3-030-03991-2_7

25. Nath, A., Arun, A.R, Niyogi, R.: A distributed approach for autonomous cooperative transportation in a dynamic multi-robot environment. In: The 35th ACM Symposium on Applied Computing, pp. 792–799 (2020)
26. Nath, A., Arun, A.R., Niyogi, R.: DMTF: a distributed algorithm for multi-team formation. In: 12th International Conference on Agents and Artificial Intelligence, vol. 1, pp. 152–160 (2020)

Source Code Recommendation with Sequence Learning of Code Functions

Erika Saito[1] and Kosuke Takano[2(✉)]

[1] Course of Information and Computer Sciences, Graduate School of Engineering, Kanagawa Institute of Technology, 1030 Shimo-ogino, Atsugi, Kanagawa 243-0202, Japan
s2185007@cco.kanagawa-it.ac.jp
[2] Department of Information and Computer Sciences, Kanagawa Institute of Technology, 1030 Shimo-ogino, Atsugi, Kanagawa 243-0292, Japan
takano@ic.kanagawa-it.ac.jp

Abstract. For finding desired source codes and articles using existing source code search engines, it is necessary to compare them in the different results to examine which source codes are the best practice for developing the target software. In this paper, we propose a method of source code recommendation based on the prediction of code function. The feature of the proposed method is that by interpreting the context of the processing procedure of the source code, it recommends source code which complements or follow the processing procedures. In the experiment using actual source codes, we evaluate the feasibility of the proposed method.

1 Introduction

For improving the efficiency in software development, there has been many studies on source code generation [1] and recommendation [2, 3]. To implement such functions, the capability of understanding source code is essentially needed. In our previous study [2], we proposed a method of suggesting a snippet of source code, where processing steps of a source code is represented as a sequence of function labels and a code function that follows the input label sequence is predicted. However, this previous method cannot predict a code function for the masked label in the input label sequence.

In this study, by extending the method in our previous study [2], we propose a method of source code recommendation based on the prediction of source code function for supporting programming. The feature of the proposed method is that by interpreting the context of the processing procedure of the source code, it recommends source code which complements or follow the processing procedures. In this study, for realizing the proposed method, we show a method for creating a corpus that is a set of sequences of function labels representing processing steps of source code based on the analysis actual source codes (hereinafter referred to as a logic corpus). Then, we show a method to train a prediction model of code function with the logic corpus for predicting a code function that complements the masked label sequences. In the prediction process, the proposed method consider the context of proceeding steps in the label sequence by interpreting it in both forward and backward directions.

© The Author(s), under exclusive license to Springer Nature Switzerland AG 2022
L. Barolli et al. (Eds.): AINA 2022, LNNS 451, pp. 314–323, 2022.
https://doi.org/10.1007/978-3-030-99619-2_30

The proposed prediction model of code function is created by applying a language model with deep learning. For example, sequence learning models such as LSTM (Long short-term memory) [4] and BERT (Bidirectional Encoder Representations from Transformers) [5] can be applied to our prediction model. In the experiment, we train a LSTM-based prediction model of source code with the logic corpus created from the actual source codes, and using the prediction model, we evaluate the feasibility of the proposed method by validating the capability of predicting a function label which complements a masked label in a input label sequence.

2   Related Work

We describe about research on several methods for recommendation of source code snippets and a neural model with distributed representation of source code as related work.

In [2], Shigeta et al. propose a method for recommending snippets of source code. In this method, processing steps are represented as a sequence of functions that are extracted from the corresponding code snippets. A sequence learning model is trained with a sequence of function labels, so that it can predict the code function that follows an input sequence of function labels. In [3], Uchiyama et al. proposes to consider the specific effect of source code by incorporating functions of removing reserved words, operators, and standard library functions and increasing neighbor words extended by word2vec.

In addition, Yamamoto proposes methods of analyzing and complementing a source code using a corpus of source codes with control statement [6] and predicting a method call statement using Recurrent Neural Network [7] by focusing on the order of the method call statements.

As for a neural model for the learning of distributed representation of source code, Alon et al. proposes code2vec that is a neural model to represent a snippet of source code as continuous distributed vector [8] by applying AST (Abstract Syntax Tree) for encoding a source code to a fixed-length code vector.

3   Proposed Method

In this study, we propose a method of source code recommendation based on the prediction of code function (Fig. 1). In the proposed method, we define a part of source code containing a function as a code snippet, a label representing the function extracted from the code snippet as a function label, and the function labels representing processing steps in the source code as a sequence of function labels, respectively.

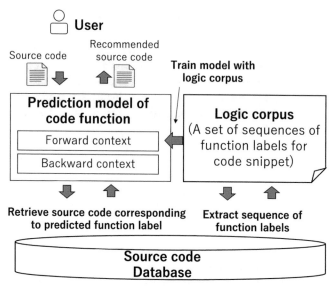

Fig. 1. Overview of proposed method

The feature of the proposed method is that by interpreting the context of the processing procedure of the source code, it recommends source code which complements or follow the processing procedures. For this, first, by analyzing a function of source code, a sequence of function labels corresponding to each code snippets are extracted to represent the processing steps in the source code. Then, a logic corpus, which is a set of sequences of function labels, to represent each processing step in the set of source codes. (Fig. 2). Here, conventional methods such as code2vec [8] can be applied for extracting a function label. Then, a prediction model of a code function is trained using the logic corpus as a learning data set.

For interpreting processing steps in a source code, sequence leaning models such as LSTM and Bi-LSTM (Bidirectional LSTM) can be applied. By inputting a sequence of function labels, the prediction model can predict a function that follows the input label sequence. In addition, for predicting a masked label in the input label sequence, the former sequence before the masked label is input in a forward direction into a prediction model, and the latter sequence after the masked label is input in a backward direction. By merging the predicted results for both inputs, the masked label can be predicted to fit the both results.

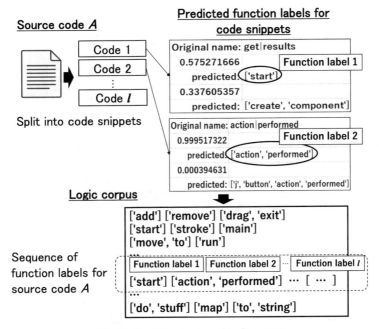

Fig. 2. Creation process of logic corpus

Then, we describe the creation process of learning data, which is used to train a prediction model of code function, from the logic corpus (Fig. 3).

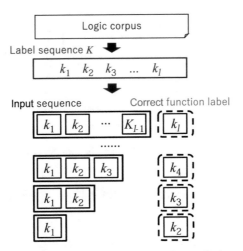

Fig. 3. Creation process of learning data for prediction model

Step-1: A source code is split into l-code snippets. The corresponding function label is extracted from each code snippets, and a sequence of function labels $K = \{k_1, k_2, ..., k_{l-1}, k_l\}$ is obtained.

Step-2: By applying Step-1 for a set of source codes, the corresponding sequences of function labels are extracted to create a logic `corpus`.

Step-3: A sequence of function labels $K = \{k_1, k_2, ..., k_{l-1}, k_l\}$ is extracted from the logic corpus.

Step-4: The sequence K is separated into two parts, the input sequence K_{input}, which is a sequence extracted from the 1st label to $(l-1)$-th label in K and the correct function label K_{label}, which is the last label of K.

Step-5: For all label sequences in the logic corpus, the pairs of input sequence K_{input} and correct label K_{label} are extracted, and used to train a sequence learning model such as LSTM.

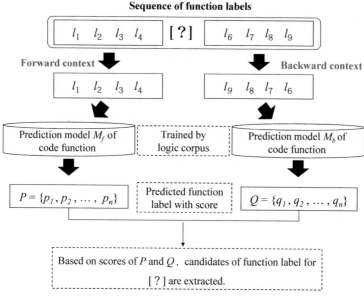

Fig. 4. Prediction process for code function

The prediction process of a function label for a input masked sequence is executed in the following steps (Fig. 4).

Step-1: An input source code is split into x code snippets, and a sequence of function labels $L = \{l_1, l_2, ..., l_x\}$ is extracted, where each label l is extracted from the corresponding code snippet.

Step-2: The label sequence L in Step-1 is split into two sequences at a masked label l_i. We call the former sequence forward context $L_f = (l_1, l_2, ..., l_{i-1})$, the latter sequence backward context $L_r = (l_{i+1}, l_{i+2}, ..., l_x)$, respectively.

Step-3: For the input of the forward context L_f, the prediction model predicts P, a set of n-function labels with score, as follows:

$$P = \{(p_1, s_1), (p_2, s_2), \ldots, (p_n, s_n)\} \tag{1}$$

The backward context L_r is sorted into $L_r' = (l_x, \ldots, l_{i+2}, l_{i+1})$. Similarly, the prediction model predicts Q, a set of n-function labels with score for the input of the backward context L_r' as follows:

$$Q = \{(q_1, s_1), (q_2, s_2), \ldots, (q_n, s_n)\} \tag{2}$$

Step-4: A set of predicted labels P and Q are merged into M. For the same label in P and Q, the prediction score is summed up. Then, m-predicted labels with top-m score are extracted from M for creating a set of predicted labels M' as follows:

$$M' = \{(m_1, s_1), (m_2, s_2), \ldots, (m_n, s_n)\} \tag{3}$$

Finally, from a source code database, m-source codes corresponding to each function label in M' is ranked in the order of the predicted score, and recommended to a user.

4 Experiment

4.1 Purpose

In the experiment, we confirm that our proposed method can be trained with a logic corpus, and predict a function label which complements a masked label in a sequence of function labels.

4.2 Experimental Environment

In order to create a logic corpus for the training, we collected actual Java source codes from a Java programming book [9] and Web sites. Out of Java source codes collected, 750 codes are used for the training of processing steps of source codes, and 878 codes are for the test of prediction.

In addition, we used code2vec [8] for splitting a source code into code snippets, and extracting the corresponding function label from each code snippet. Here, 1,636 function labels are extracted by code2vec as shown in Table 1. In Table 1, several labels inside the parentheses as ['label 1', 'label 2', ...] means one function of one snippet. Table 2 shows an example of function labels in the logic corpus extracted from 750 Java source codes for the training.

Table 1. Example of function label

['main'] ['get', 'key']
['operates', 'on', 'fact', 'handles']
['cancel', 'button', 'action', 'performed']

Table 2. Example of logic corpus

['test', 'lock', 'timeout'] ['sleep']
['main'] ['do', 'work']
['set', 'message', 'body'] ['load'] ['set', 'subject'] ['process']
['run'] ['get', 'value']

In the experiment, in order to predict processing steps of source code, we build the three prediction models of code function, which are trained with the logic corpus by bi-LSTM and LSTM, as follows:

(1) **Forward model**: Prediction model of code function with forward context
(2) **Backward model**: Prediction model of code function with backward context
(3) **Combination model**: Proposed model) Prediction model of code function with both forward context and backward context

Table 3. Test sequences of functions labels for prediction model

Number	Sequence of function labels	Correct function label
1	['start'] * ['main']	['circle']
2	['start'] * ['draw'] ['main']	['run']
3	['on', 'update'] * ['stroke'] ['main']	['draw']
4	['main'] ['print'] * ['mean'] ['max'] ['get', 'itd']	['sum']
5	['start'] ['display'] ['println'] * ['cancel'] ['main']	['connect']

Table 3 shows five test sequences of functions labels. In each sequence, '*' means the masked function label. By inputting these test sequences to the prediction models, we obtain the function labels predicted by the above three models. Then, we compare the performance of each model with the discussion focusing on top-5 predicted function labels with high prediction score.

4.3 Result

Figure 5 shows the result of number of predicted function labels for three models, forward model, backward model, and combination model. Each model is trained by applying Bi-LSTM or LSTM as a sequence model. In Fig. 5, we evaluated that if a predicted label has the adaptability with the correct function label, we defined it as high relevance. Similarly, if the predicted label is similar to the correct function label, it is defined as middle relevance, and if there is less correlation, it defined as low relevance. From the result of Fig. 5, forward model predicted more high relevance of labels in both models, Bi-LSTM and LSTM, however, backward model predicted more low relevance of labels. Since combination model depends on backward model, the low prediction accuracy of backward model caused the low prediction accuracy of combination model as well. Therefore, we have to improve especially the prediction accuracy of backward model. In addition, in the result of Fig. 5, the prediction result bi-LSTM is better that LSTM, since both forward model and backward model applied bi-LSTM predicts more high relevance of labels.

Table 4 shows the prediction result of combination model for input 1 in top-5. In Table 4, we can confirm that correct label ['circle'] for input 1 is predicted and other predicted labels have also high or middle relevance with the compatibility to the input processing steps. However, we found that the function label ['main'] tends to be predicted. This is probably because ['main'] is often appeared in almost label sequences in the logic corpus. Therefore, we have to improve to adjust the prediction output by understanding the context of proceeding steps more precisely.

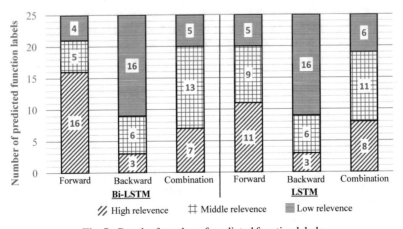

Fig. 5. Result of number of predicted function labels

Table 4. Prediction result of combination model for input 1

Bi-LSTM			LSTM		
Predicted function label	Score	Relevance	Predicted function label	Score	Relevance
Run	0.360	High	Run	0.372	High
Main	0.325	High	Main	0.273	High
Action	0.136	Middle	Action	0.136	Middle
circle	0.080	High	circle	0.095	High
Test	0.068	Middle	Create	0.077	High

5   Conclusion

In this study, we proposed a method of source code recommendation based on the prediction of source code function for supporting programming. In the experiment, we created a logic corpus, which is a set of sequences of function labels representing processing steps of source code, by analyzing actual source codes, and then, we trained a prediction model of source code with the logic corpus. Using the prediction model, we could confirm the feasibility of the proposed method by evaluating the capability of predicting a function label which complements a masked label in a sequence of function labels. However, since we used a small set of source codes in the experiment, we have to improve the experimental environment to analyze a larger set of source codes. In addition, we found that the prediction accuracy of the prediction model using only backward context was low, and as a result, it caused the low prediction accuracy of the prediction model using both forward context and backward context.

In our future work, we will improve the prediction of accuracy of that proposed method. For that, we are planning to create a logic corpus with a large set of source codes, and evaluate the effectiveness of the proposed method, for example, by comparing the proposed method with other context prediction models for the sequence data such as BERT.

Acknowledgments. This work was supported by JSPS KAKENHI Grant Number 17K00498.

References

1. Kawamura, Y., Asami, C.-C.K.K.: The production of the automatic source code generation tool from UML. In: Proceedings of the 71th Information Processing Society of Japan Annual Convention, pp. 337–338 (2010)
2. Shigeta, T., Takano, K.: A source code recommendation system based on programming logic. In: Proceedings of the 81th Information Processing Society of Japan Annual Convention, vol. 2019, no. 1, pp. 305–306 (2019)
3. Uchiyama, T., Niimi, A.: Retrieving code fragments using Word2Vec considering the lnfuence of nearby words peculiar to source code. In: Proceedings of 2017 Software Engineering Symposium (SES2017), pp. 146–154. Information Processing Society of Japan (2017)

4. Hochreiter, S., Schmidhuber, J.: Long short-term memory. Neural Comput. **9**(8), 1735–1780 (1997)
5. Devlin, J., Chang, M.-W., Lee, K., Toutanova, K.: BERT: Pre-training of Deep Bidirectional Transformers for Language Understanding. arXiv:1810.04805 (2018)
6. Yamamoto, T.: A code completion of method invocation statement using source code corpus including a consideration of contol flow. J. Inf. Process. **56**(2), 682–691 (2015)
7. Yamamoto, T.: An API suggestion using recurrent neural networks. J. Inf. Process. **58**(4), 769–779 (2017)
8. Alon, U., Zilberstein, M., Levy O., Yahav, E.: code2vec: Learning Distributed Representations of Code, arXiv:1803.09473 (2018)
9. Darwin, I.F.: The Java Cookbook Third Edition. O'Reilly Media, Sebastopol (2014)

Two-Tier Trust Structure Model for Dynamic Supply Chain Formulation

Shigeaki Tanimoto[1]([✉]), Yudai Watanabe[1], Hiroyuki Sato[2], and Atsushi Kanai[3]

[1] Faculty of Social Systems Science, Chiba Institute of Technology, Chiba, Japan
shigeaki.tanimoto@it-chiba.ac.jp, s1842135fn@s.chibakoudai.jp
[2] The University of Tokyo, Tokyo, Japan
schuko@satolab.itc.u-tokyo.ac.jp
[3] Hosei University, Tokyo, Japan
yoikana@hosei.ac.jp

Abstract. With the ongoing development of DX, the linkage between physical and cyber environments is expanding and is increasingly being applied to business aspects such as supply chains. At the same time, cyber-attacks are becoming more diversified, and information security is becoming more and more important. The Information Security White Paper 2019 reported that one of the most pressing issues in the supply chain is that the scope of responsibility between the contractor and subcontractor is unclear. In this paper, we propose a two-tier trust structure model featuring inter-organizational security policy matching for dynamic supply chains that recombines subcontractors for each contract. The proposed structure consists of a trust model at the contractor side and a zero-trust model between the contractor and the subcontractor. The results of qualitative evaluation show that this two-tier model can help clarify the scope of responsibility in dynamic supply chains.

1 Introduction

With the ongoing progress of IoT, collaborations between physical and cyber environments have been expanding, and these cyber-physical systems are increasingly being applied to business aspects. According to the IPA's Ten Major Threats to Information Security released in 2020, attacks that exploit weaknesses in the supply chain are ranked fourth among all threats to organizations [1].

Conventional supply chain risks have typically been related to the halt of business operations in the event of a disaster or accident. However, with the increase of security risks, companies in the supply chain are now facing a wide variety of new risks, such as information leakage from contractors due to internal fraud or unauthorized access, and malware contamination in deliveries from contractors [2]. The White Paper on Information Security 2019 [3] reported the results of a survey on IT supply chain risk management indicating that one of the most pressing security issues is the unclear scope of responsibility for information security between the contractor and subcontractor.

Supply chains can be classified into static supply chains and dynamic supply chains according to the relationship between the contractor and subcontractor. A static supply

L. Barolli et al. (Eds.): AINA 2022, LNNS 451, pp. 324–333, 2022.
https://doi.org/10.1007/978-3-030-99619-2_31

chain is one in which the relationship between the two parties is fixed, such as between the head office and a branch office. A dynamic supply chain is a supply chain in which the suppliers of goods are rearranged for each contract [4]. In the former case, continuous security enhancement can be expected by clarifying the boundary of responsibility, while in the latter case, things are a little more difficult because various stakeholders are assumed and the boundary of responsibility may change dynamically.

As mentioned earlier, in the DX era, the supply chain is expected to further develop and diversify, and information security measures (including the responsibility demarcation point) are becoming an urgent issue, especially in the dynamic supply chain. In this study, we propose a two-tier trust structure model for dynamic supply chains between organizations with different security policies. Specifically, we introduce Software Defined Perimeter (SDP), a zero-trust model, as the architecture of the two-tier trust structure model, and clarify its effectiveness through qualitative evaluation. Our findings show that the proposed model improves the trustworthiness of the dynamic supply chain and contributes to the formation of a safe and secure supply chain.

2 Current Status and Issues in the Supply Chain

The supply chain refers to a series of commercial flows starting from the procurement of raw materials and parts and moving on through the manufacturing, inventory management, distribution, and sales processes, as well as the group of organizations involved in these commercial flows. The supply chain is advancing rapidly with the development of IoT and Industry 4.0. At the same time, it is still fresh in our memory that the supply chain itself was paralyzed by the Great East Japan Earthquake in 2011 and more recently by the global coronavirus pandemic (COVID-19), which severely damaged the supply chain's sources of supply. In this section, we describe the current status and issues of supply chains.

2.1 Current State of the Supply Chain

Typical risks that can affect supply chains include environmental risks (e.g., natural disasters), geopolitical risks (e.g., terrorism and political instability), economic risks (e.g., economic crises and price fluctuations of raw materials), and technological risks (e.g., cyber-attacks and system failures). In a survey conducted in 2011, natural disasters, conflicts and political instability, and demand shocks were identified as risks of high importance against the backdrop of the Great East Japan Earthquake and the flooding in Thailand that occurred in the same year [5]. In a similar survey conducted in 2020, epidemic/pandemic, uncertainty of tariffs and trade restrictions, shortage of critical raw materials and components, and attacks on cyber security were rated as "major" or "moderate" risks by a high percentage of respondents [5]. Of these, the first two are considered to have stemmed from the impact of the spread of the novel coronavirus.

A supply chain is a chain of business relationships that are necessary to achieve a goal. If a problem occurs in any part of the chain, it may have a ripple effect on multiple organizations. To minimize the impact, supply chain risk management is a crucial tool to control risks and achieve objectives.

Within the same company or group, security governance can ensure that measures are implemented and managed in accordance with the security policy. However, suppliers that are not part of the same company or group may have weaker security measures due to reasons such as lack of security governance, and these suppliers may therefore be targeted in attacks. As a result, incidents such as information leakage and system shutdown have actually occurred [3]. Table 1 lists the main risk categories in the supply chain that are problematic due to the occurrence of incidents.

Table 1. Classification of incidents in the supply chain by cause [table from [3]].

Incident	Causes of incidents in the supply chain
Unauthorized access	Attacks that use business partners as stepping stones or pretend to be business partners
	Attacks that embed malware etc. in the products and systems to be procured
Mistake	Inadequate countermeasures during development
	Inadequate countermeasures during operation
Internal fraud	Internal fraud by developers, maintainers, operation staff, etc
	Transactions that deviate from the rules and norms

2.2 Supply Chain Issues

As discussed earlier, since a supply chain is composed of multiple groups of organizations, it can be classified as either a static supply chain formed within the same organization, e.g., between headquarters, branch offices, and branch offices (Fig. 1(a)), or a dynamic supply chain formed between different organizations, e.g., when outsourcing operations (Fig. 1(b)) [4, 6].

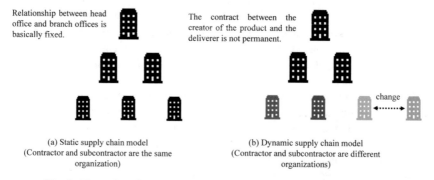

Relationship between head office and branch offices is basically fixed.

The contract between the creator of the product and the deliverer is not permanent.

change

(a) Static supply chain model
(Contractor and subcontractor are the same organization)

(b) Dynamic supply chain model
(Contractor and subcontractor are different organizations)

Fig. 1. Examples of (a) static supply chain and (b) dynamic supply chain.

In the case of a dynamic supply chain, various issues stemming from the complexity of contractual relationships generally emerge. In particular, in terms of information security measures, a pressing issue is how to clarify the security measures of the contractors, as these are between different organizations. The information security management of subcontractors is often left to the subcontractors in these cases because the contractor cannot directly manage the information security of subcontractor.

Figure 2 shows the results of a survey on the contractual issues regarding contractors and subcontractors at the time of business entrustment [7]. The largest percentage of both groups responded that they were not clear on the scope of responsibility pertaining to information security (boundary point of responsibility) in the entrustment agreement. Another pressing issue for both contractors and subcontractors was "specific information security measures to be taken are not specified."

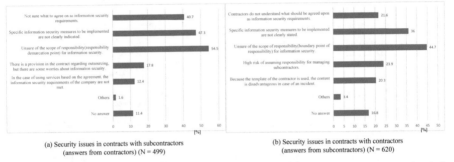

(a) Security issues in contracts with subcontractors (answers from contractors) (N = 499)

(b) Security issues in contracts with contractors (answers from subcontractors) (N = 620)

Fig. 2. Information security issues among contractors and subcontractors in the supply chain [figure from [7]].

3   Related Work

Much research has been devoted to supply chain risks. For example, C. J. Corbet et al. investigated the impact of the Customs-Trade Partnership Against Terrorism (C-TPAT) certification, a logistics security program, on international supply chain collaboration. They concluded that it is still in its early stages and there are many issues that need to be resolved before full cooperation can be achieved [8].

I. Manuj et al. proposed a comprehensive risk management and mitigation model for global supply chains that combines concepts and frameworks from multiple disciplines including logistics, supply chain management, operations management, and international business management. The specific model, shown in Fig. 3(a), provides procedures for identifying, assessing, and managing risks in global supply chains and serves as a guide for future research [9].

H. Zang et al. conducted a comprehensive review of supply chain research findings and case studies by companies that address the requirements, performance, security management, and reliability issues of digital supply chain systems [10]. Similarly, Z. Williams et al. provided empirical findings and theory building by categorizing academic papers and other literature on supply chain security (SCS), as shown in Fig. 3(b) [11, 12].

H. Aslam et al. proposed an initiative to identify the strategic orientation regarding dynamic supply chains. As shown in Fig. 3(c), they start by developing an entrepreneurial mindset internally and then encourage learning with supply chain partners to take risks and co-develop innovative ideas with suppliers [13].

M. Wang et al. investigated the complexity of dynamic supply chain formation and proposed an agent-mediated coordination approach. Each agent acts as a broker for each service and interacts with other agents to make decisions (Fig. 3(d)). Coordination among agents is focused on decision making at the strategic, tactical, and operational levels. On the basis of this approach, they implemented a prototype and reported simulation experiments that demonstrate its effectiveness [14].

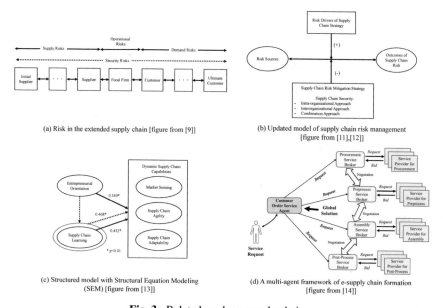

(a) Risk in the extended supply chain [figure from [9]]

(b) Updated model of supply chain risk management [figure from [11],[12]]

(c) Structured model with Structural Equation Modeling (SEM) [figure from [13]]

(d) A multi-agent framework of e-supply chain formation [figure from [14]]

Fig. 3. Related work on supply chains.

In contrast to these related studies, we propose a new approach for improving the issues related to the dynamic supply chain between contractors. We focus particularly on how to match the information security countermeasures among them.

4 Proposed Two-Tier Trust Structure Model

4.1 Dynamic Supply Chain Model

We define a dynamic supply chain as "a supply chain that does not have a fixed relationship to its contractors and recombines some of its contractors for each contract," as shown in Fig. 4.

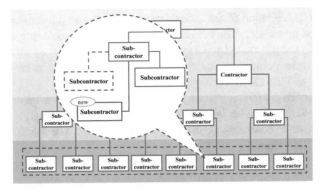

Fig. 4. Dynamic supply chain model.

4.2 Method for Establishing Reliability in Dynamic Supply Chains

For the dynamic supply chain shown in Fig. 4, we propose a secure dynamic supply chain formation method between upper-level contractors and lower-level subcontractors using a Trusted Third Party (TTP)-based reliability method. In general, TTP-based reliability methods are used as a way to build trust between different organizations. The main methods are PKI (client certificate) and IdP [15]. There is also SDP, a zero-trust model that has recently been attracting attention (Fig. 5) [16].

After comparing PKI, IdP, and SDP, we decided to adopt SDP, as we feel it will be the most effective in terms of clarifying the scope of responsibility, which is the key issue in supply chains examined in this paper. In the following, we propose a two-tiered trust structure model that introduces the zero-trust model.

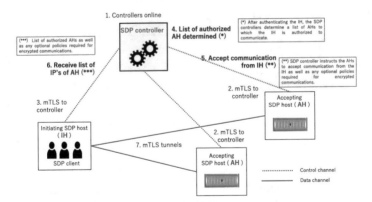

Fig. 5. Operating principles of SDP zero-trust model [figure from [9]].

4.3 Proposed Two-Tier Trust Structure Model

In the proposed model (Fig. 6(a)), we assume a two-tiered structure between contractor and subcontractor organizations in a dynamic supply chain: a trust model and a zero-trust model. For the organization of the subcontractor, as based on the zero-trust model, the subcontractor is given access privileges to the supply chain according to its trustworthiness based on predetermined security criteria of the contractor. As for the contractor, which is the contact point of the subcontractor, it is characterized as belonging to both the trust model and the zero-trust model, as shown in Fig. 6(a).

Next, the specific architecture of the two-tier trust structure model is shown in Fig. 6(b). The SDP controller is installed in a trusted provider (e.g., a Trusted Third Party (TTP)), and the contractor is installed as the SDP host. The subcontractor will be installed as an SDP client.

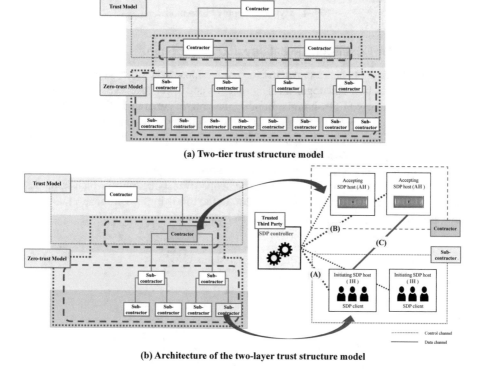

(a) Two-tier trust structure model

(b) Architecture of the two-layer trust structure model

Fig. 6. Proposed two-tier trust structure model with zero-trust model.

When the SDP client (the subcontractor) connects to the SDP host (the contractor), it follows the SDP protocol. That is, the SDP client requests the SDP controller to connect to the SDP host (Fig. 6(b)(A)), and the SDP controller checks the profile of the SDP client (in this case, the security level of the subcontractor) and uses this profile to query the SDP host (Fig. 6(b)(B)). The SDP host checks the profile of the SDP client (the security level of the subcontractor), and if it has the same security level, it grants access via the SDP controller. This completes the connection between the SDP client and the SDP host, that is, the connection between the contractor and the subcontractor (Fig. 6(b)(C)).

Thus, in the proposed two-tier trust structure model, the responsibility is concentrated on the starting point that controls the trustworthiness (the SDP controller in Fig. 6(b)), which prevents the model from being affected by issues such as the responsibility scope delineation between the contractor and the subcontractor.

4.4 Evaluation of Two-Tier Trust Structure Model

The evaluation items for the proposed two-tier trust structure model were created on the basis of the results of the survey on the scope of responsibility of the information security measures in the supply chain by IPA [7] (as discussed in 2.2; see Fig. 2). The following are the main issues concerning this scope.

(1) Clarification of the information security requirements
(2) Clarification of the specific information security countermeasures to be implemented
(3) Clarification of the scope of responsibility for the information security countermeasures.
(4) Dispelling concerns about the information security countermeasures of subcontractors
(5) Risk of being responsible for the information security countermeasures of subcontractors

Items (1), (2), and (3) are issues common to both the contractor and the subcontractor. As for (4) and (5), they are considered particular issues for the contractor.

The results of our qualitative evaluation of the effect of the proposed model regarding these five issues are as follows.

(1) **Clarification of the information security requirements**
(2) **Clarification of the specific information security countermeasures to be implemented**
These issues are considered common to both the contractor and the subcontractor organization, and both organizations feel there is a deficiency in the information security measures of the organizations they deal with.

In contrast, in the proposed two-tier trust structure model, the zero-trust model is applied to the consignee. This means the requirements for the information security measures of the contractors are explicitly stated and the issues can thus be resolved, since the contractors are required to meet the criteria defined in the zero-trust model.

(3) **Clarification of the scope of responsibility for the information security countermeasures**
As with (1) and (2), this issue is also a common problem for both the contractor and the subcontractor. It can be solved by the two-tier trust structure model because, in the zero-trust model, the responsibility is concentrated on the contractor, who controls the trust level of the subcontractor. In this way, the issue concerning the scope of responsibility can be suitably resolved.

(4) **Dispelling concerns about the information security countermeasures of subcontractors**

(5) **Risk of being responsible for the information security countermeasures of subcontractors**

These issues pertain specifically to the contractor. In the conventional supply chain, it is difficult for the contractor to manage organizations connected to the subcontractor, with whom it does not have a direct business relationship. This may be due to the fact that the contractor is forced to entrust the management to organizations only with which it has a direct business relationship. In contrast, in the two-tier trust structure model, this issue can be resolved by introducing the zero-trust model because only the subcontractor that matches the trust level indicated by the contractor can access the system.

The results of these evaluations are summarized in Table 2. As we can see, qualitatively speaking, the proposed method can improve the information security measures of the current supply chain.

Table 2. Qualitative evaluation results of proposed two-tier trust structure model.

Evaluation items	Contractor	Subcontractor
(1) Clarification of information security requirements	Improved	
(2) Clarification of specific information security measures to be taken	Improved	
(3) Clarification of the scope of responsibility for information security	Improved	
(4) Information security countermeasures for subcontractors	Improved	N/A
(5) Risks associated with the level of information security countermeasures of subcontractors	Improved	N/A

5 Conclusion and Future Work

It is now more important than ever to take information security countermeasures, including at the boundary of responsibility between the contractor and the subcontractor in a supply chain, which is expected to be further developed and diversified in the DX era.

In this paper, we proposed a two-tier trust structure model for a dynamic supply chain between the contractor and the subcontractor with different security policies. Specifically, we introduced Software Defined Perimeter (SDP), a zero-trust model, as the architecture of the two-tier trust mechanism, and clarified its effectiveness through qualitative evaluation. Our findings show that the proposed model improves the inter-organizational trust on information security countermeasures in a dynamic supply chain and helps ensure the operation of a safe and secure supply chain.

Our future work will focus on the practical implementation of the two-layer trust structure model.

Acknowledgments. This work was supported by JSPS KAKENHI Grant Number JP 19H04098.

References

1. IPA: 10 Major Threats to Information Security 2021 (2021). (in Japanese), https://www.ipa. go.jp/security/vuln/10threats2021.html
2. Kubo, T., et al.: Study of information security in the supply chain. IPSJ, DPS **161**(3), 1–8 (2014). (in Japanese)
3. IPA: Information Security White Paper 2019 (2019). https://www.ipa.go.jp/security/publicati ons/hakusyo/2019.html
4. Tanimizu, Y., et al.: A study on development of adaptive strategies in dynamic supply chains (proposal of a framework and two-layered models). Trans. JSME **75**(756), 2351–2358 (2009). (in Japanese)
5. METI: The 2021 edition of the White Paper on Trade and Commerce, p. 91 (2021). (in Japanese), https://www.meti.go.jp/report/tsuhaku2021/pdf/02-01-02.pdf
6. Sato, K., et al.: Study of dynamic supply chain model for improvement of efficiency in inter-business trading. In: Proceedings of the School of Information and Telecommunication Engineering, Tokai University, Vol. 10, No. 1, pp. 36–44 (2017). (in Japanese), https://www. ipa.go.jp/security/publications/hakusyo/2019.html
7. IPA: Research Report on Security Incidents and Management in Outsourced IT Supply Chain Operations (2018). https://www.ipa.go.jp/security/fy29/reports/scrm/index.html
8. Corbett, C.J., et al.: Designing supply contracts: contract type and information asymmetry. Manag. Sci. **50**(4), 550–559 (2004)
9. Manuj, I., et al.: Global supply chain risk management. J. Bossiness Logist. **29**(1), 133–155 (2008)
10. Zhang, H., et al.: Security and trust issues on digital supply chain. In: CyberSciTech 2019 (2019)
11. Williams, Z., et al.: Supply chain security: an overview and research agenda. Int. J. Logist. Manag. **19**(2), 254–281 (2008)
12. Juttner, U., et al.: Supply chain risk management: outlining an agenda for future research. Int. J. Logist. Res. Appl. **6**(4), 197–210 (2003)
13. Aslam, H., et al.: Determining the antecedents of dynamic supply chain capabilities. Supply Chain Manag. Int. J. **25**(4), 427–442 (2020)
14. Wang, M., et al.: Agent-based negotiation and decision making for dynamic supply chain formation. Eng. Appl. Artif. Intell. **22**(7), 1046–1055 (2009)
15. Palmo, Y., et al.: IoT reliability improvement method for secure supply chain management. In: GCCE2021, pp. 390–391 (2021)
16. CSA: Software Defined Perimeter Working Group, SDP Specification 1.0 (2014). https://clo udsecurityalliance.org/artifacts/sdp-specification-v1-0/

User Expectations When Augmented Reality Mediates Historical Artifacts

Rayed Alakhtar, Sam Ferguson, and Hada Alsobhi[✉]

UTS, 15 Broadway, Ultimo, NSW 2007, Australia
{rayed.alakhtar,hada.alsobhi}@student.uts.edu.au
Samuel.Ferguson@uts.edu.au

Abstract. Augmented reality (AR) technology has been investigated previously to enhance the experience within various tourism contexts. This paper reports on a study that aims to investigate user expectations when AR mediates exploration of historical artifacts in a museum context. The users in this study used an iOS software application to explore several historical artifacts using an Augmented Reality experience. Twenty-four participants were interviewed to evaluate their expectations and satisfaction with the experience. Analysis of the responses showed that users prioritised storytelling techniques within the AR experience to understand the history and significance of the artifacts.

1 Introduction

Tourism has become a digitized industry, meaning that tourism businesses are now employing technology to enhance tourists' experiences of locations. One technology has been implemented to serve the needs of digital tourism is Augmented Reality (AR), a technology that enables virtual knowledge to be superimposed over a real-world environment [1,2].

Tourists generally visit various countries and regions to learn and understand the cultures that they want to know more about and experience. Tourists travel to visit attractions or Points of Interest (POIs). A historical location can generally help to describe the development of countries over a long period by exploring the stories behind the artifacts. For this reason, some tourists are very excited to visit these historical tourist attractions to understand the local culture and how nations or regions within nations developed.

These historical locations comprise both natural features and human-made artifacts and architecture that generations of people from other cultures might not know about or have experienced. Visitors to these locations may want to comprehend stories of how buildings and artifacts that existed centuries ago resulted in the culture and history that permeates the country and its people now. However, these visitors can receive relevant information by checking online on Google or other websites, tour guides, or reading information platforms, books, brochures, maps. These historic locations are often the sites where unique and fragile artifacts exist, but they remain attractive to visitors who want to know what they are, how they work, and what their purpose is or was. The motivation of the study is to find the user preference of media tip of the AR technology that helps them educate and explore the history behind the museum artifacts.

L. Barolli et al. (Eds.): AINA 2022, LNNS 451, pp. 334–344, 2022.
https://doi.org/10.1007/978-3-030-99619-2_32

The use of AR technology in many countries in various ways is to improve tourists' experiences. The goal of this paper is to investigates user preferences when AR is used to facilitate the discovery of historical artifacts in a museum environment. This research involves conducting a study with tourists and exploring a user's experience, principally about what is their priority of AR media when visiting and exploring historical artifacts.

1.1 Augmented Reality in Exploring Historical Artifacts

Various studies have reported that AR technology is primarily focused on the assumption that computer-based media that are seamlessly incorporated into the surroundings will enhance real-world artifacts or activities [3–5]. The integration and interaction of virtual and physical objects are central to AR. AR was first implemented in the early 1990s, according to Dhir [6]. Mobile Augmented Reality (MAR) has become a mainstream immersive experience since the introduction of smartphones and other mobile technologies in the twenty-first century [6]. According to Rasinger [7], mobile technology in the industry include a search and browse alternative or a categorical search engine system that makes it easier for travellers to locate important information. According to [8], mobile AR tools have context-aware push alerts that allow tourists or visitors to peruse relevant data, particularly in congested urban environments. Using m-commerce or E-book functionality allows tourists to book, pay for, or cancel any services they may have subscribed to while on the road [7,9]. According to [8], MAR hosts a forum for visitors to get input. Travellers will learn from technology that makes routeing and navigating simpler, as well as the ability to get directions and navigate to the Point of Interest (POI) after selecting it in the AR show [8]. AR also enables rich travel opportunities and visitor participation, which are critical factors for the tourism industry, according to [10]. Tourists will be consciously encouraged to consider learning experiences and perceive museum spaces and objects in unique, new, and innovative ways using augmented reality.

Inside their organisations, art galleries and museums have started to incorporate augmented reality (AR) technology. In the restoration and exhibition of historical objects, AR plays a vital role [4, 11]. For example, the holographic AR app TombSeer, which is a digital cultural heritage (DCH) app, will recreate the historical inner chapel of the Egyptian Tomb of Kitines at the Royal Ontario Museum (ROM) [4]. Tomb-Seer is a website that lets people immerse themselves in ancient Egyptian history by offering meaning and knowledge about the daily lives and contributions of the civilization's inhabitants. TombSeer combines three key elements: first, 3D holograms; second, an augmented reality platform that displays computer-generated information; and third, cultural heritage objects that connect people to their ancestors [4]. AR systems are commonly used to replicate virtual and museum objects while keeping interfaces user-friendly and appealing. Currently, TombSeer is assisting in the revitalisation of an exhibition by creating computer-generated exhibits of ancient Egyptian objects that seem to be tangible to visitors of the ROM.

MAR technologies are increasingly being used to create, exchange, illustrate, and transmit historical objects, according to Akçayr [5], making cultural heritage more appealing or desirable. Numerous researchers have been drawn to the extraordinary global penetration of smartphones and tablets, according to [12]. As a result, the use of wearable and mobile devices in the field of cultural heritage can only develop [13]. Despite the fact that smartphones and tablets are simple to use, they can cause people to become distracted from their surroundings. Furthermore, according to [14], objects visualised using AR techniques lack accuracy because the emphasis is moved away from the culture and sociological elements of the artifact and towards the technologies being used. To avoid being overwhelmed by other entertainment channels, Ramly proposes that museums investigate AR as a part of their activities [15]. To build a digital museum and digital legacy that people can connect to and appreciate, the use of digital media and AR technologies must be properly expanded [16]. This will inspire more people to attend museums, where they will experience a mix of traditions and new technologies [17].

According to a study by Ridel [12], historical objects provide evidence of ancient events, allowing them to be explored and assessed, and thereby aiding in the preservation and transmission of cultural values to future generations of people, especially visitors. Geometric detail is often synonymous with museum cultural objects, and is difficult to distinguish due to ageing or deterioration [18]. AR technology such as 3D scanning and mobile computing can help save ancient objects by virtually recording them [12, 19]. Via interactive replicas, AR technologies have helped people develop their understanding of historical objects and places [20]. AR systems can provide a three-dimensional perspective of objects, allowing for the recognition of shapes and symmetrical features that would otherwise be difficult to see with the naked eye [21]. AR instruments, according to [20], provide a beneficial opportunity to comprehend and evaluate historical objects and their meanings. In comparison to virtual reality (VR), augmented reality (AR) allows museum visitors to better understand history and archaeological facts by using real-world examples [22]. For example, augmented reality allows museum visitors to see and engage with interactive objects, giving them a new experience.

2 Methodology

2.1 Research Approach

This research aims to find out what augmented reality experience will add to the experience of these artifacts and how AR technologies affect people's experience of understanding historical artifacts in general. To achieve this aim, the following question must be answered:

1. What type of media will be effective and informative to use in historical locations to explore historical artifacts?

To answer this question, a qualitative research strategy was employed. According to Rohrer [23], qualitative approaches are a better way to address the questions of 'why', 'what', and 'how'.

2.2 Participants

Participants in this study were divided into two groups of people and each group was divided in two different treatments. The first group (A) comprised 12 participants who are familiar with the area and culture of Saudi Arabia's west coast where "Old Jeddah City" is located. Meanwhile the second group (B) consisted of 12 participants who were NOT familiar with this part of the country:

- (Group A) Treatment 1: a group of 6 participants who are familiar with the artifacts and used the IOS device to explore the artifacts.
- (Group A) Treatment 2: a group of 6 participants who are familiar with the artifacts and did NOT use the IOS device to explore the artifact.
- (Group B) Treatment 1: a group of 6 participants who are NOT familiar with the artifacts and used the IOS device to explore the artifact.
- (Group B) Treatment 2: a group of 6 participants who are NOT Familiar with the artifacts and did NOT use the IOS device to explore the artifact.

This study sample comprised 24 participants. According to Guest, Bunce, and Johnson [24] a reasonable sample size for a qualitative study when interviews are being conducted is between 15 and 30 people. The sampling size here is estimated to be about 20 to 25 participants, and all the respondents were students at the Faculty of Engineering and IT, University of Technology Sydney.

2.3 Materials

This research used Unity 3D software to design an AR application that can recognise artifacts and generate information about them to users via a short and animated video. Vuforia targeted the objects in the database and offered them an augmentable rate of the picture based on the picture features (Fig. 1). Unity 3D is a cross-platform programme that can be used to create video games and smartphone apps. Vuforia, on the other hand, is an augmented reality software development kit that can be linked to Unity using a unique licence key. It recognises and tracks planar images and 3D structures in real-time using computer vision technologies.

Here are the artifacts:

Mebkhara (Incense): This is a package for burning incense. When burnt, incense is an aromatic biotic agent that emits a fragrant smoke. The word should refer to either the substance or the scent. Incense is used for a variety of purposes, including aesthetics, aromatherapy, sleep, and ritual. It's also effective as a deodorant and insect repellent.

Lightness: An old way to bright the rooms, houses, and streets

Tray: Old Fashion designed tray. the usage of the tray is to help hold things to be transferred easily to somewhere else.

Bread Holder and Keeper: This is an old way usage to keep the bread from beeng mold

Quran holder: This to be used in Mosques for those who want to read the holy Quran. The main reason for this is because prayers spend a long time reading the holy Quran and this object helps them from getting tired from holding it the whole time.

In Unity 3D, we have designed an animated video that illustrates a short story with a description of five artifacts and we took pictures of them, saving them as targets in Vuforia. When the user points the phone at an artifact, the camera will recognize the tracker of the artifact and will play a short video saved in Unity (Fig. 1).

Fig. 1. Example of one participant are pointing on the artifacts and watching the information of the artifacts.

2.4 Procedure

The five artifacts were placed on the table in a small quiet closed room (Fig. 2). After each user went through the screened questions, they entered the room and commenced the study. Based on the screening questions, participants were chosen based on their knowledge of the artifact. For example, if a participant recognized two of the five artifacts, then he/she would be tested only on the remaining three unrecognized artifacts.

Fig. 2. The displayed artifact that was used in the study.

First, we started with treatment Group 1, who were familiar with the artifact and used the IOS device. Second, we experimented with treatment Group 3, who were familiar with the artifact but did not use the IOS device. Then, we continued with treatment Group 2, who was unfamiliar with the artifact and who used the IOS device. Last, we ended the experiments with treatment Group 4, comprising people who were unfamiliar with the artifact and did not use the IOS device. Based on the experiment protocol, each participant signed the consent form and then all respondents were divided into groups for the experiment. Finally, each person was interviewed individually.

The research data were analyzed using thematic analysis. The ability to code and categorize data into themes is made possible by thematic analysis, for instance, how issues affect participants' perceptions. Processed data may be displayed and categorized according to similarities and differences in thematic analysis [25]. Therefore, we coded all the answers that employed the keywords "Story", "How it works or Information", and "Gamification or Gaming". All the coded words were documented on an Excel spreadsheet to calculate the percentages emerging from the data.

3   Results

Results show that users prefer to use media to mediate their experience of historical museums. The result came after 1 of the users who wanted to explore the historical artifact watched a short video explaining the purpose of the artifacts and how they were used in the historical sites. Meanwhile, 14 users preferred the technique of storytelling to inform and educate them about the history of the artifacts, where they came from and how they were used, while no one preferred gaming to explore the historical artifacts and obtain information in that way (Fig. 3).

Do you prefer to view a story about these artifacts, Gaming, or just to see how it works?	Percentage
Stories	59%
How it works	4%
gaming	0
Stories and Gaming	4%
stories and Information	25%
information and games	8%
Gaming, Information, and Gaming	0

Fig. 3. This is a summary of question number (17, 18) in iOS user device user questionnaire, and (13, 14) in Non iOS user device user questionnaire

After segregating the two groups, the result concerning group A who used the augmented reality technology to explore the museum artifacts revealed that six people preferred historical stories, 4 liked the option of stories and information of how they worked. Only one user preferred to receive information about how the artifacts worked, while another preferred the stories and gaming to explore the artifacts (Fig. 4).

Do you prefer to view a story about these artifacts, Gaming, or just to see how it works?	Percentage
Stories	50%
How it works	8%
gaming	0
Stories and Gaming	8%
stories and Information	34%
information and games	0
Gaming, Information, and Gaming	0

Fig. 4. The result of group A in the study.

On the other hand, the finding for group B, who did NOT use the augmented reality technology to explore the Museum Artifacts, revealed that eight people prefer stories, two like both stories and information, and two were receptive to both gamification and information of how the artifacts work. The study employed NVivo software to code participants' statements. Figure 5 shows the results concerning group B, while Table six summarizes the references to Storytelling, Information about the Artifacts, and Gamification.

Do you prefer to view a story about these artifacts, Gaming, or just to see how it works?	Percentage
Stories	67%
How it works	0
gaming	0
Stories and Gaming	0
stories and Information	16%
information and games	17%
Gaming, Information, and Gaming	0

Fig. 5. The result of group B in the study.

The study was also analyzed using NVivo software to code interesting quotations and responses from the users in the study. The table shows the references of citation, while the table shows interested quotations of the users (Figs. 6 and 7).

Query	Description	Files	References
IOS Users	Prefer Gamification	3	3
	Prefer To get information of the artifacts	3	4
	Prefer a Story	9	16
I don't Preferer to Read	Prefer Gamification	4	4
	Prefer To get information of the artifacts	5	6
	Prefer a Story	7	11

Fig. 6. The table describe the number of referencing Storytelling, Information of the Artifacts, and gamification

Query	Description
Storytelling	Story will be much better because it can relate to it. Whenever I see stories like this, it's more than just getting a passive information. for just the sake of information. So, I think the story is enough.
	stories are much better. It actually explains the things a lot more.
	showing the story in different ways could interact people for watching
	I prefer a story because it's just like giving an example. And that example includes the name in some cases, so you can fit the information inside the story that will be even better. I'm satisfied with the story that I have seen.
	Beyond knowing the history, a story is great.
Gamification	Personally, I would not use tech as little tech as possible because that's an additional overhead for the people to read and do that. Videos, gaming's, and all I think are quite intuitive and easy to follow.
	I think that would be more interactive for them to go through the history and using kind of games tracking them to too deeply to the history of that artifacts and moving them or transferring them from this year to two other years through the history to different periods. So that would be really great if you use it
	Gaming would be the new beta
Short video	I don't want a story. I just want to see how it's work. That's it. Most of the people I think they want the same thing.
	I would first I would want to know what it is. I imagine that is what would prompt me to do, a short description of this is what I want.

Fig. 7. The table is the interested quotation of participants responses.

4   Discussion

The result summarizes *"what type of media the users prefer to use when exploring the historical artifacts"*. It emerged that 59% of participants prefer the technique of storytelling to inform and educate them about the history of the artifacts, how they were used, etc. The study applied to 24 participants, but when we segregated the results into two groups, the users preferred the storytelling method to explore the artifacts. According to the data for group A which used augmented reality technology to explore the museum artifacts, it emerges that 50% of users prefer historical stories. At the same time, the other half can be broken down into 16% who prefer both stories and artifact details, 8% who like both stories and gamification, and 8% who are receptive to information about how the artifacts work. On the other hand, the data concerning group 2, those who have NOT used augmented reality technology to explore museum artifacts, indicate that 67% of users like historical stories, 34% prefer both stories and artifact details, and 17% desire to use both gamification and information about how the artifacts' function.

Moreover, exciting statements were provided by those participants who preferred storytelling to explore the museum artifacts. For example, from group 1 came the comment: *"I think it would be interesting because in gaming you still have stories"*, while another user remarked, *"Story will be much better because I can relate to it. Whenever I see stories like this, it is more than just getting passive information"*. A user from group 2 remarked, *"Beyond knowing the history, a story is great. The game is interesting. But Story is enough"* (see more responses in Fig. 5). Thus, when answering the research question, it is evident that storytelling is the preferred information technique of users to get information about the artifacts at the historic location.

5   Conclusion

This research aimed to investigate users' expectations of mediating technologies when exploring museum artifacts. To achieve this, we recruited 24 participants into this qualitative study. The participants were segregated into two groups, and the aim was to investigate the expectation and satisfaction of mediating the exploration of artifacts. This study employed a short video of the artifact explaining how it worked, playing a game to find out more about the artifact, or being informed using a storytelling technique. The investigation results demonstrate that users prioritized storytelling techniques within the AR experience to understand the history and significance of the artifacts.

References

1. Azuma, R.T.: A survey of augmented reality. Presence: Teleoperators Virtual Environ. **6**(4), 355–385 (1997)
2. Zhou, F., Duh, H.B.L., Billinghurst, M.: Trends in augmented reality tracking, interaction and display: a review of ten years of ISMAR. In: 7th IEEE/ACM International Symposium on Mixed and Augmented Reality, vol. 2008, pp. 193–202. IEEE (2008)
3. Lee, K.: Augmented reality in education and training. TechTrends **56**(2), 13–21 (2012). https://doi.org/10.1007/s11528-012-0559-3
4. Pedersen, I., Gale, N., Mirza-Babaei, P., Reid, S.: More than meets the eye: the benefits of augmented reality and holographic displays for digital cultural heritage. J. Comput. Cult. Herit. (JOCCH) **10**(2), 1–15 (2017)
5. Akçayır, M., Akçayır, G.: Advantages and challenges associated with augmented reality for education: a systematic review of the literature. Educ. Res. Rev. **20**, 1–11 (2017)

6. Dhir, A., Al-kahtani, M.: A case study on user experience (UX) evaluation of mobile augmented reality prototypes. J. UCS **19**(8), 1175–1196 (2013)
7. Rasinger, J., Fuchs, M., Beer, T., Höpken, W.: Building a mobile tourist guide based on tourists' on-site information needs. Tour. Anal. **14**(4), 483–502 (2009)
8. Yovcheva, Z., Buhalis, D., Gatzidis, C.: Smartphone augmented reality applications for tourism. E-Rev. Tour. Res. (eRTR) **10**(2), 63–66 (2012)
9. Gao, T., Deng, Y.: A study on users' acceptance behavior to mobile e-books application based on UTAUT model. In: 2012 IEEE International Conference on Computer Science and Automation Engineering, pp. 376–379. IEEE (2012)
10. Nóbrega, R., Jacob, J., Coelho, A., Weber, J., Ribeiro, J., Ferreira, S.: Mobile location-based augmented reality applications for urban tourism storytelling. In: 24° Encontro Português de Computação Gráfica e Interação (EPCGI), vol. 2017, pp. 1–8. IEEE (2017)
11. Xie, X., Tang, X.: The application of augmented reality technology in digital display for intangible cultural heritage: the case of Cantonese furniture. In: Kurosu, M. (ed.) HCI 2018. LNCS, vol. 10902, pp. 334–343. Springer, Cham (2018). https://doi.org/10.1007/978-3-319-91244-8_27
12. Ridel, B., Reuter, P., Laviole, J., Mellado, N., Couture, N., Granier, X.: The revealing flashlight: interactive spatial augmented reality for detail exploration of cultural heritage artifacts. J. Comput. Cult. Herit. (JOCCH) **7**(2), 1–18 (2014)
13. Olsson, T., Kärkkäinen, T., Lagerstam, E., Ventä-Olkkonen, L.: User evaluation of mobile augmented reality scenarios. J. Ambient Intell. Smart Environ. **4**(1), 29–47 (2012)
14. Brancati, N., Caggianese, G., De Pietro, G., Frucci, M., Gallo, L., Neroni, P.: Usability evaluation of a wearable augmented reality system for the enjoyment of the cultural heritage. In: 2015 11th International Conference on Signal-Image Technology and Internet-Based Systems (SITIS), Conference Proceedings, pp. 768–774. IEEE (2015)
15. Ramly, M.A., Neupane, B.B.: ExplorAR: a collaborative artifact-based mixed reality game. In: Proceedings of the Asian HCI Symposium'18 on Emerging Research Collection, Conference Proceedings, pp. 1–4 (2018)
16. Lee, T.-H., Hsu, K.-S., Yeh, L.-J.: Design and application of the augmented reality with digital museum and digital heritage. In: Chang, M., Hwang, W.-Y., Chen, M.-P., Müller, W. (eds.) Edutainment 2011. LNCS, vol. 6872, pp. 25–26. Springer, Heidelberg (2011). https://doi.org/10.1007/978-3-642-23456-9_5
17. Kyriakou, P., Hermon, S.: Can I touch this? Using natural interaction in a museum augmented reality system. Digit. Appl. Archaeol. Cult. Herit. **12**, e00088 (2019)
18. Manuella, K., Ovidiu, D.: ArchaeoInside: multimodal visualization of augmented reality and interaction with archaeological artifacts. In: Ioannides, M., et al. (eds.) EuroMed 2016. LNCS, vol. 10058, pp. 749–757. Springer, Cham (2016). https://doi.org/10.1007/978-3-319-48496-9_60
19. Vanoni, D., Seracini, M., Kuester, F.: ARtifact: tablet-based augmented reality for interactive analysis of cultural artifacts. In: 2012 IEEE International Symposium on Multimedia, Conference Proceedings, pp. 44–49. IEEE (2012)
20. Nofal, E., Elhanafi, A., Hameeuw, H., VandeMoere, A.: Architectural contextualization of heritage museum artifacts using augmented reality. Stud. Digit. Herit. **2**(1), 42–67 (2018)
21. Damala, A., Stojanovic, N., Schuchert, T., Moragues, J., Cabrera, A., Gilleade, K.: Adaptive augmented reality for cultural heritage: ARtSENSE project. In: Ioannides, M., Fritsch, D., Leissner, J., Davies, R., Remondino, F., Caffo, R. (eds.) EuroMed 2012. LNCS, vol. 7616, pp. 746–755. Springer, Heidelberg (2012). https://doi.org/10.1007/978-3-642-34234-9_79
22. Hu, P.-Y., Tsai, P.-F.: Mobile outdoor augmented reality project for historic sites in Tainan. In: 2016 International Conference on Advanced Materials for Science and Engineering (ICAMSE), pp. 509–511. IEEE (2016)

23. Rohrer, C.: When to use which user-experience research methods. Nielsen Norman Group, pp. 1–7 (2014)
24. Guest, G., Bunce, A., Johnson, L.: How many interviews are enough? An experiment with data saturation and variability. Field Meth. **18**(1), 59–82 (2006)
25. Miles, M.B., Huberman, A.M.: Qualitative Data Analysis: An Expanded Sourcebook. SAGE, Thousand Oaks (1994)

A Systematic Literature Review of Blockchain Technology for Identity Management

Mekhled Alharbi[1,2(✉)] and Farookh Khadeer Hussain[1]

[1] School of Computer Science, Faculty of Engineering and Information Technology, University of Technology Sydney, Sydney, Australia
Mekhled.alharbi@student.uts.edu.au, Farookh.hussain@uts.edu.au
[2] Department of Computer Science, College of Computer and Information Sciences, Jouf University, Sakaka, Saudi Arabia

Abstract. Thanks to identity management solutions, identities and operations such as access control have become increasingly common in real-world applications. The rapid growth of blockchain technology has a significant impact on various sectors like financial and healthcare institutions. Blockchain technology has attracted significant attention due to its distinctive characteristics, such as decentralization, trustworthiness, and immutability, therefore, it has recently been employed in identity management where the decentralized and distributed nature of blockchain technology eliminates the need for a trusted third party. Recently, endeavors have been made to implement blockchain-based identity management solutions that enable users to take ownership of their identities. Therefore, it is imperative to conduct an extensive review of the implementation of blockchain in identity management. This systematic review analyzes the state-of-the-art blockchain researches in the realm of identity management and provides a systematic analysis of the identified literature. After extracting relevant scientific papers, we divide the results into various categories. Finally, we highlight several research challenges and potential future research directions on blockchain in the context of identity management.

Keywords: Blockchain · Identity management · Authentication · Privacy · Trust

1 Introduction

The majority of traditional services have been migrated to online forms as a result of the internet's rapid expansion. Users must submit a valid identity to access online application services. This identity can take numerous forms, including passport, driver's license, birth certificate, etc. The term "identity management" refers to the process of identifying, authenticating, and authorizing an entity to access resources [1]. The online applications store users' personal information in a centralized fashion when users share their information to access services. This information is controlled by central authorities who access user data without their consent. Thus, users have no control over this information. Centralized identity management systems which deprive users

© The Author(s), under exclusive license to Springer Nature Switzerland AG 2022
L. Barolli et al. (Eds.): AINA 2022, LNNS 451, pp. 345–359, 2022.
https://doi.org/10.1007/978-3-030-99619-2_33

of ownership of their identity is a major concern [2]. Hence, a central authority may exploit users' trust, posing significant privacy and security concerns. This raises concerns about identity theft that necessitates robust identity management solutions. Thus, it is imperative to address these issues of identity. Efforts are underway to decentralize identity management to alleviate the aforementioned concerns.

The advent of blockchain technology is paving the way for new opportunities for resolving critical data privacy, security, and integrity challenges in identity management [3]. Over the last few years, blockchain technology has garnered substantial attention from both industry and academia. The recent introduction of blockchain and smart contracts as extensions of distributed ledger technology are redefining business models and management in different use-cases including healthcare, the Internet of Things (IoT) and smart cities. The technology is known for being a tamper-resistant and transparent ledger [4]. Thus, it can be utilized to link users' claims to their identities, thereby preventing identity fraud in identity management. Furthermore, blockchain is characterized by attractive features such as immutability, decentralized nature, and traceability, making it ideal for use in the identity management field. Blockchain technology has enormous potential in the realms of digital identity attributed to its advantages such as decentralization, tamper-resistance, and network data sharing [5].

The essential purpose of blockchain technology is to remove reliance on a third party leading to direct communication between users and stakeholders; therefore blockchain technology has emerged to establish trust between parties [6]. The intrinsic features of blockchain make it possible for both parties to communicate with each other in a trustworthy and secure way, without the disclosure of sensitive information. Blockchain technology can ensure secure and trustworthy data exchange between users and stakeholders based on its immutability and anonymity features [7]. Blockchain can bring great value in identity management by giving identity ownership to users as an identity owner. Blockchain has very substantial possibilities in the identity management area because of its characteristics, such as tamper-resistance, decentralization, and transparency. It has become a necessity to integrate identity management systems into one single system for all stakeholders to achieve transparency, security, and immutability.

There are multiple studies in the literature which discuss the use of blockchain technology in various fields, such as financial markets, the Internet of Things (IoT), smart homes, and healthcare [8–14]. A few review articles have been published on blockchain applications in the identity management domain [15, 16]. However, no systematic literature review of recent research on blockchain-based identity management applications has been conducted and no papers that address blockchain technology in identity management applications in a systematic manner have been published. We aim to overcome this shortcoming by providing a technical background in blockchain-based identity management applications which highlight recent developments in the field. This article examines the recent research on blockchain-based identity management and examines its strengths and shortcomings. The study also highlights the existing identity management research challenges.

This paper provides a systematic review of the state-of-the-art in the realm of identity management using blockchain technology and systematically gathers and categorizes blockchain-related research publications. The purpose of this study is to demonstrate the potential use of blockchain in identity management, as well as to highlight the obstacles and prospective directions of blockchain technology. We conducted a systematic literature review (SLR) on blockchain to provide valuable insights.

The contributions of this study are as follows:

- It provide a brief review and analysis of identity management and blockchain technology.
- It discusses the benefits and drawbacks of currently used blockchain-based identity management solutions.
- It investigates several identity management solutions based on blockchain technology.
- It analyses identity management solutions using blockchain technology based on a variety of factors.
- It examines the primary challenges associated with identity management solutions in the context of blockchain technology.
- It analyses the SLR's findings and makes recommendations for future research.

The rest of this paper is organized as follows: Sect. 2 briefly reviews identity management and overviews of blockchain technology. Section 3 outlines the research methodology we followed. The study's results and a discussion of the existing work is given in Sect. 4. Section 5 identifies some challenges and future research directions. Finally, Sect. 6 concludes the paper.

2 Preliminaries

This section provides the essential background for understanding the remainder of the paper, including identity management and blockchain.

2.1 Identity Management

Identity management is a mechanism by which participants are validated, recognized, and authorized to access sensitive data [17]. In the literature, identity management is often referred to as identity and access management. Identity management systems comprise three main components: a user, a service provider, and an identity provider. These three parties are interconnected entities: the user requests a service from the service provider, and the identity provider is tasked with validating the user's identity via the authentication protocol. The traditional identity management approaches are effective for service providers but ineffective for users, as they must remember numerous passwords to access various websites. In the literature [18–20], there are various identity management models that can be broadly divided into three categories which are isolated model, federated model, and centralized model.

However, the traditional method to store personal information is by adopting a centralised manner. The major challenge is to deny users legitimate ownership of their identity information online through a centralized identity management model [21]. Such architecture enables the attackers to penetrate these systems and access the information. Furthermore, a third party may violate the user's trust which elevates certain issues of security and privacy. With the aim of overcoming the concerns of centralized databases pertinent to the security and privacy aspects, a decentralized identity management approach is underway to ensure the system is robust. The emergence of the new technology which is known as blockchain, helps users to use the internet without relying on a trusted third party [22]. The major advantage of blockchain is the decentralized structure since all the nodes of the network are retained in the entire database.

2.2　Blockchain Technology

2.2.1　Overview of Blockchain

Blockchain was invented by Nakamoto in 2008 [23] and is a collection of interconnected blocks that store all transaction records. Blockchain works by storing data in distributed ledgers that are disseminated across all computing devices in a decentralized fashion. The blockchain structure is composed of a sequence of blocks. The block is comprises two major sections which are the block header and the block body. The block header contains a block version, Merkle tree, timestamp, and parent block hash. The block body consists of a transaction counter and transactions. Each block contains the prior block hash in the block header, thus it can be linked to one parent block only. This creates a connection between the blocks, resulting in the formation of a chain of blocks. The series of hash operations form an immutable chain that can be traced back to the first block produced. The genesis block is the first block on a blockchain and it does not have a parent block. The consensus of a majority of the network's participants must confirm each transaction before it can be recorded in the public ledger. Data cannot be altered or deleted once it has been entered.

Blockchain has prominent features that make it attractive as a decentralized technology due to the fact that the ledger is not controlled by a central authority. The following are some of these features:

- **Decentralization**: This is the essence of blockchain technology, as each node maintains a record of all transactions, thereby eliminating the need for a central authority. A central trusted organization should validate transactions causing performance bottlenecks at the central servers. Unlike centralized systems, blockchain eliminates the reliance on a third party.
- **Transparency**: Records are shared among all the participants in the blockchain. Each participant in the network has the same obligations and permissions to access permitted information.
- **Traceability**: The blockchain employs timestamps to identify and record each transaction, thereby reinforcing the data's time dimension. This enables the participant to maintain transaction order and to make the data traceable. Thus, every transaction can be traced back to a certain time, making it easier for participants to identify the parties involved.
- **Trust**: Data exchange between participants in the network does not require mutual trust between participants because blockchain is deployed in a decentralized manner. Therefore, the trust is shifted from a third party into the technology itself.

- **Immutability**: This feature ensures that any confirmed transaction cannot be tampered with. Hence, the data is unaltered after being stored on the blockchain.
- **Anonymity**: Every user on the blockchain has the ability to interact with an established address. The system will not reveal the user's true information; nonetheless, participants will be able to access the encrypted transaction information.

2.2.2 Smart Contracts

The notion of the smart contract was first introduced by Szabo [24]. A smart contract is a computer program that is not executed until the relevant data or action is received [25]. The smart contract is a form of electronic agreement of a legal contract between parties to the transaction. The goal of the smart contract is to eliminate the need for a trusted intermediary. A smart contract includes execution rules and execution logic. When the rule is satisfied, the execution logic is performed automatically. Data is only released by a smart contract when certain rules are satisfied. The availability of the smart contract in blockchain builds trust among participants and automatically removes the need for a trusted third party. Ethereum is the most popular blockchain platform for smart contracts.

3 Research Methodology

The primary objective of this study is to examine the existing literature that has investigated in the field of identity management within the framework of blockchain technology and to identify critical research gaps that require further investigation in future studies. We survey the existing literature to identify the relevant issues, challenges, and solutions of blockchain-based identity management. We conducted an SLR to accomplish this goal by using the procedure outlined in Kitchenham et al. [26]. The SLR is an organized and systematic approach to defining, synthesizing, and selecting recent literature related to the research objectives. This research comprises citation and evaluation procedures to complement the basic SLR approach to ensure the quality of the literature review.

The systematic approach involves the following steps:

- Data source selection and search strategies.
- Inclusion and exclusion criteria.
- Citation and inclusion decision management.
- Final selection and quality assessment.
- Data extraction and synthesis.

3.1 Data Source Selection and Search Strategies

Many sources have been explored to obtain an unbiased and comprehensive perspective, including the main online databases. The following most popular scientific databases were used to conduct this literature review:

1. IEEE Xplore (https://www.ieeexplore.ieee.org)
2. Elsevier ScienceDirect (https://www.sciencedirect.com)

3. SpringerLink (https://link.springer.com)
4. ACM Digital Library (https://dl.acm.org)
5. Google Scholar (https://scholar.google.com)

These well-known scientific databases were chosen because the related literature is sufficiently covered. The papers reviewed were chosen from industry papers, qualitative and quantitative studies, and scientific academic studies. Figure 1 shows the review process at each stage and the number of papers identified. We used the Boolean operator "AND" to search for relevant research using various combinations of items from all of the search terms. The "OR" operator is used to connect similar terms to ensure maximum coverage. The search statement is split into two major sections. The first sub-section is composed of a collection of blockchain-related phrases. The second sub-section contains a collection of phrases related to identity management. As a result, the following search string is produced:

("blockchain" OR "distributed ledger technology" OR "smart contract") AND ("identity" OR "identity management").

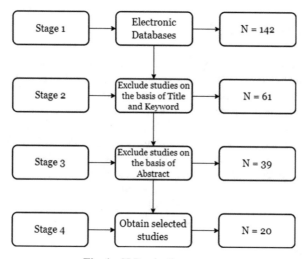

Fig. 1. SLR selection process

3.2 Inclusion and Exclusion Criteria

We include certain studies that are pertinent to blockchain-based identity management and its applications that meet a certain criterion. The following factors were taken into consideration when deciding whether to include or exclude a study:

1. The paper must be relevant to the topic of blockchain and identity management.
2. The study was conducted between 2017 and 2021.
3. The paper is written in English and the full content is available.

4. The article must have undergone a peer review process.
5. The paper must include empirical evidence relating to the use of blockchain technology for identity management.

Therefore, the study excludes papers that either do not focus on blockchain and identity management or meet the following exclusion criteria:

1. Duplicate studies.
2. Studies are not written in English.
3. Grey literature.

3.3 Citation and Inclusion Decision Management

At this stage, all 142 papers were exported and stored in EndNote, where we reviewed them using search terms in either the title or the keywords. A paper was selected if it contains at least two search terms; one from each section, in either the title or the list of keywords; otherwise, the paper was not selected for the next filtration stage. The total number of selected papers was reduced to 61 by conducting this filtering process.

3.4 Final Selection and Quality Assessment

All abstracts were thoroughly examined to ensure their relevance before being included in the final stage. The articles with pertinent abstracts were selected to go through to the next filtration stage; otherwise, the paper was excluded. The total number of selected articles was reduced to 39 by carrying out this filtration process. Table 1 describes the scientific assessment of the filtration process.

Table 1. Scientific assessment process

Filtration stage	Method	Assessment criteria
First filtration	Search keywords from scientific databases	Search terms
Second filtration	Exclude studies on the basis of titles	Title = search term Include else exclude
Third filtration	Exclude studies on the basis of abstracts	Abstract = relevant Include else exclude
Final filtration	Obtain selected papers and critically appraise studies	Discusses data relevant Yes = Accepted No = Rejected

3.5 Data Extraction and Synthesis

At this stage, the 39 papers were analyzed and their quality was ensured according to the quality criteria suggested by Dyba et al. [27] as listed in Table 2. The most relevant papers were selected after careful consideration.

An additional 19 articles were excluded at this stage by applying the criteria in Table 2 to the 39 articles, leaving 20 articles for the final data review and synthesis to address the objectives of the research. The final 20 selected articles were evaluated on the basis of the quality criteria listed in Table 2.

Table 2. Quality criteria

Quality criteria	
1	Is the paper research based?
2	Are the aims of the research clearly stated?
3	Is the context adequately described?
4	Was the design framework appropriate to address the aims of the research?
5	Was the data analysis sufficiently rigorous
6	Is there clear evidence for the findings?
7	Is the study validated or implemented?

4   Results and Discussion of Existing Work

In the literature review, we find different technical challenges in the current blockchain-based identity management systems. We grouped the papers according to the challenge they are attempting to overcome. Thus, the papers are classified into three subcategories for better presentation and to identify the natural affinity between them. This will enable us to organize the literature by grouping the papers with related themes. The classification is based on the remaining reviewed articles after the filtering processes. Of these challenges, the following stand out:

4.1   Authentication

As a decentralized distributed ledger, blockchain technology can act as a trustworthy decentralized authentication infrastructure. A number of research studies have been conducted on blockchain-based identity management for authentication. Juan et al. [28] presented an authentication model for a national electronic identity document based on blockchain. They discussed the security issues that are encountered in the Colombia's current national identities, such as the protection of citizens' information and the prevention of fraudulent transactions. Such issues can be addressed by integrating blockchain with biometric authentication technology using smart cards and leveraging the benefits of established authentication methods such as biometrics and physical security to mitigate the security concerns associated with identification documents. Hence, it helps the government to verify a document and identify counterfeit documents.

Liu et al. [29] presented a model to preserve privacy to manage identity by integrating biometrics and blockchain. A government-specified body gathers and stores the user's identity in IPFS. A smart contract on Ethereum governs the system's access control. The system's primary objective is to enhance identity management while providing data security using smart contracts. However, user registration at entry points is required to safeguard the system against any data breaches.

Zhou et al. [30] presented a self-sovereign digital identity management framework (EverSSDI) based on IPFS and smart contracts to develop a framework for decentralized identity management. Users encrypt and maintain their personal data in the IPFS system utilizing data hash fingerprints which are verified through a smart contract. Therefore, the user becomes the real dominant owner of the identity instead of just proving their digital identity. However, in the authorization procedure, users must supply identity attributes to the service providers.

Odelu [31] presented a novel biometrics-based authentication approach in which the user's identity is managed via a blockchain. The author conducts a thorough security study of the protocol, proving it is resistant to known attacks. However, the technique does not guarantee user anonymity or untraceability.

Othman and Callahan [32] proposed the Horcrux protocol, a secure decentralized authentication method which allows the end-users of a self-sovereign identity to have the control of accessing their identities through a biometric authentication that is capable of ensuring the privacy of the user. The protocol relies on decentralized identifiers and it is based on the concept of self-sovereign identity. They implement a decentralized biometric credential storage mechanism using a blockchain to store decentralized identifiers.

Xu et al. [33] developed a blockchain-based identity management and authentication mechanism based on the redactable blockchain for mobile networks, where users retain ownership over their identifying information. The blockchain stores legitimate users' self-sovereign identities (SSIs) and public keys, and the chameleon hash is utilized to remove unlawful users' data while leaving the block head unaltered.

Chen et al. [34] presented a decentralized identity management system and a cross-domain authentication method based on blockchain. The objective is to eliminate the authentication center's single point of failure and increase the cross-domain authentication performance. The uniqueness of an identifier is determined by the consensus mechanism of the consortium blockchain and anyone can request identifiers. The system uses a one-way accumulator to ensure the validity of entity identity.

Lee [35] introduced a blockchain-based solution for managing identity and authentication for mobile users and IoT devices. Their proposed approach is to generate and maintain blockchain identities as a service, without regard for interactions or messages via the blockchain. The blockchain-based identities are only intended to be used for decentralized authentication in this scenario. Thus, authentication can be accomplished without having any preregistered users' information.

Jamal et al. [36] presented a blockchain-based identity system for storing personal information. This solution makes use of blockchain features to ensure that users are aware of who has access to their personal data. The system enables third parties to access personal records while maintaining their immutability.

4.2 Privacy

Privacy is a major concern that is still being researched. Several privacy-preserving techniques have been discussed in the literature. For example, Chalaemwongwan and Kurutach [37] developed a national digital ID framework based on blockchain (NIDBC) to assist in enhancing a digital identity to a single sign-on for government services. Moreover, they affirmed that privacy is preserved by allowing users to control their data through granting permission for services to access their personal information. Furthermore, due to the inherent nature of blockchain, the system is secure since the data is distributed, which makes it difficult for attackers attack data. However, the service provider is still able to abuse users' information.

Rathee and Singh [38] proposed a blockchain-based self-sovereign identity management system. IPFS maintains users' data whereas blockchain maintains the content address of their data and the public key. Smart contracts that operate on the blockchain perform the verification process. Third parties are not permitted to access the data directly, which is exclusively accessible to the user. Data privacy is thus ensured in this manner.

Mudliar et al. [39] presented a model that utilizes blockchain technology to enable people to carry their national identity on their phones. Regime officials can verify a citizen's national identity by scanning a barcode or QR code generated automatically through the regime site. The major benefit of this approach is that communication between the regime and the citizens is transparent. However, they demonstrated the theoretical system architecture only.

Saldamli et al. [40] designed a system that uses Ethereum-based blockchain to store an identity hash on blockchain. A data hash is created by IPFS and the corresponding hash is stored on Ethereum blockchain. Documents are approved for verification of the identity by the user once the third party requests it. The user has the right to share the document or reject it. Thus, personal data can be completely controlled by the user with this system.

Faber et al. [41] proposed a conceptual design for a blockchain-based personal data and identity management system that is human-centric and GDPR compliant. They presented a framework that is transparent and gives data owners complete control over how their data is used. However, this study is still conceptual and does not provide any technical specifics or performance evaluation.

Kassem et al. [42] presented blockchain-based identity management as a means of securing personal data sharing across networks, and they emphasized the importance of the blockchain and decentralized self-sovereign identities. Moreover, the system allows users to retain their identities linked to specific attributes that can be used by service providers to authenticate the user and provide their services based on verified attributes. They attempted to leverage blockchain and its characteristics as the backbone of identity management across all the realms. The result of the security analysis demonstrated that it is possible to develop a secure and resilient identity management system that can overcome the drawbacks associated with centralized identity management systems.

Rathee and Singh [43] developed a blockchain identity management technique based on the Merkle hash digests algorithm (MHDA) that allows data sharing without being compromised by anonymous users. MHDA-based BIdM systems are efficient in terms of allowing users to maintain control over their identities and credentials. The blockchain's feature of intervening conflict ensures the integrity of the user's data.

4.3 Trust

Trust is critical when developing identity management systems. Several techniques have been proposed to provide identity in the context of mutual distrust. For example, Stokkink and Pouwelse [44] introduced a digital identity model based on blockchain which builds on a generic provable claim model using zero-knowledge proofs and the collection of third-party attestations of trust is required. The work focused on a self-sovereign identity for the Netherlands and was part of an undertaking by the government that provides identity within the context of a mutual solution. They assert that their systems are suitable for general use, however, it is not shown how the work integrates with existing IT applications.

Buccafurri et al. [45] suggested an architecture to integrate blockchain technology with identity management via identity-based encryption to achieve the trust level between users. They created a non-anonymous blockchain by binding a digital identity with a public key, which can be used to define the author of the transaction.

Hammudoglu et al. [46] developed a biometric-based authentication mechanism and blockchain storage. This enables the user to maintain personal information securely, which can be accessed upon successful biometric authentication. It combines a permissionless blockchain with identity and key attestation capabilities for use with mobile phones. However, a fully accessible blockchain is used to store the unencrypted fingerprints, which compromises both security and privacy.

Takemiya and Vanieiev [47] proposed a mobile application-based identity system that leverages blockchain technology to establish a secure protocol for storing encrypted personal data and sharing verified claims about personal data. The hash values of a user's personal information that has been encrypted using a cryptographic key are broadcast on the blockchain in this system. It is developed on top of the permissioned hyperledger "Iroha blockchain". The Sora mobile apps enable users to produce a pair of the encryption keys, insert and encrypt their data, and propagate salted hashes to the blockchain. After this, users have the option of voluntarily providing sensitive information to third parties such as institutions. The drawback is that the system cannot achieve complete decentralization if the keys are stored centrally.

5 Challenges and Future Work

Implementing blockchain technology is not a straightforward application in the realm identity management. There are a number of issues that need to be addressed in relation to the limitations of blockchain technology. We highlighted a number of research challenges in blockchain-based identity management systems that could be addressed in future research based on the findings from this systematic review. Some of these challenges are described as follows:

Identity Information Modifications: Blockchain technology featured with an immutability property which builds trust between participants. Once data is stored in a blockchain, it cannot be altered by any participant. Additionally, it eliminates the possibility of altering this information. On the other hand, this feature could pose a challenge for integrating blockchain with identity management due to the changeable nature of a user's information. As a result, it is considered a barrier to adding new information.

Cost: In light of the fact that blockchain technology is still in its infancy and many people lack the necessary training, there is a pressing demand for training to ensure a deeper understanding of the technology. The existing infrastructure of identity management has to be upgraded to support blockchain technology which will require an additional cost.

Key Management: While password-based systems include a means for recovering lost or stolen passwords, blockchain technology lacks a mechanism for resetting the private key. If a blockchain user's private key is compromised or lost, all of their data and related assets will be lost, affecting the entire system.

The focus of future efforts will involve overcoming these challenges. Therefore, before fully integrating blockchain technology in the identity management field, academics and other stakeholders should carefully evaluate the ramifications of the aforementioned research challenges.

6   Conclusion

Blockchain is a developing technology that has the potential to revolutionize the world of information technology, as an immutable ledger can be used in a wide variety of applications. In this study, a systematic literature review was carried out in order to examine the use of blockchain technology in the identity management domain, challenges and future work. The study demonstrates that utilizing blockchain in identity management has the potential to overcome the limitations of traditional identity management systems. Blockchain research trends in identity management indicate that it is mostly utilized for authentication, data sharing and data ownership, but it is rarely employed for other purposes such as supply chain management. According to our findings, the efforts to apply blockchain technology in identity management is accelerating. In addition, using blockchain, identity ownership will be controlled and owned by the legitimate users. However, certain challenges remain unresolved and need more investigations. Future research directions in identity management have been identified based on the findings, with an emphasis on resolving concerns about the use of blockchain technology in areas such as identity modifications, key management, and the cost of blockchain technology.

References

1. Zhu, X., Badr, Y.: identity management systems for the internet of things: a survey towards blockchain solutions. Sensors **18**, 4215 (2018)
2. Alharbi, M., Hussain, F.K.: Blockchain-based identity management for personal data: a survey. In: Barolli, L. (eds.) Advances on Broad-Band Wireless Computing, Communication and Applications. BWCCA 2021. LNNS, vol. 346. Springer, Cham (2022). https://doi.org/10.1007/978-3-030-90072-4_17

3. Ren, Y., Zhu, F., Qi, J., Wang, J., Sangaiah, A.K.: Identity management and access control based on Blockchain under edge computing for the industrial internet of things. Appl. Sci. **9**(10), 2058 (2019)
4. Yang, L.: The blockchain: state-of-the-art and research challenges. J. Ind. Inf. Integr. **15**, 80–90 (2019)
5. He, Z., Xiaofeng, L., Likui, Z., Zhong-Cheng, W.: Data integrity protection method for microorganism sampling robots based on blockchain technology. J. Huazhong Univ. Sci. Technol. Nat. Sci. Edn. **43**(Z1), 216–219 (2015)
6. Toth, K.C., Anderson-Priddy, A.: Self-sovereign digital identity: a paradigm shift for identity. IEEE Secur. Privacy **17**(3), 17–27 (2019)
7. Casino, F., Dasaklis, T.K., Patsakis, C.: A systematic literature review of blockchain-based applications: current status, classification and open issues. Telematics Inform. **36**, 55–81 (2019)
8. Peter, H., Moser, A.: Blockchain-applications in banking and payment transactions: results of a survey. European Finan. Syst. **141**, 141 (2017)
9. Dorri, A., Kanhere, S.S., Jurdak, R., Gauravaram, P.: Blockchain for IOT security and privacy: the case study of a smart home. In: 2017 IEEE International Conference on Pervasive Computing and Communications Workshops (PerCom workshops), pp. 618–623 (2017)
10. Banerjee, M., Lee, J., Choo, K.K.R.: A blockchain future for internet of things security: a position paper. Digital Commun. Netw. **4**(3), 149–160 (2018)
11. Andoni, M., et al.: Blockchain technology in the energy sector: a systematic review of challenges and opportunities. Renew. Sustain. Energy Rev. **100**, 143–174 (2019)
12. Alketbi, A., Nasir, Q., Talib, M.A.: Blockchain for government services—use cases, security benefits and challenges. In: 15th Learning and Technology Conference (LandT). IEEE, pp. 112–119 (2018)
13. Moniruzzaman, M., Khezr, S., Yassine, A., Benlamri, R.: Blockchain for smart homes: review of current trends and research challenges. Comput. Electric. Eng. **83**, 106585 (2020)
14. H¨olbl, M., Kompara, M., Kamiˇsali´c, A., Zlatolas, L.N.: A systematic review of the use of Blockchain in healthcare. Symmetry. **10**, 470 (2018)
15. Kuperberg, M.: Blockchain-based identity management: a survey from the enterprise and ecosystem perspective. IEEE Trans. Eng. Manage. **67**(4), 1008–1027 (2019)
16. Lim, S.Y., et al.: Blockchain technology the identity management and authentication service disruptor: a survey. Int. J. Adv. Sci. Eng. Inf. Technol. **8**, 1735–1745 (2018)
17. Domingo, A.I.A., Enr´ıquez, A.M.: Digital identity: the current state of affairs. BBVA Res., 1–46 (2018)
18. Ahn, G.J., Ko, M.: User-centric privacy management for federated identity management. In: 2007 International Conference on Collaborative Computing: Networking, Applications and Work sharing (CollaborateCom 2007). IEEE, pp. 187–195 (2007)
19. Birrell, E., Schneider, F.B.: Federated identity management systems: a privacy-based characterization. IEEE Secur. Priv. **11**, 36–48 (2013)
20. Satybaldy, A., Nowostawski, M., Ellingsen, J.: Self-sovereign identity systems, pp. 447–461. Springer, IFIP International Summer School on Privacy and Identity Management (2019)
21. Ferdous, M.S., Poet, R.: A comparative analysis of identity management systems. In: 2012 International Conference on High Performance Computing and Simulation (HPCS). IEEE, pp. 454–461 (2012)
22. El Haddouti, S., El Kettani, M.D.E.: Analysis of identity management systems using blockchain technology. In: 2019 International Conference on Advanced Communication Technologies and Networking (CommNet). IEEE, pp. 1–7 (2019)
23. Nakamoto, S.: Bitcoin: a peer-to-peer electronic cash system. Decentr. Business Rev., p. 21260 (2008)

24. Szabo, N.: Formalizing and securing relationships on public networks. First Monday (1997)
25. Lu, Y.: Blockchain: a survey on functions, applications and open issues. J. Ind. Integr. Manage. **3**, 1850015 (2018)
26. Kitchenham, B., et al.: Systematic literature reviews in software engineering–a tertiary study. Inf. Softw. Technol. **52**, 792–805 (2010)
27. Dyb °a, T., Dingsøyr, T.: Empirical studies of agile software development: a systematic review. Inf. Softw. Technol. **50**, 833–859 (2008)
28. Juan, M.D., Andres, P., Rafael, P.M., Gustavo, R.E., Manuel, P.C.: A model for national electronic identity document and authentication mechanism based on blockchain. Int. J. Modeling Optim. **8**, 160–165 (2018)
29. Liu, Y., Sun, G., Schuckers, S.: Enabling secure and privacy preserving identity management via smart contract. In: 2019 IEEE Conference on Communications and Network Security (CNS). IEEE, pp. 1–8 (2019)
30. Zhou, T., Li, X., Zhao, H.: Everssdi: blockchain-based framework for verification, authorisation and recovery of self-sovereign identity using smart contracts. Int. J. Comput. Appl. Technol. **60**, 281–295 (2019)
31. Odelu, V.: *IMBUA*: identity management on blockchain for biometrics-based user authentication. In: Prieto, J., Das, A., Ferretti, S., Pinto, A., Corchado, J. (eds.) Blockchain and Applications. Advances in Intelligent Systems and Computing, vol. 1010. Springer, Cham (2020). https://doi.org/10.1007/978-3-030-23813-1_1
32. Othman, A., Callahan, J.: The horcrux protocol: a method for decentralized biometric-based self-sovereign identity. In: 2018 International Joint Conference on Neural Networks (IJCNN). IEEE, pp. 1–7 (2018)
33. Xu, J., Xue, K., Tian, H., Hong, J., Wei, D.S., Hong, P.: An identity management and authentication scheme based on redactable blockchain for mobile networks. IEEE Trans. Veh. Technol. **69**, 6688–6698 (2020)
34. Chen, R., et al.: Bidm: a blockchain-enabled cross-domain identity management system. J. Commun. Inf. Netw. **6**, 44–58 (2021)
35. Lee, J.H.: Bidaas: Blockchain based id as a service. IEEE Access **6**, 2274–2278 (2018)
36. Jamal, A., Helmi, R.A.A., Syahirah, A.S.A., Fatima, M.A.: Blockchain-based identity verification system. In: 2019 IEEE 9th International Conference on System Engineering and Technology (ICSET). IEEE, pp. 253–257 (2019)
37. Chalaemwongwan, N., Kurutach, W.: A practical national digital id frame-work on blockchain (nidbc). In: 2018 15th International Conference on Electrical Engineering/Electronics, Computer, Telecommunications and Information Technology (ECTI-CON). IEEE, pp. 497–500 (2018)
38. Rathee, T., Singh, P.: A self-sovereign identity management system using blockchain, pp. 371–379. Springer, Cyber Security and Digital Forensics (2022)
39. Mudliar, K.E., Parekh, H., Bhavathankar, P.: A comprehensive integration of national identity with blockchain technology. In: 2018 International Conference on Communication information and Computing Technology (ICCICT). IEEE, pp. 1–6 (2018)
40. Saldamli, G., Mehta, S.S., Raje, P.S., Kumar, M.S., Deshpande, S.S.: Identity management via blockchain. In: Proceedings of the International Conference on Security and Management (SAM), The Steering Committee of The World Congress in Computer Science, Computer Engineering and Applied Computing (WorldComp), pp. 63–68 (2019)
41. Faber, B., Michelet, G.C., Weidmann, N., Mukkamala, R.R., Vatrapu, R.: Bpdims: a blockchain-based personal data and identity management system. In: Proceedings of the Annual Hawaii International Conference on System Sciences (2019)
42. Kassem, J.A., Sayeed, S., Marco-Gisbert, H., Pervez, Z., Dahal, K.: Dns-idm: a blockchain identity management system to secure personal data sharing in a network. Appl. Sci. **9**, 2953 (2019)

43. Rathee, T., Singh, P.: Secure data sharing using merkle hash digest based blockchain identity management. Peer-to-Peer Netw. Appl. **14**, 3851–3864 (2021)

44. Stokkink, Q., Pouwelse, J.: Deployment of a blockchain-based self-sovereign identity. In: 2018 IEEE International Conference on Internet of Things (iThings) and IEEE Green Computing and Communications (GreenCom) and IEEE Cyber, Physical and Social Computing (CPSCom) and IEEE Smart Data (SmartData). IEEE, pp. 1336–1342 (2018)

45. Buccafurri, F., Lax, G., Russo, A., Zunino, G.: Integrating Digital Identity and Blockchain. In: Panetto, H., Debruyne, C., Proper, H.A., Ardagna, C.A., Roman, D., Meersman, R. (eds.) OTM 2018. LNCS, vol. 11229, pp. 568–585. Springer, Cham (2018). https://doi.org/10.1007/978-3-030-02610-3_32

46. Hammudoglu, J.S., et al.: Portable trust: biometric-based authentication and blockchain storage for self-sovereign identity systems. (2017). arXiv preprint arXiv: 1706.03744

47. Takemiya, M., Vanieiev, B.: Sora identity: secure, digital identity on the blockchain. In: 2018 IEEE 42nd Annual Computer Software and Applications Conference (comp-sac). IEEE, vol. 2, pp. 582–587 (2018)

Performance Evaluation in 2D NoCs Using ANN

Prachi Kale, Pallabi Hazarika, Sajal Jain, and Biswajit Bhowmik$^{(\boxtimes)}$ (iD)

Department of Computer Science and Engineering, National Institute of Technology
Katakana, Surathkal, Mangalore 575025, India
{prachigajanankale.192is021,pallabi.192is020,
Sajj.192is024,brb}@nitk.edu.in

Abstract. A network-on-chip (NoC) performance is traditionally evaluated using a cycle-accurate simulator. However, when the NoC size increases, the time required for providing the simulation results rises significantly. Therefore, such an issue must be overcome with an alternate approach. This paper proposes an artificial neural network (ANN)-based framework to predict the performance parameters for NoCs. The proposed framework is learned with the training dataset supplied by the BookSim simulator. Rigorous experiments are performed to measure multiple performance metrics at varying experimental setups. The results show that network latency is in the range of 31.74–80.70 cycles. Further, the switch power consumption is in the range of 0.05–12.41 μW. Above all, the proposed performance evaluation scheme achieves the speedup of 277–2304× with an accuracy of up to 93%.

Keywords: Network-on-chip · Embedded system · Artificial neural network · Performance analysis and evaluation

1 Introduction

According to the requirement of highly scalable and robust multiprocessor systems, systems-on-chip (SoCs) are expected to satisfy these requirements. However, with the increase in the number of cores, the performance requirements are failed to meet by SoC architectures [1]. Alternatively, an NoC can cope with the needs and act as a promising solution. The advantage includes that the NoC provides parallelism, reusability, and scalability [2]. The analysis of the performance of NoCs is required to check the design of the architecture [3,4]. Primarily the performance evaluation is done by simulations. However, NoC evaluation via simulations is a time-consuming method [5].

Many approaches using artificial intelligence (AI) technology are discussed in the literature over the traditional evaluations. For example, Dong *et al.* [6] have proposed an approach to lower the latency in NoCs using a neural network. Authors in [7] have presented neural network-based network-on-chip interconnect designs. The advantage is a significant improvement in performance over existing

L. Barolli et al. (Eds.): AINA 2022, LNNS 451, pp. 360–369, 2022.
https://doi.org/10.1007/978-3-030-99619-2_34

architectures. Ascia *et al.* [8] have presented a deep neural network accelerator-based scheme to evaluate NoC performance. The results showed a reduction in latency by 30%. Similarly, other contributions, including [9–13] are seen in literature to address multiple NoC issues, including performance evaluation. For example, Zhao *et al.* [11] discussed a bufferless network using time-based sharing of multi-hop single-cycle paths to improve NoCs performance. The scheme lowers latency and power consumption up to 84% and 39%, respectively. Choi *et al.* [12] presented a convolutional neural network for power, network latency estimation. Experimental results show a 50% reduction in power consumption. Xiao *et al.* [13] presented a NeuronLink model that evaluates power and area of NoCs. Results show a speedup of 1.27× and 2.01× in power consumption and area reduction. The works can comparatively counter themselves, but there is a scope of improvisation, resulting in a more effective approach that can quickly evaluate NoC performance over the traditional simulation method and these existing techniques.

This paper proposes an artificial neural network (ANN) model to evaluate various performance metrics- latency, power consumption, area, etc., in NoC architectures. These metrics are a few golden metrics considered in designing on-chip networks. The proposed approach is a learning technique and is trained based on a training dataset supplied by an NoC simulator. BookSim is a popularly preferred simulator for evaluating NoCs and is used here to generate the training/testing dataset for the proposed scheme. Thorough investigations are performed at varying experimental setups with different traffic patterns. Finally, evaluation of the proposed system is done on a set of 2D mesh NoCs. A few highlights from the predicted results are as follows. For instance, the evaluation shows that the network latency is achieved in the range of 31.74–80.70 cycles. Further, the switch power consumption is in the range of 0.05–12.41 µW. Other parameters, such as the average hop count and the total area, are 2–10 and 0.08–4.66 µm², respectively. Above all, the proposed performance evaluation scheme achieves the speedup of up to 277–2304× with an accuracy of up to 93%.

The rest of the paper is organized as follows. The proposed ANN framework is described in Sect. 2. Section 3 provides the predicted results. The paper concludes in Sect. 4.

2 Proposed ANN Framework

The possible performance benefit arising out of simulating NoCs is constrained by the time of simulation limitation imposed by the simulator employed. On the contrary, AI techniques are often preferred to deal with this issue. This section presents the proposed ANN-based framework as an alternate solution.

2.1 Artificial Neural Network

An ANN is a computational algorithm of processing information. It is a layered method used to simulate biological system behavior composed of 'neurons'. ANN is a system of interconnected neurons connected with arcs that evaluate the output from the input values. Each arc has weight. The weight of arcs is used to adjust the value received as input by the node in the next layer. An ANN architecture primarily contains the following three layers:

1. Input layer - The basic information is represented in an input unit that can feed into the network.
2. Hidden layer - To check the activity of each hidden unit. It also evaluates the input unit and weights on the connection between input and hidden units. There can be one or more hidden layers in the network.
3. Output layer - The output of the ANN model depends on the hidden units and weights between remote units and output layer units.

2.2 Tunning Hyperparameters

A hyperparameter is tuned to evaluate the NoCs and improve the prediction of different evaluation metrics for the on-chip networks. Hyperparameters are the variable that decides the network's structure and how the network is trained. Several hidden layers, number of epochs, activation function, learning rate, network weight initialization, batch size are some hyperparameters. Before introducing a training of the model, the hyperparameters are chosen. Tunning means finding the optimal value that gives the test dataset the best possible result. There are different methods to find optimal hyperparameters, such as manual search, grid search, and random search. We have used a random search to find optimal hyperparameters for our model.

2.3 Hidden Layer Selection

Choosing hidden layers in the ANN model is a crucial step. If data is less complex, then 1 or 2 hidden layers are enough, but 3–5 hidden layers can be used if the information is complex. Several epochs define how many times the learning algorithm will work on the training dataset. ANN model uses an activation function to decide how the weighted sum of input is converted into an output from one node to other nodes in the network. ReLU, linear, sigmoid, leaky ReLU, softmax, etc., are the different activation functions in neural networks. As ReLU yields better results, it is used in hidden layers. A linear activation function is used in the output layer, also known as 'identity' because it returns the value of the weighted sum of the input directly and does not change the value obtained from the previous layer. Learning rate defines the size of the steps to adjust the errors in observation. It is observed that the higher the learning rate, the lower the accuracy, and the lower the learning rate, the greater accuracy. The batch size decides the number of samples on internal model parameters.

2.4 Selection of Optimizer

In ANN, there is a term called loss which tells how poorly the model is working. So the model should be trained so that the loss is minimal. For this purpose,

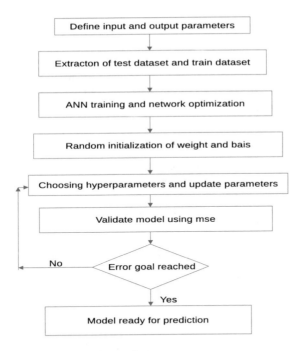

Fig. 1. Proposed ANN framework.

it becomes necessary to use a suitable optimizer that can change weights and the learning rate of the network to reduce the loss. It is impossible for a model to have correct weights right from the start but using trial and error based on loss function; then one can get accurate weights and bias. There are different optimizers such as adam, gradient descent, stochastic gradient descent, adaptive gradient. We select the adam optimizer in this work as it provides a better result.

2.5 Training and Testing ANN

The Fig. 1 gives a high-level view o the proposed ANN model for evaluating NoC performance. The working of the proposed ANN model is divided into two phases: the training phase and the testing phase. The Algorithm 1 describes these phases.

Training Phase. In the training phase, the first input and output parameters are defined, as shown in Fig. 1. Later the training and test data set are extracted from conducting a simulation of NoCs. The ANN architecture follows the network optimization by randomizing variables (weight and bias). Next, the best values of the hyperparameters are selected using the Grid Search method so that The model learns by processing the examples which have input and their corresponding output. These examples are from the probability of association of weight and store them in the net's data structure. Next, determining the difference between the obtained output of the network and the target output is used to train the ANN. The difference is known as error and measured here as mean square error (MSE). Later the network adjusts its weights using its learning rule. Successive iteration causes the network to achieve accuracy, i.e., obtain an output similar to expected.

Algorithm 1. ANN Training and Testing

1: Extraction of training and testing data.
2: ANN training and random initialization of weight and bais.
3: **For**(Selection of hyperparameters) **Do**
4:       regressor.compile(loss= mse, optimizer=adam)
5: **EndFor**
6: Train and Test the ANN
7:       regressor.fit(train data) ← train data, epoch, batch size
8:       regressor.predict(test data)
9: **return** MSE and Accuracy.

Testing Phase. Once the ANN model is trained enough, it can predict the targeted parameters, i.e., output. This prediction exploits the test dataset supplied by the NoC simulator so that the ANN can predict and validate its results. Note that simulation results from lower sized NoCs are controlled as the training dataset for the ANN while the same from the higher NoCs are used for testing the ANN, i.e., validating its predicted results. The MSE validates the predicted results.

3   Experimental Results

This section evaluates different performance metrics for NoC architectures by the proposed ANN-based framework. First, the BookSim 2.0 simulator is run at varying setups on a set of NoCs to record the dataset. The collected dataset is then used to train the proposed ANN framework and validate the predicted performance metrics.

3.1   Simulation Setup

To supply the dataset to the proposed ANN performance framework, a set of 2D mesh NoCs are simulated using the BookSim 2.0 simulator at a simulation setup that primarily includes different buffer sizes, traffic types, virtual channel (VC), etc. The proposed ANN framework evaluates standard performance parameters. The ANN framework is implemented on a large mesh topology set that varies from 2×2 to 15×15. The simulation results for the 2×2 to 8×8 are used as the training dataset. While the same from the 9×9 to 15×15 is used as the testing dataset to validate the predicted results.

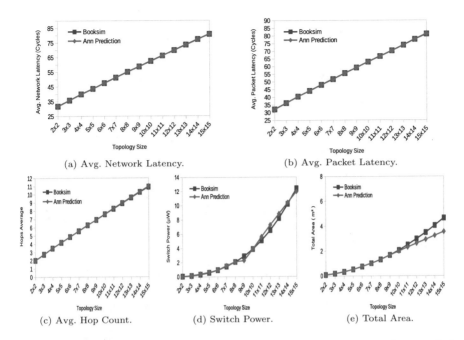

(a) Avg. Network Latency.

(b) Avg. Packet Latency.

(c) Avg. Hop Count.

(d) Switch Power.

(e) Total Area.

Fig. 2. Evaluation of ANN framework over the BookSim at VC = 4, Buffer = 8, PIR = 0.002, and Traffic = Uniform.

3.2 Result Analysis

The proposed ANN-based framework is focused on reducing the time for evaluating the performance metrics quickly over the simulation method-additionally, the proposed framework targets to predict the same metrics accurately over the previous works. The proposed learning-based framework is implemented and embedded in the BookSim 2.0 simulator that provides both training and testing datasets for the framework. We effectively exercise a set of mesh NoCs at varying experimental setups by feeding the dataset as input into our proposed approach.

Figure 2 demonstrates the learning behavior (up to 8×8 meshes) during the training phase and performance prediction on the phase for $9 \times$–11×15 meshes by the proposed ANN-based framework over the BookSim simulator with synthetic uniform traffic. Both network and packet-level latency are measured in Fig. 2a and 2b, respectively. It is observed that these metrics for meshes with size 2×2 to 15×15 lie in the range 31.74–80.70 cycles with mse = 0.0037 and 32.17–81.15 cycles with mse = 0.00097, respectively. Other metrics such as average hop count, switch power consumption, total area are provided in Fig. 2c, 2d, and 2e, respectively. The hop count is 2–11 with mse = 0.00066. A switch overage consumes the power between 0.05 to 12.41 µW with mse = 0.2931, and the total area ranges from 0.082827 to 4.65902 µm² with mse = 0.0055.

Evaluation of the proposed ANN performance framework is extended with the tornado traffic. Figure 3 shows the performance prediction of Booksim and

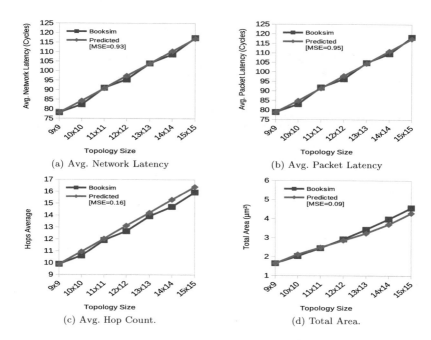

Fig. 3. Evaluation of ANN framework over BookSim at VC = 5, Buffer = 6, PIR = 0.0025, and Traffic = Tornado.

ANN-based framework. The range observed for average packet latency is 26.53–118.29 cycles with 0.95 mse. For average network latency, this range is 26–117.36 cycles with 0.93 mse. Other metrics- hop count and total area are observed as 1–15.94 with mse of 0.15 and 0.081–4.57 with mse = 0.009.

The evaluation of the proposed prediction framework further extended with the shuffle traffic pattern, virtual channel= 4, buffer size= 8, and injection rate = 0.0015. Figure 4 shows the predicted performance metrics over the BookSim simulator. The metrics-average network and packet latency observed range from 31.00–69.47 cycles with mse = 0.15 and 31.34 to 69.78 cycles with mse = 0.3. Other metric-average hop count ranges from 2–9 with mse = 0.25.

3.3   Result Validation

The predicted values must be validated so that they can be acceptable. Both simulation results by the BookSim simulator and the expected results by the ANN framework are shown in parallel in Fig. 2 through Fig. 4 for different performance parameters. Here one sees a side-by-side qualitative comparison of both results. Further, the figures ensure that both outcomes are the same in most cases and are almost identical in some instances. The correctness of the ANN-based performance evaluation is achieved through the hidden layer in the framework. We consider several hidden layers in the ANN model with the uniform, tornado, and shuffle traffic patterns. The random search method

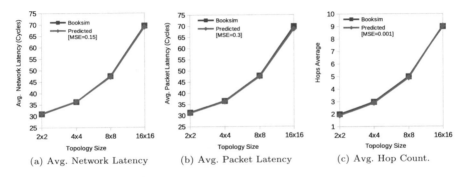

Fig. 4. Evaluation of ANN framework over BookSim at VC = 4, Buffer = 8, PIR = 0.0015, and Traffic = Shuffle.

chooses the number of hidden layers in a set. According to the random search, varying epochs are considered due to different traffic patterns. Next, the ReLU and linear activation function is used in the hidden and output layer for better results. The weights are initially randomly assigned but are updated during the training process. Adam optimizer is used to optimize the model.

3.4   Runtime Comparison

By observing the simulation results and the predicted results for multiple meshes at varying experimental setups, we demonstrate the usefulness and effectiveness of the proposed ANN-based prediction model over the conventional simulation tool in evaluating different performance parameters for NoCs. As seen, the predicted results are correct and the same in most cases. This is something that we expected. However, the accuracy in predicting a performance metric in a few instances is up to 93%. On the contrary, metrics evaluation is indicated quickly over the simulation method for an NoC. Table 1 provides a comparative study on time for evaluating the performance metrics by BookSim simulator and the proposed ANN framework on mesh NoCs with the uniform traffic pattern. The range of execution time of the Booksim simulator is 56.42–516 s, and the same for the proposed ANN framework is 0.2035–0.2419 s. Thus, the later technique quickly returns a performance parameter, resulting in a speedup of 277–2304×. The proposed ANN framework thus delivers both the accuracy and speedup parameters far better than the traditional simulation method and many existing ML-based techniques, increasing more acceptability of the proposed model.

Table 1. Timing comparison between BookSim and proposed model.

Network size	Booksim time (Sec)	ANN time (Sec)	Speedup (×)
9 × 9	56.42	0.2035	277
10 × 10	60.81	0.2164	280
11 × 11	84.5	0.2419	349
12 × 12	209.77	0.2156	972
13 × 13	282.3	0.2181	1294
14 × 14	318.3	0.2201	1446
15 × 15	516	0.2239	2304

4   Conclusion and Future Work

We have proposed a learning-based technique for NoC performance evaluation in this work. The proposed method is an artificial neural network targeted to reduce the time needed to evaluate various performance parameters by simulating an NoC and accurately predicting the parameters by the proposed technique in place of the simulation method. A thorough evaluation of the proposed framework has ensured that it can predict the performance metrics in a speedup up to 2304×. Also, the expected results are 93% accurate, as seen by the simulation method. Future works include a comparison study with the existing techniques and reframing the proposed ANN model to improve the accuracy. Additionally, it can be employed for other types of NoCs such as torus, octagon, and spidergon.

References

1. Bhowmik, B.: Dugdugi: an optimal fault addressing scheme for octagon-like on-chip communication networks. IEEE Trans. Very Large Scale Integr. (VLSI) Syst. **29**(5), 1009–1021 (2021)
2. Bhowmik, B., Deka, J.K., Biswas, S.: Reliability monitoring in a smart NoC component. In: 2020 27th IEEE International Conference on Electronics, Circuits and Systems (ICECS), pp. 1–4 (2020)
3. Zhang, W., Hou, L., Zuo, L., Peng, Z., Wu, W.: A network on chip architecture and performance evaluation. In: 2010 Second International Conference on Networks Security, Wireless Communications and Trusted Computing, vol. 1, pp. 370–373 (2010)
4. Foroutan, S., Thonnart, Y., Petrot, F.: An iterative computational technique for performance evaluation of networks-on-chip. IEEE Trans. Comput. **62**(8), 1641–1655 (2013)
5. Bhowmik, B., Hazarika, P., Kale, P., Jain, S.: AI technology for NoC performance evaluation. IEEE Trans. Circuits Syst. II Express Briefs **68**(12), 3483–3487 (2021)
6. Dong, Y., Wang, Y., Lin, Z., Watanabe, T.: High performance and low latency mapping for neural network into network on chip architecture. In: 2009 IEEE 8th International Conference on ASIC, pp. 891–894 (2009)

7. Theocharides, T., Link, G., Vijaykrishnan, N., Invin, M., Srikantam, V.: A generic reconfigurable neural network architecture as a network on chip. In: IEEE International SOC Conference. Proceedings, vol. 2004, pp. 191–194 (2004)
8. Ascia, G., Catania, V., Monteleone, S., Palesi, M., Patti, D., Jose, J.: Networks-on-chip based deep neural networks accelerators for IoT edge devices. In: 2019 Sixth International Conference on Internet of Things: Systems, Management and Security (IOTSMS), pp. 227–234 (2019)
9. Silva, J., Kreutz, M., Pereira, M., Da Costa-Abreu, M.: An investigation of latency prediction for NoC-based communication architectures using machine learning techniques. J. Supercomput. **75**, 11 (2019)
10. Choi, W., et al.: On-chip communication network for efficient training of deep convolutional networks on heterogeneous manycore systems. IEEE Trans. Comput. **67**(5), 672–686 (2018)
11. Daya, B.K., Peh, L.S., Chandrakasan, A.P.: Low-power on-chip network providing guaranteed services for snoopy coherent and artificial neural network systems. In: 2017 54th ACM/EDAC/IEEE Design Automation Conference (DAC), pp. 1–6 (2017)
12. Zhao, Y., Ge, F., Cui, C., Zhou, F., Wu, N.: A mapping method for convolutional neural networks on network-on-chip. In: 2020 IEEE 20th International Conference on Communication Technology (ICCT), pp. 916–920 (2020)
13. Xiao, S., et al.: NeuronLink: an efficient chip-to-chip interconnect for large-scale neural network accelerators. IEEE Trans. Very Large Scale Integr. (VLSI) Syst. **28**(9), 1966–1978 (2020)

Security, Power Consumption and Simulations in IoT Device Networks: A Systematic Review

Roland Montalvan Pires Torres Filho, Luciana Pereira Oliveira[✉],
and Leonardo Nunes Carneiro

IFPB Campus João Pessoa, Av. Primeiro de Maio,
720 - Jaguaribe, João Pessoa, PB, Brazil
{roland.montalvan,leonardo.nunes}@academico.ifpb.edu.br,
luciana.oliveira@ifpb.edu.br

Abstract. Research efforts have been directed to propose security solutions with low energy for Internet of Things (IoT) device, which have peculiar similarities with Wireless Sensor Network (WSN). This work used the Systematic Literature Review to identify, analyze and classify 5,671 papers in these context. The results about papers with simulations, energy and security contain COOJA the most used in explicit IoT papers and NS2 in WSN. MATLAB is the second prefer simulator for both. Few papers explore software programmed networks and artificial intelligence as a low-power security solution for studies with explicit and implicit IoT.

1 Introduction

Internet of Things (IoT) constitutes a network of physical tools capable of accumulation and sending of data collected in sensors arranged in them. Its areas of operation enable a wide spectrum of possibilities, such as smart cities, agriculture, public safety, healthcare, to train dog and others [1].

The work present by [2] detailed the infrastructure of networks composed of sensors. They have the necessary peculiarities for its validation as an IoT network. Among the attributes presented by these devices, safety and energy consumption are key factors for their proper functioning [2–4]. The IoT infrastructure consists of a set of devices connected to the internet network, to collect and transmit information about objects around people, enabling a faithful capture of signals by sensors and expanding the interaction with the environment [3]. Another characteristic point is its similarity with WSN, which has been used in various aspects of daily life, with IoT being one of the accentuating factors in its use [5].

In both IoT and WSN, planning is a crucial phase so that there can be implemented with less possibility of critical failures. The tools used for this type of planning are network simulator, representing the structure that will be created. In the present study, a total of 5,671 documents were analyzed and 448 paper referenced some simulator in context of the IoT, energy and security.

The purpose of this research is based on the Systematic Literature Review (SLR) method in order to select papers that respond a set of research questions.

L. Barolli et al. (Eds.): AINA 2022, LNNS 451, pp. 370–379, 2022.
https://doi.org/10.1007/978-3-030-99619-2_35

The results of this work present information extracted from the analyzed papers, gaps and recommendations for future studies.

Study [6] mad [7] make use of SLR method to plan, test, measure or validate idealizations of security solutions with low energy, considering formal methods and real experiments. So, they did included paper with simulations.

This research organization was based on systematic rationale as follows. On Sect. 2 related works are described, with specif subtopics about Security, Power Consumption and Simulations. The research methodology (SRL process) is informed in Sect. 3. The discussion of the papers selected by SRL is described in Sect. 4. Conclusion and future works are in Sect. 5.

2   Related Works

This section presents some peculiarities of the baseline studies that describe information about security, energy consumption and simulation, with a focus on segments on WSN and IoT.

2.1   Security

The Wireless Sensor Networks are the direct basis for IoT networks with different application fields ranging from healthcare, industry, home automation, military, monitoring and tracking of environment/objects [8]. Applications with this profile demand enhanced security services such as mutual authentication, data integrity and end-to-end data confidentiality, making security requirements relevant to many factors such as node power status, delay, bandwidth and data packet delivery rate.

There are several artifices adopted by the authors to maintain security levels, the experiments [9] stands out intrusion detection and how it plays an important role in many IoT applications for detecting malicious intruders. They introduce a novel deployment strategy to overcome the limitations of homogeneous and heterogeneous devices deployments for energy-efficient and quick detection in both IoT networks. Initially, investigate the problem of physical intrusion detection, furthermore, examined the effects of different network parameters elevating the intrude detection probability. But they can't extend they framework for a real-time intruders detection.

The method presented by [14] can be used in a large scale sensor network, and it only uses the symmetric cryptography avoiding the use of the public key cryptography. Simulation results show that the proposed method only needs a small memory (below 140 KB) for each node. They use identity certificates to defend against Sybil attacks, the energy consumption by new identity generations within each group is low (below 60 mJ), which is much lower than the available energy at each sensor node on scenario of experiment.

2.2   Power Consumption

Energy consumption in WSN or IoT devices is a factor that can be explored from different perspectives, but always with a common goal, which is consumption efficiency. Communication security is normally provided by encryption, data is encrypted before transmission and will be decrypted first on reception, with a focus on the cost of processing. The paper [10] proposes a mechanism for sensors to operate at an energy-efficient fashion using the dynamical voltage scaling technique. The experiments showed the sensor's lifetime can be extended substantially.

The biggest current difficulties faced in WSN are the large number of nodes, heterogeneity of data and the low energy power of the devices. A major highlight to ensure the reduction of energy consumption in devices deployed in these networks are artificial intelligence and deep learning technologies. Using Mobile Agent (MA) concepts for WSN, [4] develops a system in which deep reinforcement learning is aimed at WSN flow control. The processing energy consumption of each transmitted and received bit is estimated at 50 nJ/bit, from which a comparison is made with other algorithms already consolidated in the market. During the learning phase, energy consumption is higher, after consolidating the absorption of a certain amount of data from the network, the algorithm promoted a significant reduction in energy consumption, proving to be an excellent solution.

The performance of energy cost calculations in the sensor nodes is described in the formula below, where are assumed that a sensor node consumes e_{out} when sending out one bit of data, and consumes e_{in} when receiving one bit of data. The transmitter amplifier's energy is $e_{amp} \times dk$, d means the distance and k is the propagation loss exponent. P_{in}^i means the packet size received by the node i, use P_{out}^i to mean the packet size of the sent data, then the total energy consumption of the sensor node is:

$$E(P_{in}^i, P_{out}^i) = e_{in} \times P_{in}^i + (e_{out} + e_{amp} \times dk) \times P_{out}^i$$

From this basic calculation model expose before, the authors start their specific measurement processes of each study. In [4,10] based on this logic it is possible verify the total consumption of the network, including the calculation of the consumption of MA moving within the network.

2.3   Simulations

Network simulation allow to study the behavior under different traffic and topology conditions, obtain performance measures, and evaluate its sensitivity in relation to operational parameters. With efficient results, the algorithm developed by [4] utilizes deep neural network for learning, inputs the state of WSN as well outputs the optimal route path. The simulation experiments demonstrate that the algorithm is feasible to control traffic in wireless sensor network and can reduce the energy consumption. The results shows contrast when the increase in size of mobile agents packet will cost higher energy and the security its not a priority.

Some researchers prefer to create algorithms to describe the simulation environment. However, the algorithm is not a standard simulator, making it difficult to compare studies. Therefore, several studies choose an open source (OMNet++, NS2, NS3, GTSNetS, SIDnet, TOSSIM for TinyOS devices, Cooja for Contiki devices, Verilog and others) or property simulator (for example, OPNeT, Altera Quartus II and MATLAB).

However, the reference [11] presented limits in NS2, NS3 and TOSSIM when needing more robust ad hock simulator networks. The main problem found was not being able to simulate networks containing hundreds of nodes for periods of months, with very detailed modeling characteristics of the lower layers (physical and MAC), and they chose to implement the simulator itself that idealizes the physical and MAC layers.

When working on problems with a focus on energy saving, security and network efficiency in the control of topology for WSN, [10] presents a mobile agent based topology control (MATC) algorithm that could solve three main issues in WSN topology control: routing void, isolated node and sleeping control. Simulation results shows that, compared with the algorithm with no agent, MATC could save more energy, prolong network lifetime and reduce packet loss rate.

An specific case [12] of wireless multimedia sensor networks (WMSNs), where given rise to intelligent transportation systems as a mobile data-sharing model. With focus on IoT to include versatile resources such as mobile devices and sensors on the roads for emergency and security conditions.

3 Methodology

Among the key points that make up a systematic review, the ordering of steps in order to make possible reproduce the process has great weight in this type of study. Through a set of well-defined steps to select and analyze the studies to extract their most relevant characteristics when compared in sets. This work followed the method performed by [6] which used three steps: definition of research questions; execution; and analyzes that generate the results of graphs, tables and descriptions. Below are presented a description of each process.

3.1 Research Questions

In the search for studies that address security, energy an simulations in IoT, initially five questions were listed:

   I. If the work were to guide the choice of a protocol (Simulators and parameters) for the IoT environment, what would it be like?;

  II. What is the result of comparing simulators used in documents that have explicit and implicit IoT?;

 III. When I look at WSN and IoT networks, are the characteristics similar?;

 IV. Which papers use AI?;

  V. What research gaps in the area of the data present?

3.2   Search Process

The search synthesis had the choice of the keywords "Wireless Sensor Networks, IoT, Simulation, Security and Energy Consumption", by searching the library of Institute of Electrical and Electronics Engineers (IEEE), Science Direct directories and on the Digital Library of Association for Computing Machinery (ACM). The total of 5,671 articles were found and 448 documents selected from the reading of keywords and abstract. Therefore, the following inclusion (IC) and exclusion (EC) criteria were important to identified a viable number of papers to analyze: EC1 (works that not written in English); EC2 (Papers does not contain in the title or in the abstract the words in the context of security, power consumption or simulation; EC3 (Duplicate articles); EC4 (Secondary work identified after to read it, because relevant words [security, power consumption or simulation] were not in title or abstract); EC5 (Incomplete work, because the paper has no value related to measurement experiments); EC6 (Papers with irrelevant content, because after reading, they did not present measurement description in terms of the security, energy consumption and simulation) and IC1 (papers did not exclude by any of the exclusion criteria).

3.3   Studies Selection

This SLR was performed manually in two phases. In the first step, an analysis of the title and abstract of all works was performed using the exclusion criteria (EC1, EC2 and EC3). In a second moment, there was a total reading, exclusion of works to apply EC4, EC5 and EC6. The papers selected at this time are classified by IC1.

As result, in the first moment, 448 studies were selected and 4384 excluded. At the second moment, the full text of each primary study included as IC1 was read to extract information related to RQs addressed by this SLR.

4   Research Results

After performing the chronological analysis regarding the implicit (IIT) or explicit (EIT) reference of the IoT theme in the documents found, it is possible to observe an interesting pattern. As the years go by, the Fig. 1 shows a higher frequency of publications addressing EIT (more than two-thirds of papers) since 2017 in contrast to the years 2012 and 2016.

4.1   RQ1: If the Work were to Guide the Choice of a Protocol (Simulators and Parameters) for the IoT Environment, What Would it Be Like?

The Fig. 2 presents 40 distinct simulators. Most use open source solutions. COOJA is the first place of use to simulate explicit IoT scenarios. The explanation for this must be in [6] which identified the Tmote Sky and TesloB hardware

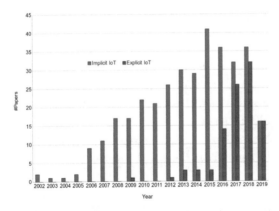

Fig. 1. Years distribution of selected papers

are most used by IoT papers with real experiments. Both support the Contiki as the operating system which is also used by COOJA simulator. Implicit IoT contains several older papers and this justifies NS2 as the usually simulator.

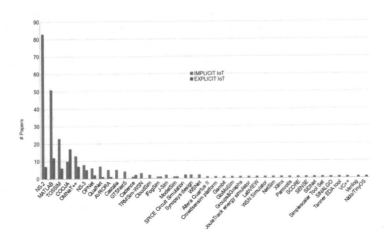

Fig. 2. Frequency of explicit and implicit IoT simulators

Observing infrastructure characteristics found (Table 1), WSN networks presents a greater variation in the number of nodes chosen as parameters for simulations with experiments reaching up to 10,800 nodes. However, the numbers are inverted in third line, where the number of nodes in IoT simulations remains higher than WSN.

Table 1. Simulations parameters on IoT networks.

Simulations parameters	Implicit IoT	Explicit IoT
Min. node number	10	10
Max. node number	10,800	600
Most used node number	100	500
IEEE standard	802.11	802.11
Covered layer	Routing	Routing
Communication range	10–300 m	20–500 m

4.2   RQ2:What is the Result of Comparing Simulators Used in Documents that have Explicit and Implicit IoT?

The studies were classified as "Explicit IoT" when the study directly references IoT and "Implicit IoT" when it refers only to WSN. So, the Fig. 2 exposes NS2 (old open source simulator) as the most used simulator, because WSN was more referenced than explicit IoT, which is a new research area.

Among a wide variety in the types of simulated network technologies in the explicit IoT category, it is possible to find works with Mobile Ad Hoc Net-works (MANETs) [15] and Vehicular Ad-Hoc Network (VANET) [13]. The related fields of application may involve numerous areas such as Medicine, Physics, Astronomy and others, as the studies act directly to improve networks, enabling safer connections, being processed with the best levels of energy consumption.

4.3   RQ3:When I Look at WSN and IoT Networks, are the Characteristics Similar?

The percentage of works with MATLAB is the similarity between WSN and IoT, because this simulator most used for both explicit IoT and WSN (implicit IoT).

In Table 1, IoT and WSN showed some similarities in simulations parameters: 10 as minimum node number, 802.11 as usually technology and several work with routing simulation. In Table 2, WSN and explicit IoT had more intervention papers (new proposal) than observation papers.

However, in Table 2, an accumulation of 99 documents pointing directly to networks with IoT devices, contrasting with the incidence of 349 studies addressing WSN. A factor to be considered for sequencing the data cited is the fact that the first studies with IoT started in 2009, showing a theme with more recent approaches compared to the WSN that show first studies in 2002.

4.4   RQ4: Which Papers Uses AI?

The possibility to implement a solution that enables sensor network to behave as an intelligent multi-agent are presented by [4]. Consist in a novel traffic-control system based on deep reinforcement learning, which regards traffic control as

Table 2. Characteristics of studies divided by type of IoT network.

Characteristics of studies	Implicit IoT	Explicit IoT
Numbers of studies	349	99
Years that publications covered	2002 to 2019	2009 to 2019
Analytical intervention papers	236	52
Analytical observation papers	12	8

a strategy-learning process, to minimize energy consumption. The algorithm utilizes deep neural network for learning, inputs the state of wireless sensor network as well as outputs the optimal route path. The simulations demonstrate that the algorithm is feasible to control traffic in WSN reducing the energy consumption.

Showing one promising technology [13] introduces a Intelligent Transportation Systems (ITS) to support the increasing number of Intelligent vehicles on roads what leads to congestion and safety problems. Additionally, this SRL found sixty-six papers with AI algorithms tags to provide a low-power security solution. There are a greater number of papers with AI and IoT implicit. However, the total number of papers with implicits is large. Thus, considering the percentage, learning, game theory and intelligent tags were found more in recent work with explicit IoT.

4.5   RQ5: What Research Gaps in the Area do the Data Present?

The Table 3 shows an overview of the studies that referred to the protocol used in the simulations. It is evident that the most chosen protocols were those of the routing layer, as this is where there is more freedom to work with algorithms or modifications in the protocol structure.

Table 3. Protocols used in simulations

Worked protocol	Implicit IoT	Explicit IoT
Routing	144	20
Link layer security	8	0
Data aggregation	8	0
Own	4	4
Authentication	4	0

Although the Table 3 does not contain algorithms for virtualization, this SRL found only five papers based on virtualization that can be implemented by virtual machine, container and software defined networks (SDN). Moreover, only two papers used SDN. For example, in [16] the authors decouple the control plane and data plane of routers enabling a more flexible and efficient configuration of security policies. With focus on security and low consumption, the Privacy-aware, Secure and energy-aware Green Routing was developed, which can defend against internal attacks, achieve stronger privacy protection and reduce energy consumption. So, it is important to highlight the absence of studies that work with AI and SDN. Furthermore, this paper suggests investigating how AI and SDN can collaborate to reduce energy consumption in IoT security solution.

5   Conclusion

The results presented in this SLR aim to investigate the characteristics of simu-lations in context of IoT (implicit or explicit), energy and security. This review helped to understand current trends and the state-of-the-art to identify research gaps and future directions. After reviewing each selected paper, they are ranked according to their goals and the way they work.

Thus, this SLR suggests COOJA or MATLAB as tool to simulate explicit IoT scenarios, because they are most used with a number of nodes between 10 and 500 nodes using 802.11 technology. There are several studies considering routing problems, but there are gaps for other network layers.

In addition, few studies have addressed studies of artificial intelligence and software-defined networks. So, another recommendation for future work is the investigation of how such areas can collaborate to reduce energy consumption in IoT security solutions.

Acknowledgments. The authors would like to thank the Federal Institute of Paraíba(IFPB)/Campus João Pessoa for financially supporting the presentation of this research and, especially thank you, to the IFPB Interconnect Notice.

References

1. Menezes, A.H.S., et al.: IoT environment to train service dogs. In: IEEE First Summer School on Smart Cities (S3C) (2017). https://doi.org/10.1109/S3C.2017.8501386
2. Asim, M., et al.: A self-configurable architecture for wireless sensor networks. Dev. E-Syst. Eng. DeSE (2010). https://doi.org/10.1109/DeSE.2010.20
3. Lata, B.T., et al.: SGR: secure geographical routing in wireless sensor networks. In: 9th International Conference on Industrial and Information Systems ICIIS (2015)
4. Lu, J., et al.: Artificial agent: the fusion of artificial intelligence and a mobile agent for energy-efficient traffic control in wireless sensor networks. Future Gener. Comput. Syst. **95**, 45–51 (2019). https://doi.org/10.1016/j.future.2018.12.024
5. Mbarek, B., Meddeb, A.: Energy efficient security protocols for wireless sensor networks: SPINS vs TinySec. In: ISNCC (2016). https://doi.org/10.1109/ISNCC.2016.7746117
6. Oliveira, L.P., Vieira, M.N., Leite, G.B., de Almeida, E.L.V.: Evaluating energy efficiency and security for Internet of Things: a systematic review. In: Barolli, L., Amato, F., Moscato, F., Enokido, T., Takizawa, M. (eds.) AINA 2020. AISC, vol. 1151, pp. 217–228. Springer, Cham (2020). https://doi.org/10.1007/978-3-030-44041-1_20

7. Oliveira, L.P., da Silva, A.W.N., de Azevedo, L.P., da Silva, M.V.L.: Formal methods to analyze energy efficiency and security for IoT: a systematic review. In: Barolli, L., Woungang, I., Enokido, T. (eds.) AINA 2021. LNNS, vol. 227, pp. 270–279. Springer, Cham (2021). https://doi.org/10.1007/978-3-030-75078-7_28

8. Zeng, G., Dong, X., Bornemann, J.: Reconfigurable feedback shift register based stream cipher for wireless sensor networks. IEEE Wirel. Commun. Lett. **2**, 559–562 (2013)

9. Halder, S., Ghosal, A., Conti, M.: Efficient physical intrusion detection in Internet of Things: a node deployment approach. Comput. Netw. **154**, 28–46 (2019)

10. Hong, L.: Mobile agent based topology control algorithms for wireless sensor networks. In: IEEE Wireless Communications and Networking Conference Workshops WCNCW (2013)

11. Sichitiu, M.L.: Cross-layer scheduling for power efficiency in wireless sensor networks. In: IEEE INFOCOM (2004)

12. Al-Turjman, F.: Energy-aware data delivery framework for safety-oriented mobile IoT. IEEE Sens. J. **18**, 470–478 (2018). https://doi.org/10.1109/JSEN.2017.2761396

13. Gaber, T., et al.: Trust-based secure clustering in WSN-based intelligent transportation systems. Comput. Netw. (2018). https://doi.org/10.1016/j.comnet.2018.09.015

14. Yin, J., Madria, S.K.: Sybil attack detection in a hierarchical sensor network. In: SecureComm (2007). https://doi.org/10.1109/SECCOM.2007.4550372

15. Silva, A.A.A., et al.: Predicting model for identifying the malicious activity of nodes in MANETs. In: Proceedings - IEEE IEEE Symposium on Computers and Communication (2015). https://doi.org/10.1109/ISCC.2015.7405596

16. Lin, H., et al.: A trustworthy and energy-aware routing protocol in software-defined wireless mesh networks. Comput. Electr. Eng. **64**, 407–419 (2017)

Real Time Self-developing Cybersecurity Function for 5G

Maksim Iavich[1](✉), Razvan Bocu[2], and Avtandil Gagnidze[3]

[1] Caucasus University, P. Saakadze Street 1, 0102 Tbilisi, Georgia
`miavich@cu.edu.ge`
[2] Transilvania University of Brasov, Bulevardul Eroilor 29, 500036 Brașov, Romania
`razvan@bocu.ro`
[3] East European University, Shatili Street 4, 0178 Tbilisi, Georgia

Abstract. The telecommunications industry is undergoing a significant transformation towards 5G networks. Research has shown that 5G has security issues. In the paper, we present a new machine learning-based cybersecurity model that also includes a firewall and IDS/IPS. The attack patterns are created in the simulated laboratory and the IDS is trained by means of these patterns. We offer the model where the cyber security function, will be trained in real time and will be improved based on its mistakes. As a part of the experiment, we integrated this model into the existing 5G architecture. The study was carried out in a laboratory consisting of a server and 20 raspberry pi systems that simulated attacks on the server. The strategy proposed in the document will also be useful in the future versions of the system.

1 Introduction

The challenge for 5G wireless networks is to provide very high data rates (QoS), very low latency and high coverage by deploying high-bandwidth stations over short distances. 5G service delivery requires new data processing and storage technologies, new network architectures and service delivery models. With the introduction of these technologies, we will face new challenges related to new 5G cybersecurity systems. Soon, all critical infrastructure will be completely dependent on 5G networks, so it is necessary to create a reliable security system to ensure the safety of infrastructure and the public. For example, security breaches in power systems can have devastating consequences for an electronic system that is critical to all communities. Thus, to ensure the security of 5G networks, it is necessary to study their weaknesses and outline ways to eliminate them. It is also important to compare the security of 4G and 5G systems and analyze the differences between them [1–5].

2 Notes on 5G Security Features

It should be noted that 4G and 5G have similar security architectures. The LTE/4G and 5G network nodes that are providing security and the communication connections that need to be protected. These nodes are similar. Therefore, security mechanisms of 4G and 5G can be divided into two categories. The first category includes network access security devices that are used to securely access various services for users using certain devices, as well as to protect the radio system interface located between a specific device and the radio node from attacks (Fig. 1).

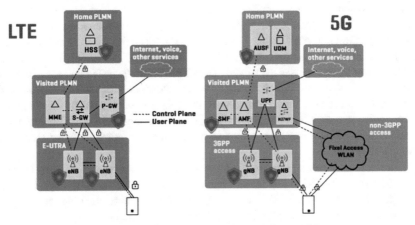

Fig. 1. Security architecture of LTE/4G and 5G

5G is considering a new authentication framework. Access authentication is the main security procedure in 3GPP networks, and it is considered as the main authentication related to 3GPP 5G security standards. This procedure is performed at the initial stage of device authentication. Upon successful completion of this procedure, the session keys must be determined. The keys are used to ensure communication between the device and the network. This procedure is designed in accordance with 3GPP 5G security to support the Extended Authorization Protocol (EAP). EAP is widely used in various IT fields. This allows to use different types of credentials stored on SIM cards, such as certificates, public keys, certificates, usernames, and passwords. The design of authentication and key agreement mechanisms allow authorization procedures to be performed between the network and user equipment needed for future communication between them. The keys that are generated as a result of the initial authentication framework and the key agreement procedure form the binding key are called (KSEAF). After the key is transferred by the Home Network Authorization Server (AUSF) function to the SEAF Service Network, it is used as the secure channel requested by the Subscriber and Service Networks (SEAF) [6–8].

So, KSEAF can be used in a variety of security contexts where re-authentication procedures will no longer be required. For example, the 3GPP access network uses an authentication process that generates keys to establish secure channels between the UE and the 3GP function. This feature is used for unreliable 3GPP access. In a 5G network, each subscriber must be assigned a globally unique permanent subscriber identifier (SUPI). Permanent caller ID formats include Network Access Identifier (NAI) and International Mobile Caller subscriber identity (IMSI). It is significant SUPI not to be revealed when connecting to a specific mobile device, as it happened in case of 3G and 4G networks. Where, during authentication process of particular device to the new network IMSI is disclosed. It is notable that the Subscription Concealed Identifier (SUCI) must be used before the authentication process of the network and device. Just after authentication process is finished home network SUPI can be exposed to the serving one. This procedure is used to ensure that international mobile subscriber identity catchers (IMSI) do not steal the subscriber's identity. This can be avoided by using a device that can be attached to the Rogue Base Station (RBS) or to the operator base station when an attempt is made to sniff unencrypted traffic. The Subscription Identifier De-Concealing Function (SIDF) function should be used to hide SUPI from SUCI. The Subscription Identifier De-Concealing Function uses the privacy key of the public/private key pair of the home network related to privacy, which is safely stored by the operator's home network. SIDF access rights should be defined so that only the network element of the home network has the right to request SIDF. Disclosure must pass at the UniFi Dream Machine (UDM). The SIDF access rights must be defined such that only the home network element can request the SIDF.

To protect messages sent via the N32 interface on the perimeter of a public terrestrial mobile network (PLMN) in the 5G architecture Proxy Server Security Edge Protection Proxy (SEPP) is embedded. All the service layers of the network function (NF) send messages to the SEPP which protects them until sending them over the network through the N32 interface. In addition, the N32 interface sends all messages to SEPP, the last one verifies its security data and resends them to the appropriate function layer of the network. SEPP implements the security of all application layer information that is interleaved between two NFs in two different PLMNs. To reach highest security level in 5G the network slicing method is used [9, 10]. Network slice is the separate permeable logical network that operates on a common physical infrastructure that can provide an appropriate level of quality of service. Therefore, in 5G whole network can be divided into pieces so that they do not intersect. This mechanism secures management of related services. Therefore, the system will be more secure, well structured, and easy to manage.

3   Security Concerns of the 5G Standard

Based on the studies carried out, it can be concluded that the security of 5G is not yet perfect and faces big challenges. The analysis presented in this document identifies the following security issues:

1. The 5G network is constantly susceptible to software attacks and has many more penetration points for hackers, since the software configuration of 5G network lets hackers to target on various security vulnerabilities, flaws and bugs that can affect the performance of the 5G network.
2. As the introduction of the new 5G network architecture brings new functionality, we get much more vulnerable points in the form of some network equipment parts and functions. Hackers may target the base stations and primary management functions.
3. Since all mobile network operators are dependent on suppliers, it can inspire new directions of hacking attacks in addition to being able to significantly increase the severity of the impact of such attacks.

4. 5G network architecture will be used by almost all vital IT applications, so attacks on the integrity and availability of these types of applications can also pose a security problem.
5. The 5G network will include many devices that can provoke various types of attacks such as DoS and DDoS.
6. Fragmentation of the network can also lead to security problems, as attackers can try to connect a specific device to the desired network segment.

The number of vulnerabilities have already been identified in 5G security systems, with the help of which hackers can inject malicious code and perform illegal actions [10–14]. These attacks can be grouped as follows:

• MNmap
The team of researchers sent the information in plain text to the network, which they used to outline the map of the network. They set up a fake base station and determined all devices connected to the network and their capabilities. They found out the manufacturer, model, operating system, version, and type of the device.

• MiTM
Among others 5G infrastructure also allows Man-in-the-Middle (MiTM) attacks. So, with the implementation of MiTM, it is possible to carry out bidding-down and battery drain attacks. Attackers can disable Multiple-Input Multiple-Output (MIMO) enabling, the physical part that provides high speeds. Without this component, 5G speeds will drop to the speeds of 2G/3G/4G networks. MIMO enabling is a feature that significantly increases the data transfer rate on a 5G network.

• Battery Drain Attack
Battery drain attack targets NB-IoT devices. Here low-power sensors rarely send out small information packages that drain the battery even in power saving mode. Attacker adapts the PSM to keep devices running and trying to connect to the network. Thus, the attacker can offer the victim the desired networks, and then the equipment can be used at the attacker's discretion.

It is important to note that a new architecture for 5G and future 6G networks needs to be defined in order to create new AI/ML-based algorithms that should ensure the highest level of cybersecurity and adequate protection for mobile subscribers, industry and government.

4 Description of the Approach

We propose to apply the cybersecurity module to each base station as an additional server that integrates IDS/IPS and firewall (Fig. 2).

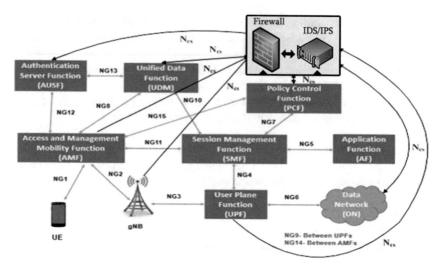

Fig. 2. Cybersecurity module

Based on the work presented in this article, we can conclude that 5G networks are assailable to software-based, Probe and Dos attacks. The primary version of our IDS (Invasion Detection System) is designed on the basis of M/L algorithms. It is noteworthy that resistance to these types of attacks can be achieved through the use of various sets of data in the training process. One of the most common among scientists is the NSL-KDD dataset. It is actively used in academic research and for prototyping intrusion detection systems [15–19]. A lot of attacks in these datasets are old and not relevant. We offer to create our own pattern and train IDS by means of these patterns. Our laboratory contains 20 RASPBERRY PI devices and the server with network protocol analyzer TShark. By means of Tshark we create four datasets, every dataset contains from 1 to 2 Gb of data of different patterns. We have implemented DoS/DDOS (SYN flood, UDP flood, slowloris, ICMP, Ping of Death and Volume Based Attacks) and brute force attacks. The first dataset contains the following attacks: SYN flood, UDP flood. The second dataset contains: Slowloris and ICMP attacks. The third one contains Ping of Death and Volume Based Attacks. Finally, the fourth block contains brute force attacks.

We have divided each dataset into training and test sets. The first one includes 90% of the information, while the test data set contains the remaining 10% of it. This selected method for data sharing, gave us the best results in terms of accuracy after training. The IDS is trained separately by means of these patterns, using Convolutional Neural Network model. Each dataset was trained in the separate thread in order to increase the efficiency. The accuracy of training each dataset was 0.9937894736842106, 0.9912156123148208, 0.9977654435232112, and 0.9932111212578854 correspondingly.

We propose the following model of IDS. On the main server, we install network protocol analyzer TShark and our novel clever IDS. After completing the training process, the system waits for data entry. One more server will be added to the network - the log server. On the log server, we will save all the traffic patterns, if some of the attacks is still successful it will be analyzed manually and added to the dataset. Our IDS will be periodically trained with the collected datasets, like it was mentioned above, and the trained model will be transferred to the main server. Therefore, we can divide our model into the following stages:

1. Training the data on the main server.
2. Analyzing the traffic on the main server
3. Sending the data to the log server
4. Training data on the log server
5. Updating the trained model on the main server
6. In the case of attack the information is sent to IPS
7. In the case of successful attack, the data is analyzed manually and sent to the log server.

The process is infinitely repeated. The model of scheme is described on the Fig. 3.

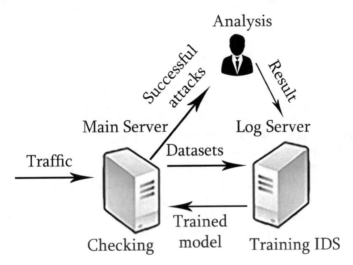

Fig. 3. The novel model

5 Experiments

We have checked our IDS in our simulated laboratory. The laboratory contained 20 RASPBERRY PI devices and the server with network protocol analyzer TShark. By means of Tshark we created four datasets, every dataset contained from 1 to 2 Gb of data of different patterns. We have implemented DoS/DDOS (SYN flood, UDP flood, slowloris, ICMP, Ping of Death and Volume Based Attacks) and brute force attacks. The first dataset contains the following attacks: SYN flood, UDP flood. The second dataset contained: Slowloris and ICMP attacks. The third one contains Ping of Death and Volume Based Attacks. Finally, the fourth block contained brute force attacks.

We have installed our IDS on the server and trained it by means of the received attack patterns.

Table 1. The results of experiment

Attack name	Number of attacks	Number of identified attacks
SYN flood	1000	998
UDP flood	1000	996
Slowloris	1000	1000
ICMP, Ping of Death	1000	970
Volume Based Attacks	1000	940
Brute force attacks	1000	992

We have received the following results (Table 1).

Obtained results show that proposed Intrusion Detection System is very useful and can be used as a prototype for a real IDS system.

6   Conclusions and Future Plans

The scheme described in this paper covers most of the attacks that pose a threat to 5G. The described approach is fundamentally different from the works done in this direction. Existing strategies typically train the IDS model using the NSL-KDD data set. However, these approaches are only academicals and can not be used in practice. Experiments have shown that the model can successfully detect most of the mentioned attacks and will be more and more precise. Consequently, IDS's real-time detection results will be improved. In addition, more targeted and multipurpose attack vectors will be created. Our system is always updated and is in the constant training mode. Accordingly, the new IDS will be implemented in the real world 5G data networks. In future it would be interesting to automate the manual analysis of the identified attacks.

References

1. Schneider, P., Horn, G.: Towards 5G security. IEEE Trustcom/BigDataSE/ISPA **2015**, 1165–1170 (2015). https://doi.org/10.1109/Trustcom.2015.499
2. Ji, X., et al.: Overview of 5G security technology. Sci. China Inf. Sci. **61**(8), 1–25 (2018). https://doi.org/10.1007/s11432-017-9426-4
3. Jover, R.P., Marojevic, V.: Security and protocol exploit analysis of the 5G specifications. IEEE Access **7**, 24956–24963 (2019). https://doi.org/10.1109/ACCESS.2019.2899254
4. Dutta, A., Hammad, E.: 5G Security challenges and opportunities: a system approach. In: 2020 IEEE 3rd 5G World Forum (5GWF), pp. 109–114 (2020). https://doi.org/10.1109/5GWF49715.2020.9221122
5. Bocu, R., Iavich, M., Tabirca, S.: A real-time intrusion detection system for software defined 5G networks. In: Barolli, L., Woungang, I., Enokido, T. (eds.) AINA 2021. LNNS, vol. 227, pp. 436–446. Springer, Cham (2021). https://doi.org/10.1007/978-3-030-75078-7_44
6. Iwamura, M.: NGMN view on 5G architecture. In: 2015 IEEE 81st Vehicular Technology Conference (VTC Spring), pp. 1–5 (2015). https://doi.org/10.1109/VTCSpring.2015.7145953

7. Agyapong, P.K., Iwamura, M., Staehle, D., Kiess, W., Benjebbour, A.: Design considerations for a 5G network architecture. IEEE Commun. Mag. **52**(11), 65–75 (2014). https://doi.org/10.1109/MCOM.2014.6957145

8. Ahmed, I., et al.: A survey on hybrid beamforming techniques in 5G: architecture and system model perspectives. IEEE Commun. Surv. Tutor. **20**(4), 3060–3097 (2018). https://doi.org/10.1109/COMST.2018.2843719. Fourthquarter

9. Foukas, X., Patounas, G., Elmokashfi, A., Marina, M.K.: Network slicing in 5G: survey and challenges. IEEE Commun. Mag. **55**(5), 94–100 (2017). https://doi.org/10.1109/MCOM.2017.1600951

10. Zhang, S.: An overview of network slicing for 5G. IEEE Wirel. Commun. **26**(3), 111–117 (2019). https://doi.org/10.1109/MWC.2019.1800234

11. Yao, J., Han, Z., Sohail, M., Wang, L.: A robust security architecture for SDN-based 5G networks. Future Internet **11**, 85 (2019). https://doi.org/10.3390/fi11040085

12. Yang, Y., Wei, X., Xu, R., Peng, L., Zhang, L., Ge, L.: Man-in-the-middle attack detection and localization based on cross-layer location consistency. IEEE Access **8**, 103860–103874 (2020). https://doi.org/10.1109/ACCESS.2020.2999455

13. Kang, J.J., Fahd, K., Venkatraman, S., Trujillo-Rasua, R., Haskell-Dowland, P.: Hybrid routing for man-in-the-middle (MITM) attack detection in IoT networks. In: 2019 29th International Telecommunication Networks and Applications Conference (ITNAC), pp. 1–6 (2019). https://doi.org/10.1109/ITNAC46935.2019.9077977

14. Iavich, M., Gnatyuk, S., Odarchenko, R., Bocu, R., Simonov, S.: The novel system of attacks detection in 5G. In: Barolli, L., Woungang, I., Enokido, T. (eds.) AINA 2021. LNNS, vol. 226, pp. 580–591. Springer, Cham (2021). https://doi.org/10.1007/978-3-030-75075-6_47

15. Iavich, M., Akhalaia, G., Gnatyuk, S.: Method of improving the security of 5G network architecture concept for energy and other sectors of the critical infrastructure. In: Zaporozhets, A. (ed.) Systems, Decision and Control in Energy III, pp. 237–246. Springer International Publishing, Cham (2022). https://doi.org/10.1007/978-3-030-87675-3_14

16. Rezvy, S., Luo, Y., Petridis, M., Lasebae, A., Zebin, T.: An efficient deep learning model for intrusion classification and prediction in 5G and IoT networks. In: 2019 53rd Annual Conference on Information Sciences and Systems (CISS), pp. 1–6 (2019). https://doi.org/10.1109/CISS.2019.8693059

17. Kumudavalli, T.R., Sandeep, S.C.: Machine learning IDS models for 5G and IoT. In: Velliangiri, S., Gunasekaran, M., Karthikeyan, P. (eds.) Secure Communication for 5G and IoT Networks, pp. 73–84. Springer International Publishing, Cham (2022). https://doi.org/10.1007/978-3-030-79766-9_5

18. Hu, N., Tian, Z., Lu, H., Du, X., Guizani, M.: A multiple-kernel clustering based intrusion detection scheme for 5G and IoT networks. Int. J. Mach. Learn. Cybern. **12**(11), 3129–3144 (2021). https://doi.org/10.1007/s13042-020-01253-w

19. Ghosh, P., Mitra, R.: Proposed GA-BFSS and logistic regression-based intrusion detection system. In: Proceedings of the 2015 Third International Conference on Computer, Communication, Control and Information Technology (C3IT), pp. 1–6 (2015). https://doi.org/10.1109/C3IT.2015.7060117

Analysis of A-MPDU Aggregation Schemes for HT/VHT WLANs

Kaouther Mansour[1(⊠)] and Issam Jabri[2]

[1] IResCoMath Laboratory, ENIG, University of Gabès, Gabès, Tunisia
mansour.kaouther@enit.utm.tn
[2] CoEA, Yamamah University, Riyadh, Kingdom of Saudi Arabia
issam.jabri@enig.rnu.tn

Abstract. Frame aggregation technique is a milestone in 802.11-based Wireless Local Area Networks (WLANs) progress. Firstly defined by the 802.11n amendment for High throughput (HT) WLANs, this technique supports two forms of aggregation; MAC service Data unit aggregation (A-MSDU) and MAC Protocol Data Unit Aggregation (A-MPDU). Regarding its great impact on performance enhancement, this technique is also supported by the upcoming amendments. Indeed, the use of A-MPDU aggregation is compulsory in 802.11ac-based WLANs. In spite that the standard provides a detailed specification of A-MPDU technique implementation, an alternative approach is adopted by the majority of 802.11 card drivers. However, no investigations on the reasons behind the resort to an alternative implementation have been provided in the research literature. In this paper, we conduct a series of assessments to compare the performance of the conventional A-MPDU aggregation scheme called the greedy scheme and the alternative scheme known as the conservative scheme for multimedia traffics of different types in terms of throughput and delay. Next, guidelines on the choice of the most suitable scheme for a given scenario are outlined.

1 Introduction

Contemporary applications such as live streaming, video gaming and video conferencing present strict requirements in terms of Quality of service (QoS) [9]. Thus, in spite of the high capacities of contemporary WLANs, fine tuning of the network settings is crucial to guarantee better performance [11]. Different settings should be considered. Such as the choice of the suitable data rate, the channel width, the number of transmitting antennas. Studies have shown that such adjustments have great influence on network performance enhancement. In this works, we are interested in the frame aggregation technique, in particular A-MPDU aggregation. This technique was defined for the first time by the 802.11n amendment [8]. The key idea is to transmit multiple frames at a single access to the channel at the aim of optimizing channel usage efficiency. Two forms of aggregation are defined: A-MPDU aggregation and A-MSDU aggregation. The former consists in accumulating and encapsulating multiple MSDUs of the same

L. Barolli et al. (Eds.): AINA 2022, LNNS 451, pp. 388–398, 2022.
https://doi.org/10.1007/978-3-030-99619-2_37

access category addressed to the same destination. An MSDU is composed of a Link Layer Control (LLC) header, an IP header and the data payload. The latter offers the possibility to deliver up to N mpdus to the same destination using a unique PHY header. N presents the size of the Block Ack window and corresponds to 64 and 128 with 802.11n and 802.11ac amendments, respectively. A-MPDU aggregation is accomplished at the MAC layer. Firstly, MAC Packets Data Units are logically aggregated. Next, padding bits are added such that the length of each subframe is a multiple of four bytes. An MPDU delimiter is inserted at the end of each subframe. The latter is necessary for aggregated subframes depiction by the receiving node. The Block Acknowledgement technique is used, jointly with A-MPDU aggregation technique, to ensure individual error control over the aggregated subframes. Indeed, the sequence number of the oldest mpdu in the aggregated frame is recorded in the Starting Sequence Number subfield. Then each bit in the bitmap subfield ensures the acknowledgement of a specific mpdu in the aggregated frame such that the bit in position k is assigned to the subframe of sequence number $(SSN + k - 1)$ (Fig. 1).

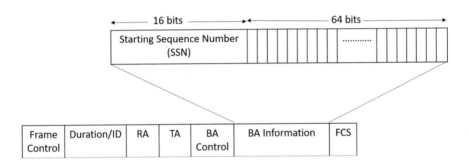

Fig. 1. Block acknowledgement frame format

In this paper we focus on the influence of the selected A-MPDU aggregation technique implementation scheme on network performance. Indeed, two different implementations are considered: the greedy scheme and the conservative scheme. the former corresponds to the conventional scheme. The latter is supported by most of wireless devices in spite of being different from standard specifications [1]. A comparative study is conducted to evaluate the performance of each of these schemes for different traffic types under various network conditions.

The paper is structured as follows. Preliminaries are given in Sect. 2. The most relevant related works to the area of our concern are briefly reviewed in Sect. 3. Section 4 is dedicated to analysis and discussions. And finally the paper is concluded in Sect. 5.

2   Background

2.1   The Conservative A-MPDU Aggregation Scheme

According to this scheme, once an aggregated frame composed of N new subframes is transmitted, no new subframes can be added until the initial N subframes are either delivered with success or dropped [2]. Figure 2 illustrates an example of A-MPDU aggregation transmission using the conservative scheme. The transmission of an aggregated frame composed of $N = 64$ subframes is initiated. 61 subframes are delivered with success however 3 subframes are corrupted. In the $i^{th} + 1$ transmission attempt, the aggregated frame is composed only of the 3 corrupted subframes transmitted with errors in the ith transmission attempt. As it has been claimed earlier, even though that this scheme is different from standard specifications, it is supported by most of WLAN card drivers.

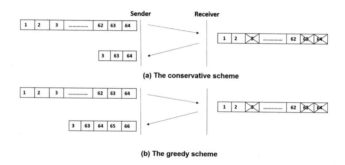

Fig. 2. A-MPDU aggregation schemes

2.2   The Greedy A-MPDU Aggregation Scheme

Presents the conventional scheme since it is conform to standard specifications [4]. According to this scheme, an aggregated frame composed of N subframes is initially transmitted. In case of mpdus corruption, the next transmitted A-MPDU will include the corrupted mpdus followed by eligible mpdus from the MAC queue. We note that the supported number of new subframes to be added depends on the sequence number of the first lost packet in the previous transmission attempt. Indeed, the conventional acknowledgement scheme requires that the sequence number of aggregated mpdus be in the interval $[SSN--SSN+N-1]$, here SSN refers to the sequence number of the first mpdu in the aggregated frame. Otherwise, the subframe can not be acknowledged and should be retransmitted [6]. Figure 2 illustrates an example of A-MPDU aggregation according to the conservative scheme. The transmission of a new aggregated A-MPDU composed of $N = 64$ is initiated. Feedback reveals that mpdus with sequence numbers $(3, 63$ and $64)$ are corrupted whereas the remaining subframes are transmitted with success. The following aggregated frame is composed of the 3 corrupted mpdus followed by 2 new ones. As it has been explained earlier, the last aggregated subframe is of sequence number $66 = 3 + N - 1$ which corresponds to $(SSN + N - 1)$.

3   Related Works

Regarding its importance, great research activities on frame aggregation technique have been conducted. Many area have been explored: performance evaluation, analytical modelling and improvement for QoS support. In this paper, we will review research activities in the first area. Namely, works that focused on the evaluation of the performance of packet aggregation technique on network performance. We focus, in particular, on multimedia traffic since this type of traffic presents strict requirements in terms of QoS.

Several works have demonstrated that frame aggregation technique improves considerably the achieved throughput since it permits to minimize channel access and control overhead. In [10], authors have shown that enabling packet aggregation permits to improve the number of calls supported by a VoIp application by a factor of 7 times. In [5], simulations have revealed that thanks to frame aggregation the network throughput shows an increase by above 2.5%. More than 200% improvement in terms of throughput is achieved due to packet aggregation in [12].

From another hand, research works have demonstrated that frame aggregation technique may cause performance degradation in terms of delay. In [17], authors have conducted extensive analysis to investigate the impact of frame aggregation on packet delay. Analysis results revealed that packet aggregation introduces excessive delays and thus it causes dramatic raise of packets transmission delays. Authors have concluded that this technique is not recommended for delay-sensitive applications. In [13] authors have examined the impact of packet aggregation for some particular applications. They claimed that this technique is not suitable for applications with data packets of small sizes because of the excessive introduced overhead. As a solution, authors have designed a new aggregation scheme called mA-MSDU. This new aggregation scheme aims to minimize the overhead introduced by the legacy A-MSDU and A-MPDU aggregation techniques. Hence, it is suitable for applications with small headers and permits individual MSDUs subframes error control. Authors have conducted extensive simulations to prove the efficiency of this scheme especially for VoIP applications. The same topic was tackled in [14] to demonstrate the efficiency of packet aggregation in saturated network conditions.

Authors in [16] have proposed a QoS-aware MPDU aggregation scheme adequate for delay-sensitive applications. In [3], a simulation-based study has been conducted to investigate the impact of packet aggregation for voice and video traffic. Simulation results showed that the impact of aggregation on delay depends on the codec rate. Thus, it shows high efficiency with high rate traffic. whereas, it causes severe performance degradation for low-rate applications.

The major deficiency of these works is that they suppose that A-MPDU aggregation is always performed at the maximum supported aggregation level N. Whereas this can not be implemented in practice due to the limit imposed by the conventional acknowledgement method (as explained earlier in Sect. 1)

Contrarily to previous analytic studies, in this paper, we consider the two schemes adopted for A-MPDU aggregation implementation. Hence, we conduct a comparative study to investigate the efficiency of each of these schemes for multimedia traffic transport.

For our analysis, we use the model proposed in [2] for the conservative scheme. Compared to other existing models, this model presents high accuracy since it takes into account the anomalous slots and the freezing of the backoff counter. For the greedy scheme, we use the model proposed in [7]. Contrary to the most of models found in literature that suppose that A-MPDU aggregation is always performed at the maximum aggregation size, this model respects the standard specifications. Thus, it takes into account the impact of the Block Ack window limit on the aggregation size variation. It is worthwhile to note that the issue of the variation of the aggregation size due to the Block Ack Window limit is a contemporary issue. Thus this point was not considered by previous research activities on aggregation. However recent studies have demonstrated that this problem has a great impact on A-MPDU aggregation technique performance especially in error prone environments [4,7,15].

In this study, three different multimedia applications are considered: voice traffic, video traffic and streaming traffic. A-MPDU aggregation behaviour variation face to channel quality degradation with both conservative and greedy schemes is studied. Performance in terms of throughput and delay are further analyzed under different channel conditions in the context of both 802.11n and 802.11ac networks.

4   Analysis and Discussions

4.1   Analysis Setup

We consider an infrastructured WLAN, composed of n mobile nodes transmitting saturated uplink traffic to the AP. We suppose that A-MPDU aggregation is enabled for all the transmissions in this network. For simplicity reasons, we suppose/assume that all the exchanged data packets over this network are of the same length. As recommended by the standard, we consider that the Request To Send/ Clear To Send (RTS/CTS) handshake mechanism is used for all transmissions. We assume that preambles and control frames are transmitted without errors, since they are transmitted at the lowest basic rate, furthermore they are very short. Finally, the binary symmetric channel model with independently distributed bit errors is considered. Simulation parameters are presented in Table 1.

Analysis are carried on for three types of traffic: Voice traffic, Video traffic and Streaming traffic in the context of both 802.11n and 802.11ac-based WLANs. Table 2 summarizes the characteristics of each of these traffic types.

Table 1. Analysis parameters.

Parameter	Value	Parameter	Value
Time slot	9 µs	PLCP	6 bytes
$SIFS$	16 µs	MAC Header	34 bytes
$DIFS$	34 µs	MPDU delimiter	4 bytes
$EIFS$	89 µs	Data rate	300/866 Mbps
RTS	208 bits	Basic rate	54 Mbps
CTS, BACK	160 bits	L_{subf}	160; 660 and 1500 bytes
W	15	m	7
N_s	20	N	64/128

Table 2. Traffic characteristics.

	Voice	Video	Streaming
Frame size (B)	160	660	1500
Delay target (ms)	50	150	250

4.2 Average Aggregation Size

Table 3 depicts the variation of the average estimated aggregated frame size in respect to the measured Packet Error Rate (PER). We note first that compared to simulation results, our model shows high degree of accuracy. Even in very lossy environments, the estimated values are extremely near to those resulting from ns-3 simulations. Regarding A-MPDU aggregation schemes behavior, as

Table 3. The average number of aggregated mpdus within an A-MPDU frame (802.11n) for different $PERs$.

PER	The greedy scheme	The conservative scheme
0.00	64	64.00
0.05	6.03	39.06
0.10	4.69	34.82
0.15	3.89	31.76
0.20	2.82	29.04
0.25	2.44	27.83
0.30	2.10	26.01
0.35	1.91	24.96
0.40	1.74	23.78
0.45	1.60	22.97
0.50	1.47	22.16

expected, in ideal channel conditions aggregation is accomplished at the maximum aggregation size N since no subframes losses are recorded. A very low PER ($PER = 0.01$) triggers a total change in the aggregation technique functioning process. The number of aggregated subframes falls to 39 and 5 with the greedy scheme and the conservative scheme respectively. For the greedy scheme, this is explained by the impact of the blocked bins in the bitmap of the Block Ack frame because of the legacy acknowledgement policy. Regarding the conservative scheme, this dramatic deterioration is due to the expensive adopted retransmission policy. Besides, for higher PER values, the conventional scheme maintains a relatively acceptable aggregation level (*about* 23 *mpdus*). However, the conventional scheme shows severe degradation. The average number of mpdus within an aggregated frame is below 2 mpdus.

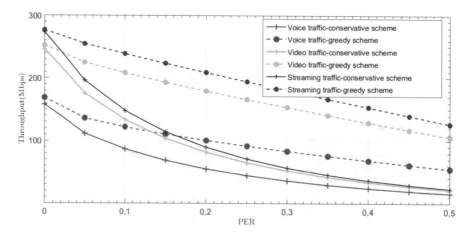

Fig. 3. Throughput against Packet Error Rate for voice, video and streaming traffic for 802.11n WLAN

Finally, even though that the conventional A-MPDU aggregation scheme presents some deficiencies in presence of bit errors because of the problem of the Block Ack window limit, it shows a potential benefit compared to the conservative scheme.

4.3   Throughput Analysis

Figure 3 and 4 depict the results of throughput analysis for Voice, video and streaming traffic for 802.11n and 802.11ac-based networks respectively. Firstly, we remark that the impact of A-MPDU aggregation on WLANs performance increases greatly with higher data rates. We notice also that for both schemes the achieved throughput depends on the considered application. Streaming traffic shows the best performance; The achieved throughput raises up to 270 Mbps

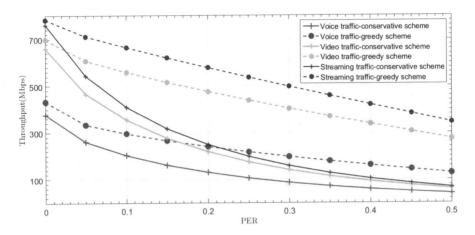

Fig. 4. Throughput against Packet Error Rate for voice, video and streaming traffic for 802.11ac WLAN

in absence of errors, whereas voice traffic presents the poorest output; The realized throughput is limited to 170 Mbps (see Fig. 3). The gap is much more important when we consider a 802.11ac-based WLAN. Compared to voice traffic, streaming traffic is almost the double. In accordance with previous studies on A-MPDU aggregation, these results prove that inspite of their great impact on MAC efficiency enhancement for traffic with large payload size, the gain achieved by aggregation is limited for applications with packets of small sizes [3].

In ideal channel conditions. The two schemes present equal throughput values. Thus both of them perform A-MPDU aggregation at the maximum supported aggregation size N. However, a low packet error rate provoques a dramatic difference between the two schemes. The greedy scheme outperforms the conservative scheme for the three considered types of traffic.

Compared to the conservative scheme, the greedy scheme presents an improvement that raises up to 130%, 110% and 40% for streaming, video and voice traffic respectively in the context of a 802.11n/ac-based Wlan for a $PER = 0.2$. Indeed, the greedy scheme presents higher efficiency level since it allows the insertion of new subframes in addition to the retransmitted ones. Hence, it permits to carry on larger frames. However with the conservative scheme, in presence of errors, the aggregation gain becomes limited since much time is devoted to the retransmission of corrupted mpdus.

4.4 MAC Access Delay Analysis

Figure 5 and 6 draw the average access delay performance analysis. Firstly we notice that the estimated delay depends on the size of the aggregated subframes; larger mpdus show longer delays. Indeed, streaming traffic presents the highest delay; above 0.4 s when $PER = 0.5$ for 802.11ac (Fig. 6).

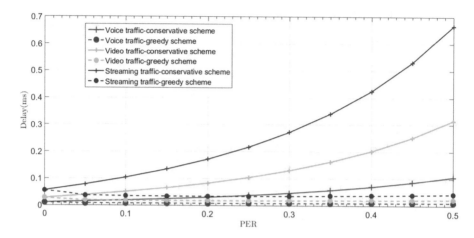

Fig. 5. Average Delay against Packet Error Rate for voice, video and streaming traffic for 802.11n WLAN

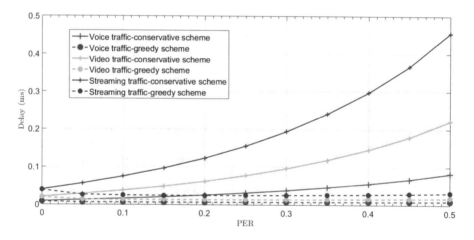

Fig. 6. Average Delay against Packet Error Rate for voice, video and streaming traffic for 802.11ac WLAN

Regarding the performance of the two considered A-MPDU aggregation schemes, we remark that in an error free channel they present almost equal average delays. Thus both of them perform aggregation at the maximum supported aggregation level N. However, they show extremely different behavior in presence of channel errors. From one hand, with the conservative scheme, the estimated delays raise up significantly. This is due to the adopted retransmission policy which causes a dramatic decrease of the size of the aggregated frame in presence of errors. As it has been introduced earlier (Table 3), the average number of subframes within an aggregated mpdu is above 2. From another hand, analysis results show that the greedy scheme show high ability to resist to channel quality degradation. Indeed, even in very lossy environments, a slight increase of the estimated delays is observed; The measured delay for voice traffic in the context of a 802.11ac WLAN increases from 0.007 s when $PER = 0.05$ to 0.009 s when $PER = 0.5$ (Fig. 6 (b)). Thus, the greedy scheme approach guarantees a relatively higher efficiency level inducing thus better performance. It is important to note that with the greedy scheme the delays estimated in error free channel conditions are higher than those estimated in presence of errors; In the context of 802.11ac networks, the delay estimated in ideal conditions for streaming traffic is about 0.04 s, whereas it is below 0.025 s when the $PER = 0.15$ (Fig. 5). This is explained by the impact of the frame size on delay. Thus in absence of errors, the aggregated frame includes N subframes which badly affects delay performance, whereas the decrease of the number of aggregated subframes in lossy environments plays a positive role on delays minimization. The slight increase of the estimated delays for high $PERs$ is due to the raise of the number of required retransmission attempts. These results prove that the conventional scheme shows high ability to guarantee good performance in terms of delay under different channel conditions.

5 Conclusion

In this paper, we have conducted a comparative study between two different A-MPDU aggregation schemes: the conventional scheme, specified by the standard, and the conservative scheme, implement on the most of WiFi card drivers. The study is based on analytic models proposed in [2] and [7]. Analysis of the behavior of the two considered schemes in terms of average frame size, throughput and delay are done.

Analysis results show that the two schemes present similar performance in error-free environments, whereas, in presence of channel errors they behave differently. The performance of the conservative scheme is significantly influenced by channel quality degradation. However, the greedy scheme shows certain robustness. Thus, even though that the achieved throughput has decreased, delay performance are improved. Indeed, the decrease of the maximum aggregation size in error prone environments because of the problem of the Block Ack window limit permits to maintain good delay performance. To conclude, in spite of its higher computational complexity compared to the conservative scheme, the greedy scheme presents better efficiency, especially in erroneous environments.

Currently we are working on an intelligent A-MPDU aggregation scheduler for 802.11ac WLANs based on the greedy aggregation scheme.

References

1. Atheros ath9k Wireless Driver. http://linuxwireless.org/en/users/Drivers/ath9k
2. Hajlaoui, N., Jabri, I., Jemaa, M.B.: An accurate two dimensional Markov chain model for IEEE 802.11n DCF. Wirel. Netw. **24**(4), 1019–1031 (2018)

3. Hajlaoui, N., Jabri, I., Taieb, M., Benjemaa, M.: A frame aggregation scheduler for QoS sensitive applications in IEEE 802.11n WLANs. In: Proceedings of the ICCIT, Tunisia, pp. 221–226 (2012)

4. Jabri, I., Mansour, K., Al-Oqily, I., et al.: Enhanced characterization and modeling of A-MPDU aggregation for IEEE 802.11n WLANs. Trans. Emerg. Telecommun. Technol. **33**, e4384 (2021)

5. Jibukumar, M.G., Datta, R., Biswas, P.K.: New frame aggregation schemes for multimedia applications in WLAN. In: Proceedings of the NOMS, Japan, pp. 424–431 (2010)

6. Inamullah, M., Raman, B.: 11 ac frame aggregation is bottlenecked: revisiting the block ACK. In: International ACM Conference on Modeling, Analysis and Simulation of Wireless and Mobile Systems, 22nd edn., pp. 45–49 (2019)

7. Mansour, K., Jabri, I., Ezzedine, T.: Revisiting the IEEE 802.11n A-MPDU retransmission scheme. IEEE Commun. Lett. **23**(6), 1097–1100 (2019)

8. IEEE 802.11-2020 - IEEE Approved Draft Standard for Information Technology - Telecommunications and Information Exchange Between Systems Local and Metropolitan Area Networks - Specific Requirements - Part 11: Wireless LAN Medium Access Control (MAC) and Physical Layer (PHY) Specifications (2007)

9. International Telecommunication Union: Global Information Infrastructure, Internet Protocol Aspects and Next-Generation Networks; ITUT Y.1541; Series Y. International Telecommunication Union (ITU), Geneva (2011)

10. Kim, K., Ganguly, S., Lzmailov, R., Hong, S.: On packet aggregation mechanisms for improving VoIP quality in mesh networks. In: IEEE Vehicular Technology Conference, Victoria, 63rd edn., pp. 891–895 (2006)

11. Khalil, N., Najid, A.: Performance analysis of 802.11ac with frame aggregation using NS-3. Int. J. Electr. Comput. Eng. **10**(5), 5368 (2020)

12. Majeed, A., Abu-Ghazaleh, N.B.: Packet aggregation in multi-rate wireless LANs. In: Annual IEEE Communications Society Conference on Sensor, Mesh and Ad Hoc Communications and Networks (SECON), Korea, 9th edn., pp. 452–460 (2012)

13. Saif, A., Othman, M., Subramaniam, S., Hamid, N.A.W.A.: An enhanced A-MSDU frame aggregation scheme for 802.11n wireless networks. Wirel. Pers. Commun. **66**, 683–706 (2011)

14. Selvam, T., Srikanth, S.: A frame aggregation scheduler for IEEE 802.11n. In: Proceedings of the NCC, Chennai, pp. 1–5 (2010)

15. Seytnazarov, S., Choi, J.G., Kim, Y.T.: Enhanced mathematical modeling of aggregation-enabled WLANs with compressed blockACK. IEEE Trans. Mob. Comput. **18**(6), 1260–1273 (2018)

16. Seytnazarov, S., Kim, Y.T.: QoS-aware MPDU aggregation of IEEE 802.11n WLANs for VoIP services. In: CNSM and Workshop, Brazil, 10th edn., pp. 64–71 (2014)

17. Teymoori, P., Yazdani, N., Hoseini, S.A., Reza, M.: Analyzing delay limits of high-speed wireless ad hoc networks based on IEEE 802.11n. In: Proceedings of the IST, USA, pp. 489–495 (2010)

An Implementation of V2R Data Delivery Method Based on MQTT for Road Safety Application

Akira Sakuraba[1]([⊠]), Yoshitaka Shibata[2], and Mamoru Ohara[1]

[1] IoT Technology Group, Information Systems Technology Division, Tokyo Metropolitan Industrial Technology Research Institute, Koto-ward, Tokyo-met., Japan
{sakuraba.akira,ohara.mamoru}@iri-tokyo.jp
[2] Regional Cooperative Research Division, Iwate Prefectural University, Takizawa-city, Iwate-pref., Japan
shibata@iwate-pu.ac.jp

Abstract. Vehicle-to-Road (V2R) communication is a part of the essential elements to realize near-the-future safer driving technologies. There is a challenge for temporal restriction of data delivery when a vehicle communicates on the move, due to requirements for speedy data exchange to complete the operation within a quite short time. Hence, designing V2R wireless communication also requires consideration that there is disruption of the wireless link, much latency, loss of data, or other challenging issues. This paper reports a new implementation for the V2R wireless communication designed to exchange road surface weather information. This new implementation considers a principle of minimum data delivery to avoid redundant data exchanging. The implementation utilizes a messaging system based on Message Queueing Telemetry Transport (MQTT). The implementation provides a point-of-interest (PoI) based message extraction method to reduce the number of RSI records to deliver. We have evaluated the performance of data delivery while exchanging information between a roadside unit and an onboard unit. The result suggested that the proposed system can deliver RSI correctly in case of the connection disruption, however, it revealed additional requirements for redesign messaging between nodes.

1 Introduction

Wireless communication surrounding road vehicles is becoming one of the commodity elements of the modern automotive technology. There are many inter-vehicular networking installations on the street-legal vehicles sold today. The vehicles equip public mobile networking modules, such as 4G cellular standards, to connect roadside units (RSUs) or other vehicles. The modern vehicles can provide road safety applications to indicate dynamic information on the road to their drivers, such as existing of road-crossing pedestrians, the tail of traffic jams frontward located at blind corners, etc. Vehicle-to-Road (V2R), Vehicle-to-Vehicle (V2V), or Vehicle-to-Everything (V2X) wireless communication are essential technical elements to realize the near-the-future safer driving

L. Barolli et al. (Eds.): AINA 2022, LNNS 451, pp. 399–410, 2022.
https://doi.org/10.1007/978-3-030-99619-2_38

environments, in which road traffic safety information is exchanged and shared among road infrastructures and vehicles as well as drivers on the vehicle.

Meanwhile, the data rate of vehicular wireless connectivity is getting broader in the context of introducing high data rate wireless communication standards. For example, C-V2X based wireless communication enables to delivery of about 2 to 50 times the data amount at the same unit time compared to IEEE 802.11p based WAVE system. In addition, the future 5G based V2X communication can provide 10 Gbps data rate with multimedia and cloud support [1]. Thus, the onboard vehicular application can deal with multiple types of data and exchange information with multiple vehicular/infrastructure nodes.

On the other hand, public wireless access services are not necessarily available on every road, especially in mountainous or rural areas. According to an investigation by the Japanese government, the coverage of public mobile networks accounts for only 24.7% to 36.4% of the national land area for 2 GHz band which is the most spread public mobile RF band in Japan [2]. Road safety application is required to be available in any environment; we believe there is a requirement for considering and building private wireless communication methods over the direct link topology to provide inter-node communication in the non-service area in which the mobile network operators do not provide their service. Therefore, there is a motivation to develop a private V2X wireless communication method in the real road environment.

Private networks based on the unlicensed RF bands usually cover very smaller area compared to the cellular networks. The smaller coverage of the wireless links rises the challenges for availability. For instance, limited effective throughput and disruption of connection are often observed issues in the actual V2X communication. In addition, the impact of these challenges could vary according to the quality of service in the actual road environment. To develop a V2X communication system, there is a requirement for efficient and reliable design in order to complete data delivery within a short period, e.g., during the vehicle is passing by the RSU.

This paper presents a V2R wireless communication method using two different wavelength wireless links on unlicensed bands to enable both long-distance node discovery and exchanging road safety information over high data rate wireless links. We have built a Message Queueing Telemetry Transport (MQTT) based road state information (RSI) exchange system. The system delivers limited information containing only RSI records requested by the vehicle in order to reduce delivery time.

We have evaluated the new RSI delivery system on the actual devices. The result shows improvement of availability to some extent in a realistic scenario in V2R communications with TCP connection disruptions. We also figure out that there is a temporal overhead in the messaging during RSI delivery over the MQTT-based method.

2   Related Works

Vehicular ad-hoc networking (VANET) is a very interesting topic for researchers. There are many implementations of multi-hop inter-vehicle networking method. Ducourthial et al. observed an increase in the packet loss rate corresponding to increasing of the hopping nodes in the stable traveling convoy [3]. Moreover, a multi-hop messaging can deliver data to nodes, where the opposite node is out of the line of sight, however, it has difficult to discover other nodes due to breaking out of the retransmission of the beacon signal [4].

MQTT is utilized for data delivery methods over inter-vehicular networking. Vilalta et al. investigated the feasibility of a hierarchical vehicular message exchange solution [5]. They evaluated the inter-broker and cross-border message exchanging delay between the vehicle and the public cloud resource. The delay was about 260 ms and the overhead during exchanging was 0.2 ms to deliver 100 samples of data.

In our past work, we developed a file-based vehicular data exchanging system to realize V2X communication [6]. Our system was intended to operate in rural and mountainous areas, in which there is a limitation of the connectivity to the public cellular network services. Thus, the onboard system also connected to the private V2X system. The V2X communication system exchanged the categorized RSI between the nodes by transferring a file over SSH secure channels. The past system was difficult to recover or lose the RSI from disruption of the connection.

3 Design of V2R Wireless Network

3.1 Overview

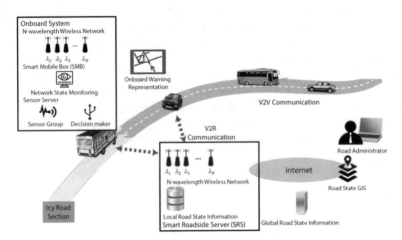

Fig. 1. Overview of the proposed system.

The proposed system is a single-hop opportunistic delay-tolerant vehicular networking system on the sensing vehicle. Figure 1 describes the overview of this proposed system.

The V2R communication system is connected to an onboard road surface weather categorizing system. The vehicle analyzes the road surface weather state while the vehicle is traveling. The categorized road surface weather state is associated with geocoordinate data to the RSI and stored into the storage on the vehicle when the vehicle travels outside of the RSU coverage. When a vehicle is entering the coverage of wireless links to a RSU, the vehicular sends RSI records in a burst. Finally, the onboard unit provides RSI to visualize road state for a GIS application.

Similarly, the RSU merges the received RSI records into a consistent database called Zone DB to manage RSI in wide area. The RSU connects to the cloud-computing resource to obtain and submit RSI records over the region. Because the RSU may issue a huge number of datasets at once, the system needs to exchange only the essential part of them in order to reduce the amount of data delivery. Our new approach extracts the necessary part corresponding to the Point-of-Interest (PoI) from the consistent database into the storage of RSUs so that the vehicle can receive only necessary RSI records along to require the length of road sections. We expect that the system could deliver RSI even when available time of the V2R network is very short time.

A vehicle initiates the communication when approaching an RSU. Both nodes exchange the control messages over a LPWA link before they come in their coverage of WLAN links. The LPWA link has wider coverage than the WLAN link. The vehicular and the RSU nodes can discover vehicular nodes and initiate the connection When a vehicle is approaching an RSU, which has both LPWA and WLAN links, the vehicle can discover the LPWA link establish a control link over the LPWA earlier than the WLAN becomes available.

We have developed an MQTT-based RSI message delivery method on the WLAN datalink. MQTT is a message-oriented publisher/subscriber-model data delivering protocol. We consider that the attractive point of MQTT on data delivery links is the quality of service (QoS). The quality level of MQTT can be designated in the following three different levels when a publisher sends a topic: AT_MOST_ONCE, AT_LEAST_ONCE, and EXACTLY_ONCE. We can assure the exchange of RSI record sby utilizing the QoS mechanism, for example by the system delivering RSI records in the dangerous road section such as icy or snowy roads with EXACTLY_ONCE QoS level. Whereas the RSI in the safer road section is categorized as dry or damp, the system can exchange with AT_MOST_ONCE in QoS topics over the challenged network environments. The system uses these different QoS levels depending on the road state to be at risk of traffic collisions.

3.2 PoI and the Consistent Database

Figure 2 illustrates the PoI based data management in our system. PoI is a distance which the driver desired to obtain the frontward RSI. The vehicular node sends the PoI distance and the traveling direction of the vehicle to the RSU node during the control link is initiating.

The RSU node extracts RSI records including the circle of requested PoI from the Zone DB in response. Each RSI record in the Zone DB is associated with a Geohash code-based zones and the RSU extracts the corresponding records in the internal zones of the circle centered at the provided vehicle's geolocation with a radius of the PoI distance. Geohash represents coordinates as the surface on the corresponding grid. Geohash based management of RSI can associate with the mesh meteorological data that public meteorological authorities provide, to predict the forecast road surface weather state.

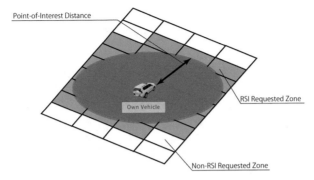

Fig. 2. Data requesting method for PoI delivery based on PoI distance.

3.3 LPWA Link to Initiate Communication

LPWA enables long-distance packet delivery even when using unlicensed band standards. We use it as the control message link on which both onboard/roadside units exchange control messages with each other.

Generally, the 1-hop coverage of the typical implementation of private LPWA standards is several to several ten kilometers in radius [7]. We investigated several field experiments and obtained the result that the effective coverage radius has remained at approximately a kilometer with any spreading factors on 200 kHz bandwidth on 920 MHz band LoRaWAN [8]. We determined that with this coverage we can deliver control messages for discovering other nodes, obtaining the PoI distance, and acquiring the geolocation of the vehicle.

3.4 WLAN Links to Deliver RSI

The proposed system adopts high data rate WLAN standards for dedicated data deliver links. On these links, the vehicular node transmits categorized RSI recorded by the vehicle to the RSU node. Conversely, the RSU node selects the RSI records from the Zone DB and sends them to the vehicular node.

4 RSI Data Delivery Method on MQTT

4.1 Brokers and Clients

Our system is composed of MQTT brokers and clients in both the vehicular and RSU nodes as illustrated in Fig. 3. The Communication Unit (Comm Unit) works as an MQTT broker for the local Database Unit (DB Unit) and connects to the broker in the opposite node as a client when the WLAN link is available. A Comm Unit also relays between the local DB Unit and the opposite Comm Unit. DB Unit issues the local RSI records in response to the request message from the local broker and receives the relayed RSI records from the opposite node via the remote and the local brokers.

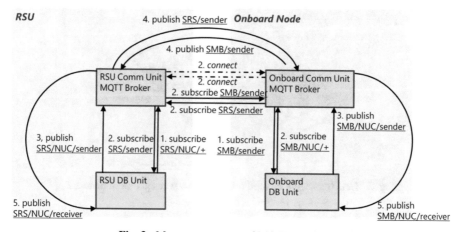

Fig. 3. Message sequence of MQTT messaging

4.2 Sequence of Messaging

The system establishes a messaging session over MQTT after the reachability of the WLAN datalink between the vehicular and RSU nodes is confirmed. Figure 3 also describes the message sequence over MQTT among the components of a pair of RSU and vehicular nodes. Firstly, the DB Unit attempts connecting to the local broker. The DB Unit immediately subscribes to the relayed topics sent by the local broker. The relayed topics are forwarded messages originated from the opposite nodes. Next, the Comm Unit connects to the broker in the opposite node and immediately subscribes sender topic to obtain RSI records from the opposite node. After establishing a mutual subscription between both brokers, the Comm Unit publishes a *start* topic to initiate a data delivery session and the local DB Unit receives it.

On receiving the *start* topic, the DB Unit publishes one or more *sender* topics, whose payload contain RSI records, toward the local broker. Then, the Comm Unit relays *sender* topics to the remote broker. Finally, the local broker publishes *receiver* topics according to the received *sender* topics.

4.3 Payload of Sender/Receiver Topics

The sensor vehicle collects road surface weather state, to which the system associates with geolocation, air/road temperature, the humidity of the outside air, and other environmental sensor values in the constant interval Each *sender* or *receiver* topic contains an RSI record formatted in a plain JSON string. An example is described in Fig. 4.

"{\"id\":\"12623\",\"timestamp\":\"2020-02-25 08:12:45.109\",\"latitude\":\"39.645245\",
\"longitude\":\"141.3683\",\"state\":\"Dry\",\"friction\":\"0.9593998164442743\",
\"air_temp\":\"6.321174399713035\",\"road_temp\":\"8.340193838654487\",\"humidity\":
\"38.505669202189566\",\"roughness\":\"0.9827495334726336\"}"

Fig. 4. An example of the RSI record payload format in sender/receiver topic

5   Evaluation

This section reports the performance evaluation of the WLAN link in the proposed system in an indoor experiment with the actual device setup. This experiment evaluates the capability of data delivery under an emulated V2X wireless communication environment. We performed the evaluation under an indoor configuration with the network performance emulation, in which a scenario emulates an end-to-end performance over the V2R network between the RSU and traveling vehicular nodes on the road. This evaluation focused on WLAN link performance.

5.1   Scenario of Experiment and Deliver RSI

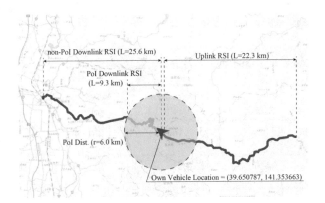

Fig. 5. Visualized RSI resources to deliver in the experiment

The evaluation was performed based on the data exchanging while the vehicle is passing by an RSU to measure the capability of data delivering and availability of our prototype while the vehicular node is on-the-move. We set an actual RSI resource which is obtained along Japan National Route 106 in the northern part of mainland Japan in February 2020.

RSI records collected by a vehicle node were sent to the RSU via "uplink". Reversely, from the RSU to the vehicle direction, RSI records describing the forward in the direction of travel to the vehicular node were sent via "downlink" in Fig. 5. In the experiment, both nodes published all RSI records with the default QoS = 1 (AT_LEAST_ONCE) for MQTT.

The evaluation system emulated the throughput limitation of moving vehicles. We measured the end-to-end throughput while V2R wireless communication in the past work and have implemented a throughput limitation script to emulate the vehicular node traveling at 50 km/h described in Fig. 6.

We measured the data delivery performance of the proposed system and compared to our previous file transferring-based exchanging system [9]. Our new approach delivered downlink a smaller number of records of RSI resource than that of the previous work

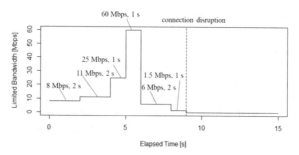

Fig. 6. Network emulation in the experiment trials

due to designating of PoI. The previous system delivered whole of the accumulated RSI records as a file in the SQLite format over SFTP on the WLAN link. Details of the delivery RSI resources are shown in Table 1.

Table 1. Specification of delivery data over SFTP/MQTT trials

RSI resource	Number of records	Size of RSI resource [bytes]	Duration of traveling	Distance of traveling [km]
MQTT, PoI-sent, downlink	651	163,163	12 m 45 s	9.3
SFTP, non-PoI, downlink	3,051	290,816	59 m 57 s	25.6
MQTT, uplink	1,508	378,322	29 m 56 s	22.3
SFTP, uplink	1,508	151,552	29 m 56 s	22.3

5.2 Prototype for Evaluation

We built a prototype environment to evaluate our new data delivery method as shown in Fig. 7. The prototype consists of a pair of vehicular nodes and an RSU node. Both types of nodes have one ARM-based single-board computer Raspberry Pi 3 with Raspberry Pi OS Buster for the communication unit and x86_64 based Intel NUC for the DB Unit with Microsoft Windows 10. We installed Eclipse Mosquitto 1.5.7 as the MQTT broker on the communication units. MQTT clients and the evaluation module are written by Ruby 2.5.

The control message link is based on ARIB STD-T108 LoRa standard, we use Oi Electric OiNET-923/928 for the single-hop connection.

We choose IEEE 802.11ac for a data delivery link. We connected a Ruckus Wireless ZoneFlex T300 access point to the RSU node. On the other hand, a PLANEX GW-900D USB-WLAN adaptor provides WLAN connectivity in the vehicular node.

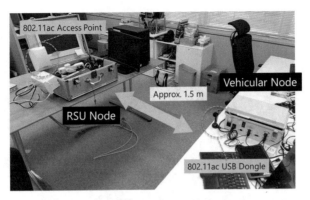

Fig. 7. Prototype system in the experiment environment.

5.3 Results

Figure 8 represents the amount of data transfer by the timeline. We observed both SFTP and MQTT data delivery methods in 20 trials for each. Data deliveries showed a linear increase in the lapse of time in both data delivery methods. Both data delivery methods required frequent messaging while in the initialization of the connection to secure the session such as encryption processing. The emulated throughput was increased to 11 Mbps, even when a SFTP session sent a file in the constant delivery rate.

The MQTT-based method completed RSI records transmissions via the uplink in almost all trials. However, the downlink transferring over MQTT did not complete to deliver the opposite node the entire sender RSI records until the link disruption broke out. In contrast, the SFTP based per-file delivery was completed in both directions with very few deviations of completion time. We found just a failure to delivery of an RSI database file over SFTP. This failure means partial reception of the RSI resource file ends up in complete loss in the receiver node because the RSI resource is a compressed binary file.

The statistics of the experiment are shown in Table 2. Surprisingly, SFTP based implementation delivered RSI records to the opposite node with better performance. Despite the uplink payload having the same number of records, MQTT requires 3.6 times longer time to deliver the RSI.

We also focused on the completion ratio of the total number of trials to the number of trials that the system finished delivering all of the prescribed data. The result shows that over 95% of the records were delivered properly. Exceptionally, it got worse in the MQTT data delivery method on the uplink, it totally failed to complete to deliver the entire RSI resource. We observed explicitly poor delivery performance over MQTT-based one compared to the file-based transferring. However, when the MQTT session encountered TCP session disruptions, it succeeded to publish 53.5% of RSI records of the entire uplink payload on average.

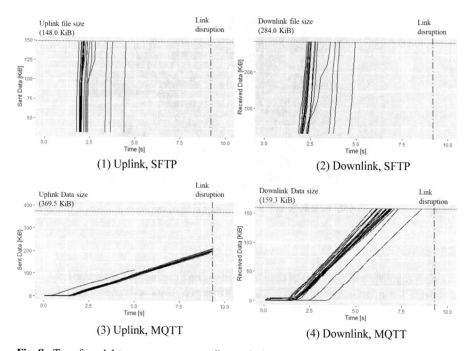

Fig. 8. Transferred data amount corresponding to the lapse of time. Each line denotes the amount in the individual trial.

Table 2. Statistics of the results.

	SFTP uplink	SFTP downlink	MQTT uplink	MQTT downlink
Mean time required to deliver Data [s]	2.49 ($\sigma = 0.66$)	2.84 ($\sigma = 0.71$)	9.05 ($\sigma = 0.94$)	6.43 ($\sigma = 1.57$)
Mean delivered data amount [Byte]	151,552	287,475	202,282	155,018
Completion ratio to the entire trials [%]	100	95	0	95

5.4 Discussion

The result of the evaluation suggests that MQTT can deliver the fragments of the whole RSI resource as many as possible. In case the event of a disruption has occurred, the file-based RSI exchanging could fail and lose the entire RSI resources. We confirmed the partial delivery of RSI resources even when the TCP connection was disrupted during the exchange. This result suggested that RSI transportation by the unit of records should be considered for availability on the challenged V2X networks.

Over the MQTT message links, publishers publish the RSI records one by one and receive the subscribed topics. In the experiment, the payload length of messages was only about 330 octets, therefore the system could send them in about 10 ms. These results suggests that the design of message leads very large temporal overhead for publishing, due to, e.g., waiting for receiving PUBACK messages corresponding to the published message.. We can also consider redesigning of topic payload to resolve this. For example, a record can be bonded with other records existing in the same zone represented in the same Geohash code. It is possible to reduce waiting time for ACK due to decrease of the number of messages to publish.

Another consideration is the priority for RSI delivery based on QoS and ordering of the topic publishing. As we discussed in Sect. 3.1, the MQTT-based system can vary the QoS level in RSI records. To decrease the temporal overhead, the publisher can send RSI records in the low-risk road section with the lower "fire-and-forget" QoS level so that it needs not wait for the PUBACK message. In addition, the system can begin to publish RSI records in the high-risk road section independently from the temporal sequence in the original RSI resource. The experiment results show the possibility of improvement in performance by dynamic prioritization of data delivering.

6   Conclusion

This paper reported a new approach for a single-hop and delay-tolerant MQTT-based V2R wireless communication system. The design considers a challenging situation ascribed to the disruption of the connection while the vehicular node is on-the-move. We have presented the message sequence in the system consisting of components in/between vehicular/RSU nodes.

We evaluated the end-to-end performance of the system on actual V2R communication nodes. As the result, we confirmed that the V2R communication system succeeded to exchange 53.5% of the RSI records at the minimum in a practical scenario even when a connection loss was broken out during data delivery.

In our future work, to improve the data delivery rate between the nodes, we will work to reconsider the MQTT messaging between the components to reduce temporal overhead in the message exchanging with dynamic QoS designation. We are also planning to measure the performance between the nodes in the actual public road environment.

Acknowledgments. This work was supported by JSPS KAKENHI JP20K19826, JSPS KAKENHI JP20K11773, and Japan Keiba Association Grant 2021M-198.

References

1. Usman, M., et al.: A business and legislative perspective of V2X and mobility applications in 5G networks. IEEE Access **8**, 67426–67435 (2020)
2. Draft: Outline of Evaluation Result of RF Utilization Status for Mobile Phone and BWA in the Nation (in Japanese), Electromagnetic Environment Division, Radio Department, Telecommunications Bureau, Ministry of Internal Affairs and Communications (2020). https://www.soumu.go.jp/main_content/000720959.pdf. Accessed 07 Jan 2022
3. Ducourthial, B., Khalfallah, S.: A platform for road experiments. In: 2009 IEEE 69th Vehicular Technology Conference, pp. 1–5 (2009)

4. Mittag, J., et al.: A comparison of single- and multi-hop beaconing in VANETs. In: Proceedings of the 6th ACM International Workshop on VehiculAr InterNETworking, pp. 69–78 (2009)
5. Vilalta, R., et al.: Vehicular message exchange in cross-border scenarios using public cloud infrastructure. In: 2020 IEEE 3rd 5G World Forum, pp. 251–256 (2020)
6. Leontiadis, I., et al.: On the effectiveness of an opportunistic traffic management system for vehicular networks. IEEE Trans. Intell. Transp. Syst. **12**(4), 1537–1548 (2011)
7. Shinha, R.S., Wei, Y., Huang, S.: A survey on LPWA technology: LoRa and NB-IoT. ICT Express **3**(1), 14–21 (2017)
8. Sakuraba, A., Shibata, Y., Tamura, T.: Evaluation of performance on LPWA network realizes for multi-wavelength cognitive V2X wireless system. In: 10th International Conference on Awareness Science and Technology, pp. 434–440 (2019)
9. Sakuraba, A., Shibata, Y., Sato, G., Uchida, N.: Evaluation of end-to-end performance on N-wavelength V2X cognitive wireless system designed for exchanging road state information. In: Barolli, L., Okada, Y., Amato, F. (eds.) EIDWT 2020. LNDECT, vol. 47, pp. 277–289. Springer, Cham (2020). https://doi.org/10.1007/978-3-030-39746-3_30

Smart Metering Architecture for Agriculture Applications

Juan C. Olivares-Rojas[1]([⊠]) [iD], José A. Gutiérrez-Gnecchi[1] [iD], Wuqiang Yang[2] [iD], Enrique Reyes-Archundia[1] [iD], and Adriana C. Téllez-Anguiano[1] [iD]

[1] Tecnológico Nacional de México/I. T. de Morelia, Av. Tecnológico 1500, Lomas de Santiaguito, Morelia, Mexico
{juan.or,jose.gg3,enrique.ra,adriana.ta}@morelia.tecnm.mx
[2] Department of Electrical and Electronic Engineering, The University of Manchester, Manchester M13 9PL, UK
wuqiang.yang@manchester.ac.uk

Abstract. The development of technology has made agriculture more and more efficient, being more productive with less resources consumed, such as water and fertilizers. Our objective is to develop more efficient measurement systems for agricultural applications, e.g. to be more intelligent measurement and data processing and to make forecasts, helping decision-making. In this work, an architecture based on distributed edge-fog-cloud computing is presented. Through low power consumption and wide-area communications, this architecture can be used for measurement of climate, energy, and soil variables to achieve more productive crops.

List of Abbreviations

AI	Artificial intelligence
AMI	Advanced measurement infrastructure
CAN	Campus area network
DC	Data concentrator
DL	Data lake
DER	Distributed energy resource
GSM	Global system for mobile communications
IoT	Internet of Things
LAN	Local area network
LPWAN	Low power wide area network
MAN	Metropolitan area network
SBC	Simple card computer
SM	Smart meter
WAN	Wide
WP	Water pumps
WSAN	Wireless sensor and actuator network

L. Barolli et al. (Eds.): AINA 2022, LNNS 451, pp. 411–419, 2022.
https://doi.org/10.1007/978-3-030-99619-2_39

1 Introduction

Electricity is an essential commodity to support our daily life and activities, in particular those activities involving in agriculture, which requires extensive vital resources, such as water and electricity mainly. Because the costs of these resources are eventually transferred to final consumers, it is essential to adopt more efficient measurement to reduce the consumption of resources.

On the other hand, because clean energy micro-generation technologies, such as photovoltaic panels and wind turbines, become cheaper and cheaper, electrical energy is increasingly accessible by the final consumers, which is called now prosumer (producer/consumer). Notably, solar panels have begun to be widely used in the agricultural sector for self-consumption. This entire area of study is called agrovoltaics.

One of the main problems with agrovoltaic systems is that although they can efficiently measure the production of electrical energy, agrovoltaic systems must be integrated into the traditional measurement of electrical consumption due to their distance. On the other hand, it is necessary to measure other resources and variables, such as water, climate, wind, and soil, for agricultural products.

This work aims to develop a smart metering system that remotely integrates electrical consumption/production and climatic variables, allowing agricultural producers to be more efficient in consuming key natural resources. The system is intended to integrate a multi-level architecture that allows remote measurement of data for subsequent visualization, processing, and analysis of information.

The development of new technologies, such as the Internet of Things (IoT), cloud computing, robotics, artificial intelligence (AI), autonomous vehicles, among others, has resulted in great changes in production systems, facilitating people's life quality. In agriculture, IoT can carry out much more meticulous control of its production. Sensors can be used to acquire soil conditions in real-time, such as humidity, temperature, and chemical indicators [1]. Other technologies, such as geolocation systems and wireless sensor networks, have brought ubiquity to the field.

Excessive population growth and ongoing climate change put high pressure on natural resources and require new approaches to planning and managing crops [2]. On the other hand, agricultural producers are forced to increase productivity and quality while preserving the environment. Unfortunately, only farmers in developed countries have the technology required to increase yield, ensure food security, and increase the profitability of the field [3].

Farming systems are enhanced with technology and data analytics to implement smart farming systems. For example, these systems can continuously monitor the water requirements of the land and crops in real-time through a network of wireless sensors while saving and managing the collected data in a cloud database [4].

Social, ecological, economic, and health incentives favor the growing adoption of technology in agriculture and highlight its importance for comprehensive national development in the short and medium term. Through the 2019 Agriculture Promotion program in the "2019 Agricultural Research, Innovation, and Technological Development Component," the Mexican government emphasizes research and technological development to solve production problems [5]. On the other hand, most smart agricultural systems have not considered the efficient use of electrical energy. With the proliferation of solar panels for self-generation, these are beginning to be feasible.

To address the need mentioned above, we propose an architecture of an intelligent measurement system for agrovoltaic applications that considers a multi-level scheme to obtain electrical energy consumption/production variables and climatic variables, allowing the information in a data lake repository for later use for further data processing and visualization. The multi-tier architecture is based on the edge-fog-cloud computing paradigm and the Advanced Metering Infrastructure (AMI). The communication between the various components is intended to be done through wireless wide area networks and low power consumption, such as LoRa, which is a long-range proprietary low-power wide-area network modulation technique and LoRaWAN. While LoRa defines the lower physical layer, LoRaWAN is one of several protocols to define the upper layers of the network.

2 State of the Art

Recent developments of technologies have made it possible to deploy low-cost, miniaturized electronics in a wide variety of sectors, such as agriculture. The use of information and communication technologies in the agricultural sector has been used since the very beginning after the Second World War. Existing work on intelligent agricultural systems ranges from precision agriculture, irrigation systems, and, more recently, agrovoltaic systems. Smart agricultural systems require construction of technological infrastructure in layers [3–6]. Infrastructure can include IoT equipment, software, embedded operating systems, network communications, databases, and cloud data analytics. Agrovoltaic systems consist of the efficient use of land to install photovoltaic panels [7]. These systems are integrated with intelligent agricultural systems derived from precision agriculture and intelligent irrigation systems.

A review of the existing literature is presented to lay the technical foundations of the state of art. Sales *et al.* [3] proposed a wireless sensor and actuator system for intelligent irrigation in the cloud. The proposed architecture is modular to adapt to each scenario's requirements. It is divided into three main components: a Wireless Sensor and Actuator Network (WSAN) component, a cloud platform, and a Web application. The WSAN component contains three different types of nodes: a collector node, a sensor node, and an actuator node.

Srivastava *et al.* [6] developed an irrigation system based on an ESP8266 WiFi module using IoT and demonstrated the efficiency of using IoT for traditional agriculture. They showed a smart irrigation system controlled and monitored by an Arduino and an ESP8266 WiFi module, which is of low cost and simple to implement.

Tajwar *et al.* [8] designed and implemented an automated agriculture monitoring and control system that implements a point-to-point connection between individual subsystems, which are connected using a common router via WiFi. The number of individual systems is solely up to the user, as it is only limited by the number of systems present on the farm. The proposed architecture is composed of a central control/monitoring system connected to a network of microcontrollers through the cloud. For the core system, they used a Raspberry Pi 3, which is the core of the entire system as it is used to create an

MQTT server, which is a lightweight, publish-subscribe network protocol, transporting messages between devices and connecting all the microcontrollers of the individual systems. The microcontrollers used are NodeMCU nodes because they support the MQTT IoT protocol, allowing access to MQTT. The user can access the information in the cloud through an application or a web browser connected to the same WiFi network.

Munir *et al.* [9] proposed a smart irrigation system using edge computing and IoT. The proposed architecture consists of four layers: application, processing, transport, and perception, contrary to the basic IoT architecture that consists of three layers. The perception layer is the physical layer, which refers to the sensors that collect data. They measure temperature, soil moisture, and moisture in the air. The transport layer is how the data obtained previously in the physical layer is transmitted through wireless, 2G, 3G, and LAN networks. The processing layer stores and processes large amounts of data from the transport layer. It uses technologies, such as databases, cloud computing, and border computing. The application layer is to provide specific services to the end-user. Its system manages sensors, GSM (i.e. Global System for Mobile Communications) modules, border server + IoT server, and an Android application.

While the use of wireless networks with low energy consumption that reaches long distances is just beginning to be used in the agricultural sector, it will become more and more extensive [10–13]. On the other hand, agrovoltaic systems are also beginning to proliferate as useful solutions to make more efficient use of natural resources, such as electricity, water, and fertilizers [14, 15].

Smart metering systems based on AMI consist of smart metering equipment, data concentrators, wired and wireless telecommunications networks, and a metering database management system [16]. The new distributed edge and fog computing paradigms complement the functionality of cloud computing, including applications in the agricultural sector. The edge computing paradigm allows computing processes to be executed directly on IoT sensors and devices whose computing capacity grows day by day. In contrast, fog computing corresponds to an intermediate computation between end devices (edge) and cloud computing (data center), achieving improved performance for queries closer to where the data is generated [17, 18].

Until recently, wireless sensor networks were of a tiny area (personal or local networks). Great power was required to cover large geographical distances, and therefore, it is difficult to reduce electrical consumption in embedded systems. Today, thanks to advances in modulation techniques, it has been possible to obtain extensive geographical distances that, in many cases, include locations of several kilometers. These new networks have been called Low Power Wide Area Network (LPWAN) and have been used in various fields of knowledge, including agriculture [19–22].

As can be seen, there are still no solutions to integration of electricity consumption/production resources with the climatic part in a single, smart metering system. Our proposed architecture integrates the best smart electric metering systems with the agricultural sector.

3 Proposed Architecture

Figure 1 shows in the proposed smart metering architecture for agricultural applications. It comprises three tiers representing the edge, fog, and cloud levels. In turn, the various components can form data networks across their geographic reach: short (LAN, Local Area Network), medium (CAN, Campus Area Network), long (MAN, Metropolitan Area Network), and wide (WAN, Wide Area Network). The edge tier comprises distributed energy resource (DER) systems, where photovoltaic panels stand out, but wind turbines or some other system for micro-generation of electrical energy through renewable sources are also included. It consists of equipment that consumes electrical energy, such as water pumps (WP), although it can be other equipment, such as sensors (S) and actuators (A). The S allows the measurement of various climate and soil variables. In contrast, the actuators allow actions on the crops, such as turning on a sprinkler to irrigate the field. The essential element is the smart meter (SM), which is composed of a simple card computer (SBC) with telecommunications connectivity through wired and wireless means. The SM is responsible for measuring mainly the electrical energy consumption of the devices interconnected to it, but it is also responsible for measuring the electrical energy production of the DERs, while this part is what an SM typically does in a smart grid. Part of the new proposal is that this SM for agricultural applications can collect data on climatic and soil variables to be processed in the same device, helping make local decisions. With the increase in SBCs, it is now possible to carry out data analytics and machine learning processes directly on the SM, giving rise to the concept of edge computing.

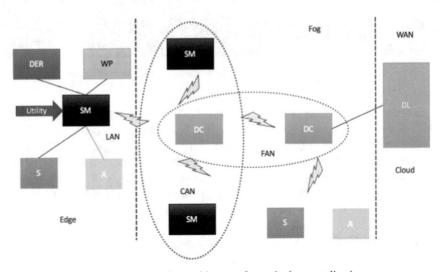

Fig. 1. Smart metering architecture for agriculture applications.

The fog layer is formed by the connection between various SMs with a data concentrator (DC). A DC is intermediate computing equipment made up of a higher capacity SBC that allows the data of various SMs to be concentrated so that they can be worked on as a block. While our SM is made up of a Raspberry Pi 4 SBC with 2 GB of RAM, the DC is made up of a LattePanda Alpha 864s with 8 GB of RAM and 64 GB of internal storage. In turn, the DC can receive data from sensors and actuators that could be isolated from an SM. Due to their location or low density of nodes and its cost, however, it is not feasible to have an SM. DC can run more complex data analytics and machine learning models capable of making regional decisions.

The last tier is the cloud. Here, the data coming from the DCs and in some cases remote SMs. Due to their location or density, however, it is not feasible for them to communicate with another DC. Sensors and actuator's data are stored in a data lake (DL) capable of storing structured and semi-structured information from databases. of relational and non-relational data, files in CSV, XML, JSON format, among others. In this layer there are quite high computational capacities coming from a group of servers, allowing to process all the information available in the entire crop area.

Figure 2 shows part of the data lake schema corresponding to a relational database. There is the management of users and experiments, as well as the configuration and measurement of various sensors, such as moisture front detector, infiltrometer, and hygrometer.

Fig. 2. Part of the data lake structure of a relational database model.

Data communication is done through wired and wireless networks. For example, at the edge tier, communication between various components can be done over wired media, such as Ethernet or over wireless sensor networks (WSN), such as ZigBee and 6LoPAN. In the fog tier, communication is done through wireless WiFi or Sub-1GHz radio networks between SM and DC, while communication from DC to DC, which is generally over larger areas, is done through LoRa networks and LoRaWAN. Finally, the communication between the final DCs with the cloud is generally done through fiber optics, although it can be done through cellular networks due to its remote location.

A fundamental part of any smart metering system is the user interface. Generally, these interfaces are through Web or mobile applications, which allow end-users to see the energy consumption/production values, see statistics, and graphically consult the history of climatic and soil variables. Figure 3 show the implementation of the SMS connecting DERs and consumption loads.

More advanced systems can use data analytics and machine learning to determine patterns of electrical energy consumption/production, weather forecasting, or classification of various soil phenomena.

Fig. 3. Smart metering system user interface: part energy.

4   Testing the Architecture

The proposed architecture was tested using the McCall software quality factors [23] and the testing was focused on operation, revision, and product transition (software). Each of the tests is shown below. 86% of farmers consider that this work will allow agricultural producers to carry out more efficient energy and climate measurements, allowing the best use of natural resources and the technification of the Mexican countryside.

(1) Correctness. We conducted a survey of 30 potential architecture clients, being a representative sample of farmers in the state of Michoacán. We concluded that the proposed architecture satisfies its specification and fulfills the customer's mission objectives. By improving the consumption of electricity and climatic resources, such as water, the prices of agricultural products are lower for the entire population, with better care for the environment and therefore healthier.

(2) Reliability. We compared the measurements from our sensors with others in a backup system and compared whether the data transmitted over the network made any difference. The system reports the measurements 100% correctly. On some occasions, there is an average latency of approximately 0.3 s due to the characteristics of the wireless networks and the distance.

(3) Efficiency. It is the amount of computing resources and code required by a program to perform its function. In our case, a comparison of processes in the SM, DC and measurement data server was made, measuring the CPU time with the htop tool, obtaining a metric of 27.31%, 15.26% and 9.95% respectively. This is largely due to the processing capabilities of the hardware in each type of device. Regarding RAM memory, an average of 233.47 Mb, 1.49 Gb and 4.56 Gb respectively was obtained, because the data load increases in each device containing more information.

(4) Integrity. Data security is a critical aspect in any software architecture today. We use the AES 128 symmetric encryption algorithm to cypher and decrypted the measurements.

(5) Usability. We assessed the effort required to learn and operate the user interface with the 30 farmers. 96.66% considers the prototype very useful and friendly. In general, user experience was satisfactory.

(6) Maintainability, flexibility, and reusability. The architecture is developed with Python in object-oriented paradigm. The software can be modified in an easy way and reuse these modules to other systems.

(7) Portability. One of the advantages of this architecture is that it can be ported to other platforms because it is constructed in Python and can be implemented in several hardware architectures.

(8) Interoperability. The proposed architecture can be integrated with other measurement systems including sensors and actuators. We developed an API to export the metering data and can be used in other systems.

5 Conclusions

Agrovoltaic systems are increasingly necessary to guarantee food security of the world population. For its good performance, it is necessary to have a measurement architecture and infrastructure that allows to effectively measure the variables of electrical energy, climate, and soil conditions. In turn, this architecture needs to be highly scalable and interoperable with a wide variety of sensors, actuators and IoT devices for agriculture.

This paper presents a multi-level architecture for intelligent measurement systems for agricultural applications based on the edge-fog-cloud computing paradigm that allows segmentation of the computing capabilities present in the various sensors/actuators, as well as such as being able to achieve communications over long distances through wireless wide-area networks with low power consumption such as LoRaWAN.

Acknowledgments. The authors thank the Tecnológico Nacional de México for partial financial support through grant 13537-22.P and the British Council Newton Fund grant 540323618.

References

1. Guerrero-Ibañez, J.A., et al.: SGreenH-IoT: Plataforma IoT para Agricultura de Precisión. In: CISCI 2017 - Decima Sexta Conferencia Iberoamericana en Sistemas, Cibernetica e Informatica, Decimo Cuarto Simposium Iberoamericano en Educacion, Cibernetica e Informatica, SIECI 2017 - Proceedings, pp. 315–320 (2017)

2. Ragab, R., Prudhomme, C.: Climate change and water resources management in arid and semi-arid regions: Prospective and challenges for the 21st century. Biosyst. Eng. **81**(1), 3–34 (2002)

3. Sales, N., Remedios, O., Arsenio, A.: Wireless sensor and actuator system for smart irrigation on the cloud. In: Proceedings of the IEEE World Forum on Internet of Things, WF-IoT 2015, pp. 693–698. Institute of Electrical and Electronics Engineers Inc. (2015)

4. Khriji, S., El Houssaini, D., Kammoun, I., Kanoun, O.: Precision irrigation: an IoT-enabled wireless sensor network for smart irrigation systems. In: Hamrita, T.K. (ed.) Women in Precision Agriculture. WES, pp. 107–129. Springer, Cham (2021). https://doi.org/10.1007/978-3-030-49244-1_6

5. Delegación SADER Querétaro: Convocatoria Componente Investigación, Innovación y Desarrollo Tecnológico Agrícola. Programa de Fomento a la Agricultura (2019). https://rb.gy/f6incd

6. Srivastava, P., Bajaj, M., Rana, A.: Overview of ESP8266 Wi-Fi module based smart irrigation system using IOT. In: Proceedings of the 4th IEEE International Conference on Advances in Electrical and Electronics, Information, Communication and Bioinformatics, AEEICB 2018. Institute of Electrical and Electronics Engineers Inc. (2018)

7. Agrovoltaic. https://agrovoltaic.org/

8. Tajwar, M., et al.: Design and implementation of an IoT based automated agricultural monitoring and control system. In: 1st International Conference on Robotics, Electrical and Signal Processing Techniques, ICREST 2019, pp. 13–16. Institute of Electrical and Electronics Engineers Inc. (2019)

9. Munir, M., et al.: Intelligent and smart irrigation system using edge computing and IoT. Complexity **2021**, 1–16 (2021)

10. Sophocleous, M., Karkotis, A., Georgiou, J.: A versatile, stand-alone, in-field sensor node for implementation in precision agriculture. IEEE J. Emerg. Sel. Top. Circ. Syst. **11**(3), 449–457 (2021)

11. Shanmuga, J.P., et al.: A survey on LoRa networking: research problems, current solutions, and open issues. IEEE Commun. Surv. Tut. **22**(1), 371–388 (2020)

12. Lloret, J., et al.: Cluster-based communication protocol and architecture for a wastewater purification system intended for irrigation. IEEE Access **9**, 142374–142389 (2021)

13. Di Renzone, G., et al.: LoRaWAN underground to aboveground data transmission performances for different soil compositions. IEEE Trans. Instrum. Meas. **70**(2021), 1–13 (2021)

14. Willockx, B., Herteleer, B., Cappelle, J.: Theoretical potential of agrovoltaic systems in Europe: a preliminary study with winter wheat. In: 2020 47th IEEE Photovoltaic Specialists Conference (PVSC), pp. 0996–1001 (2020)

15. John, R., Mahto, V.: Agrovoltaics farming design and simulation. In 2021 IEEE 48th Photovoltaic Specialists Conference (PVSC), pp. 2625–2629 (2021)

16. Huang, C., et al.: Smart meter pinging and reading through AMI two-way communication networks to monitor grid edge devices and DERs. IEEE Trans. Smart Grid (2021)

17. Alharbi, H., Aldossary, M.: Energy-efficient edge-fog-cloud architecture for IoT-based smart agriculture environment. IEEE Access **9**(2021), 110480–110492 (2021)

18. Taneja, M., et al.: Connected cows: utilizing fog and cloud analytics toward data-driven decisions for smart dairy farming. IEEE IoT Mag. **2**(4), 32–37 (2019)

19. Habibi, M., et al.: A comprehensive survey of RAN architectures toward 5G mobile communication system. IEEE Access **7**, 70371–70421 (2019)

20. Liya, M., Arjun, D.: A survey of LPWAN technology in agricultural field. In: 2020 4th International Conference on I-SMAC (IoT in Social, Mobile, Analytics and Cloud) (I-SMAC), pp. 313–317 (2020)

21. Valecce, G., et al.: NB-IoT for smart agriculture: experiments from the field. In: 2020 7th International Conference on Control, Decision and Information Technologies (CoDIT), pp. 71–75 (2020)

22. Chew, K., et al.: Fog-based WSAN for agriculture in developing countries. In: 2021 IEEE International Conference on Smart Internet of Things (SmartIoT), pp. 289–293 (2021)

23. Pressman, R., Maxim, B.: Software Engineering: A Practitioner's Approach, 9th edn. McGraw Hill (2020). ISBN 9781259872976

Apple Brand Texture Classification Using Neural Network Model

Shigeru Kato[1]([⊠]), Renon Toyosaki[1], Fuga Kitano[1], Shunsaku Kume[1], Naoki Wada[1], Tomomichi Kagawa[1], Takanori Hino[1], Kazuki Shiogai[1], Yukinori Sato[2], Muneyuki Unehara[3], and Hajime Nobuhara[4]

[1] Niihama-College, National Institute of Technology, Niihama, Japan
s.kato@niihama-nct.ac.jp
[2] Hirosaki University, Hirosaki, Japan
[3] Nagaoka University of Technology, Nagaoka, Japan
[4] University of Tsukuba, Tsukuba, Japan

Abstract. This paper describes a texture measurement method for apples and a neural network model to infer the texture classification in apple brands. The features such as load and sound measured by the texture inspection equipment are input to the neural network model. The model identifies the corresponding apple brand. The authors had measured the apple texture effortful manner in our previous paper. For instance, we had to hollow out the sample of apple flesh using a stainless-steel pipe, and then we cut the flesh in the same size for texture inspection. This paper describes an improved method, where the inspector will slice the entire apple into the same thickness to examine the texture of flesh. We experimented with the proposed method to obtain the load and sound signals for three brands of apples: Sun-Fuji, Orin, and Shinano-Gold. The load and sound features are input to the neural network model to classify three apple specimens. Even though the features were complicated to distinguish the specimens, the neural network model could identify the corresponding specimens. This paper shows the validation result of the proposed method and the neural network model and describes future works.

Keywords: Apple · Texture · Neural network model

1 Introduction

Apples are rich in nutrients such as polyphenols [1], minerals, vitamins, and dietary fiber [2], and apple consumption has been reported to be effective in preventing cancer [3] and maintaining cardiovascular [4] and liver functions [5]. Studies have shown that apple consumption by the diseased and elderly effectively supports their health [6, 7] and that apple consumption strengthens the immune system [8, 9].

On the other hand, filtered and pure apple juice has a high fructose concentration, and excessive consumption of apple juice may increase the risk of diabetes [4]. Therefore, to maintain nutrient requirements, it is necessary to consume apple pulp orally and digest and absorb dietary fiber in a balanced manner. The "texture" quality is significant for the regular consumption of apples. Therefore, it is necessary to develop a system to manage

© The Author(s), under exclusive license to Springer Nature Switzerland AG 2022
L. Barolli et al. (Eds.): AINA 2022, LNNS 451, pp. 420–430, 2022.
https://doi.org/10.1007/978-3-030-99619-2_40

the texture quality. Therefore, this study aims to construct a texture measurement system for widely distributed apples such as Sun-Fuji, Orin, and Shinano-Gold and develop an intelligent information processing system to estimate the "texture" from the obtained physical data such as load and sound.

We have been developed a method to measure the texture of apples [10]; however, it was very time-consuming to hollow out the apples with stainless-steel pipes and shape them to the same length. Therefore, we adopted the measurement method described in the research paper [11]. Furthermore, we found that it is not easy to estimate the apple brands from the measured load and sound signals. Therefore, we decided to employ the Neural Network Model (NN) [12] to classify individuals with complex characteristics. This paper describes the original equipment, the experiment, and the validation of the proposed NN.

The latest study [11] has analyzed the acoustic vibration of apple pulp when pierced with a probe to estimate the texture. Several studies systematically classified the texture and chemical composition of various apple varieties based on the features of sound and load [13, 14]. Note many of study related to apple texture mainly summarize the findings in chemical and physical characteristics. On the other hand, few studies aim to construct a system that can automatically evaluate the texture quality of apples employing machine learning techniques.

2 Improved Method

As shown in Fig. 1(a), in the previous method of our paper [10], the examiner had to hollow out the apple with a stainless-steel pipe. The examiner shapes the hollowed pulp into the same length as Fig. 1(b), and then the many sample pulps are preserved in the refrigerator as Fig. 1(c). The sample is set under the blade of the texture measurement equipment as Fig. 1(d), and the load and sound are measured when the blade cuts the sample (Fig. 1(e)).

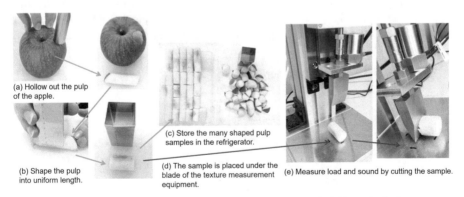

(a) Hollow out the pulp of the apple.

(b) Shape the pulp into uniform length.

(c) Store the many shaped pulp samples in the refrigerator.

(d) The sample is placed under the blade of the texture measurement equipment.

(e) Measure load and sound by cutting the sample.

Fig. 1. Previous method to measure the load and sound of the apple flesh.

As shown in Fig. 2(a), an apple is sliced horizontally at a predetermined height. The sliced pulp is then placed under the wedge-shaped probe of the texture measurement equipment as shown in Fig. 2(b). The probe is used to pierce the pulp and measure the load and sound signals (Fig. 2(c)).

(a) Slice the apple horizontally at the same height.

(b) Place the apple flesh under the wedge-shaped probe.

(c) Measure the load and sound when the probe stabs the pulp.

Fig. 2. Improved method to measure the load and sound.

The method in Fig. 2 is more convenient and saves time and workload. It is practical and minimizes oxidation and moisture evaporation of the apple. The technique in Fig. 1 used a blade; on the other hand, the method in Fig. 2 employed a wedge-shaped probe resembling a human tooth. This technique was proposed by Sakurai [11]. The experiment detail in Fig. 2 is described in the later section.

3   System Configuration

As shown in Fig. 3(a), the air cylinder rod goes down, and then the wedge-shaped probe pierces the apple's flesh. The sensors detect the load and sound, and each amplifier magnifies the sensor signal. The AD converter converts the amplified signals to digital ones and inputs them to the PC. PC performs signal processing to calculate the load and

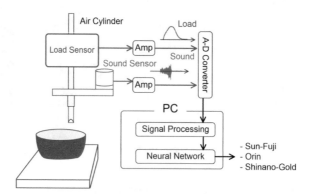

Fig. 3. System configuration.

sound parameters and inputs them to the NN, which outputs the results of the texture classification of the apples such as Sun-Fuji, Orin, and Shinano-Gold.

The calculation of input parameters to the neural network is illustrated in Fig. 4. The load and sound signals for 2.0-[s] are automatically extracted. The signals are extracted automatically from 1.0-[%] of maximum load point, as shown in Fig. 4. The average load value for 2.0-[s] period is W1, ones for the first and last half are W2 and W3, respectively. The sound is transformed into a frequency spectrum by Fast Fourier Transform (FFT). The integrated values of the frequency intensity in each region divided into five equal-divided ranges are F1 to F5, and one of the whole ranges is F6.

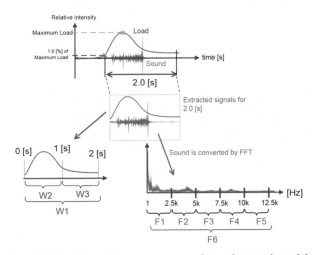

Fig. 4. Calculation of input parameters of neural network model.

Figure 5 shows the neural network model. The input parameters are the values calculated in Fig. 4. The output layer (Softmax layer) outputs the probability of specimens A (Sun-Fuji), B (Orin), and C (Shinano-Gold). The one with the highest probability is the classification result. Two hidden layers comprise 30 nodes, respectively, and each node

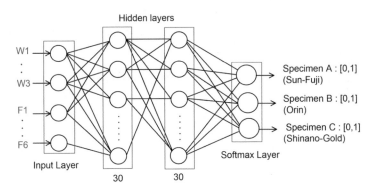

Fig. 5. Neural network model.

has a sigmoid transfer function. The connection weights between layers are adjusted by the backpropagation algorithm [12].

4   Experimental Settings and Results

4.1   Apple Specimens

We carried out load and sound measurements for specimens A (Sun-Fuji), B (Orin), and C (Shinano-Gold) displayed in Fig. 6 using the texture measurement method in Fig. 2. We purchased apples at a local supermarket. The size and weight of specimens A and B were almost the same, while C was slightly smaller.

(a) Specimen A (Sun-Fuji)   (b) Specimen B (Orin)   (c) Specimen C (Shinano-Gold)

Fig. 6. Apples used in experiment.

We measured the width and height shown in Fig. 7(a) of the three apples for each specimen. Before the texture inspection, each apple is trimmed in the 40-[mm] thickness as shown in Fig. 7(b). Table 1 shows the maximum and minimum width, height, and weight of each brand.

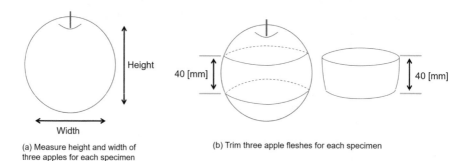

(a) Measure height and width of three apples for each specimen

(b) Trim three apple fleshes for each specimen

Fig. 7. Measured size.

Table 1. Size and weight of the apples.

Specimen	Brand	Width (Max/Min) [mm]	Height (Max/Min) [mm]	Weight (Max/Min) [g]
A	Sun-Fuji	85 / 74	86 / 81	397 / 289
B	Orin	84 / 79	97 / 78	358 /303
C	Shinano-Gold	76 / 74	72 / 60	240 /201

4.2 Equipment for Texture Measurement

Figure 8 shows the details of the probe configuration. As shown in Fig. 8(a), the probe tip's height is 21-[mm] when the probe tip gets down lowest. Therefore, the probe is inserted into the 40-[mm] thick apple flesh up to a depth of 19-[mm].

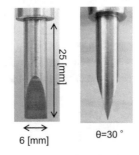

(a) Height of the probe tip when the probe is at lowest point.

(b) Size and form of probe

(c) The probe is pierced around the center between the core and peel.

Fig. 8. Probe configuration.

As shown in Fig. 8(b), the probe length, width, and angle are 25-[mm], 6-[mm], and 30-[degree] angle, respectively. The probe is stabbed along the centerline between the core and peel, as shown in Fig. 8(c). Table 2 shows the experimental conditions, such as probe speed, pressure in the air cylinder, the AD converter's sampling frequency, observation time, and so on.

Table 2. Experimental condition.

Item	Value
Probe speed	approx.30 [mm/s]
Air cylinder pressure	0.3 [MPa]
Sampling rate	25 [k Samples/s]
Observation time	10 [s]
Date	December 20, 2021
Climate	Clear
Temperature / Humidity	22.7 [°C] / 37 [%]

4.3 Texture Measurement

We carried out texture measurements of three brands of apples under the condition shown in Table 2. We obtained 25 load and sound signal data for each specimen A (Sun-Fuji), B (Orin), and C (Shinano-Gold). The labels A1–A25, B1–B25, and C1–C25 were assigned for each specimen data, respectively. Figures 9(a), (b), (c) show examples of signal data for specimens A, B, and C. The red curve indicates the load in the top graph, and the blue line is the sound. The middle graph shows 2.0-[s] signals automatically extracted. The bottom graph shows the FFT result on the extracted 2.0-[s] sound. Comparing the three graphs in Figs. 9(a), (b), and (c), we can see they slightly differ. The load curve of C1 is smaller than the others.

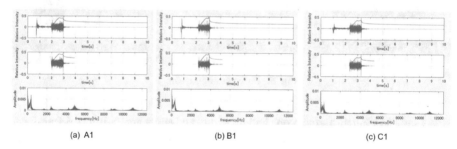

(a) A1   (b) B1   (c) C1

Fig. 9. Signal examples for each specimen.

Figure 10(a) shows the mean of W1 expressing the average load for 2.0-[s] period, and Fig. 10(b) shows the mean of F6 expressing the sound intensity. Figure 10(c) shows the plot with W1 on the horizontal axis and F6 on the vertical axis. The red asterisks "*" indicate A1–A25, the green plus "+" and blue circles "o" are B1–B25 and C1–C25, respectively.

As seen in Fig. 10(c), it is not easy to classify specimens A, B, and C based on load and sound intensity because individuals widely spread and overlap among specimens.

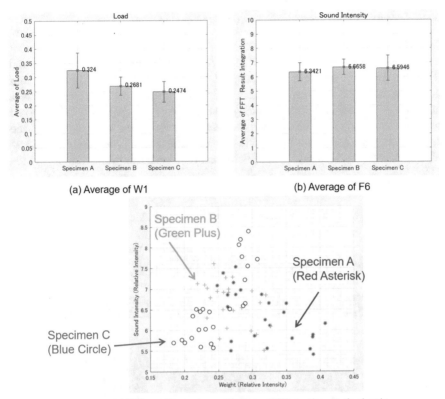

(a) Average of W1

(b) Average of F6

(c) W1 on the horizontal axis and F6 on the vertical axis

Fig. 10. Graphs of load (W1) and sound intensity (F6).

Figures 11(a), (b), and (c) show all load curves for each specimen. This result is consistent with Fig. 10(a). The load strength is higher for specimens A, B, and C in order. Specimen A, in particular, has the highest load strength and dispersion.

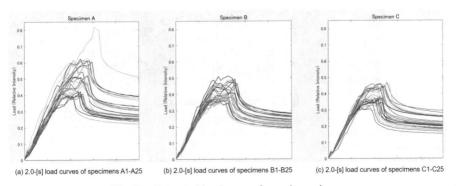

(a) 2.0-[s] load curves of specimens A1-A25   (b) 2.0-[s] load curves of specimens B1-B25   (c) 2.0-[s] load curves of specimens C1-C25

Fig. 11. Extracted load curves for each specimen.

4.4   Texture Classification Using Neural Network Model

We validated the classification capability of the neural network model in Fig. 5. The total data was 75, including A1–A25, B1–B25, C1–C25. To evaluate the performance of the NN with a small amount of data, we adopted the Bootstrap method [15]: we randomly chose five data for each specimen A, B, and C, and thereby 15 data were left as test data not used for training. The remaining 60 data were used for training the NN. And then, the 15 test data were input to the NN after the training and accuracy was verified. This Bootstrap method was conducted in 10 trials. The training conditions are shown in Table 3, and the MSE (Mean Squared Error) curve in each trial while training is shown in Fig. 12.

Table 3. Neural network training condition.

	Method / Value
Method	Levenberg-Marquardt Back Propagation
Max Epochs	300
Number of Training Data	60 (Specimens A:20, B:20, C:20)

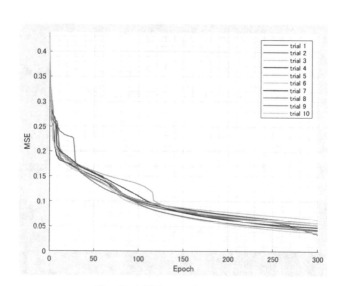

Fig. 12. MSE curve in each trial.

The validation result is shown in Table 4. The average classification accuracy was 76.67%.

Table 4. Validation of NN result.

Trial	Test for Specimen A					Test for Specimen B					Test for Specimen C					Accuracy
1	3	10	13	15	16	4	5	9	10	18	4	8	20	21	23	0.8667
2	2	10	15	17	19	11	15	19	20	25	8	9	13	20	24	0.8000
3	4	12	17	18	19	6	7	10	15	17	3	7	13	18	25	0.6667
4	4	8	12	23	25	5	6	9	16	18	2	13	16	24	25	0.9333
5	5	8	13	18	24	2	14	16	19	22	2	4	15	16	19	0.7333
6	2	3	5	7	24	5	7	8	17	22	2	4	10	17	22	0.8667
7	1	8	10	18	25	1	2	5	19	23	5	7	12	17	24	0.6667
8	1	4	6	13	18	3	4	12	13	14	5	9	10	17	19	0.7333
9	6	15	18	19	20	7	15	17	23	24	8	12	13	14	16	0.5333
10	2	6	8	10	23	4	5	16	19	25	7	9	10	14	21	0.8667
—	—					—					Mean Accuracy					0.7667

5 Conclusions

This paper describes a texture measurement method less burdensome. The method of piercing the flesh with a tooth-shaped probe was based on the paper by Sakurai [11]. The probe was controlled with an air cylinder in the experiment, and the load and sound signals were measured simultaneously. Using the data obtained in the experiment, we attempted to classify the three brands of apples using a neural network model. The classification accuracy was 76.67%. In the future, we will work on the practical application of the proposed method to apples' texture quality control system.

Acknowledgments. This work was supported by a Grant-in-Aid from Nagaoka University of Technology (Collaborative research grant for national college and Nagaoka Univ. of Technology). This work was supported by JSPS Grant Numbers 20K06116.

References

1. Wruss, J., et al.: Differences in pharmacokinetics of apple polyphenols after standardized oral consumption of unprocessed apple juice. Nutr. J. **14**, 32 (2015)
2. Manzoor, M., et al.: Variations of Antioxidant characteristics and mineral contents in pulp and peel of different apple (Malus domestica Borkh.) cultivars from Pakistan. Molecules **17**, 390–407 (2012)
3. Gerhauser, C.: Cancer Chemopreventive potential of Apples, Apple juice, and Apple components. Planta Med. **74**(13), 1608–1624 (2008)
4. Bondonno, N.P., et al.: The cardiovascular health benefits of apples: whole fruit vs. isolated compounds. Trends Food Sci. Technol. **69**(Part B), 243–256 (2017)
5. Skinner, R.C., et al.: Apple pomace improves liver and adipose inflammatory and antioxidant status in young female rats consuming a Western diet. J. Funct. Foods **61**, 103471 (2019)
6. Giaretta, A.G., et al.: Apple intake improves antioxidant parameters in hemodialysis patients without affecting serum potassium levels. Nutr. Res. **64**, 56–63 (2019)

7. Avci, A., et al.: Effects of Apple consumption on plasma and erythrocyte antioxidant parameters in elderly subjects. Exp. Aging Res. **33**(4), 429–437 (2007)
8. Pires, T.C.S.P., et al.: Antioxidant and antimicrobial properties of dried Portuguese apple variety (Malus domestica Borkh. cv Bravo de Esmolfe). Food Chem. **240**, 701–706 (2018)
9. Martin, J.H.J., Crotty, S., Warren, P., Nelson, P.N.: Does an apple a day keep the doctor away because a phytoestrogen a day keeps the virus at bay? A review of the anti-viral properties of phytoestrogens. Phytochemistry **68**(3), 266–274 (2007)
10. Kato, S., et al.: Apple brand classification using CNN aiming at automatic apple texture estimation. In: Barolli, L., Hellinckx, P., Natwichai, J. (eds.) 3PGCIC 2019. LNNS, vol. 96, pp. 811–820. Springer, Cham (2020). https://doi.org/10.1007/978-3-030-33509-0_76
11. Sakurai, N., Akimoto, H., Takashima, T.: Measurement of vertical and horizontal vibrations of a probe for acoustic evaluation of food texture. J. Texture Stud. **52**(1), 25–35 (2020)
12. Rumelhart, D.E., Hinton, G.E., Williams, R.J.: Learning representations by back-propagating errors. Nature **323**, 533–536 (1986)
13. Costa, F., et al.: Assessment of apple (Malus × domestica Borkh.) fruit texture by a combined acoustic-mechanical profiling strategy. Postharvest Biol. Technol. **61**(1), 21–28 (2011)
14. Corollaro, M.L., et al.: A combined sensory-instrumental tool for apple quality evaluation. Postharvest Biol. Technol. **96**, 135–144 (2014)
15. Priddy, K.L., Keller, P.E.: Artificial neural networks - an introduction, chap. 11. In: Dealing with Limited Amounts of Data, pp. 101–105. SPIE Press, Bellingham (2005)

Adaptive Analysis of Electrocardiogram Prediction Using a Dynamic Cubic Neural Unit

Ricardo Rodríguez-Jorge[⊠], Paola Huerta-Solis, Jiří Bíla, and Jiří Škvor

Jan Evangelista Purkyně University,
Pasteurova 3632/15, Ústí nad Labem, Czech Republic
Ricardo.Rodriguez-Jorge@ujep.cz

Abstract. In this work, the implementation of a dynamic cubic neural unit for the prediction of heartbeats using a wireless method is presented. The data were recorded with the BITalino biomedical acquisition card using its ECG input and output module via Bluetooth. This paper aims to predict a prediction horizon according to the learning rate, the number of samples used to train the model, and the specified times required for training. The signal (input) was acquired from electrodes, which were placed on the surface of the chest near the heart. The signal was visualized and presented through a graphical interface. For the interface evaluation, tests are performed using the obtained signal in real time.

1 Introduction

Embedded systems have certain requirements, such as the integration of hardware and software components, and their hierarchical relationship with a super system is responsible for controlling communication between systems of the same level. Integrated systems now dominate different aspects of life with portable phones, mobile devices, tablets, and smart devices, which allow the transmission of physiological data via Bluetooth. Thanks to the development of mobile Internet and wireless networks, monitoring systems, such as cardiac arrhythmia monitoring systems, have emerged, that allow detection and transmission of a signal to a smartphone via a wireless transmission platform, such as Bluetooth [1].

Affordable and reliable mobile health systems have increasingly become a basic need of society. The ECG records the electrical impulses of the heart and transmits them through a wireless channel, thus representing the electrical activities of the heart, that normally occur between two successive heartbeats. During normal heart function, each ECG cycle represents an orderly progression [2].

Cardiovascular disease is among the top 10 causes of death according to the World Health Organization (WHO). Most of the risk factors for cardiovascular diseases are related to the lifestyle of the individual. The corresponding percentages of the following factors attributable to mortality have been reported: 40.6% from high blood pressure; 13.7% for smoking; 13.2% for a poor diet; 11.9% for insufficient physical activity; and 8.8% for abnormal glucose levels. Other factors account for the remaining percentage points [3,4].

L. Barolli et al. (Eds.): AINA 2022, LNNS 451, pp. 431–440, 2022.
https://doi.org/10.1007/978-3-030-99619-2_41

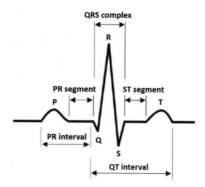

Fig. 1. Typical ECG signal waveform [4].

A gradual increase in heart-related problems, including ischemia, arrhythmias, and heart attack, distort the consistency of cardiac sequences and alter the progression of heart disease. Abnormal rhythms of the heart are known as arrhythmias, and the most common include premature ventricular contraction (PVC) and atrial premature contraction (APC). Prediction of cardiac arrhythmias helps professionals respond early to possible cardiovascular disease [1].

Optimization algorithms and machine learning algorithms are basic tools that can benefit healthcare applications. Therefore, researchers have put great effort into proposing new algorithms and architectures for healthcare. Given the rapid increase in computing power and the availability of health data, health care applications can include artificial intelligence algorithms and thus provide intelligent health care [3].

An electrocardiogram (ECG) represents cardiovascular activities in the context of cardiomyopathy. However, ECGs can only measure cardiac activities, which offer little knowledge about various cardiac mechanical activities, such as the movement of the heart valves, the blood circulation towards the ventricles, the suppression and relaxation of the ventricular walls [5]. The Fig. 1 presents a typical ECG signal waveform.

2   Prediction of Electrocardiograms via Wireless Communication

In this section, the steps to be followed are described. The system was developed in stages, and the implementation of the dynamic cubic neural unit that will result in the prediction of heartbeats is explained in each of the stages.

Next, Fig. 2 shows the block diagram. The general architecture of the methodology of this work is presented, and the main stages of the diagram are shown.

As the input of the system, ECG signals are acquired through sensors called electrodes. In the signal acquisition process, there are five different phases where each of the points that were made were developed. Figure 3 shows the 5 stages for the acquisition of ECG signals.

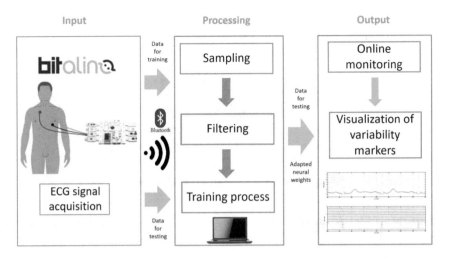

Fig. 2. Proposed methodology for the ECG prediction method.

Fig. 3. Stages for the signals acquisition.

ECG Recording. ECG signals are captured by means of electrodes placed on the surface of the skin in the area close to the heart.

Identification of the Electrode Placement. The electrodes for cardiac sensors are an ideal monitoring element for reading biomedical signals, and 3 electrodes were used in this system. The BITalino (r)evolution kit includes standard pre-gelled and self-adhesive disposable electrodes, which are presented in Fig. 4. The kit also includes the three-lead electrode cable, see Fig. 5.

Fig. 4. Standard pregelled and self-adhesive electrode.

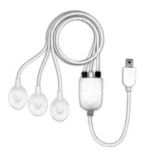

Fig. 5. Three lead cable for electrodes [4].

Cleaning of the Identified Area. Clean and disinfect the area with alcohol where the electrodes will be located to achieve effectiveness and thus obtain a better connection.

Electrode Placement. The electrodes were placed based on investigations and recommendations to acquire the ECG signal successfully. The electrodes were placed. So they were located in the area where the heart is. The steps to follow for the electrode placement process is shown bellow.

- Set the sensor patches to the cables before applying them to the body.
- The individual should not move during data capture.
- Avoid using the same patches for different measurements.
- Adjust the sensors for each person.
- Clean and, if necessary, shave the area where the patches are to be attached.

Circuit Power. The electrodes were fed through the connectivity in the BITalino platform.

BITalino is a low-cost biomedical data acquisition development board that allows you to create projects using sensors and physiological tools. The BITalino platform is equipped with a Microcontroller Block (MCU), Bluetooth Block (BT), Power Block (PWR), Prototyping Block, Electromyography Block (EMG), Electrodermal Activity Block (EDA), Block electrocardiogram (ECG), electroencephalography (EEG) block, light block (LUX), accelerometer block (ACC), push button block (BTN), light-emitting diode (LED) block, buzzer, and a digital to analog translation device. The ATMega328 microcontroller, is located at the center of the board. The microcontroller can be configured with a sample rate of up to 1,000 Hz and is capable of supporting six analog inputs (four at 10 bits, two at 6 bits) as well as four input pins. Digital input and four digital output pins [7].

2.1   ECG Signal Acquisition

Data acquisition was performed using the BITalino (r)evolution card, which has all the necessary components to start working with physiological data

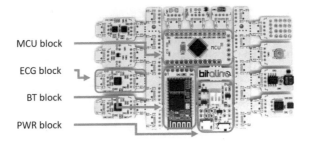

Fig. 6. BITalino (r)evolution card.

(see Fig. 6). In the signal acquisition stage, ECG signals are captured through patches placed on the surface of the skin on the correct sides of the heart. The Bitalino (r)evolution card has been used for the identification of electrodes and their positions. BITalino's biomedical development kit is an easy-to-use, low-cost toolkit for the development of learning applications and prototypes using body signals [4]. The small dimensions of this system and the wireless acquisition system allow their implementation in a greater number of experiments. The different modules and sensors are described in the following sections.

2.1.1   Microcontroller

The block of the microcontroller unit (MCU) converts the analog signals from the sensors to a digital format and samples all the channels. The MCU provides access to the analog and digital channels of the BITalino card, as well as the peripherals. The module is presented in Fig. 7. The MCU block is designed for reliable data transmission in real time over any UART compatible interface (for example: Bluetooth and FTDI, among others). The MCU block also has important features, such as high-performance open-source firmware, a crystal oscillator for maximum precision, status and low battery indicators, the capacity to cancel interference (cross-talk), and the ability to sample and output raw data.

Fig. 7. MCU module (ATMEL Processor ATMEGA328P-AU).

2.2 Sampling

The analog filter processes the analog input to obtain the band-limited signal, which is sent to the analog-to-digital conversion unit (ADC). The ADC unit samples the analog signal, quantizes the sampled signal, and encodes the quantized signal levels into the digital signal. Therefore, the analog signal contains an infinite number of points.

2.3 Filtering

A digital filter is applied to facilitate prediction of the ECG signal.

Butterworth Filter. The frequency response in the passband and in the stopband is as flat as possible; therefore, this filter is also called the maximally flat magnitude filter. This filter has a minimum phase change over the passband compared to other conventional IIR filters. On the other hand, the output decreases at a rate of -20 dB per decade per pole, and the phase response becomes less linear with the increasing order of the filter. Compared to Chebyshev and elliptical filters, the Butterworth filter has a slower output signal but a more linear phase response in the passband.

2.4 Training Process

The dynamic cubic neural unit is known as type higher order unit (HONU). Compared with the linear neural unit, the HONU provides a good quality non-linear approximation. The input-output relationship of a dynamic CNU is presented in the following equation [6,8].

$$\widetilde{y}(k+h) = \sum_{i=0}^{nx-1} \sum_{j=i}^{nx-1} \sum_{l=j}^{nx-1} x_i \cdot x_j \cdot x_l \cdot w_{i,j,l} \tag{1}$$

where $\mathbf{x}(k)$ and $\widetilde{y}(k+h)$ are the input vector and the neuron output, respectively; $\mathbf{W} = \{w_{i,j,l} : i,j,l \in N\}$ are the weights of the DCNU. The weight matrix $\mathbf{W}$ is an upper triangular tridimensional matrix with neural bias $w_{0,0,0}$. The structure of the DCNU for prediction of electrocardiogram signals is shown in Fig. 8 [8].

The input $\mathbf{x}(k)$ is the augmented vector defined by the next equation.

$$\mathbf{x}(k) = \begin{bmatrix} 1 \ \widetilde{y}(k+h-1) \ \widetilde{y}(k+h-2) \cdots \widetilde{y}(k+1) \ y(k) \ y(k-1) \cdots y(k-n+1) \end{bmatrix}^T \tag{2}$$

where n is the number of samples of the real signal that feeds the neural input, $\mathbf{y}(k)$ is the vector of real values, k is the variable that describes the discrete time, h is the prediction horizon and ^{T} represents transposition. The DCNU calculates the output neuron $\widetilde{y}(k+h)$ when the input vector $\mathbf{x}(k)$ is provided. The DCNU

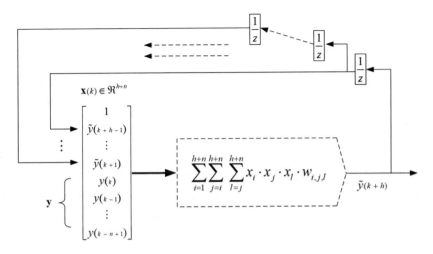

Fig. 8. Dynamic cubic neural unit architecture for prediction [8].

learns by applying the real time recurrent learning technique, sample-by-sample, to adjust the weight matrix **W** [8].

The back-propagation technique as a learning rule is applied in a sample-by-sample adaptation. It follows the convergence criterion of the training performance which is given as the sum of squared error in each iteration time *epoch* for N training samples as shown below.

$$Q(epoch) = \sum_{k=1}^{N} e(k+h) \tag{3}$$

where $e(k+h) = y(k+h) - \tilde{y}(k+h)$. The neural weights adaptation $w_{i,j,l}(k+1)$ in the new adaptation time of the model is presented as follows:

$$w_{i,j,l}(k+1) = w_{i,j,l}(k) + \Delta w_{i,j,l}(k) \tag{4}$$

The value corresponds to the sum of each individual neural weight increment $\Delta w_{i,j,l}(k)$ from the adaptation of the model with each individual neural weight $w_{i,j,l}(k)$ of each k discrete value. The increase of each individual weight based on the gradient descent rule is described in the next equation [8,9].

$$\Delta w_{i,j,l}(k) = \mu \cdot e(k+h) \cdot \frac{\partial \tilde{y}(k+h)}{\partial w_{i,j,l}} \tag{5}$$

where μ is the learning rate which determines the velocity of the learning process multiplying the error and the partial derivative of the neural output with respect to the weights $w_{i,j,l}$.

2.5 Online Monitoring

In this step, an online recording is fed into the DCNU, which uses the neural weights optimized from the training process. Then, after running the DCNU once with the new data, the output data are fed back again into the DCNU along with the following portion of the data. This process allows real-time predictions [8,9].

2.6 Visualization of Variability Markers

In this step, the predicted signal is plotted along with the real signal. In addition, the visualization of the variability markers is shown. The variability markers are plotted according to the evaluation of the weight increments. Thus, when the current weight increments are greater than a weight increment threshold, the markers will be plotted [8,9].

3 Experimental Results

The input configuration of the model was $(nx + 1)$, where $nx = h + n$ represents the number of signal samples used to feed the model, and a bias is used as $\mathbf{x}(k = 1) = 1$. The number of training times has been set at $epochs = 300$, whereas the learning rate is set as $\mu = 0.1$. Figure 9 shows the sum of squared error performance of the DCNU training. The ECG results from the dynamic CNU predictive model implemented in Python language are shown in Fig. 10. The superimposition on the ECG prediction horizon is set as $h = 10$ steps ahead. The model shows that the prediction for the ECG signal is a highly time variant and complex signal due to its dynamics changes based on amplitude. The proposed dynamic CNU predictive model allows capturing the current dynamics of the data using a sample-by-sample adaptation. The architecture of the predictive

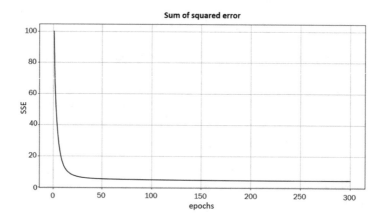

Fig. 9. Training performance (sum of squared error vs epochs) of the DCNU model with RTRL technique.

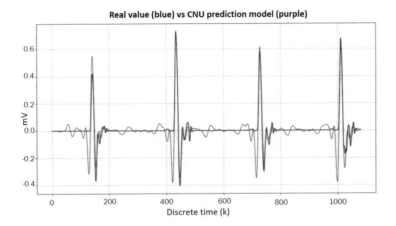

Fig. 10. Prediction of electrocardiogram signals by the DCNU model.

model is computationally time-consuming due to the dynamic CNU, a type of HONU. Nevertheless, the computing time can be reduced even further using parallel computing or specialized supercomputing.

4   Conclusion

The present work proposed the implementation of a cubic neural unit for the prediction of heartbeats via a wireless unit. Heart signals were obtained in real time with the help of electrodes, Python programming and the ECG acquisition module of the BITalino platform. Different methods, such as data acquisition, sampling and the implementation of a band-pass filter, were employed in the development of the dynamic cubic neural unit. Finally, training and real-time prediction were involved. For medical care applications could be created with artificial intelligence and thus provide intelligent medical care. The checkups would be performed in the comfort of the home and wirelessly, helping to solve some existing problems such as the lack of medical staff.

Our future work will be aimed at communicating via WI-FI, to have greater reach and to be able to monitor remotely with the specialist physician. The variability markers of the neural weights should be printed to identify patterns of the different cardiac arrhythmias. An Android App should be developed for monitoring and prediction, so that an appointment can be scheduled automatically with the specialist cardiologist in case of any anomaly. Tests should be performed with different types of filters to attenuate different frequencies. Tests should be performed on a server so that the response of the prediction is faster at the time of compilation. Tests should be performed with cloud computing so that the prediction response is faster.

Acknowledgements. This project is supported by Jan Evangelista Purkyně University. Title of the project - Detection of Cardiac Arrhythmia Patterns through Adaptive Analysis.

References

1. Yang, Z., Zhou, Q., Lei, L., Zheng, K., Xiang, W.: An IoT-cloud based wearable ECG monitoring system for smart health care. J. Med. Syst. **40**, 1–11 (2016)
2. Sahoo, P., Thakkar, H., Yih Lee, M., Auslander, D.: A cardiac early warning system with multi channel SCG and ECG monitoring for mobile health. Sensors **17**, 1–28 (2017)
3. Chui, K., Alhalabi, W., Pang, S., Pablos, P., Liu, R., Zhao, M.: Disease diagnosis in smart health care: innovation, technologies and applications. Sustainability **9**, 1–23 (2017)
4. Rodríguez-Jorge, R., De León-Damas, I., Bila, J., Škvor, J.: Internet of Things-assisted architecture for QRS complex detection in real time. Internet Things **14**, 100395 (2021)
5. Ibarra-Fierro, G.I., Rodriguez-Jorge, R., Mizera-Pietraszko, J., Martinez-Garcia, E.A.: Design and implementation of a data acquisition system for R peak detection in electrocardiograms. In: Proceedings of the 11th International Joint Conference on Biomedical Engineering Systems and Technologies-AI4Health, pp. 715–721. SciTePress, INSTICC (2018)
6. Rodriguez-Jorge, R., Bila, J., Mizera-Pietraszko, J., Martínez-Garcia, E.A.: Weight adaptation stability of linear and higher-order neural units for prediction applications. In: Choroś, K., Kopel, M., Kukla, E., Siemiński, A. (eds.) MISSI 2018. AISC, vol. 833, pp. 503–511. Springer, Cham (2019). https://doi.org/10.1007/978-3-319-98678-4_50
7. Vieira, M., Costa, N., Fonseca, R., De Lima, S.: Robotic Hand. School SENAI (2016)
8. Rodriguez-Jorge, R., Bila, J., Mizera-Pietraszko, J., Loya Orduño, R.E., Martinez Garcia, E., Torres Córdoba, R.: Adaptive methodology for designing a predictive model of cardiac arrhythmia symptoms based on cubic neural unit. Front. Artif. Intell. Appl. **295**, 232–239 (2017)
9. Rodríguez-Jorge, R., Bila, J.: Cardiac Arrhythmia prediction by adaptive analysis via Bluetooth. MENDEL **26**(2), 29–38 (2020)

Evaluation of the Crack Severity in Squared Timber Using CNN

Shigeru Kato[1]([⊠]), Naoki Wada[1], Kazuki Shiogai[1], and Takashi Tamaki[2]

[1] Niihama College, National Institute of Technology, Niihama, Japan
s.kato@niihama-nct.ac.jp
[2] Ehime Forest Research Center, Ehime Research Institute of Agriculture,
Kumakougenchou, Japan

Abstract. Natural woods for building materials, such as cedars used in wooden construction, have high moisture. It is essential to dry them to use as building materials. Currently, high-temperature drying is commonly employed. However, this process sometimes results in internal cracks that are not apparent on the surface. It takes considerable time and effort to measure the length and area of the cracks in the cross-section of square timber using calipers.

A system to automatically estimate the crack severity is required to facilitate researchers to derive the conditions and methodology so that internal cracks do not occur while drying. Therefore, we tried to rank the severity of internal cracks in this study using CNN. The authors experimented with verifying CNN's ability to evaluate the severity of cracks in cross-sectional images. The classification accuracy was 90%. In this paper, we describe the experimental detail and future works.

Keywords: Timber · Internal cracks · High temperature drying · CNN

1 Introduction

Timber building materials such as cedar require a drying process after being harvested from the forest. In general, low-cost high-temperature drying is adopted [1]. However, internal cracks sometimes occur. This internal cracking causes a reduction in structural performance, such as buckling strength and joint durability [2]. Accordingly, a high temperature drying condition that does not cause internal cracking has been investigated [3]. To evaluate the severity of internal cracks, humans currently measure the length and area of the cracks using calipers. However, this process is time-consuming and effort intensive. Therefore, we decided to apply CNN (CNN: Convolutional Neural Network) [4] to estimate the severity of internal cracks more easily. We conducted an experiment to validate CNN's performance to evaluate the severity level of cracks in cross-sectional images. In next section, the method to estimate the crack severity and experiment details are described.

There are studies on the automatic detection of surface timber cracks using CNN [5]. As well, automatic detection of inside faults of woods such as cracks, knots, and mildew, using CNN [6, 7] are investigated. On the other hand, a study on the application of CNN to evaluate the internal cross-section crack-severity of dried timber is not found within the scope of our investigation.

© The Author(s), under exclusive license to Springer Nature Switzerland AG 2022
L. Barolli et al. (Eds.): AINA 2022, LNNS 451, pp. 441–447, 2022.
https://doi.org/10.1007/978-3-030-99619-2_42

2   Timber Crack Pictures

We prepared 64 cuboid timber pieces illustrated in Fig. 1(a). Figure 1(b) displays the top and middle cross-sections of the timber.

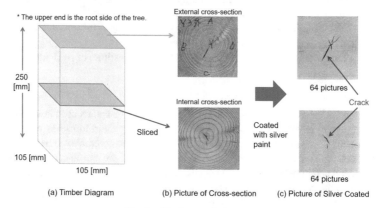

Fig. 1. Timber and pictures.

As shown in Fig. 1(c), we coated them with silver paint so that the cracks on the surface become visible. And then we took their pictures. In the following experiment, a total of 128 silver paint pictures in Fig. 1(c) were used.

We classified 128 pictures into three levels according to the degree of cracks. The "bad" label was assigned to the pictures with large or many cracks as shown in Fig. 2(a). In contrast, we assigned a label named "good" to pictures with shallow cracks and smooth surface as Fig. 2(b) shows. Pictures middle level between "good" and "bad" were assigned with labels "neutral" as shown in Fig. 2(c).

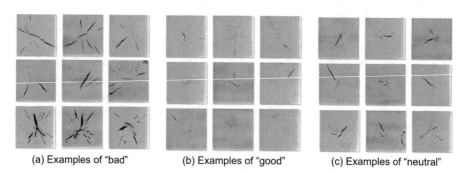

Fig. 2. Image classification depending on the cracks.

As a result of our classification, 44 images were assigned to "bad,", 38 to "good," and 46 to "neutral," respectively.

3   CNN Configuration and Validation

Figure 3 shows the schematic of the proposed CNN overview. The detailed configuration is enumerated in Table 1. The input to the CNN is a 500 × 500-pixels grayscale image. The output layer of CNN comprises three nodes that output the probabilities of bad, good, and neutral respectively. Each node outputs a real number between 0 and 1. The label of the node with the highest value is the evaluation result.

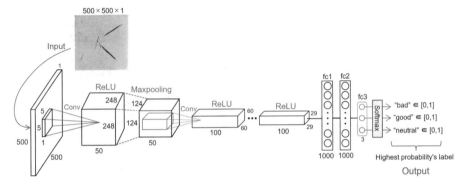

Fig. 3. Overview of proposed CNN.

Firstly, 5 × 5 convolution operation is applied to the input image, resulting in a 248 × 248 × 50 feature map. For each channel of 248 × 248 × 50, a batch normalization operation performs normalization across a mini-batch training data [8]. The normalized feature map 248 × 248 × 50 is input to the ReLU layer [9], and then negative values are set to zero. Max pooling operation with 2 × 2 filter is performed on 248 × 248 × 50 ReLU feature map [10]. The dropout operation from conv3 to fc2 is adopted as shown in Table 1 to avoid overfitting [11, 12].

Table 1. CNN configuration.

Layer name	Operation	Filters Num	Filter	Stride	Output
Input	-	-	-	-	500 × 500 × 1
conv1	Conv + BatchNorm + ReLU	50	5 × 5 × 1	2 × 2	248 × 248 × 50
pool 1	Maxpooling	-	2 × 2	2 × 2	124 × 124 × 50
conv2	Conv + BatchNorm + ReLU	100	5 × 5 × 50	2 × 2	60 × 60 × 100
conv3	Conv + BatchNorm + ReLU + dropout (50%)	100	3 × 3 × 100	2 × 2	29 × 29 × 100
fc1	BatchNorm + ReLU + dropout (50%)	-	-	-	1000
fc2	BatchNorm + ReLU + dropout (50%)	-	-	-	1000
fc3	softmax	-	-	-	3

The proposed CNN was validated with the bootstrap method [13, 14], which is reliable for validating when a small number of data are available. This method is commonly used to evaluate the performance of machine learning. In this method, the test data is randomly selected, and the training data other than the test data is used to train the CNN. After training, the test data is fed into the trained CNN to validate its accuracy. The images are flipped left and right with random probability during the training phase to avoid overfitting. The training conditions are listed in Table 2.

Table 2. CNN training condition.

	Method / Value
Solver	SGDM(Stochastic Gradient Descent with Momentum)
Learn Rate	10^{-4}
Max Epochs	500
Mini batch Size	32
Total Iterations	1500
Augmentation	Left (50%) and Right (50%) Reflection
Number of Data	118 data (augmented data = 118*3 = 354)

Overall, we used 44 bad (named as bad 1 ~ bad 44), 38 good (good 1 ~ good 38), and 46 neutral images (neutral 1 ~ neutral 46); thus, a total of 128 images were used for the validation.

In the proposed bootstrap validation, a total of 10 images (5 "bad" and 5 "good") are randomly selected for the test. Ten test images will not be used for training CNN. However, they will be input to the CNN after training to verify the accuracy. The remaining 118 images will be used for training CNN on the condition in Table 2.

The above evaluation was performed for 15 trials. Figures 4 (a) and (b) show the mini batch loss and accuracy in the training progress in each trial. As the iterations progressed, we can see that the mini-batch loss decreased, and the accuracy for the training data improved. Therefore, it is found that the training proceeded properly.

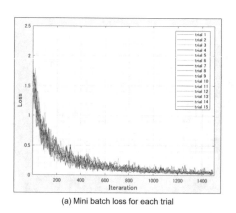

(a) Mini batch loss for each trial

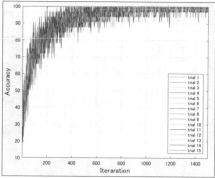

(b) Mini batch accuracy for each trial

Fig. 4. Training progress.

The validation results are shown in Table 3. The accuracy is illustrated in Fig. 5.

Table 3. CNN validation result.

Trial	"bad" image for test					"good" image for test					Accuracy [%]
1	22	31	32	37	41	1	7	13	16	23	1
2	11	18	20	21	42	6	29	33	34	35	0.9000
3	7	10	13	25	42	8	10	19	21	35	1
4	1	6	20	23	32	1	3	13	25	30	0.8000
5	1	3	8	27	35	3	5	18	29	35	0.7000
6	8	17	18	21	25	2	23	26	32	34	0.9000
7	2	3	12	21	23	8	12	16	29	37	0.9000
8	19	21	23	30	39	1	5	13	33	36	1
9	5	11	13	20	25	2	8	13	25	37	0.9000
10	16	20	23	28	44	1	12	23	26	28	0.9000
11	17	18	27	33	43	3	26	27	28	31	0.9000
12	18	35	36	38	44	1	13	14	25	37	0.9000
13	4	6	16	19	39	7	8	28	29	34	0.9000
14	10	11	22	32	36	7	9	10	19	26	0.9000
15	2	9	23	42	44	6	9	15	26	34	0.9000
—	—					**Mean Accuracy**					0.9000

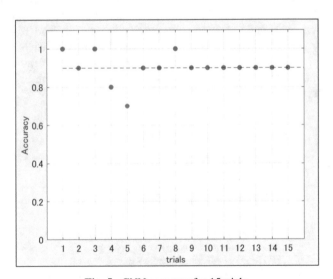

Fig. 5. CNN accuracy for 15 trials.

Since the mean accuracy was 90%, CNN recognized the crack features. However, in trial 5, the accuracy was 70%. Hence, we analyzed the inference results in detail, and Fig. 6 shows the confusion matrix and misjudged images.

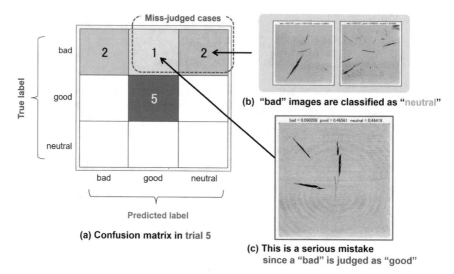

Fig. 6. Detailed result in trial 5.

As Fig. 6(a) illustrates, five "good" images were judged correctly. In contrast, two "bad" images were evaluated as "neutral," as Fig. 6(b) shows. In addition, one "bad" image shown in Fig. 6(c) was regarded as "good." Table 4 shows the probability of CNN output in trial 5. For "bad 1" and "bad 8," the probability of "neutral" was the highest; hereby, they were evaluated as "neutral" incorrectly. The image "bad 3" was judged as "good" because the probability of "good" is highest. In this case, CNN was wavering between "good" and "neutral." This fault was serious because "bad" was confused with "good." To improve such mistakes, it is necessary to prepare many data and to categorize the training data correctly so that CNN is appropriately trained.

Table 4. CNN output probability value in trial 5.

probability value	bad	good	neutral	correct label
bad 1	0.0127	0.0012	0.9861	
bad 3	0.0902	0.4656	0.4442	
bad 8	0.0613	0.0683	0.8704	bad
bad 27	0.9782	0.0153	0.0065	
bad 35	0.9997	0.0000	0.0003	
good 3	0.0001	0.9981	0.0018	
good 5	0.0004	0.9946	0.0050	
good 18	0.0001	0.9994	0.0005	good
good 29	0.0001	0.9989	0.0011	
good 35	0.0004	0.9908	0.0088	

4 Conclusions

Research has been conducted to investigate the optimal drying conditions [3] for timbers. However, internal cracks occur that are not visible on the surface. They have been overburdened, measuring the cracks in the cross-section of the square timber using calipers. This study is the first attempt applying CNN to evaluate the internal cracks' severity automatically to facilitate and promote above studies.

In the experiment, silver paint was coated on the cross-section to make the cracks more visible. The validation was conducted to verify the CNN's ability to evaluate the severity of the cracks from the silver-painted pictures. The classification accuracy was 90%. In the future, we will improve the accuracy by adding more image data.

Acknowledgments. The authors would like to thank Ueno in MathWorks for technical advice. This work was supported by JSPS Grant Numbers 20K06116.

References

1. Bergman, R.: Drying and control of moisture content and dimensional changes. In: Wood Handbook, Wood as an Engineering Material, Chap. 13, pp. 1–21 (2021)
2. Yamashita, K., Hirakawa, Y., Saito, S., et al.: Internal-check variation in boxed-heart square timber of sugi (Cryptomeria japonica) cultivars dried by high-temperature kiln drying. J. Wood Sci. **58**, 375–382 (2012)
3. Yin, Q., Liu, H.H.: Drying stress and strain of wood: a review. Appl. Sci. **11**(11), 5023 (2021)
4. LeCun, Y., Bottou, L., Bengio, Y., Haffner, P.: Gradient-based learning applied to document recognition. Proc. IEEE **86**(11), 2278–2324 (1998)
5. Liu, Y., Hou, M., Li, A., Dong, Y., Xie, L., Ji, Y.: Automatic detection of timber-cracks in wooden architectural heritage using Yolov3 algorithm. Int. Arch. Photogramm. Remote Sens. Spatial Inf. Sci. **XLIII-B2–2020**, 1471–1476 (2020)
6. He, T., Liu, Y., Yu, Y., Zhao, Q., Hu, Z.: Application of deep convolutional neural network on feature extraction and detection of wood defects. Measurement **152**, 107357 (2020)
7. Pan, L., Rogulin, R., Kondrashev, S.: Artificial neural network for defect detection in CT images of wood. Comput. Electron. Agric. **187**(106312), 1–7 (2021)
8. Ioffe, S., Szegedy, C.: Batch normalization: accelerating deep network training by reducing internal covariate shift. arXiv:1502.03167 (2015)
9. Nair, V., Hinton, G.E.: Rectified linear units improve restricted boltzmann machines. In: Proceedings of the 27th International Conference on Machine Learning (ICML-10), pp. 807–814 (2010)
10. Nagi, J., et al.: Max-pooling convolutional neural networks for vision-based hand gesture recognition. In: IEEE International Conference on Signal and Image Processing Applications (ICSIPA), pp. 342–347 (2011)
11. Srivastava, N., Hinton, G., Krizhevsky, A., Sutskever, I., Salakhutdinov, R.: Dropout: a simple way to prevent neural networks from overfitting. J. Mach. Learn. Res. **15**, 1929–1958 (2014)
12. Krizhevsky, A., Sutskever, I., Hinton, G.E.: ImageNet classification with deep convolutional neural networks. Adv. Neural. Inf. Process. Syst. **25**, 1–9 (2012)
13. Priddy, K.L., Keller, P.E.: Artificial neural networks - an introduction. In: Chapter 11, Dealing with Limited Amounts of Data, pp. 101–105. SPIE Press, Bellingham (2005)
14. Ueda, N., Nakano, R.: Estimating expected error rates of neural network classifiers in small sample size situations: a comparison of cross-validation and bootstrap. In: Proceedings of ICNN 1995 - International Conference on Neural Networks, vol. 1, pp. 101–104 (1995)

Information Security Fatigue in Visually Impaired University Students

Masataka Kakinouchi[1(✉)] and Kazumasa Omote[2]

[1] Tsukuba University of Technology, 4-12-7 Kasuga, Tsukuba, Ibaraki 305-8521, Japan
kakinouchi@cs.k.tsukuba-tech.ac.jp
[2] University of Tsukuba, 1-1-1 Tennoudai, Tsukuba, Ibaraki 305-8577, Japan
omote@risk.tsukuba.ac.jp

Abstract. Information security measures for ICT users are always issues that must be considered. This may lead to a sense of mental exhaustion (information security fatigue) owing to the troublesome and time-consuming measures. It has already been shown that information security fatigue prevents the effectiveness of information security measures, and some necessary measures have been proposed. The visually impaired have some problems that cannot be dealt with by the measures considered for the sighted in the previous studies. In addition, considering studies on usable security for the visually impaired, although there are studies on authentication technology and methods, few studies on information security fatigue have been conducted. In this study, we have conducted a survey among only visually impaired university students and have analyzed the results. We found that the distribution of the condition matrix of the sighted students was almost even among the six groups, whereas the distribution of visually impaired students was very different. We also observed that the higher the level of implementation of the security measures is, the higher the level of security fatigue. In addition, we have clarified the countermeasures for the visually impaired because the countermeasures proposed in the existing studies for the sighted cannot fully compensate for their effects.

1 Introduction

Recently, the threats related to information security, such as information leakage and cyber-attacks, have diversified with the remarkable advancement of ICT technology. Considering a countermeasure against these threats, users are required to take information security measures. The Information Security Site for Citizens of the Ministry of Internal Affairs and Communications lists updating software, installing anti-virus software, and proper management of ID and password as the three principles for beginners [16]. The number of Web services and software used by users varies, and the necessary countermeasures increase and become more detailed every day. In addition, it is important to pay attention to the latest information because the attacker's technology advances daily.

A 2016 survey by the US National Institute of Standards and Technology (NIST) [20] revealed that people got fed up with daily updates and took reckless actions owing to fatigue. This behavior is known as security fatigue [4].

L. Barolli et al. (Eds.): AINA 2022, LNNS 451, pp. 448–458, 2022.
https://doi.org/10.1007/978-3-030-99619-2_43

Hatashima et al. explored the relationship between the degree of security fatigue and the degree of implementation of security measures [7–11,21] and proposed measures for each group after organizing them into a classification known as the information security condition matrix as a model for visualization. However, this is only for the sighted, and it has not been clarified whether the distribution is similar for the visually impaired or whether the same measures can be taken.

In this study, we survey only visually impaired university students and analyze the results. The distribution of the condition matrix of the visually impaired students was quite different from that of the sighted students, whereas the distribution was approximately uniform among the six groups. The distribution of the visually impaired was quite different from that of the six groups. This study makes two major contributions.

1. The classification by the information security condition matrix, was conducted based on a survey of the visually impaired. A positive correlation was found for the visually impaired, whereas the distribution was uniform for the sighted. Alternatively, the higher the level of implementation of information security measures is, the higher the level of information security fatigue.
2. We propose two new measures for screen readers, which have an inseparable relationship with the visually impaired who use ICT, especially in the group with the most respondents.

2   Preparation

2.1   Security Fatigue

Security fatigue was proposed by Furnell and Thomson in 2009 [4], and in 2016, Stanton et al. analyzed the actual situation and presented three recommendations [20] to reduce security fatigue. In addition, Hatashima et al. summarized information security fatigue and its measures to reduce it [7].

Table 1. Information security condition matrix.

High-fatigue state → F3	F3-Im1 Group	F3-Im2 Group (5.2)
Ideal-fatigue state → F2	F2-Im1 Group	F2-Im2 Group (Ideal state)
Low-fatigue state → F1	F1-Im1 Group (5.1)	F1-Im2 Group
	Low implementation Group → Im1	High implementation Group → Im2

F: Scale of Information Security Fatigue

Im: Scale of Implementation of Information Security Measures

2.2 Assessment by Information Security Condition Matrix

Hatashima et al. [11] presented the information security condition matrix as shown in Table 1, and defined F2-Im2 as the ideal state. This matrix is classified into six groups based on the combination of three ranks for security fatigue (F group) and two ranks for the implementation of security measures (Im group).

2.3 Screen Reader

The use of a screen reader is essential for visually impaired people when using a personal computers (PC) or smartphone. Based on a survey [23] conducted by Accessibility in mind (WebAIM) in September 2018 among 248 people with low vision, 59.9% of them vision use a screen reader on a PC, and 58.1% of them use a screen reader on a mobile tablet.

In this study, we identify the increased safety of screen readers for the visually impaired and risk owing to their use. Therefore, providing them as part of information security education will help create an appropriate and safe ICT environment.

3 Related Work

3.1 Actions on Information Security Measures

Koppel et al. clarified that the perception of passwords was very different between general users of ICT and cyber security experts. They asserted that general users have misunderstandings and wrong approaches to improve security [14]. Therefore, it is necessary to reduce such misunderstandings and wrong approaches to enable users to take appropriate security measures.

3.2 Information Security for the Visually Impaired

There have been several studies on information security for the visually impaired, including a study that has developed an original system to replace the completely automated public Turing test to tell computers and humans apart (CAPTCHA) [24]. There is another study that has developed an improved version of the existing audio CAPTCHA [3]; nonetheless, there are few studies on security fatigue.

On the contrary, the visually impaired and training materials on the accessibility features of smartphones need to be made easily available to schools and all young people with disabilities worldwide [18]. Regarding the information security fatigue of the visually impaired, it is considered significant to enhance ICT education, including information security education by presenting such training materials.

Furthermore, considering studies [2] on the visually impaired cannot directly target the visually impaired. Nonetheless, they often target blindfolded sighted people, and there are few studies that limit the target to the visually impaired. The purpose of this study is to investigate the differences between blind and visually impaired individuals.

4   Previous Research and Analysis

In this section, we summarize the scales and questionnaires that Hatashima et al. have developed in their previous studies [7–11]. In this study, these have been adapted, and the questions described in Subsubsect. 4.1.3 have been added as an improvement. The analysis is the same as that conducted in [11], and is described in Subsect. 4.3.

4.1   Preparing a Questionnaire

4.1.1   Information Security Fatigue Scale

The information security fatigue scale was developed using (1) a questionnaire consisting of 23 items with five choices in a previous study [8], extracted from the English version of the MBI-GS published in [13] and Kubo's (Japanese version) Burnout Scale [15]. (2) A questionnaire survey was conducted on university students using the questionnaire in (1) of the previous study [10], which was summarized to 13 items (Question 2) through factor analysis, reliability, and validity studies.

4.1.2   Information Security Measures Implementation Scale

This section describes the measurement of the degree of implementation of information security measures. Hamatsu et al.'s [6], experience of virus infection, IT knowledge, and IT skills created as the experience of information security measures and knowledge about security. These questions were adapted by referring to the ten most socially influential information security threats [12] announced by the Information-Technology Promotion Agency (IPA) in 2015 and the measures to be taken by teleworkers in the Telework Security Guidelines [17] by the Ministry of Internal Affairs and Communications in 2017. After preparing 17 questions [9], Hatashima et al. prepared a questionnaire (Question 3) consisting of 11 items with five choices, excluding questions on corporate policies to include university students and others who do not work for companies [11].

4.1.3   Open-Ended Questions About the Safety and Dangers of Information Security for the Visually Impaired

In a previous study, the scale of information security fatigue and implementation of countermeasures were surveyed only for the sighted, whereas in this study, the survey was conducted for the visually impaired. To extract their unique experiences, we set two questions as follows. (Questions 1 and 4)

1. Which do you rely on more when using a PC or smartphone, audio, or visual? Please check one of them (Audio/Visual).
2. In relation to information security measures, if you have ever felt that you are vulnerable to fraud because of your visual impairment or less vulnerable to fraud because of your visual impairment, please describe your thoughts freely.

4.1.4   Open-Ended Question About Information Security

In addition to the questions in Subsubsect. 4.1.3, the question "Please freely describe your thoughts on the information security measures, if any," was included to let the respondents freely answer their opinions on the information security for extracting the characteristics of each group and conducting the assessment (Question 5).

4.2   Administering the Questionnaire

A survey was conducted at a national university in the Tokyo metropolitan area. Prior to the implementation, the survey was reviewed and approved by the Institutional Research Ethics Review Board. The survey was conducted from September 2 to 16, 2021, and 11 first-year students attending a national university in the Tokyo metropolitan area participated. Written and oral consents were obtained from the parents of nine minors through e-mail before the survey. No honorarium was offered. Eleven valid responses were collected, and there were no missing values in the data. Eleven participants is a small number for a normal survey but is not small for a survey of the visually impaired[1].

4.3   Classification of Respondents by Latent Rank Theory

In this study, to compare the results with those of the previous study [11], we use the software exametrika [19] by Shojima, which is a similar analysis. The analysis was conducted in the following order: (1) determination of the number of latent ranks to classify the respondents; (2) execution of exametrika with the number of latent ranks determined in (1); discussion of the analysis results. Considering step (1), the number of latent ranks is increased from (2), and exametrika is run each time, and the method of adopting the number of latent ranks that minimizes the calculated information criterion is adopted. Nevertheless, practically, the analyst may decide the number of latent ranks based on the analysis results [22].

Latent rank theory was applied to the results of the questionnaire survey to classify the respondents. Specifically, exametrika was run for the responses to the information security fatigue scale and those to the information security measures implementation scale. The respondents are classified into three and two groups for the F and Im groups, respectively, as shown in Subsect. 2.2. Default values were used, excluding the selection of the self-organizing map (SOM), which is the analysis option that could be set during these runs. The respondents are classified into six groups, and the number of respondents belonging to each group is shown in Table 2.

Table 2. Number of respondents on information security condition matrix.

Number of respondents (N = 11)		Im Group	
		Low implementation group [Im1]	High implementation group [Im2]
F Group	High-fatigue state [F3]	0	4
	Ideal-fatigue state [F2]	1	1
	Low-fatigue state [F1]	4	1

---

[1] For example, the number of participants in Studies [1] and [5] are similar.

5   Our Assessment Results for Each Group

In this section, we first present the results of the assessment for each of the groups shown in Table 1, based on the results of the questionnaire survey in Sect. 4. Table 3 summarizes the characteristics of the respondents in each group, the proposed measures, and their evaluations. Subsequently, we propose two new countermeasures, (⋆1) and (⋆2) for the two groups, F1-Im1 and F3-Im2, respectively, which showed characteristics unique to the visually impaired. Considering these two new countermeasures, the authors decided on the assessments to extract the features from the responses to the open-ended questions in Subsubsects. 4.1.3 and 4.1.4. Regarding the 14 measures proposed in the previous studies, those that seemed less applicable and the assessments of the measures for the groups that have no respondents, we've written "-".

5.1   Low Level of Information Security Fatigue and Implementation (F1-Im1 Group)

Regarding this group, the degree of information security implementation is low, similar to the degree of information security measures implemented. Four respondents belonged to this group (Table 2). The information security fatigue and measures implementation scale scores are 48.25 and 27.5 with standard deviations of 5.74 and 4.51, respectively.

Considering this group, some respondents answered that they had insufficient understanding of the countermeasures as an answer indicating a low level of implementation.

Considering the information security measures unique to the visually impaired, "I feel that I am more susceptible to fraud owing to my visual impairment because I use my voice to control my PC, and I access different links when exploring websites (Respondent ID4)" is one of the responses. Some respondents mentioned that they accessed more sites because of the use of screen readers. On the contrary, one respondent mentioned that the accuracy of security measures was improved owing to supplementary tools such as screen readers, saying, "I can read the fine print with a screen reader; therefore, I do not miss any warning signs that I will normally miss (Respondent ID9)." One respondent took both aspects into consideration by indicating, "I think it is sometimes difficult to get caught because the voice sometimes reads out things that are not displayed. I think it is sometimes easier for me to get caught than for sighted people because I sometimes click on links without fully recognizing where I am going (Respondent ID5)." It can be observed that both aspects are considered; nonetheless, considering the low level of implementation while being aware of the capabilities and limitations of screen readers, some measures are required.

Table 3. Results of assessing open-ended questions by group for information security condition matrix, countermeasures and evaluation.

Group name	Assessment results (Characteristics of respondents belonging to each group)	Measurements Move to the ideal state Maintain the ideal state	Evaluation of countermeasure effectiveness (5.3)
F2-Im1	Rely on Sharepoint and other tools to avoid the hassle of setting passwords Be aware that vulnerable to shoulder hacking because enlarging the screen	1. Presentation of recommended environment for information security solutions 2. Self-check using a checklist of information security measures 3. Provision of information security solutions	Rising (to Im2) - Rising (to Im2)
F2-Im2 (Ideal State)	Be hesitant to update because worrying that the operability will change	4. Presentation of more advanced information security measures, such as case studies of information security incidents and their countermeasures 5. Let people know that they are in an ideal state and motivation to maintain it	No change No change
F1-Im1 (5.1)	Even with the use of a screen reader, I sometimes click on links without knowing where they will take me, so I feel that I am more likely to be caught by fraudulent activities than sighted people	6. Conducting education to raise awareness of the parties involved 7. Specific presentation of information security measures such as manualization (⋆1)**Promote understanding of security incidents that can be avoided by using screen readers**	Rising (to F2-Im2) Rising (to F2-Im2) Rising (to F2-Im2)
F1-Im2	It all feels so natural	8. Stimulate individuals to take active measures without relying on solutions 9. Implementation of motivation recovery measures (education, etc.) after the occurrence of incidents	Rising (to F2) -
F3-Im1	N/A	10. Exercises to simulate information security incidents	-
F3-Im2 (5.2)	I feel that there is no advantage in information security for the visually impaired Feeling anxious about accessing unauthorized sites that they have no experience with	11. Hearing on information security measures (Letting out stress) 12. Extending the interval for information security education 13. Simplification of information security education 14. Development of a systematic education system (⋆2)**Provide education on the limitations of screen reader functionality**	Falling (to F2) Falling (to F2) Falling (to F2) Falling (to F2) Falling (to F2)

To move this group to the ideal state, in addition to the two countermeasures proposed in the previous study, the screen reader can prevent access to illegal links by reading out fine text. This is a very important aspect because the screen reader can assist, considering the sighted people. Therefore, it will be effective to promote the understanding of security incidents that can be avoided by using screen readers (a new countermeasure proposal (⋆1)).

5.2 High Level of Information Security Fatigue and Implementation (F3-Im2 Group)

This group has a high level of information security fatigue and implementation of information security measures. Four respondents belong to this group (Table 2). The information security fatigue and measures implementation scale scores were 44.50 and 47.25 with standard deviations of 9.64 and 4.92, respectively.

Regarding this group, there were responses that considered the burden of security measures for ICT novices, and responses that felt the inconvenience in accessibility of screen readers related to authentication. This included: "I think it would be nice if the visually impaired could use authentication, such as 'I'm not a robot' authentication or image authentication when registering on a website. I think it is safe; however, I do not like it because it does not read out loud. I found it inconvenient, especially in online stores (Respondent ID6)." This response indicates that there is a lack of situations to educate people about the limitations of screen readers.

In addition, we received the following responses: "I will like to see security measures that are safe for the visually impaired to recognize because there are times when I cannot register or log in to recent sites because I cannot recognize the 'I am not a robot' image. Recently, pop-up advertisements have appeared on major companies' websites, and I feel this is a problem because it sometimes opens links that are different from the ones I focus on. This is especially annoying when I am viewing a site with mixed advertisements on my smartphone or tablet (Respondent ID7)." Regarding the security measures, the respondents felt inconvenienced by those that were difficult for the visually impaired to access and considered the difficulties of password management.

In addition, about information security measures for the visually impaired, another respondent recognized the aspect of safety maintained by Voiceover, a screen reader included in iOS saying, "Especially, using the iPhone, I can read e-mails and messages without opening them; therefore, I can distinguish suspicious e-mails and messages and feel less likely to get caught (Respondent ID7)." Another respondent mentioned actual damages caused by visual impairment saying "I was charged approximately 1,500 yen per month to read a QR code in an app in the App Store. The reason for this was that the font was too small, and the explanation for use was written in English. We should definitely read the explanation, no matter how small it is. (Respondent ID11)."

To shift this group to the ideal state, in addition to the four measures presented in the previous study, education on the limitations of the functions of screen readers used in smartphones and tablets is also necessary (new measure proposal (⋆2)).

5.3 Evaluation of Proposed Measures to Be Considered as Unique Measures for the Visually Impaired

In addition to the countermeasures proposed in the previous studies, two new countermeasures are proposed in Table 3. In this section, we evaluate the effect of the transition of the rank of the group to which they belong on the scales of information security fatigue (F group) and implementation of information security measures (Im group). The evaluations in this subsection (including the countermeasure proposals of the previous studies) are independent considerations for the F and Im groups. They do not necessarily show the countermeasure effects for the F and Im groups at the same time. Considering the evaluation of the rise and fall of the ranks, there is a shift of one level before and after unless otherwise mentioned.

Countermeasure ($\star$1) of the F1-Im1 group (Subsect. 5.1) is considered to have an educational effect that makes it more proactive about security measures by correctly understanding the risks that can be prevented by screen readers.

The proposed countermeasure ($\star$2), which is a countermeasure for the F3-Im2 group (Subsect. 5.2), has also been evaluated. Although many visually impaired people use screen readers on smartphones and tablets, education about the limitations of screen readers remains insufficient as pointed out by Prakash et al. [18]. A proper understanding of the situations in which screen readers do not read out loud and the limitations of their functions will encourage more appropriate use. If appropriate use can be made based on a proper understanding, countermeasures can be prepared in advance against the screen reader not reading out loud, and security fatigue can be reduced.

By combining these countermeasure plans, it is considered that information security measures to approach the ideal state of the F2-Im2 group can be systematically established.

6   Discussion

Regarding the tendency of the visually impaired, there is a positive correlation between the scale of security fatigue and that of the implementation of security measures. Alternatively, (1) the higher the level of security fatigue, the higher the level of implementation of security measures is, and (2) the opposite is true. Considering (1), the visually impaired are more stressed than the sighted in taking information security measures because of the lack of information caused by (partial) reliance on visual information and the screen reader does not read out. On the contrary, as shown in (2), those who are less likely to implement security measures may lack an understanding of the risks that can be prevented by using a screen reader. Regarding both of them, it is necessary to provide security education to facilitate the consideration of security measures with a sense of tension. In addition, the results might be biased owing to the small sample size, and the questionnaire survey was conducted only for students in the Department of Information Systems. Verifying whether the same conclusions can be drawn for the visually impaired generally by expanding the target departments and grades is yet to be studied.

7 Conclusion

The distribution of the condition matrix proposed by Hatashima et al. and the measures for each condition cannot be considered similarly for the visually impaired. In this study, as a characteristic of the distribution, we observed that the correlation between the scale of security fatigue and implementation of security measures is positive for the visually impaired, whereas it is uniform for the sighted. In addition, we proposed specific measures for security incidents that are unique to the visually impaired.

In the future, we would like to continue our study on the effectiveness of the proposed measures and to propose measures that can be correlated more generally with the visually impaired and those that can be adapted to the problems specific to the visually impaired by expanding the survey targets. Therefore, we would refer to the previous studies that were considered for people with sighted people to propose measures that would be more individually optimized, leading to a better enhancement of security education.

References

1. Bai, J., Liu, Z., Lin, Y., Li, Y., Lian, S., Liu, D.: Wearable travel aid for environment perception and navigation of visually impaired people. Electronics **8**(6), 697 (2019)
2. Brulé, E., Tomlinson, B.J., Metatla, O., Jouffrais, C., Serrano, M.: Review of quantitative empirical evaluations of technology for people with visual impairments. In: CHI 2020: Proceedings of the 2020 CHI Conference on Human Factors in Computing Systems, April 2020, pp. 1–14 (2020)
3. Fanelle, V., Karimi, S., Shah, A., Subramanian, B., Das, S.: Blind and human: exploring more usable audio CAPTCHA designs. In: Sixteenth Symposium on Usable Privacy and Security (SOUPS 2020) (2020)
4. Furnell, S., Thomson, K.-L.: Recognizing and addressing security fatigue. Comput. Fraud Secur. **2009**(11), 7–11 (2009)
5. Götzelmann, T.: Visually augmented audio-tactile graphics for visually impaired people. ACM Trans. Access. Comput. **11**(2), 1–31 (2018)
6. Hamatsu, S., Kurino, S., Yoshikai, N.: Proposal and application of a model of intention to implement information security measures based on collective protective motive theory. Trans. Inf. Process. Soci. Jpn. **56**(12), 2200–2209 (2015). (in Japanese)
7. Hatashima, T., et al.: Evaluation of the effectiveness of risk assessment and security fatigue visualization model for internal E-crime. In: 2018 IEEE 42nd Annual Computer Software and Applications Conference (COMPSAC) (2018)
8. Hatashima, T., Nagai, K., Tanimoto, S., Kanai, A.: A study on visualization of information security fatigue of college students. In: Proceedings of Computer Security Symposium 2017, pp. 888–895 (2017). (in Japanese)
9. Hatashima, T., Sakamoto, Y.: Analysis of teleworker's propensity for information security anxiety behavior. Trans. Inf. Process. Soc. Jpn. **58**(12), 1912–1925 (2017). (in Japanese)
10. Hatashima, T., Tanimoto, S., Kanai, A.: Proposal of information security fatigue measurement scale (University Student Version) - design and evaluation of measurement method by using burnout scale. IEICE Trans. **J101-D**(10), 1414–1426 (2018). (in Japanese)
11. Hatashima, T., Tanimoto, S., Kanai, A., Fuji, H., Okubo, K.: Proposal of information security fatigue countermeasure for university students using Improved information security condition matrix. Trans. Information Process. Soc. Jpn. **59**(12), 2105–2119 (2018). (in Japanese)

12. Information-technology Promotion Agency: 10 major threats to information security (2016). (in Japanese), https://www.ipa.go.jp/security/vuln/10threats2016.html. Cited 29 Sep 2021

13. Itakura, H.: Burnout and project management. J. Project Manage. Soc. Jpn **11**(1), 17–19 (2009). (in Japanese)

14. Koppel, R., Blythe, J., Kothari, V., Smith, S.: Beliefs about cybersecurity rules and passwords: a comparison of two survey samples of cybersecurity professionals versus regular users. In: Twelfth Symposium on Usable Privacy and Security (SOUPS 2016) (2016)

15. Kubo, M.: Burnout: stress in human service workers. Jpn. J. Labor Res. **558**(1), 54–64 (2007). (in Japanese)

16. Ministry of Internal Affairs and Communications: Information security site for citizens to use the Internet with peace of mind. (in Japanese), https://www.soumu.go.jp/main_sosiki/joho_tsusin/security/intro/beginner/index.html. Cited 29 Sep 2021

17. Ministry of Internal Affairs and Communications: Telework Security Guidelines. (in Japanese), https://www.soumu.go.jp/main_sosiki/cybersecurity/telework/. Cited 29 Sep 2021

18. Prakash, S., Frode, E.S.: A glimpse into smartphone screen reader use among blind teenagers in rural Nepal. Disability Rehabil. Assist. Technol. **12**, 1–7 (2020)

19. Shojima, K.: Exametrika. http://antlers.rd.dnc.ac.jp/shojima/exmk/index.htm. Cited 29 Sep 2021

20. Stanton, B., Theofanos, M.F., Prettyman, S.S., Furman, S.: Security fatigue. IT Prof. **18**, 26–32 (2016)

21. Tanimoto, S., Nagai, K., Hata, K., Hatashima, T., Sakamoto, Y., Kanai, A.: A concept proposal on modeling of security fatigue level. In: 5th International Conference on Applied Computing and Information Technology (ACIT) (2017)

22. Ueno, M., Shojima, K.: New Trends in Learning Evaluation, Asakura Shoten (2010). (in Japanese)

23. WebAIM: Survey of Users with Low Vision # 2 Results. https://webaim.org/projects/lowvisionsurvey2/#mobilescreenreader. Cited 13 Oct 2021

24. Zhang, Z., Zhang, Z., Yuan, H., Barbosa, N.M., Das, S., Wang, Y.: WebAlly: making visual task-based CAPTCHAs transferable for people with visual impairments. In: Seventeenth Symposium on Usable Privacy and Security (SOUPS 2021) (2021)

Privacy and Security Comparison of Web Browsers: A Review

R. Madhusudhan[(✉)] and Saurabh V. Surashe

Department of Mathematical and Computational Sciences,
NIT - K Surathkal, Karnataka, India
`madhu@nitk.edu.in`

Abstract. In today's digital world, mobile phones, computers, laptops and other digital devices are the most important things. The global pandemic like, Covid-19 has drastically changed the whole world, and in the post Covid world, most of the businesses are switching to online business, online marketing, online customer service, etc. Due to this, the usage of web browsers has increased exponentially. So, when we are using the internet to a very much great extent then, there are also chances of getting tracked, hacked, or cyber-bullying, etc. Hence, there comes the need for privacy protection while internet surfing in web browsers. In this paper, we have surveyed different research papers. This paper focuses on the popular desktop browsers on Windows such as Google Chrome, Mozilla Firefox, Microsoft Edge, Apple Safari, Brave, Tor, etc. This work studies the different parameters considered for privacy leakage; different methods used for evaluation both in normal browsing mode and in private browsing mode. The work proposed that the documentation given for the private mode for popular browsers is not complete, and also that users' privacy, can be leaked.

Keywords: Privacy · Web browsers · Private browsing · Security · Third-party tracking · Machine learning

1 Introduction

In the 21st century, the use of the internet is increasing exponentially and the global pandemics like Covid, played a very important role in extending this reach of internet to each and every industry, school, college and even house. Because of that, the use of web browsers is also increasing rapidly. Even though, internet technology may have reached most people on the earth, but most of these internet users, or the average users are not technical nor security-savvy. They don't try new security and privacy techniques available and rely on the default security measures that are provided by the web browsers, such as protection from sites having malware or hosting phishing attacks [1]. They don't even check what they are allowing to the websites, such as collecting user information, location, saving passwords, etc. and they end up following the default browser settings set by

L. Barolli et al. (Eds.): AINA 2022, LNNS 451, pp. 459–470, 2022.
https://doi.org/10.1007/978-3-030-99619-2_44

web browsers [2]. Here, web privacy refers to the right of web users to conceal their personal information and have some degree of control over the use of any personal information disclosed to others [3]. Similarly, web security means the protective measures and protocols that users/organizations adopt to protect the user/organization from, cyber criminals and threats that use the web channel [4].

As per the information reported to and tracked by the Indian Computer Emergency Response Team (CERT-In), 3,94,499 and 11,58,208 cyber security incidents have been observed during 2019 and 2020, respectively [5]. That is nearly 3 times more cyber attacks were reported in 2020 than in 2019 and 20 times more than in 2016 [6]. So, we can clearly see that pre-pandemic data and post-pandemic data are a lot different and that the increase is very much higher and dangerous and this also highlights the importance of web security.

Private browsing is the generalized word used for the method, by which users can prevent themselves from having evidence of their web browsing behavior stored on the local device [7,8]. Depending on the browser in use, private browsing facility is referred to with different names in different terminology like 'Incognito mode' in Chrome, 'InPrivate' in Microsoft Edge, 'Private window' in Firefox and Brave as shown in Table 1. According to Browser market share statistics 2019, in the United States, 46% of peoples have decided to give private browsing a shot at least once. On the other hand, 33% of people haven't even heard about this mode in browsers [9].

Table 1. Private browsing modes in web browsers

Browser	Synonym	Date of introduction
Safari 2.0	Private browsing	April 29, 2005
Google Chrome	Incognito	December 11, 2008
Internet Explorer	Private browsing	March 19, 2009
Mozilla Firefox	InPrivate browsing	June 30, 2009
Opera	Private tab/window	March 2, 2010
Microsoft Edge	InPrivate browsing	July 29, 2015
Brave	Private browsing	November 13, 2019

Private browsing mode does not necessarily protect users from being tracked by websites. It is a common misconception that private browsing modes can protect users from being tracked by other websites or their internet service provider (ISP). Such entities can still use information such as IP addresses and user accounts to uniquely identify users [9]. Some browsers have partly addressed this shortcoming by offering additional privacy features that can be automatically enabled when using private browsing mode, such as Firefox's 'Tracking Protection' feature to control the use of web trackers [10]. In the documentation on private browsing of browsers, it says that they don't save cookies; it gets deleted as a session gets terminated. But in actual practice, various forensic tools have successfully found these saved cookies.

A cookie is a textual piece of data sent to the browser whenever a user visits a website [11]. A cookie makes a site to remember the information about user visit, user data and make it faster to visit the site again and again [12]. Cookies are of different types, but the most used classification is first-party cookies and third-party cookies. First-party cookies are stored under the same domain user is currently visiting [13]. For example, if a user is on example1.com, all cookies stored under this domain are considered first-party cookies. Those cookies are usually used to identify a user between pages, remember selected preferences, or store your shopping cart, or store settings [12]. Third-party cookies are cookies that are stored under a different domain than user is currently visiting. They are mostly used to track users between websites and display more relevant ads between websites [11]. There are a lot of examples where we can see that third-party cookies are used by different websites either to track the user or to provide the ads to him. Sometimes there are also chances of web session linking or cross-site, cross-browser user tracking by using personal information collected from cookies. So, in order to avoid this, there is need to improve privacy and security of a user in web browsers.

This work does a survey of the privacy and security of web browsers when a user is surfing in private mode. This work also talks about the changes made in the filesystem of browsers while using private browsing mode. This paper also states what information in public mode is available in private mode even after closing public/normal mode and vice versa. It also talks about third-party tracking used in websites, how it is done, and different machine learning methods and machine learning models used to avoid third-party tracking.

The rest of the paper is organized as follows: Sect. 2 discusses the literature review; Sect. 3 presents the detailed description of the method used in the referred papers, and Sect. 4 gives insight to the result obtained. Finally, we conclude the paper in Sect. 5.

2   Related Work

The privacy and security of browsers have been the subject of substantial literature. This can be broad as concerned with online tracking by websites, attacks on browsers themselves, such as SQL injection attacks, and phishing attacks on the users [14]. Now-a-days, many researchers are working on privacy tools for web browsers, increasing privacy and security in web browsers and also on private browsing mode and its vulnerabilities, third-party cookies, its use and how to prevent them from getting tracked, etc.

Private browsing is a security control tool used by browsers that aim to protect users' data which is generated during a private browsing session by not storing them in the file system [1,7]. In [1], author examines the protection that is offered by the private browsing mode of the most popular browsers on Windows. It is a common misconception that private browsing modes can protect users from being tracked by other websites or their ISP [9]. This paper discovers occasions in which even if users browse the web in a private session, privacy violations exists contrary to what is documented by the browsers. To raise the bar of privacy protection offered by web browsers, the author proposes the use of virtual filesystem as the storage medium for browsers' cache data [1].

In [7,15] and [16], authors did a study on forensic analysis of the private browsing mode of common web browsers. Many private browsing modes are designed to be 'locally private', preventing data denoting a user's browsing actions from being stored on their device. In [7], author examines nearly 30 web browsers to determine the presence of a 'private mode', and where available, the 'privateness' of said mode. The paper concludes that out of 30 browsers 5 leaked users' data, namely, Avant, Comodo Dragon, Edge, Epic, and Internet Explorer. This happens in two ways; a flaw in browser design and development leading to data being leaked outwards from within, and an operating system is taking more control over the browser than it should, leading to data being extracted from within. In [16], author conducted a digital forensic examination, to determine the recoverable artifacts of three enhanced privacy web browsers i.e., Doodle, Comodo Dragon, and Epic, and three commonly used web browsers i.e., Chrome, Edge and Firefox, in private browsing mode. For the comparisons of recoverable artifacts, the author used two digital forensic tools namely, Forensic Toolkit imager lite (FTK) and Autopsy, commonly used by law enforcement. The author proposes FTK recovered more artifacts than Autopsy [16].

In [14], author provide an insight into the data exchange done by different web browsers with their back-end servers and also identifying what kind of identifiers are used or sent to back-end servers while browsing. This work aims to assess the privacy risks associated with this data exchange between web browsers and their back-end servers. The author proposes that Chrome, Firefox, and Edge share details of web pages visited with backend servers, and they also send long-lived identifiers such as IP addresses, that can be used to link connections together and potentially allow tracking over time. The author also proposes that Brave doesn't use any identifiers allowing tracking of IP address over time, and it doesn't send any details to its backend server [14].

A common practice for websites to rely on services provided by third-party sites is to make websites easy and fast to access. An attacker can use this information for the reconstruction of personal profiles and misuse those data. To avoid this, [17] and [18] proposes the use of a machine learning (ML) technique for classifying websites as trackers or non-trackers. While, examining the websites authors found that length of malicious URLs is shorter than the benign one, also numbers of JavaScript tags are more in benign than of malicious website. In [17], author used a hybrid mechanism that is, blacklisting and machine learning (ML) for the automatic identification of privacy-intrusive services while browsing web pages. The author makes use of the blacklisting technique in combination with the machine learning model, which distinguishes between malicious and functional resources and hence updates the blacklists. They were able to classify JavaScript programs with an accuracy of 91% and HTTP requests with 97% accuracy. The author proposes, by using this hybrid mechanism called GuardOne, that it is possible to filter out malicious resources from users' request without performance degradation [17].

Similarly, in [18] also, author assumed that the most effective defense against the third-party trackers is blocking them, and as a part of blocking, the blacklist is manually created and is difficult to maintain. This work proposes an effective system with high accuracy, called TrackerDetector to detect third-party trackers automatically. The behaviors of trackers and non-trackers are different; thus, an incremental classifier is trained from JavaScript files crawled from a large number of websites to detect whether a website is a third-party tracker. A high accuracy of 97.34% is obtained with training data set and that of 93.56% is obtained within 10-fold cross-validation [18].

3   Discussion

The scope of this analysis includes the popular web browsers for Windows such as Chrome, Edge, Firefox, Brave, Comodo Dragon, Internet Explorer, TOR, etc. This study not only talks about the privacy protections in the private browsing mode of different desktop browsers, but also to detect third-party tracking and avoid it.

3.1   Private Browsing Mode

Private browsing is a privacy feature in web browsers; when operating in such modes, the browser creates a different temporary session which is isolated from the browser's main session by using sandboxing techniques. There is a possibility that identifiable traces of activity could be leaked from private browsing sessions utilizing the operating system, security flaws in the browser, or via malicious browser extensions. It is a common misconception that private browsing modes can protect users from being tracked by other websites or their Internet service provider [9]. It is important to note that; this can be done by the transmission of user data to backend servers [14]. This data can be found out by using test setup such as starting the browser from fresh install/new user profile, pasting a URL into the browser to search bar, press enter and recording the network activity, closing and restarting the browser, and recording the network activity during both events and lastly, starting the browser from fresh install/new user profile and pasting a URL into the browser to search bar and keeping it untouched for next 24 h [14]. In this process, privacy evaluation of backend servers is also done.

Safe browsing API is one of the techniques used by almost all browsers, to maintain and update a list of web pages or URLs associated with phishing and malware attacks [19]. There is a second potential privacy issue associated with the use of this service. Since requests carry the IP address as a value to a parameter, then linking requests together would allow tracking of web browsing behavior and session on cross-site, cross-browser user tracking. This work identified that browsers typically contact the Safe Browsing API roughly every 30 min to request updates. Similarly, the chrome extension update API is used to check for updates in chromium-based browsers. In each round of update checking, it typically generates multiple requests to the update server. The header of this request contains cup2key and cup2hreq values [14]. If any of these values are dictated by the server, then they can potentially be used to link requests by the same browser instance together over time, and hence link the device IP addresses over time. Public documentation for this chrome extension updates API is lacking, but inspection of the Chromium source provides some insight. Firstly, the cup2key value consists of a version number before the colon and after a colon, a new random value is generated for each request [14]. So, IP address which is used by chrome extension update API may result in the user getting tracked.

For checking of private browsing sessions, i.e., to identify whether any traces of the user's online activities remain after a private session; each browser was executed in private mode and online activities were performed. More specifically, we assume that the user in a private session performs online activities that will create the artifacts. For example, when he visits a website, bookmark it, and downloads a file, then those actions will create artifacts regarding bookmarking, browser cache memory, browsing history, cookies, download list, and downloaded files [1]. The online activities that were performed aimed to create data that is typically left behind during a normal browsing session. More specifically the artifacts that were generated can be classified as (a) generic artifacts, which is a set similar to the set of protected data sources compiled from the browser's documentation pages and includes simple browsing activities (e.g., bookmarking a webpage, downloading a file, etc.), (b) browser artifacts, which describe changes in the browser itself (e.g., installing a digital certificate, modifying browser settings) and (c) website artifacts, which include per site configurations, such as website translation or website zoom [20].

Similarly, in [16], author used this method to monitor the changes made in the filesystem, when a user does a web browsing session in private browsing mode. Whenever user browse in normal browsing mode, different folders or files and cache data is created and this data can be saved in the filesystem of the local device and can be used for later use, like cookies can be used later to load the websites faster if a user visits that website again. But when users are using the private mode then these changes must not happen in the filesystem, as occurring of this change shows that even in private mode, browsers are saving user data and this is dangerous. For the monitoring of these changes, FTK imager and Autopsy are used, and this is done by data population method [7]. The data population is a data acquisition method used to collect data related to visiting websites, creating bookmarks, searching words, sending e-mails and messages, and surfing videos and images, simply saying the data population method was designed to create browser artifacts. For this data acquisition method, the top ten websites from a list of the top 500 most viewed sites in the USA as ranked in a 2015 Alexa poll (Alexa, 2016) [7].

According to StatCounter, in last 12 months Google Chrome has highest usage share with 66.6% while Apple Safari has 9.56% share, Microsoft Edge has 9.22% share, Mozilla Firefox has 8.49% share and Brave has lowest usage share with 1.02% [21]. Although Brave browser has the lowest usage share, but lately it is gaining so much popularity and as of December 2021, Brave has more than 50 million active monthly users [22]. So, we decided to use all these desktop browsers, namely Google Chrome, Mozilla Firefox, Microsoft Edge, Apple Safari, and Brave and did a comparison of private browsing modes in these browsers. During this comparison we found out that:

1. In Firefox and Safari, users can open 'Preferences/Settings' page in private mode.
2. Data in public mode like history, download list, bookmarks, password database, etc. are available in private mode even after users close their sessions in public mode. Such information could allow websites to link users' sessions in public mode to those in private mode, as shown in Table 2.
3. All browsers delete history, cookies, password database, and auto-completed forms in private mode after the user closes private sessions as shown in Table 3.
4. Firefox provides Mozilla VPN in private mode, while other browsers do not provide any VPN service. This service is provided in Europe, New Zealand, and the US. But it is yet to be introduced in India.
5. Only Desktop Brave browser provides two types of private browsing mode. The first is 'New private window' which uses 'Google' as search engine, and the second 'New private window with TOR' which uses 'DuckDuckGo' as search engine with TOR network.

Table 2. Data in public mode which is accessible in private mode

Data	Chrome	Firefox	Safari	Edge	Brave
History list	No	Yes	Yes	No	No
Cookies	No	No	No	No	No
Download list	Yes	Yes	Yes	Yes	Yes
Bookmarks	Yes	Yes	Yes	Yes	Yes
Passwords	Yes	Yes	Yes	Yes	Yes
Form autocomplete	Yes	Yes	Yes	Yes	Yes

3.2 Third-Party Tracking

A cookie is a small piece of text sent to a web browser by a website user visits [23]. When a user visits a website, first-party cookies are directly stored by the website (or domain) visited [24]. These cookies allow website owners to collect analytics data, remember language settings, and perform other useful functions that provide a good user experience [12]. While third-party cookies are created by domains that are not the website (or domain) that you are visiting. These are usually used for online advertising purposes and placed on a website through a

Table 3. Data in Private mode which is accessible even after Private mode ends/terminates

Data	Chrome	Firefox	Safari	Edge	Brave
History list	No	No	No	No	No
Cookies	No	No	No	No	No
Download list	No	No	No	No	No
Bookmarks	Yes	Yes	Yes	Yes	Yes
Passwords	No	No	No	No	No
Download item	Yes	Yes	Yes	Yes	Yes
Form autocomplete	No	No	No	No	No

script or tag or tracking image. A third-party cookie is accessible on any website that loads the third-party server's code [25]. Simply put, a first-party cookie is set by the publisher's web server or any JavaScript loaded on the website. A third-party cookie can be set by a third-party server, or via code loaded on the publisher's website [26]. For example, Facebook is a first-party website, and it is also a third-party website for it can embed a 'Like' button into other websites [18]. However, all third-party websites don't track users; some are used to provide faster and efficient user experience while some are used to track users, and this tracking third-party websites are called as third-party trackers and these are harmful websites.

The goal of work in [17] is to design a machine learning approach, GuardOne, that allows us to provide a valid and effective instrument to protect privacy without performance degradation with respect to the existing privacy tools that exclusively rely on blacklisting mechanisms. The idea is to verify whether machine learning can be efficiently employed in the real-time application of filtering unwanted content, without impacting on the user experience and also want to overcome the problems of having manually update blacklists and reducing the overhead associated with the use of a high number of regular expressions to apply filtering mechanisms. To collect raw data for the machine learning model, they used labelling provided by Ghostery and Disconnect for JavaScript programs. For HTTP requests, they developed an auxiliary add-on for Chrome named WatchHTTP, to log HTTP headers for each issued web request and to indicate if that request is a first-party one or a third-party one. Finally, developed a crawler based on the Selenium webdriver and crawled the Alexa top 8000 sites, randomly following 3 links on each page, for a total amount of 32,000 downloaded websites. Crawling has been performed in parallel, by envisioning three different instances of Selenium webdriver, one for HTTP requests and two for JavaScript programs [17].

In [18], author worked on the third-party tracker detection system called TrackerDetector. For that purpose, they collected data on a large number of websites on whether those sites are trackers or not; whether any first-party site is embedded with a third-party website or not. And all this data is collected from Ghostery. For machine learning modelling, feature extraction is done. While training and classification of the model, model contains two parts; a manual label and incremental learning. Manual labelling consists of classifying code by using the following aspects: i) Stateful tracking: Whether the code contains tracking, HTTP Cookies, Local Storage, or Etag to record information. ii) Stateless tracking: Whether the code calls JavaScript APIs on the window, navigator, screen objects, font family, and soon, to obtain fingerprint information and send this information back through the 'document.write' function, URL parameters, or web bugs. iii) Sharing tracking id with other trackers. iv) Loading the JavaScript code, iframes, or advertising contents of trackers. If the code loads JavaScript code or iframes, we also check whether the loading content satisfies any of the above rules [18]. If the code satisfies even one of the above rules, we label it as a tracker; otherwise non-tracker.

GuardOne and TrackerDetector both make use of different ML models like Random Forest (RF), multilayer perceptron (MLP), support vector machine (SVM), k-nearest neighbor, linear discriminant analysis (LDA), BTree, etc. After using these ML models, the third-party tracker detection systems were successfully able to classify sources/websites as trackers or non-tracker. GuardOne and TrackerDetector classified websites with an accuracy upto 97% and 98%, respectively.

4   Result

All common web browsers such as Chrome, Firefox, Safari, Brave and Edge are equipped with private browsing mode. The artifacts created during private browsing sessions are not erased completely by the browsers. Some data can be retrieved by using some tools, while some data can be directly found in the filesystem of a browser. Basically, the data which should be deleted after terminating a private session can be found on the filesystem even after the session is terminated. So to avoid this, the study proposes the use of a virtual filesystem. The virtual filesystem is stored in volatile memory i.e., RAM instead of a hard disk. So, when someone terminates a private session, it's all data will get erased as soon as the private session is terminated.

For private browsing, it is found that some browsers provide adequate information such as, Firefox and Explorer, while some don't, such as Chrome. Firefox and Brave do not document what happens to passwords created and used in private sessions. Also, there is no similarity in what data is to be protected. Each browser tries to protect different artifacts and documentation, says all. The tools like FTK and Autopsy show that artifacts like history, cookies and bookmarks are recoverable. Some of these artifacts contain all the data which users input while some artifacts when recovered do not contain data, but that data can be easily recovered by using some tools.

In our comparison of private browsing mode, we found that data in public mode like history, download list, bookmarks, password database, etc. are available in private mode even after users close their sessions in public mode. Such information could allow websites to link users' sessions in public mode to those in private mode for a cross-site, cross-browser user tracking as shown in Table 2. All browsers delete history, cookies, password database, and auto-completed forms in private mode after the user closes private sessions as shown in Table 3. Only Firefox provides VPN in private mode, while other browsers do not provide any VPN service. And we also found that in the default settings of web browsers, third-party cookies are not blocked by default, but they provide that option to the user. So the user has to start that manually. Like in Chrome default setting provided is Standard Protection, but user can change it to Enhanced Protection to avoid third-party cookies and trackers, similarly for Edge and Firefox user can use Strict tracking prevention to avoid tracking.

While in case of third-party tracker detection by using machine learning algorithms, it is found that Guardone gives better results and also, it does not hamper web page rendering while effectively accomplishing filtering task. As evaluation metrics for ML models, confusion matrix/error matrix is used. It is particularly used for binary classification problems. It consists of True Positive (TP) samples, True Negative (TN) samples, False Positive (FP) samples and False Negative (FN) samples. Here TP, TN means the output of ML model and original value/target value is the same, while FP and FN means both output and original value/target value are opposite and different. So, by using this values accuracy, precision, false positive rate and false negative rate of the model can be predicted. This matrix can also be used to calculate the area under ROC curve (AUC) score.

GuardOne makes use of different models such as RF, MLP, SVM, k-nearest neighbor, LDA, etc. These models classify third-party tracking with the accuracy of 97%, 87.17%, 86.76%, 95.20%, 75.39% respectively, as shown in Table 4. Whereas, TrackerDetector uses the process of incremental learning of classifiers, that is, trying different ML algorithms and selecting one which gives higher accuracy for both training sets and for a 10-fold cross-validation set. TrackerDetector uses Naïve Bayes, LR, RF and BTree for incremental learning of classifiers. Based on all this data, BTree is finalized as final model, as its training and classification model and has an accuracy up to 98.70% with training set and 97.10% with 10-fold cross validation as shown in Table 5.

Table 4. Different ML models used and their result

Classifier models	Accuracy obtained in JavaScript Programs (%)	Accuracy obtained in HTTP Programs (%)
Random Forest	90.60	97
MLP	87.17	87.06
SVM	84.20	86.76
k-nearest neighbor	85.19	95.20
LDA	75.39	74.50

Table 5. Different ML models used and their accuracy

Classifier models	Accuracy obtained in training set (%)	Accuracy obtained in 10-fold validation set
Naive Bayes	62.15	61.85
Logistic Regression	100	82.95
Random Forest	100	92.60
BTree	98.70	97.10

5   Conclusion

In this paper, we have done a survey of different research papers related to web browser privacy and security and found that all the web browsers make use of safe browsing services to protect the user from phishing attacks. But this method has few privacy concerns. Similarly, for chrome update extension, it also contacts chrome update API every 30 min and the documentation for this is lacking. Also, common web browsers send details of a web page visited by a user and user details to their respective back-end servers in connections made to back-end servers by browsers. This can all be disabled, but this is enabled by default. This work identifies whether the documentation of each browser is either inadequate or inconsistent, as artifacts that were documented to be deleted after the private session were found. So, there is a possibility of increasing privacy in private modes of browsers.

We found that in browsers except Safari and Edge, users can directly start private mode, but for Safari and Edge, users have to start a private mode from existing public mode sessions. All browsers except Safari allow pages from the same website in different tabs to share cookies. Therefore, only Safari prevents a web attacker from linking sessions in different tabs using cookies. Similarly, history lists, bookmarks, page autocomplete, etc., in public mode are available in private mode even after users close their sessions in public mode. Such information could allow websites to link users' sessions in public mode to those in private mode. It is also found that 'Do Not Track' feature of web browsers is not that powerful; even after using that feature websites are able to track the users. So, we can conclude that web browsers are providing privacy and security, but it are not up to the mark.

References

1. Tsalis, N., Mylonas, A., Nisioti, A., Gritzalis, D., Katos, V.: Exploring the protection of private browsing in desktop browsers. Comput. Secur. **67**, 181–197 (2017)
2. Virvilis, N., Mylonas, A., Tsalis, N., Gritzalis, D.: Security busters: web browser security vs. rogue sites. Comput. Secur. **52**, 90–105 (2015)
3. Bouguettaya, A., Rezgui, A., Eltoweissy, M.: Privacy on the web: facts, challenges, and solutions. IEEE Secur. Priv. **1**(6), 40–49 (2003)

4. Mimecast: What is web security? (2021). https://www.mimecast.com/content/web-security/

5. Indian Computer Emergency Response Team (2021). https://www.cert-in.org.in/s2cMainServlet?pageid=PUBANULREPRT

6. The Hindu: More than 6.07 lakh cyber security incidents observed till June 2021: Government (2021). https://www.thehindu.com/business/cert-in-observed-more-than-607-lakh.-cyber-security-incidents-till-june-2021-government/article35726974.ece

7. Horsman, G., et al.: A forensic examination of web browser privacy-modes. Forensic Sci. Int. Rep. **1**, 100036 (2019)

8. Korniotakis, J., Papadopoulos, P., Markatos, E.P.: Beyond black and white: combining the benefits of regular and incognito browsing modes. In: ICETE, pp. 192–200 (2020)

9. Private browsing (2018). https://en.wikipedia.org/wiki/Private_browsing

10. CodeDocs: Private browsing (2021). https://codedocs.org/what-is/private-browsing

11. Kerschbaumer, C., Crouch, L., Ritter, T., Vyas, T.: Can we build a privacy-preserving web browser we all deserve? XRDS: crossroads. ACM Mag. Stud. **24**(4), 40–44 (2018)

12. Cookie Script: All you need to know about third-party cookies (2021). https://cookie-script.com/all-you-need-to-know-about-third-party-cookies.html

13. The end of third-party cookies and what to focus on now (2021). https://www.mightyroar.com/blog/third-party-cookies

14. Leith, D.J.: Web browser privacy: what do browsers say when they phone home? IEEE Access **9**, 41615–41627 (2021)

15. Jadoon, A.K., Iqbal, W., Amjad, M.F., Afzal, H., Bangash, Y.A.: Forensic analysis of Tor browser: a case study for privacy and anonymity on the web. Comput. Secur. **299**, 59–73 (2019)

16. Gabet, R.M., Seigfried-Spellar, K.C., Rogers, M.K.: A comparative forensic analysis of privacy enhanced web browsers and private browsing modes of common web browsers. Int. J. Electron. Secur. Digit. Forensics **10**(4), 356–371 (2018)

17. Cozza, F., et al.: Hybrid and lightweight detection of third party tracking: design, implementation, and evaluation. Comput. Netw. **167**, 106993 (2020)

18. Wu, Q., Liu, Q., Zhang, Y., Wen, G.: TrackerDetector: a system to detect third-party trackers through machine learning. Comput. Netw. **91**, 164–173 (2015)

19. Google safe browsing (2021). https://en.wikipedia.org/wiki/Google_Safe_Browsing

20. Kim, H., Kim, I.S., Kim, K.: AIBFT: artificial intelligence browser forensic toolkit. Forensic Sci. Int. Digit. Invest. **36**, 301091 (2021)

21. StatCounter GlobalStats: Desktop browser market share worldwide (2021). https://gs.statcounter.com/browser-market-share/desktop/worldwide

22. Brave (web browser) (2022). https://en.wikipedia.org/wiki/Brave_(web_browser)

23. Cookie policy - Arad Group (2021). https://arad.co.il/app/uploads/Cookie-policy.pdf

24. Malandrino, D., Scarano, V.: Privacy leakage on the web: diffusion and countermeasures. Comput. Secur. **57**(14), 2833–2855 (2013)

25. Mazel, J., Garnier, R., Fukuda, K.: A comparison of web privacy protection techniques. Comput. Commun. **144**, 162–174 (2019)

26. CookiePro: What's the difference between first and third-party cookies? (2021). https://www.cookiepro.com/knowledge/whats-the-difference-between-first-and.-third-party-cookies/

Blockchain Search Using Searchable Encryption Based on Elliptic Curves

Marius Iulian Mihailescu[1] and Stefania Loredana Nita[2(✉)]

[1] Spiru Haret University, Scientific Research Center in Mathematics and Computer Science, Bucharest, Romania
m.mihailescu.mi@spiruharet.ro
[2] Military Technical Academy, Department of Computers and Cyber Security, Bucharest, Romania
stefania.nita@mta.ro

Abstract. Nowadays there are more key technologies, such as cloud computing, big data, blockchain, and different techniques that process the data via a well-established purpose. The security of data should be a significant aspect of any type of system, therefore the systems should implement security techniques that solve this aspect. A relatively new encryption technique is searchable encryption, which enables the user to submit search queries based on chosen criteria (for example, keywords) to the cloud server on which the encrypted data (in the form of encrypted documents) is stored. Another emerging technology is blockchain, a decentralized approach for data sharing, based on a direct communication between two peers. To protect the data and its integrity, blockchain uses cryptographic mechanisms and strong mathematics computations as challenges that need to be solved by the nodes that adheres to the blockchain network. In this paper, we propose a searchable encryption scheme, which can be used in the cloud environment, based on elliptic curves and symmetric bilinear mapping for its algorithms and blockchain to store the indexes structure and to process the search query. Using elliptic curves, the proposed scheme is efficient, as it is proven in the literature that elliptic curve cryptography is usually faster than other encryption systems with the same level of security.

1 Introduction

In the last decade, cloud computing was adopted by an increasing number of companies, becoming today an important part for many enterprises, even if it is about storing some documents on the cloud or working with complex services such as platforms or infrastructures. A recent report from 2021 [1] shows that the interest for the infrastructure as a service (IaaS) and platform as a service (PaaS) has drastically increased, taking into consideration the pandemic pressure for moving as many as possible processes in the digital world. It is predicted that 40% of the computation work of enterprises will be moved to the cloud until 2023, comparing with 20% in 2020, and moreover, the cloud providers will embrace Edge computing in their services [2]. A solution for the security concerns of the cloud environment can be searchable encryption. This is an encryption

L. Barolli et al. (Eds.): AINA 2022, LNNS 451, pp. 471–481, 2022.
https://doi.org/10.1007/978-3-030-99619-2_45

technique that allows the users to submit search queries based on pre-defined criteria to a cloud server that performs the search process directly on the encrypted data.

Another technology that has drawn attention lately is blockchain. It was introduced firstly in [3] as a solution for electronic coins. Although the first application of blockchain was in the financial field (more specifically, for cryptocurrencies), it can be applied in numerous other domains such as law (distributed notary), trade, supply chain, healthcare, etc. A key characteristic of blockchain (that can be seen a distributed ledger) is decentralization, where the network uses direct communication between each two peers (or nodes) of the network (called peer-to-peer communication). When a node tries joining the network it should prove some specific things depending on the approach used by the blockchain network. There are two common approaches based on which a network accepts a node: proof-of-work (PoW) and proof-of-stake (PoS). PoW requires that the node that adheres to the network to solve a mathematical function of high difficulty, in this way showing that it owns required computational capabilities, therefore PoW is a zero-knowledge protocol. PoS is a consensus technique that allows a peer to validate transactions based on the number of coins that owns [4]. In general a blockchain network starts with a PoW approach, and at some point, it migrates to PoS for providing an even stronger security. The data in the blockchain is stored into blocks that are chained together and when a new block is added to the chain, every node record this transaction, therefore, a copy of the chain is stored on each node of the network. There are situations in which multiple nodes validate a transaction at the same time, and here the question would raise: which of the two transactions is recorded into the chain? Of course, such situations are treated in the blockchain networks, usually resulting in more branches of the main chain.

From the blockchain description, it can be concluded that this technology has more advantages. One of the most important of them is *decentralization*, which eliminates the need for a central /trusted authority, because when the transaction is initiated between any two nodes, it is not required to be authenticated by a central/trusted authority. Another advantage of blockchain is *persistency*, because a transaction must be confirmed and registered by the nodes of the network, a fact that makes it almost impossible to falsify. On the list of advantages *anonymity* and *audibility* can be added. Eliminating the need for a central/trusted authority, sensitive data of users are not required anymore, therefore the network generates an address that represents the identity when a user asks to join the network. Further, when a preceding record needs to be reviewed, it is an easy process, because timestamps are used when a transaction is completed, which makes all records traceable.

In this paper, we present a public-key searchable encryption scheme that uses symmetric bilinear pairing for elliptic curves. These primitives are used in all algorithms of the searchable encryption scheme. The proposed system is based on the system from [7], in which the authors use a similar approach, involving bilinear pairing between two multiplicative cycle groups, while our scheme uses a symmetric bilinear pairing within an additive group of an elliptic curve. The motivation for choosing elliptic curves lies in the fact that usually, these are faster than other encryption systems with the same level of security. For example, in [8], there is a comparison of elliptic curve cryptography (ECC) and RSA in which is proved that ECC is faster. The indexes for the encrypted documents within the proposed system are stored in a blockchain network, while the encrypted documents are stored on a cloud server. The trapdoor process triggered by the data user is submitted to the blockchain network as a transaction, and then mainly the blockchain will run the search process. However, the blockchain network will need a validation from the cloud server during the search. Our scheme is based on the discrete logarithm problem and we will prove that it is secure against keyword guessing attack (KGA), ensuring the indistinguishability of the trapdoors and indexes. Moreover, we will show that the scheme has security properties due to using blockchain. The remaining of the paper is organized as follows: in the next section (Sect. 2), we present more deeply the technologies and the tools that we will use in our proposed security mechanism, and also we will present some related work. Further (Sect. 3), we will describe our solution and discuss the security and performance. Lastly, we will present the conclusion of the paper and will provide further research directions (Sect. 4).

2 Background and Preliminaries

In this section, we present briefly the main concepts used further in the paper and notable results in each research topic.

2.1 Elliptic Curves and Symmetric Billinear Pairing

The equation from below gives an elliptic curve E which is defined over a finite field G:

$$y^2 = x^3 + ax + b, \tag{1}$$

In Eq. (1) the coefficients a, b are constant values that must fulfill the following condition: $\Delta \neq 0$, where Δ is defined as $\Delta = 4a^3 + 27b^2$. The restriction for delta being different of 0 it is ensured that there are no multiple roots.

There is an additional important point contained by the elliptic curve, called the *point at infinity* (∞). Having these components, the elliptic curve E defined over G can be expresses as:

$$E(G) = \{\infty\} \cup \{(x, y) \in G \times G | y^2 = ax^3 + b\} \tag{2}$$

The set $E(G)$ accepts an operation with points, namely *addition*, which is defined and described in details in [5]. With this operation, $(E(G), +)$ becomes a group, with the identity (∞). The hardness assumption used in cryptography regarding the elliptic curves is the *discrete logarithm problem* (DLP), which states that given a point P and a point Q in the form $Q = kP$, it is hard to find the integer k that satisfies the preceding equality. Here the term "hard" means that the problem cannot be solved in polynomial time regarding something (usually the security parameter).

Given an additive cyclic group G, whose order is the prime number q, a *symmetric bilinear pairing* represents a map from $G \times G$ to a group G_T, $e : G \times G \to G_T$ (with $(G_T, \cdot)$ a cyclic group with order q), which meets the following requirements:

- **(bilinearity)** $\forall a, b \in \mathbb{F}_q^*, \forall P, Q \in G$ we have $e(aP, bQ) = e(P, Q)^{ab}$

- **(nondegeneracy)** $e(P,P) \neq 1$, where $1 \in G_T$ is the neutral element of G_T
- **(computability)** $\forall P, Q \in G \exists A()$ such that e is easy computable. Here $A()$ is a polynomial-time algorithm.

To secure the proposed scheme we will use the settings for the parameters as they are described in [6]. In this work, the authors provide a comprehensive analysis of the implications of using bilinear maps (Weil pairing and Tate pairing) for elliptic curves in cryptography.

According [6], to achieve a high level of security, the curve E should be defined over a finite field of the form $\mathbb{F}_{p^k}$, where p is a large prime and k represents the embedding degree of the curve. Here $\mathbb{F}_{p^k}$ must be large enough to make the discrete logarithm problem hard to compute. For a basepoint P its order expressed as a prime number n should be also large enough to resist a Pollard attack launched for the DLP in the group $\langle P \rangle$. For example, to achieve a security level of 192 bits, the minimum number of bits for n should be 384 and for p^k the minimum number of bits should be 8192.

2.2 Searchable Encryption

As it is mentioned above, searchable encryption (SE) is an encryption technique that enables a user to submit search queries based on some chosen criteria. Note that searchable encryption follows two directions: (1) the queries are submitted for encrypted databases [9, 10] and (2) the queries are submitted on a server to search for encrypted documents [11, 12] that meet the search criteria. In this paper, we use the second approach.

Regarding the number of keys, there are two types of SE schemes: *Symmetric Searchable Encryption* (SSE), in which a single private key (k_{prv}) is used for both encryption and decryption, and *Public-key Searchable Encryption* (PKSE), in which the public key (k_{pub}) is used for encryption and the private key (k_{prv}) is used for decryption. SSE was introduced in [13] and since then important progress has been made. For example, [14] represents an improvement of [13] securing the indexes of the documents, while [15] introduced inverted indexes that provided stronger security and introduced some technical aspects that can be seen as a standard for this research topic. Recent works are focusing on dynamic SSE, a variant that allows updates for keywords or documents directly on the server [16, 17]. On the other hand, PKSE was introduced in [18], under the name Public Key Encryption with Keyword Search (PEKS) derived from an identity-based encryption (IBE) scheme. In [19], the authors designed the proposed scheme to enable search operation based on multiple keywords once. Recent works introduced an additional phase for the keywords, namely authentication [20–22]. Here, the keywords that describe the documents are encrypted and authenticated by the data owner, resulting in resistance against keyword guessing attacks (KGA). In general, a searchable encryption scheme has the following algorithms: key generation, build index, encryption, trapdoor, search, decryption. All these algorithms will be explained in details in Sect. 3.

2.3   Blockchain

An important element of a blockchain structure is the *smart contract*. The smart contract was introduced in [23] based on protocols that were securing the relationships between the public networks. The advantages of smart contracts are in the first place trust and transparency resulted from the disappearance of a third party and the recording of all encrypted transactions in all nodes within the network. In the blockchain, these are included from the beginning and describe rules that must be followed when during all processes in the network. When the smart contract is defined, it cannot be modified, therefore, in this way, trust is built. It is actually a computer program that makes sure the rules are followed and it has not legal aspects.

The proposed scheme is uses a PoW approach of blockchain network (as presented in [3]), but it can be transitioned to a PoS approach when it reaches a proper amount of blocks in the chain. Therefore, a block of the chain will contain the following elements:

- The hash of the previous block;
- The nonce resulted through computations in the PoW process;
- The timestamp on which the block has been published into the network.

3   The Proposed Solution

In this section, we present the algorithms from the proposed scheme and then, we will discuss the security of our solution. A general view of the system is presented in Fig. 1.

The entities involved in the proposed SE system are:

- The *data owner* (Usr_O) possesses the data, encrypts it, and sends it to be stored on a cloud server.
- The *data user* (Usr_U) initiates the search process. For this, firstly, chooses the keyword based on which the search will be made, then computes a value called trapdoor, and finally, the trapdoor is sent to the server. Another competence for the data user is decrypting the encrypted data sent by the server.
- The *server* (Srv) stores the encrypted documents. Other important things that are made by the server are performing the search process and sending the result to the data user.
- The *trusted authority* (TA) in general it is an entity that generates the pair of public and private keys for the previous participants.
- The *blockchain network* (BN) that stores the index structure and performs a part of the search process.

Note that a data owner may be a data user.

3.1   The Searchable Encryption Scheme

Our scheme in composed from the algorithms from below.

Setup $(\lambda) \to PS$. *TA* runs this algorithm in order to generate the parameters of the system. *Input*: λ - the security parameter. *Output*: $PS = (E(\mathbb{F}_{p^k}), G, \text{hf}, \text{Hf})$ - the system parameters, where $E(\mathbb{F}_p)$ is the elliptic curve defined over $\mathbb{F}_p$ and $G \in \mathbb{F}_p$

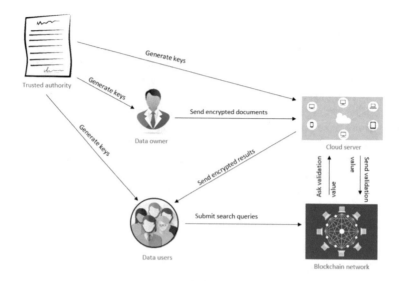

Fig. 1. The system for the proposed SE scheme.

is a generator point of $E(\mathbb{F}_p)$ of order n, $hf : \{0,1\}^* \to \mathbb{F}_{p^k}$ is a hash function and $Hf : \{0,1\}^* \times \{0,1\}^* \to \{0,1\}^*$ is also a hash function.

KeyGeneration $(PS) \to ((PK_{Usr_O}, SK_{Usr_O}), (PK_{Usr_U}, SK_{Usr_U}), (PK_{Srv}, SK_{Srv}))$. TA runs this algorithm from which the public and the private keys for Usr_O, Usr_U and Srv are generated. *Input:* PS - the parameters of the system.

Output: $((PK_{Usr_O}, SK_{Usr_O}), (PK_{Usr_U}, SK_{Usr_U}), (PK_{Srv}, SK_{Srv}))$ - the pairs of the public key PK and secret key SK for Usr_O, Usr_U and Srv, which are computed as follows:

(a) Usr_O: picks an integer $a \leq \#E(\mathbb{F}_{p^k})$ and computes the point $P = aG$, then $PK_{Usr_O} = P$ and $SK_{Usr_O} = a$.

(b) Usr_U: picks an integer $b \leq \#E(\mathbb{F}_{p^k})$ and computes the point $Q = bG$; the public key of the data user has two components, namely $PK_{Usr_U 1} = e(G,G)^{1/b}$ and $PK_{Usr_U 2} = Q$; therefore, $PK_{Usr_U} = (PK_{Usr_U 1}, PK_{Usr_U 2})$ $SK_{Usr_O} = b$.

(c) Srv: picks an integer $c \leq \#E(\mathbb{F}_{p^k})$ and computes the point $R = cG$, then $PK_{Srv} = R$ and $SK_{Srv} = c$.

BuildIndex $(PS, PK_{Usr_O}) \to Idx$. Usr_O runs this algorithm through which the indexes are encrypted and then stored in an index structure. *Input:* PS, PK_{Usr_O}, Kw - the parameters of the system PS, the Usr_O's public key PK_{Usr_O}, and the set of keywords $Kw = \{kw_1, \ldots, kw_n\}$. *Output:* Idx - the structure of the encrypted indexes. In order to construct the index structure based on each keyword $w_j \in W = \{w_1, \cdots, w_\gamma\}$, where γ is the total number of the keywords used in the index structure, he/she proceed as follows:

(a) Usr_O picks randomly a value $r \in \mathbb{F}_{p^k}$ then computes the following components of the index Idx_i for the current keyword w_j:

(b) $Idx_{i0} = e(G,G)^r$; $Idx_{i1} = rPK_{Usr_U 2} = rQ = rbG$; $Idx_{i2} = e(G,G)^{arb^{-1}}$; $Idx_{ij} = rhf(w_j)G$

(c) The index structure Idx contains all tuples of the form $(Idx_{i0}, Idx_{i1}, Idx_{i2}, Idx_{ij})$, where j is the index of the current keyword w_j. The index structure is sent to the smart contract to be registered.

(d) Usr_O uses a secret sharing scheme to send the value r to Usr_U.

Encryption has more steps. This algorithm is called by the Usr_O to encrypt the documents and send them to the server cloud Srv. *Input*: $PS, PK_{Usr_U}, SK_{Usr_O}, D$: the system parameters PS, data user's public key PK_{Usr_U}, data owner's private key SK_{Usr_O}, and the set of the documents D. *Output*: C - the set of the encrypted documents. Firstly consider $D = \{d_1, \cdots, d_n\}$ the set of the documents that Usr_O owns and wants to encrypt.

(a) Usr_O computes the following value:
$K_{Enc} = SK_{Usr_O} PK_{Usr_U} 2 = aQ = abG$. This represents the encryption key used to encrypt the documents.
(b) Usr_O uses a symmetric encryption mechanism based on elliptic curves to encrypt the documents $C_i = Encryption_{K_{Enc}}(d_i), \forall i \in \{1, \cdots n\}$, and the index of each document $c_i = Encryption_{K_{Enc}}(i), \forall i \in \{1, \cdots n\}$. After obtaining each pair, Usr_O applies the hash function Hf, over it, obtaining $C_i' = Hf(C_i, c_i)$ then C_i' is sent to Srv to be stored. The last step regarding documents' encryption is that Usr_O sends the key K_{Enc} using a secret sharing scheme to Usr_O that will use it to decrypt the documents resulted from the search process.

Trapdoor$(PK_{Srv}, PK_{Usr_O}, Qw) \rightarrow T_{Qw}$. Usr_U runs this algorithm when wants to submit search queries to Srv, based on a set of keywords. *Input*: PK_{Srv}, PK_{Usr_O}, Qw - the public key of the server PK_{Srv}, the public key of the data owner PK_{Usr_O}, and the set of query keywords Qw. *Output*: T_{Qw_1}, T_{Qw_2} - the computed trapdoor has two components $T_{Qw} = (T_{Qw_1}, T_{Qw_2})$. To receive the documents that matches query criteria $Qw = \{qw_1, \cdots, qw_\gamma\}$, Usr_U proceeds needs to compute the trapdoor based on these criteria and proceeds as follows for every query keyword qw_i:

(a) Usr_U picks randomly an integer $x \in \mathbb{F}_{p^k}$ and sets $T_{Qw_1} = x$.
(b) Usr_U computes the value $T_{Qw_2} = r[(-xPK_{Srv} + b^{-1}PK_{Usr_O}) - (b - hf(qw_i))G] = r[(-xcG + b^{-1}aG) - (b - hf(qw_i))G]$
(c) The trapdoor value is the following pair: $T_{Qw} = (T_{Qw_1}, T_{Qw_2})$.

Search$(Idx, T_{Qw}, SK_{Srv}) \rightarrow C_{Qw}$. This algorithm is performed mainly by BN in collaboration with Srv. *Input*: T_{Qw} - the trapdoor value computed by Usr_U, Idx - the index structure, SK_{Srv} - the secret key of the server. *Output*: C_{Qw} - the set of the encrypted documents that match the query criteria.

(a) BN computes the value $V_2' = Idx_{i0}^{T_{Qw_1}}$ and sends it to Srv. Here, Srv computes the value $V_2 = (V_2')^{SK_{Srv}} = (Idx_{i0}^{T_{Qw_1}})^{SK_{Srv}} = ((e(G,G)^r)^x)^c = e(G,G)^{rxc}$. Then, Srv sends the value V_2 back to BN.
(b) Further, BN computes the value $V_1 = e(G, T_{Qw_2}) \cdot e(G, Idx_{i1} - Idx_{i\psi(k)})$, where ψ is a indexing function such that $\psi(k) = j$ and $V_3 = Idx_{i2} = e(G,G)^{arb^{-1}}$, and then

verifies whether $V_1 \cdot V_2 = V_3$. If the equality occurs, then BN records the transaction and sends the index $\psi(k)$ for which the equality is true to Srv, otherwise a proper message is sent and the transaction is dismissed and the search process is ended.
(c) For each $\psi(k)$ received, Srv extracts the corresponding document.

Decryption. The data user Usr_U calls this algorithm to decrypt the documents resulted from the search process. *Input*: PS, SK_{Usr_U}, C_q - the system parameters SP, the private key of the data owner SK_{Usr_U}, and the set of the encrypted documents resulted from the search query C_q. *Output*: D_q - the set of the decrypted documents.

3.2 Security

Correctness. The correctness of the proposed solution lays in the verification of the equality $V_1 \cdot V_2 = V_3$, namely, for each valid query keyword that generates the trapdoor pair, the result of the search process always returns the corresponding document(s) based on this equality. Indeed:

$$
\begin{aligned}
V_1 \cdot V_2 &= e(G, T_{Qw_2}) \cdot e(G, Idx_{i1} - Idx_{i\psi(k)}) \cdot e(G,G)^{rxc} \\
&= e(G, r[(-xcG + b^{-1}aG) - (b - \mathtt{hf}(qw_i))G]) \cdot \\
&\quad \cdot e(G, rbG - r\mathtt{hf}(w_{\psi(k)})G) \cdot e(G,G)^{rxc} \\
&= e(G, r(-xc + b^{-1}a - b + \mathtt{hf}(qw_i))G) \cdot \\
&\quad \cdot e(G, (rb - r\mathtt{hf}(w_{\psi(k)})G) \cdot e(G,G)^{rxc} \\
&= e(G,G)^{r(-xc + b^{-1}a - b + \mathtt{hf}(qw_i))} \cdot \\
&\quad \cdot e(G,G)^{rb - r\mathtt{hf}(w_{\psi(k)})} \cdot e(G,G)^{rxc} \\
&= e(G,G)^{-rxc + rb^{-1}a - rb + r\mathtt{hf}(qw_i) + rb - r\mathtt{hf}(w_{\psi(k)}) + rxc} \\
&= e(G,G)^{rb^{-1}a} = V_3
\end{aligned}
$$

Controlled Searching. The search process is approved the blockchain network, which verifies if the user that initiates the search has the proper computational power to support this operation.

Encrypted Queries. To generate a trapdoor pair, Usr_U uses a randomly chosen value $x \in \mathbb{F}_{p^k}$ and the public keys of Usr_U and Srv to "hide" the query keywords.

Query Isolation. As the search process is made by the blockchain network, Srv will learn nothing during the search process. However, Srv cannot generate a valid trapdoor pair, because it needs data user's private key SK_{Usr_U} and the secret value $r \in \mathbb{F}_{p^k}$. Because, $\mathbb{F}_{p^k}$ has p^k elements, the probability that Srv to guess SK_{Usr_U} is $1/p^k$. Similarly, the probability of guessing r is $1/p^k$. Therefore, the probabilities that Srv guesses the values x, r are negligible.

Security Against CKA2. The security against KGA is proven using a security game between a challenger $\mathscr{C}$ and an outsider adversary $\mathscr{A}$. For this, we use the method described in [18], as follows:

1. The challenger $\mathscr{C}$ generates the key, and makes public all three public keys, namely $PK_{Usr_O}, PK_{Usr_U}, PK_{Srv}$.
2. The attacker $\mathscr{A}$ may ask in an adaptive way for the trapdoor T_{qw} of any query keyword qw.
3. After receiving some trapdoor values, A chooses two query keywords qw_0, qw_1 with which it will be challenged. Further, the two values are sent to the challenger $\mathscr{C}$. However, the trapdoors should be fresh, meaning that these were not chosen in previous rounds.
4. The challenger $\mathscr{C}$ selects randomly an element $b \in \{0,1\}$. Further, $\mathscr{C}$ calls the trapdoor algorithm to achieve T_{qw_b}, then the result that was obtained is sent to $\mathscr{A}$. On its side, the attacker may continue to ask for trapdoor values with the condition these were not asked in previous rounds.
5. When $\mathscr{A}$ considers, it outputs value $b' \in \{0,1\}$ and wins if $b = b'$. otherwise it looses.

In order to be secure against KGA, the advantage of the attacker in this searchable encryption scheme should be negligible. In order to obtain the trapdoor value, $\mathscr{A}$ should actually determine the two components that resembles the trapdoor value, in particular, it should determine the value picked randomly $x \in \mathbb{F}_{p^k}$. The chances to get this value is $1/p^k$, which is a negligible quantity.

As we used blockchain in the proposed scheme, it preserves the blockchain security properties, namely, decentralization for the search process, persistency, anonymity, and audibility.

The overall time complexity of the proposed scheme consists of the maximum complexity within the algorithms of the scheme. Each algorithm involves operations of scalar multiplication with points on an elliptic curve, for which the complexity is proven to be $O(log_2 k)$, where k is the scalar [5]. However, it also depends on the bilinear pairing used. For example, the Tate pairing is more efficient than Weil pairing, but it needs a careful choice of the elliptic curve [24].

4   Conclusion and Further Research

In the last few years, the blockchain technology has evolved rapidly, being adopted in different domains, even if it was initially proposed for the financial sector. In this paper, we proposed a searchable encryption scheme that includes in its workflow a blockchain network used in the search process. The scheme is based on symmetric bilinear maps for elliptic curves, with the hardness assumption of discrete logarithm problem. We have chosen an approach based on elliptic curves, because these are faster than other cryptographic mechanisms that provide the same security level. We have seen that the proposed scheme preserves standard security requirements, namely controlled searching, encrypted query, and query isolation. Additionally, we proved the scheme is KGA-secure. As further research direction, we will implement the proposed scheme firstly in a laboratory environment in order to make tests and simulations for real-life use regarding practical security and performance, then the implementation will be extended to real scenarios of usage.

References

1. 2021 State of the Cloud Report —Flexera. https://info.flexera.com/CM-REPORT-State-of-the-Cloud?lead_source=Website%20Visitor&id=Blog
2. Gartner Predicts the Future of Cloud and Edge Infrastructure. www.gartner.com/smarterwithgartner/gartner-predicts-the-future-of-cloud-and-edge-infrastructure
3. Nakamoto, S.: Bitcoin: a peer-to-peer electronic cash system. In: Decentralized Business Review, p. 21260 (2008)
4. Li, X., Jiang, P., Chen, T., Luo, X., Wen, Q.: A survey on the security of blockchain systems. Futur. Gener. Comput. Syst. **107**, 841–853 (2020)
5. Washington, L.C.: Elliptic Curves: Number Theory and Cryptography. CRC Press, New York (2008)
6. Koblitz, N., Menezes, A.: Pairing-based cryptography at high security levels. In: Smart, N.P. (ed.) Cryptography and Coding 2005. LNCS, vol. 3796, pp. 13–36. Springer, Heidelberg (2005). https://doi.org/10.1007/11586821_2
7. Chen, Z., Wu, A., Li, Y., Xing, Q., Geng, S.: Blockchain-enabled public key encryption with multi-keyword search in cloud computing. In: Security and Communication Networks (2021)
8. Lauter, K.: The advantages of elliptic curve cryptography for wireless security. IEEE Wirel. Commun. **11**(1), 62–67 (2004)
9. Shang, Z., Oya, S., Peter, A., Kerschbaum, F.: Obfuscated access and search patterns in searchable encryption. arXiv preprint arXiv:2102.09651 (2021)
10. Wang, H., Sui, G., Zhao, Y., Chen, K.: Efficient SSE with forward ID-privacy and authentication in the multi-data-owner settings. IEEE Access **9**, 10443–10459 (2020)
11. Khazaei, S.: Fuzzy retrieval of encrypted data by multi-purpose data-structures. Signal Data Process. **17**(4), 123–138 (2021)
12. Patranabis, S., Mukhopadhyay, D.: Forward and backward private conjunctive searchable symmetric encryption. Cryptology ePrint Archive (2020)
13. Song, D.X., Wagner, D., Perrig, A.: Practical techniques for searches on encrypted data. In: Proceeding 2000 IEEE Symposium on Security and Privacy, S&P 2000, pp. 44–55 (2000)
14. Goh, E.J.: Secure indexes. IACR Cryptol. ePrint Arch. (2003). https://eprint.iacr.org/2003/216
15. Stefanov, E., Papamanthou, C., Shi, E.: Practical dynamic searchable encryption with small leakage. Cryptology ePrint Archive (2013)
16. Etemad, M., Kupcu, A., Papamanthou, C., Evans, D.: Efficient dynamic searchable encryption with forward privacy. arXiv preprint arXiv:1710.00208 (2017)
17. Liu, Z., et al.: Eurus: towards an efficient searchable symmetric encryption with size pattern protection. IEEE Trans. Depend. Secure Comput., 1 (2020). https://doi.org/10.1109/TDSC.2020.3043754
18. Boneh, D., Di Crescenzo, G., Ostrovsky, R., Persiano, G.: Public key encryption with keyword search. In: Cachin, C., Camenisch, J.L. (eds.) EUROCRYPT 2004. LNCS, vol. 3027, pp. 506–522. Springer, Heidelberg (2004). https://doi.org/10.1007/978-3-540-24676-3_30
19. Park, D.J., Kim, K., Lee, P.J.: Public key encryption with conjunctive field keyword search. In: Lim, C.H., Yung, M. (eds.) WISA 2004. LNCS, vol. 3325, pp. 73–86. Springer, Heidelberg (2005). https://doi.org/10.1007/978-3-540-31815-6_7

20. Huang, Q., Li, H.: An efficient public-key searchable encryption scheme secure against inside keyword guessing attacks. Inf. Sci. **403**, 1–14 (2017)
21. He, D., Ma, M., Zeadally, S., Kumar, N., Liang, K.: Certificateless public key authenticated encryption with keyword search for industrial internet of things. IEEE Trans. Industr. Inform. **14**(8), 3618–3627 (2017)
22. Liu, Z.-Y., Tseng, Y.-F., Tso, R., Chen, Y.-C., Mambo, M.: Identity-certifying authority-aided identity-based searchable encryption framework in cloud system. IEEE Syst. J., 1–12 (2021). https://doi.org/10.1109/JSYST.2021.3103909
23. Szabo, N.: Formalizing and securing relationships on public networks. First Monday (1997)
24. Enge, A.: Bilinear pairings on elliptic curves. Enseign. Math. **61**(1), 211–243 (2016)

Ensuring Data Integrity Using Digital Signature in an IoT Environment

Nadia Kammoun$^{(\boxtimes)}$, Aida ben Chehida Douss, Ryma Abassi,
and Sihem Guemara el Fatmi

Innov'COM, Sup'com, University of Carthage, Tunis, Tunisia
{Nadia.Kammoun,Ryma.Abassi,Sihem.Guemara}@supcom.tn

Abstract. The Internet of Things (IoT) devices produce large amounts of data which increase their susceptibility to cyber attacks. Data Integrity attack is a serious limitation to the evolution of IoT as it leads to bad financial impact or even human fatality. In this work, we will focus on ensuring exchanged data integrity using digital signature for public key cryptography. Elliptical Curve Digital Signature Algorithm (ECDSA) is gaining more attention than RSA in IoT for its same level of security with smaller keys. ECDSA is applied on our predefined Software Defined Architecture (SDN).

1 Introduction

Internet of Things (IoT) is broadly available nowadays in different forms of devices such as smartphones, medical sensors, smart home security systems, etc. These devices are interconnected to each other and to the internet by different forms of communications technologies. According to users demand, different IoT applications are implemented like smart homes, smart cities, crowd sensing, etc. which has led to a spectacular increase of big data issued from IoT devices. In fact, big data exchanged between devices as well as between devices and Data Owners(DO) and stored in cloud, is not far from security attacks that some of them affect data integrity. This latter is the way towards protecting the precision, consistency and the reliability of information during its whole life cycle from external influences or attacks [1]. Data should maintain its originality during transmission, reception and storage [2]. In IoT, the distortion of basic information could prompt considerable issues, e.g., in smart healthcare systems, it could even cause the death of the patient [1]. In this paper we concentrate on guaranteeing the integrity of data while transferring over the network.

In addition, security measures should not be heavy in IoT networks seeing that IoT devices have limited capacities with respect to processing time and bandwidth consumption. Incorporating data security solutions to protect network resources and devices from several types of attacks should be lightweight and cover different goals like authentication, access control, Data Integrity. We already implemented in previous works security models for IoT network based on an SDN architecture. This latter can be adopted by different kinds of network

L. Barolli et al. (Eds.): AINA 2022, LNNS 451, pp. 482–491, 2022.
https://doi.org/10.1007/978-3-030-99619-2_46

including IoT. SDN alleviates nodes workload by limiting their task on forwarding data between each other across controllers. An authentication mechanism was required to establish trusted communication inside the network. In addition, controlling the access to limit users action and access to services is crucial for IoT environments. For ensuring malicious nodes elimination, we had also proposed a trust management model. These security systems cannot ensure that data received by a user or an object is conformed with the original data. That is why an Integrity verification system is required in such IoT networks. Data integrity checking could be in different levels on the network. Remote data integrity verification is a challenging security task that protects cloud-based information systems. Many researchers are interested on this theme include trendy technologies such as Blockchain [4]. Otherwise, ensuring end-to-end data integrity is very crucial. There are many ways to ensure in transit data integrity. Even that we already implement an authentication scheme for our IoT environment, including digital signatures will enhance the authentication process and ensure data integrity by preventing IoT devices from being compromised and data from being tampered. We chose to use digital signatures based on Elliptic Curve Cryptography ECC. In fact, digital signatures gained legal acceptance just like the traditional hand-written signatures. It uses public-key cryptography which is conceived to protect the authenticity of a digital document. Digital signatures has become an essential part of IoT to keep under control illegal users. Digital signature schemes have been recommended to overcome the security problems in confidentiality, integrity and authentication related problems in IoT [10]. Nowadays Elliptical Curve Digital Signature Algorithm (ECDSA) is gaining more popularity than other digital signature algorithms like RSA because of its better performance [11]. In fact, the main advantage of ECC is that it provides the same grade of security with less overhead and key size than RSA [9]. [9,11] show the superiority of ECDSA comparing to RSA for resource-constrained IoT devices which support our choice. In addition, using SDN architecture is a favor for IoT networks, where it alleviate nodes workload by implicating controllers in computation tasks. Rare researchs include ECDSA in IoT where an SDN architecture is applied [12]. This paper contains five other parts. Section 2, gives some related works of data Integrity for IoT networks. Section 3, presents an overview of Security Challenges in IoT Environments. We present our previous work in Sect. 4. Section 5 is dedicated to the System Model. Finally,

2 Related Works

Data integrity techniques are used for verification and protection of the integrity of:

- Data generated by a device,
- Stored data (e.g. in cloud)

- Data Generated by a Device
Some examples of integrity checking of data generated by a device (or data in communication) are presented as follows:

Dua, A. and Chaudhary, R. proposed a novel scheme which prevents data from unauthorized production and preserves the integrity of data in e-healthcare. Integrity is ensured by a trusted third party (proxy server). A tag is assigned to each block (after patient's data fragmentation) using the lightweight approach (ECC). This block-tag pair, after being uploaded on the data server, is used for integrity checking. Authors also proposed a lightweight lattice-based authentication model for users authentication [5].

Dubal, M. and Deshmukh, A. proposed an algorithm based on Elliptic Curve Integrated Encryption Scheme (ECIES) for achieving secrecy and hashing function for authentication. The hashed text is encrypted with RSA then decrypted at the receiver end. The hash value of the decrypted text is compared to the latter hash. The integrity is assured if they are found equal [6].

Some other authors propose digital signature algorithms to provide not only generated data integrity but also authenticity and non-repudiation of the message. Toradmalle, D., Singh, R. and al proposed a review on the performance of RSA and ECDSA to prove the efficacity of ECDSA over RSA in favor of smaller keys [9]. Suarez-Albela, Manuel and al proposed a practical performance comparison of RSA and ECC for resource-constrained IoT devices. They prove by analysing all tests that ECDSA is a better alternative than RSA in term of data throughput values and energy consumption in order to secure resource-constrained IoT devices [11].

- Stored Data in Cloud

Some other works concentrate on ensuring integrity of outsourced data such as by using Provable Data Possession (PDP). Y Liu and S Xiao proposed an improved scheme for a work made by other authors that secure outsourced data in cloud computing based on PDP and deletion. They prove flaws of the other's scheme by applying an attack on it.

In this work, we focus on ensuring the integrity of data in communication using digital signatures. The particularity of this proposition is using ECDSA on an SDN architecture which alleviate nodes workload during signature computation.

3   Security Challenges in IoT Environments

We mention in this section the major security challenges in IoT environments.

3.1   Authentication

Authentication Keys deployment and management are a real issue for IoT devices. Any cryptographic key generation and exchange will produce an important overhead on IoT nodes. Moreover, if the Certificate Authority (CA) is absent, other mechanisms ensuring validation of cryptographic keys and key transfer are required.

3.2   Authorisation and Access Control

Authorisation implies the specification of access rights for different resources in the network while Access Control mechanisms guarantee an access right for only authorised resources. IoT nodes, with their constrained memories, may only support lightweight mechanisms for access attribution and verification, which could variate from a node to another. Thus, deployment and management of different access control and authorisation mechanisms is a challenge for devices in an heterogeneous IoT environment.

3.3   Integrity

Data integrity perform a vital role in IoT security systems to check if an attacker has modified the data or not while the data was transmitted to the destination or stored on the data server.

3.4   Secure Architecture

Construct an architecture that overcomes above-mentioned security challenges in IoT is not trivial. These architectures should inevitability include devices that inherited IoT devices process such as cloud infrastructure and SDN which will create other challenges by deploying IoT devices over them [4].

4   Previous Work

One of the most crucial requirements for IoT networks is taking into consideration security measurements under limited capacities. Our present contribution completes our previous work on ensuring security in an IoT network based on SDN Architecture. We already proposed an authentication and access control mechanisms to authorise and limit the access to the network as well as a trust management to exclude malicious nodes. In this paper, we used our predefined SDN architecture for an IoT network, which aims to secure it by employing Attribute-based access control (ABAC) as an access control paradigm, a trust management mechanism and a lightweight authentication scheme based on ECC. In fact, ECC is a public key encryption technique based on elliptic curve theory that is used to make faster in favour of smaller and more effectual cryptographic keys comparing to other keys like RSA. In addition, common SDN architectures are based on 3 layers: Application layer consists on SDN applications which are set by network administrators. Control layer contains a central(s) controller(s) which are responsible(s) of all control tasks, whereas data forwarding between IoT nodes and controllers is ensured by network devices (switches, routers, etc.) in infrastructure layer.

In our security model, using ECC for authentication represents a crucial measurement to authorise users and devices communication without using a heavy system. In fact, ECC is a public key encryption technique based on elliptic curve theory that is used to make faster, shorter, and more effectual cryptographic keys. In addition, implementing an access control mechanism limits users and devices access under predefined attributes in ABAC paradigm. For excluding malicious nodes from the network, using trust management is an efficient way. A variable trust level is assigned to each node to attest its trustworthiness. Figure 1 shows our previous SDN architecture. It presents an infrastructure layer containing 3 under-layers:

- Bases Stations (BSs), switches and routers.
- Cluster Heads (CHs): which are some IoT objects possessing high Energy and Trust levels calculated in [7].
- Cluster Members (CMs): which are most of network's IoT objects. CMs are divided into clusters according to a clustering algorithm already defined in a previous work [7].

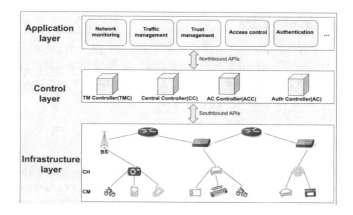

Fig. 1. DN architecture based on authentication, access control and trust management

For better security, integrity is the guarantee that data received by a node has not been changed during transit either via a planned tampering by an untrusted node or through collision and the received data was as originally sent [3].

5   System Model

This network model includes the following entities on a communication:

- The user: any entity seeking for a service from nodes in the network.
- BS: an entity which is responsible of forwarding data from IoT objects (precisely CH) to controllers and vice versa.
- CH: is an IoT node elected as a cluster head on a group of nodes in favor of its high levels of energy and trust.
- Thing: its the IoT device which offers a service for a user or other devices.

- The authentication controller (AC): Whenever a new request is launched for data accessing or storing, AC authenticates the user who raises this request. It also generates all ECC parameters for authentication and integrity.
- The central Controller (CC): responsible of standard tasks in the network and forwarding messages to the right destination.

We aim in this paper to ensure integrity during transit. In fact, after authentication process, when a user or thing request a service from a specific thing, this request passes by several equipment cited above to reach its destination as well as the response back. This circuit increase tampering risks exercised by internal and external attackers. For this reason integrity checking is a crucial security measurement. Figure 2 resume messages exchange model.

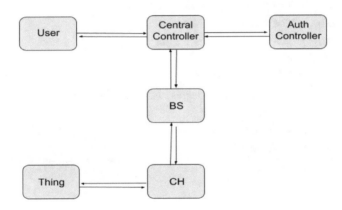

Fig. 2. Entities contributing in a communication

Data integrity verification could be only when users demand []. It can also made automatically when a user request a service from the network. After entity authentication, the proposed scheme verifies data integrity of exchanged messages using digital signature. For gaining in term of process and storage, we chose to ensure data integrity using ECC considering that it is already used for authentication in previous work. In fact, Elliptical Curve Digital Signature Algorithm (ECDSA) uses an elliptical curve to encrypt a message then it attaches the signature over it to secure the document from any unauthorized access. ECDSA is an approach that works likewise RSA digital signature, but key size is lesser. Digital signatures are made for exchanged messages between nodes and users or nodes and other nodes. To simplify the process, we propose digital signature steps on message sent by user. This algorithm is made according to 3 steps:

1. Key pair generation
2. Signature Generation
3. Signature verification

The following table includes all notations that are used in forthcoming steps:

Table 1. Table of notation

AC	:	Authentication Controller
U	:	User
T	:	Thing
ID_{AU}	:	The user's ID
ID_{AT}	:	The thing's ID
Pub_{U_i}	:	The public key of the user
Pub_{AC}	:	The public key of AC
Pub_{T_i}	:	The public key of the thing
Sec_{U_i}	:	The secret key of the user
Sec_{T_i}	:	The secret key of the thing
F_q	:	A finite field
P	:	A base point on EC
EC	:	An elliptic curve defined F_q over a finite field
n	:	Integer order of P
m	:	message sent by a user or a thing
$h()$	:	Hush function

1) Key Pair Generation

We already used key pairs for authentication. In fact, AC generates a couple of public keys (Pub_{U_i}, Pub_{AC}) for users when they join the network; Pub_{U_i} is a unique public key for each user and Pub_{AC} is AC's public key. The hash function h as well as the elliptic curve and its parameters are also generated by AC. Each user evaluate its private key $Sec_{U_i} = a * h(Pub_{U_i})$ such that a $\in F_q$. The hash function h of Pub_{U_i} was sent by AC during the authentication process [8]. Then, each user has a couple of keys (Pub_{U_i}, Sec_{U_i}).

2) Signature Generation

In addition to the finite field F_q and equation of the curve, a base point P of large order p on the curve is assigned; p is the multiplicative order of the point P (where q and p are primes). To sign a message, a user (or a thing) follows these steps:

STEP 1: calculate $e = h(m)$ which is the hash function of the message m.
STEP 2: Select a random integer k in the range $[1, n-1]$.
STEP 3: Calculate the curve point $(x_1, y_1) = k \times P$ where x_1 is an integer
STEP 4: Determine $R = x_1 mod n$; If $R = 0$, repeat steps from step 2 (to choose a different value of k)
STEP 5: Determine $S = k^{-1}(e + RSec_{U_i}) mod n$; If $S = 0$, then repeat from step 2.
STEP 6: The signature is the pair (R, S)

3) Signature Verification

To verify sender's signature, there are some steps to evaluate by the receiver. This process is a time and battery consumer which is not advisable in IoT. Applying SDN in our network can relieve objects from additional workload. In fact, before attending IoT objects, messages coming from users pass inevitably by AC.

STEP 1: AC obtains Pub_{U_i} and checks that is unique and lies on the curve. It verifies also ID_{AU} the user ID.
STEP 2: AC also verifies Pub_{T_i} and ID_{TU} of the destination.
For ensuring the secrecy of a message, next steps of signature verification and message recovery are made by the destination itself.
STEP 3: Verify that R and S are in the interval $[1, n-1]$. The signature is invalid if they are not.
STEP 4 : Calculate $W = S^{-1} mod n$
STEP 5: Resolve $e = h(m)$, where H is the hashing function used by the sender(user).
STEP 6: Determine $u1 = eW mod n$
STEP 7: Determine $u2 = RW mod n$
STEP 8: Determine $u1P + u2Pub_{U_i} = (x2, y2)$
STEP 9: Determine $V = x2 mod n$
STEP 10: Accept if and only if $V = R$

In this step, the receiver (in this case the thing) will accept the message if $V = R$ and will respond by repeating the same steps using its own public and secret keys. Figure 3 simplifies ECDSA algorithm using for IoT.

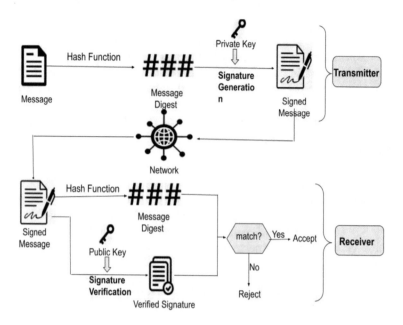

Fig. 3. ECDSA algorithm for IoT

6   Conclusion

IoT, the rapidly growing network of things, attracts many malicious users that try to gain access for exchanged data within the network. Security measurements are obviously required to protect the network from attackers. We already proposed an SDN architecture provided by an access control and an authentication systems for both users and things. This paper presents a data integrity assurance mechanism based on digital signatures. ECDSA permits to users and nodes to insert a digital signature after hashing the message to provide the message with a high degree of security. This algorithm's security analysis and implementation will be made within our IoT network in the incoming work.

References

1. El-hajj, M., Fadlallah, A., Chamoun, M., Serhrouchni, A.: A survey of Internet of Things (IoT) authentication schemes. Sensors **19**(5), 1141 (2019)
2. Lu, Y., Xu, L.D.: Internet of Things (IoT) cybersecurity research: a review of current research topics. IEEE IoT J. **6**(2), 2103–2115 (2019)
3. Aman, M.N., Sikdar, B., Chua, K.C., Ali, A.: Low power data integrity in IoT systems. IEEE IoT J. **5**(4), 3102–3113 (2018)
4. Liu, B., Yu, X.L., Chen, S., et al.: Blockchain based data integrity service framework for IoT data. In: 2017 IEEE International Conference on Web Services (ICWS), pp. 468–475. IEEE (2017)

5. Dua, A., Chaudhary, R., Aujla, G.S.: LEASE: lattice and ECC-based authentication and integrity verification scheme in E-healthcare. In: IEEE Global Communications Conference (GLOBECOM), vol. 2018, pp. 1–6. IEEE (2018)
6. Dubal, M., Deshmukh, A.: Achieving authentication and integrity using elliptic curve cryptography architecture. Int. J. Comput. Appl. **69**(24), 11–15 (2013)
7. Kammoun, N., Abassi, R., Guemara, S.: Towards a new clustering algorithm based on trust management and edge computing for IoT. In: 15th International Wireless Communications & Mobile Computing Conference (IWCMC), vol. 2019, pp. 1570–1575. IEEE (2019)
8. Kammoun, N., Abassi, R., Guemara El Fatmi, S., Mosbah, M.: A lightweight authentication scheme for SDN-based architecture in IoT. In: Barolli, L., Woungang, I., Enokido, T. (eds.) AINA-2021, pp. 336–345. Springer, Cham (2021). https://doi.org/10.1007/978-3-030-75078-7_34
9. Toradmalle, D., Singh, R., Shastri, H., et al.: Prominence of ECDSA over RSA digital signature algorithm. In: 2018 2nd International Conference on I-SMAC (IoT in Social, Mobile, Analytics and Cloud), pp. 253–257. IEEE (2018)
10. Gupta, P., Sinha, A., Srivastava, P.K., Perti, A., Singh, A.K.: Security implementations in IoT using digital signature. In: Favorskaya, M.N., Mekhilef, S., Pandey, R.K., Singh, N. (eds.) Innovations in Electrical and Electronic Engineering: Proceedings of ICEEE 2020, pp. 523–535. Springer, Singapore (2021). https://doi.org/10.1007/978-981-15-4692-1_40
11. Suárez-albela, M., Fernández-caramés, T.M., Fraga-Lamas, P.: A practical performance comparison of ECC and RSA for resource-constrained IoT devices. In: Global Internet of Things Summit (GIoTS), vol. 2018, pp. 1–6. IEEE (2018)
12. Chattaraj, D., Saha, S., Bera, B., et al.: On the design of blockchain-based access control scheme for software defined networks. In: IEEE Conference on Computer Communications Workshops (INFOCOM WKSHPS), IEEE INFOCOM 2020, pp. 237–242. IEEE (2020)

Beaver Triple Generator from Multiplicatively Homomorphic Key Management Protocol

Huafei Zhu[1]([✉]) and Wee Keong Ng[2]

[1] Institute for High Performance Computing, Agency for Science and Technology,
Singapore, Singapore
`zhu_huafei@ihpc.a-star.edu.sg`
[2] School of Computer Science and Engineering, Nanyang Technological University,
Singapore, Singapore
`awkng@ntu.edu.sg`

Abstract. SPDZ, the nickname of secure multi-party computation (MPC) protocol of Damgård et al. from Crypto 2012, is an elegant secret value splitting technique which enjoys highly scalable and efficient multiplications with the help of off-line Beaver (multiplicative) triples. Since an arithmetic circuit defined over a finite field can be implemented by addition and multiplication gates, and also since an implementation of addition gate is almost cost free within the SPDZ framework, a design of efficient secure multi-party computations within the SPDZ framework is reduced to that of Beaver (multiplicative) triples. This paper studies Beaver (multiplicative) triple generators and our contributions are two-fold:

- First, an efficient construction of Beaver triple generators (BTGs), formalized within the framework of 3-party computation leveraging multiplicatively homomorphic key management protocols (mHKMs) is presented and analyzed. We show the equivalence between BTGs and mHKMs. As such, we are able to provide an alternative construction method for BTGs beyond the state-of-the-art solutions.
- Second, a new notion which we call blind triple dispensation protocol is introduced for securely dispensing the generated Beaver triples. We then propose an implementation of blind triple dispensation protocols leveraging mHKMs. As such, we are able to demonstrate the power of mHKMs by showing that it is a useful notion not only for securely generating Beaver triples, but also for securely dispensing the generated triples as well.

1 Introduction

Since the introduction of secure multi-party computation ([13,16,24] and references therein), numerous practical yet provably secure multi-party computation (MPC) protocols have been developed. The notable developments among them are ring-based zero-splitting solutions [1,6–8] and SPDZ-based solutions [4,9–11,14,17]. Each development has its own pros and cons. While a ring-based zero-splitting solution is efficient for 3-party computations, its scalability is problematic. On the other hand, while SPDZ provides high scalability, the efficiency of multiplication (or Beaver) triple generations is a bottleneck [3]. Since a computation of arithmetic addition within the SPDZ framework is as efficient as that within the ring-based zero-splitting framework and also since a multiplication can be computed efficiently in the SPDZ framework if Beaver triples used to assist the multiplications are available during an on-line processing, an efficient Beaver triple generator results in an efficient implementation of SPDZ.

The Beaver triples deployed in the SPDZ framework are originally constructed from somewhat fully homomorphic encryptions [11]. Since computation cost of a ciphertext multiplication is high if one leverages somewhat fully homomorphic encryptions, more and more researchers are considering alternative constructions such as oblivious-transfers [14], additively homomorphic encryptions [15,18] and multiplicatively homomorphic encryptions [26]. An interesting question is that *can we construct Beaver triple generators beyond the above mentioned methods?*

To assist secure computations, we need to dispense generated Beaver triples to a set of predefined computing nodes. Since a double spending of a Beaver triple leaks private input data, it is paramount importance to design a security mechanism to prevent double spending of multiple triples. Very recently, a blockchain based solution called TaaS [19] for dispensing Beaver triples leveraging commodity-based cryptography (CBC) first introduced by Beaver [4] has been proposed. In the TaaS (triple as a service) framework, concept of a ledger (in the context of blockchain) is introduced to make sure that MPC service providers do not reshare the same Beaver triple twice, and that every transaction must be logged in the ledger. While the TaaS solution is useful to blockchain-based applications such as on-chain and off-chain computations, we aware that many use cases are non-blockchain oriented. For example, in the GWAS computing [2], a case-control study may run in the federated computing model [20,21], where a privacy-preserving computation of a case-control classifier could invoke large of numbers of multiplications. As such, large numbers of blockchain transactions will be created to prevent double sharing. This could render TaaS inefficient for certain industrial scale deployments. An interesting research problem thus is − *how to construct efficient dispensation protocols for preventing double resharing within the commodity-based cryptography (CBC) framework for industrial-scale deployment?*

1.1   The Related Work

In Zhu et al. [25], an interesting notion called multiplicatively homomorphic key management (mHKM) protocol is introduced and formalized in the context of multi-party computation. In their model, data owner O and a set of key management (KM) servers $S_1, \cdots, S_n$ are introduced to collaboratively dispense master key k of data owner O. A KM scheme is multiplicatively homomorphic if $k = k_O k_1 \cdots k_n$, where multiplicatively homomorphic encryptions are performed over the underlying cyclic group. The definition of k_O benefits O controlling the dependency of shares $(k_1, \cdots, k_n)$ for cloud key management. If we remove k_O and restrict $n = 2$, then a secure computation for the relationship $k = k_1 \times k_2$ can be derived, which in essence, is an instance of Beaver triples, where k_j $(j = 1, 2)$ is privately computed by each Beaver triple generators BTG_i $(i \in \{A, B\})$ and k by BTG_C [26]. Interestingly, we are able to show that the notions of BTGs and mHKMs are equivalent.

1.2   Our Contributions

In this paper, an efficient construction of Beaver triple generators (BTGs) leveraging multiplicatively homomorphic key management (mHKM) system is presented and analyzed. We are able to show that given an mHKM protocol, one can efficiently construct BTGs and vice versa. We further show that BTGs can be used to blindly dispense Beaver triples as well and present two solutions for data owners blindly dispensing the generated Beaver triples, which benefit us to construct lightweight BTGs for industrial scale deployments. Consequently, mHKMs are backbone not only for constructions of BTGs but also for secure triple dispensations.

The rest of this paper is organized as follows: Notions and notations of MPC is sketched in Sect. 2. A rigorous proof of equivalence between mHKMs and BTGs is presented in Sect. 3. We provide blind triple dispensation protocol in Sect. 4. We test the efficiency of the proposed Beaver triple generator in Sect. 5 and conclude our work in Sect. 6.

2   Notions and Notations

In this section, the notions and notations of secure multi-party computations are sketched. We refer the reader to Goldreich [13] for more details.

2.1   Secure Multi-party Computation

A multi-party computation securely computes an m-ary functionality (i.e., secure multi-party computation, SMPC) if the following simulation-based security definition is satisfied.

Let $[m] = \{1, \cdots, m\}$. For $I \in \{i_1, \cdots, i_t\} \subseteq [m]$, $f_I(x_1, \cdots, x_m)$ denote the subsequence $f_{i_1}(x_1, \cdots, x_m)$, $\cdots$, $f_{i_t}(x_1, \cdots, x_m)$. Let Π be an m-party protocol for computing f. The view of the i-th party during an execution of Π on $\overline{x} := (x_1, \cdots, x_m)$ is denoted $\text{View}_i^{\Pi}(\overline{x})$. For $I = \{i_1, \cdots, i_t\}$, we have $\text{View}_I^{\Pi}(\overline{x}) := (I, \text{View}_{i_1}^{\Pi}(\overline{x}), \cdots, \text{View}_{i_t}^{\Pi}(\overline{x}))$.

- If f is a deterministic m-ary functionality, we say Π privately computes f if there exist a probabilistic polynomial-time algorithm S, such that for every $I \subseteq [m]$, it holds that $S(I, (x_{i_1}, \cdots, x_{i_t}), f_I(\overline{x}))$ is computationally indistinguishable with $\text{View}_I^{\Pi}(\overline{x})$.

- In the general case, we say Π privately computes f if there exist a probabilistic polynomial-time algorithm S, such that for every $I \subseteq [m]$, it holds that $S(I, (x_{i_1}, \cdots, x_{i_t}), f_I(\overline{x}), f(\overline{x}))$ is computationally indistinguishable with $\text{View}_I^\Pi ((\overline{x}), f(\overline{x}))$.

2.2 Oracle-Aided Multi-party Computation

An oracle-aided protocol is a read and write mechanism augmented by a pair of oracle types, per each party. An oracle call is defined as follows: A party writes an oracle request on its own oracle tape and then sends it to other parties. In response, each of the other parties writes its query on its own oracle tape and responds to the first party with an oracle call message. The oracle is invoked and the oracle answer is written by the oracle on the read-only oracle tape of each party.

Definition 1. An oracle-aided protocol is said to privately reduce g to f if it securely computes g when using the oracle-functionality f.

3 Equivalence of BTGs and mHKMs

In this section, we show that the two interesting notions BTGs and mHKMs are equivalent. That is, given an mHKM protocol, we can construct BTGs and vice versa.

3.1 Beaver Triple Functionality

We write $[x]$ to mean that each party P_i holds a random, additive sharing x_i of x such that $x = x_1 + \cdots + x_n$, where $i = 1, \cdots, n$. The values are stored in the dictionary Val defined in the functionality $\mathcal{F}_{\text{Triple}}$ [14]. We refer to the reader Table 1 for more detail on the Beaver triple functionality.

Table 1. Beaver triple functionality

The functionality of Beaver triple generator $\mathcal{F}_{\text{BTG}}$

- The functionality maintains a dictionary, Val to keep track of assigned value, where entry of Val lies in a fixed field.
- On input (Triple, id_A,id_B, id_C) from all parties, sample two random values $a, b \leftarrow F$, and set [Val[id_A], Val[id_B], Val[id_C]] $\leftarrow (a, b, ab)$.

3.1.1 Multiplicatively Homomorphic Key Management Functionality

Each party P_i holds a random, multiplicative sharing x_i of x such that $x = x_1 \cdots x_n$, where $i = 1, \cdots, n$. The values are stored in the dictionary Val defined in the functionality $\mathcal{F}_{\mathrm{mHKM}}$. We refer to the reader the Table 2 for more details.

Table 2. Description of multiplicatively homomorphic key management functionality

---

Functionality of multiplicatively homomorphic key management $\mathcal{F}_{\mathrm{mHKM}}$

- The functionality maintains a dictionary, Val to keep track of assigned value, where entry of Val lies in a fixed field.
- On input $(\mathrm{mHKM}, \mathrm{id}_1, \cdots, \mathrm{id}_{n-1}, \mathrm{id}_n)$ from all parties, sample random values $a_1, \cdots, a_{n-1} \leftarrow F$, and set $a_n \leftarrow a_1 \cdots a_{n-1}$, [Val[$\mathrm{id}_1$], $\cdots$, Val[id_n]] $\leftarrow (a_1, \cdots, a_n)$.

Definition 2 ([25]). A key management (dispensation) protocol is a multi-party computation consisting of data owner O and a set of servers $S_1, \cdots, S_n$ who collaborate with each other to dispense master key k by running the following algorithms:

- key generation algorithm: on input of security parameter 1^κ, it generates a public and secret key pair for participants:
  - on input of security parameter 1^κ, an instance of semantically secure encryption scheme is invoked. The public key of instance is denoted pk_E (note that secret sk_E is not an output; hence our protocol is defined in the common public key model);
  - on input of pk_E, a data owner generates a pair of public and secret key pair (pk_O, sk_O);
  - on input of pk_E, individual server S_i generates a pair of public and secret key pairs (pk_i, sk_i) independently, $i = 1, \cdots, n$.
- on input of (pk_E, pk_O, pk_i), a common reference public key pk that will be used to dispense the master key of the data owner is generated, $i = 1, \cdots, n$.
- on input of secret keys k and pk, a multi-party processing executes among the data owner and the n servers. The output of each server is a random key k_i that is secretly selected by each server S_i.

We say a key management scheme is multiplicatively homomorphic if $k = k_O k_1 \cdots k_n$ (a multiplication procedure is performed over a cyclic group).

3.2 Construction of BTG Using mHKM

In this section, a new construction of BTGs different from our previous result [26] is proposed and the details are depicted in Table 3. We remark that our previous

Table 3. Construction of BTG protocol based on mHKM

BTGs based on mHKM protocols

- Step 1. BTG_A performs the following computations:
  - computing $h_{AC} = h_A \times h_C \bmod p$;
  - randomly selecting $r_A \in Z_q$, $a \in Z_p^*$ and then computing $u_A = g^{r_A} \bmod p$ and $v_A = a \times h_{AC}^{r_A} \bmod p$;
  - sending (u_A, v_A) to BTG_B;
- Step 2. Upon receiving (u_A, v_A) from BTG_A, BTG_B performs the following computations:
  - randomly selecting $r_B \in Z_q$, $b \in Z_p^*$ and computing $u_B = u_A g^{r_B}$ and $v_B = v_A \times b \times h_{AC}^{r_B} \bmod p \ (= a \times b \times h_{AC}^{r_A} \times h_{AC}^{r_B} \bmod p)$;
  - sending (u_B, v_B) to BTG_A;
- Step 3: Upon receiving (u_B, v_B) from BTG_B, BTG_A performs the following computations:
  - $v_C \overset{def}{=} v_B / u_B^{x_A} \ (= a \times b \times h_{AC}^{r_A} \times h_{AC}^{r_B} / u_B^{x_A} = a \times b \times (h_A \times h_C)^{(r_A + r_B)} / u_B^{x_A}$ $= a \times b \times h_C^{(r_A + r_B)})$;
  - computing $u_C \overset{def}{=} u_B \times g^{r'_B} \ (= g^{(r_A + r_B + r'_B)} \bmod p)$ and $v_C \leftarrow v_C \times h_C^{r'_B} \bmod p$, and then sending (u_C, v_C) to BTG_C.
- Upon receiving (u_C, v_C), BTG_A performs the following computations:
  - computing $c \overset{def}{=} v_C / u_C^{x_C} \bmod p$.

---

solution [26] is constructed leveraging El Gamal encryption scheme by a random selection of c and a and then computing $b = ca^{-1}$. In this paper, we propose a concise procedure to compute Beaver triple by starting from the computations of a and b and then computing $c = ab$. Consequently, the inversion procedure required in the current solution [26] can be eliminated.

Let p be a large safe prime number, $p = 2q + 1$, p and q are prime numbers. Let $G \subseteq Z_p^*$ be a cyclic group of order q and g be a generator of G. Let $h_i = g^{x_i} \bmod p$, where $x_i \in_R [1, q]$ is randomly generated by the Beaver multiplication triple generator BTG_i $(i \in \{A, B, C\})$. By applying the same proof technique [26] and assuming that the underlying El Gamal encryption [22] is semantically secure, we claim that the proposed BTG protocol is secure against semi-honest adversary.

3.3 Construction of mHKM Using BTG

In this section, we provide a construction of multiplicatively homomorphic key management protocols based on the Beaver triple generators depicted above. The idea behind our construction is that for each group of triple generators $(BTG_n, BTG_{n-1}, BTG'_{n-2})$, we generate $a_n = a_{n-1} a'_{n-2}$. This a'_{n-2} will be viewed as a master key and the owner of this value invokes an instance of BTGs as a party computation $(BTG'_{n-2}, BTG_{n-2}, BTG'_{n-3})$ to generate a triple $(a'_{n-2}, a_{n-2}, a'_{n-3})$ such that $a'_{n-2} = a_{n-2} a'_{n-3}$. The procedure continues until we

Table 4. Construction of mHKMs based on BTGs

Description of mHKM protocol leveraging BTG

- Invoking an instance of BTG protocol with three parties BTG_n, BTG_{n-1} and BTG'_{n-2} such that $a_n = a_{n-1} \times a'_{n-2}$, assuming that BTG_i generates its private value a_i and relaying BTG'_i generates a'_i, $(i = 1, \cdots, n)$;
- Invoking an instance of BTG protocol with three parties BTG'_{n-2}, BTG_{n-2}, BTG_{n-3} such that $a'_{n-2} = a_{n-2} \times a'_{n-3}$.
- The invocation procedure of BTG instances continues until it reaches final result $a'_2 = a_2 \times a'_1$ $(:= a_2 \times a_1)$, where $a_1 \leftarrow a'_1$.

---

obtain $a_n = a_{n-1} \cdots a_1$. Details are depicted in Table 4. One can verify that from $(BTG_n, BTG_{n-1}, BTG'_{n-2})$, $(BTG'_{n-2}, BTG_{n-2}, BTG'_{n-3})$, $\cdots$, (BTG_3, BTG_2, BTG_1), we know that $a_n = a_{n-1} \cdots a_1$. A rigorous proof of our construction is presented in Theorem 1. Consequently, we are able to show the equivalence between two different cryptographic primitives.

3.4 Proof of Security

Theorem 1. *Let* g_{mHKM} *be the multiplicatively homomorphic key management functionality. Let* $\Pi^{g_{mHKM}|f_{BTG}}$ *be an oracle-aided protocol that privately reduces* g_{mHKM} *to* f_{BTG} *and* $\Pi^{f_{BTG}}$ *be a protocol privately computes* f_{BGT}. *Suppose* g_{mHKM} *is privately reducible to* f_{BTG} *and that there exists a protocol privately computing* f_{BTG}, *then there exists a protocol for privately computing* g_{mHKM}.

Proof. We construct a protocol Π for computing g_{mHKM} by replacing each invocation of the oracle f_{BTG} with an execution of protocol $\Pi^{f_{BTG}}$. Note that in the semi-honest model, the steps executed $\Pi^{g_{mHKM}|f_{BTG}}$ inside Π are independent of the actual execution of $\Pi^{f_{BTG}}$ but depends only on the output of $\Pi^{f_{BTG}}$.

Let $S_i^{g_{mHKM}|f_{BTG}}$ and $S_i^{f_{BTG}}$ be the corresponding simulators for the view of party P_i (either BTG_i or BTG'_i). We construct a simulator S_i for the view of party P_i in Π. That is, we first run $S_i^{g_{mHKM}|f_{BTG}}$ and obtain the simulated view of party P_i in $\Pi^{g_{mHKM}|f_{BTG}}$. This simulated view includes queries made by P_i and the corresponding answers from the oracle. Invoking $S_i^{f_{BTG}}$ on each of partial query-answer (q_i, a_i), we fill in the view of party P_i for each of these interactions of $S_i^{f_{BTG}}$. The rest of the proof is to show that S_i indeed generates a distribution that is indistinguishable from the view of P_i in an actual execution of Π.

Let H_i be a hybrid distribution representing the view of P_i in an execution of $\Pi^{g_{mHKM}|f_{BTG}}$ augmented by the corresponding invocation of $S_i^{f_{BTG}}$. That is, for each query-answer pair (q_i, a_i), we augment its view with $S_i^{f_{BTG}}$. It follows that H_i represents the execution of protocol Π with the exception that $\Pi^{f_{BTG}}$ is replaced by simulated transcripts. We show the following facts, from which our claim is proved.

- The distributions between H_i and Π are computationally indistinguishable: Note that the distributions of H_i and Π differ from $\Pi^{f_{BTG}}$ and $S_i^{f_{BTG}}$ which is computationally indistinguishable assuming that $\Pi^{f_{BTG}}$ securely computes f_{BTG}.

- The distributions between H_i and S_i are computationally indistinguishable: Note that the distributions between $(\Pi^{g_{mHKM}|f_{BTG}}, S_i^{f_{BTG}})$ is computationally indistinguishable from $(S_i^{g_{mHKM}|f_{BTG}}, S_i^{f_{BTG}})$. The distribution $(S_i^{g_{mHKM}|f_{BTG}}, S_i^{f_{BTG}})$ defines S_i. That means H_i and S_i are computationally indistinguishable.

4   Decoupling Model for Beaver Triple Dispensation

In this section, we provide a decoupling model for Beaver triple dispensation. Different from the single/centric Beaver triple generator case [5], our model works in the three Beaver triple generators setting, where each of BTGs holds its own private value a or b or c such that $c = a \times b$. We assume that there are m-MPC servers $(MPC_1, \cdots, MPC_m)$ running in the federated computing architecture, where each private data silo plays the role of a MPC server as well. We separate roles of BTGs from that of MPCs and assume that the role of each BTG is to generate and dispense triples while the role of each MPC is to process data leveraging the available triples. Borrowing the notion of random processing, we propose a new protocol, which we call blind triple dispensation (BTD). BTD in essence is a randomized processing for Beaver triple dispensation, which benefits MPCs protecting their private data if multiplications are processed in the framework of SPDZ. The details are described below.

4.1   Blind Triple Dispensation

Let F be a finite field with a predefined security parameter. To multiply $x \in F$ and $y \in F$, the data owner O_x of x calls for the Beaver triple component a while data owner O_y of y calls for the Beaver triple component b. Different from the state-of-the-art sharing-before-processing solutions, our solution is based on processing-before-sharing paradigm. That is, we allow data owners to first make requests for share entire Beaver triple components to the corresponding BTGs and then process the gained components for later sharing. Our solution saves communication bandwidth for blind dispensations for off-line computations. Below, two randomized splitting solutions are provided to dispense the generated Beaver triples.

4.1.1   Single-Randomness Solution

A splitting of the private value a is defined by $[a] = (a_1, \cdots, a_n)$, where $a = a_1 + \cdots + a_n$ over F. To compute a multiplication xy in F, O_x and O_y collaboratively generate a random value $r \in F$, which plays the role of blind factor to protect the called triple components. O_x then sends a request to BTG_A who holds private triple component a while O_y sends a request to BTG_B who holds private triple component b. Let $a' = r \times a$ and $b' = r \times b$, it follows that $[a'] = r[a]$ and $[b'] = r[b]$. The details of processing is depicted below.

- O_x (for simplicity, we assume that O_x is MPC_1) selects $a'_2, \cdots, a'_m \in Z_p^*$ uniformly at random, and then sends a'_2 to MPC_2, $\cdots$, a'_m to MPC_m via private channels shared between O_x and MPC_j, $j = 1, \cdots, m$;
- O_x computes $a'_1 = ra - a'_2 - \cdots - a'_m$ mod p and keeps a'_1 locally.

Since there are m-MPC (as above, we simply assume that MPC_1 holds x and MPC_2 holds private data y), an interesting question is who is a good candidate making request c to BTG_c. Here we apply the committee selection technique presented in [23] to the set $\{\mathrm{MPC}_3, \cdots, \mathrm{MPC}_n\}$. The selected committee leader (say, MPC_3) will make a quest to BTG_C and then computes $[c'] = r^2[c]$, where r is shared with committee leader MPC_3. One can verify that $[c'] = [a'] \times [b']$. We emphasize that the shares of randomized triple $([a'], [b'], [c'])$ will be used as Beaver triples in the context of SPDZ (instead of $[a], [b], [c]$).

4.1.2 Two-Randomness Solution

Let $a' = r_a \times a$ and $b' = r_b \times b$. It follows that $[a'] = r_a[a]$ and $[b'] = r_b[b]$. To compute $r_c = r_a \times r_b$, we apply three-party computation protocol as described in Sect. 3, where O_x, O_y and committee leader (say, MPC_3) are involved and the shares $r_a[a]$, $r_b[b]$ and $r_c[c]$ are dispensed respectively.

4.2 Multiplication

Assuming that MPC_1 holds private data x and MPC_2 holds private data y. Suppose m parties MPC_1, $\cdots$, MPC_m wish to compute $[xy]$ collaboratively given $[x]$ and $[y]$. We assume MPC_1 makes a request of a to BTG_A, and then applies the depicted blind triple dispensation procedure to get a share $[a]$. The same procedure applies to MPC_2 to obtain a share $[b]$. Borrowing the notation from SPDZ, by ρ, we denote an opening of $[x] - [a]$ and by ϵ, an opening of $[y] - [b]$. Given ρ and ϵ, each party can compute his secret share of $[xy] = (\rho + [a])$ $(\epsilon + [b]) = [c] + \epsilon[a] + \rho[b] + \rho\epsilon$ locally.

5   Experiments

In this section, we test the efficiency of Beaver triple generators constructed from multiplicatively homomorphic key management protocol. We use Python 3.8.1 in a Window 10 system with processors: Intel(R) Core(TM) i7-8665U CPU 1.90 GHz 2.11 GHz; installed memory (RAM) 16.0 GB (15.8 GB usable) and system type: 64-bit operating system, x64-based processor.

19. Smart, N.P., et al.: TaaS: commodity MPC via triples-as-a-service. In: ACM SIGSAC Conference on Cloud Computing Security Workshop, London, UK, pp. 105–116 (2019)
20. Hardy, S., et al.: Private federated learning on vertically partitioned data via entity resolution and additively homomorphic encryption. CoRR abs/1711.10677 (2018)
21. Nock, R., et al.: Entity resolution and federated learning get a federated resolution. CoRR abs/1803.04035 (2018)
22. ElGamal, T.: A public-key cryptosystem and a signature scheme based on discrete logarithms. IEEE Trans. Inf. Theor. **31**(4), 469–472 (1985)
23. Chen, J., Micali, S.: Algorand: a secure and efficient distributed ledger. Theor. Comput. Sci. **777**, 155–183 (2019)
24. Yao, A.C.-C.: How to generate and exchange secrets (extended abstract). In: FOCS 1986, pp. 162–167 (1986)
25. Zhu, H., et al.: Sustainable data management strategies and systems in untrusted cloud environments. In: 6th International Conference on Information Technology: IoT and Smart City, Hong Kong, China. December 2018, pp. 163–167 (2018)
26. Zhu, H., et al.: Privacy-preserving weighted federated learning within the secret sharing framework. IEEE Access **8**, 198275–198284 (2020)

Highly Scalable Beaver Triple Generator from Additively Homomorphic Encryption

Huafei Zhu[1]([✉]) and Wee Keong Ng[2]

[1] Institute for High Performance Computing, Agency for Science and Technology, Singapore, Singapore
zhu_huafei@ihpc.a-star.edu.sg
[2] School of Computer Science and Engineering, Nanyang Technological University, Singapore, Singapore
awkng@ntu.edu.sg

Abstract. In a convolution neural network, a composition of linear scalar product, non-linear activation function and maximum pooling computations are intensively invoked. As such, to design and implement privacy-preserving, high efficiency machine learning mechanisms, one highly demands a practical crypto tool for secure arithmetic computations. SPDZ, an interesting framework of secure multi-party computations is a promising technique deployed for industry-scale machine learning development if one is able to generate Beaver (multiplication) triple offline efficiently. This paper studies secure yet efficient Beaver triple generators leveraging privacy-preserving scalar product protocols which in turn can be constructed from additively homomorphic encryptions (AHEs). Different from the state-of-the-art solutions, where a party first splits her private input into a shared vector and then invokes an AHE to compute scalar product of the shared vectors managed MPC servers, we formalize Beaver triple generators in the context of 2-party shared scalar product protocol and then dispense the generated shares to MPC servers. As such, the protocol presented in this paper can be viewed as a dual construction of the state-of-the-art AHE based solutions. Furthermore, instead of applying the Paillier encryption as a basis of our previous constructions or inheriting from somewhat fully homomorphic encryptions, we propose an alternative construction of AHE from polynomial ring learning with error (RLWE) which results in an efficient implementation of Beaver triple generators.

1 Introduction

In a convolution neural network (CNN for short), a party first performs a filtering computation leveraging the proposed kernel, then applies an activation function (say, relu) to the output of convolution layer and finally applies a sub-sampling/pooling strategy (say, max-pooling) to the output of the relu function. This procedure repeats several times (depending on the number of layers defined

© The Author(s), under exclusive license to Springer Nature Switzerland AG 2022
L. Barolli et al. (Eds.): AINA 2022, LNNS 451, pp. 504–514, 2022.
https://doi.org/10.1007/978-3-030-99619-2_48

by the CNN structure) before a sequence of full connection procedures will be applied to the output of the convolution neural network. Suppose Alice who holds a private CNN image classification model, where each layer can be abstracted as a weight vector $w = (w_1, \cdots, w_n)$ and a bias b, collaboratively makes a prediction on a private input $x = (x_1, \cdots, x_n)$ from by the player Bob. A secure computation of CNN can thus be reduced to that of $(b, w) \cdot (1, x)^T$ for filtering and full connection computations together with a comparison protocol for the relu procedure. Since different techniques will be applied for implementing secure scalar product and secure comparison protocols, we will focus on the secure scalar product protocol throughout the paper and leave the development of secure yet highly scalable comparison protocol for the future research.

We remark that a scalar product protocol can be implemented using different ways. One may choose homomorphic encryption based solutions, or differential privacy based solutions or secure multi-party computations. Each of method has its own pros and cons. While the efficiency of fully homomorphic encryption has been improved dramatically in the recent years, computational complexity of multiplication of two ciphertexts is still a bottleneck. For a differential privacy based computation, the raw data is polluted. While a secure computation within the framework of SPDZ provides an efficient way to compute arithmetic additions and multiplications, the generation of offline multiplication (Beaver) triples is challenging on which this paper is focusing.

Since the introduction of secure multi-party computation ([11,14,20] and references therein), many practical yet secure two-party and multi-party computation (MPC) protocols have been developed. The notable developments among them are ring-based zero-splitting [1,2,5,6] and SPDZ-based solutions [3,4,7–9,12,15,18]. Since a computation of arithmetic addition within the SPDZ framework is as efficient as that within the ring-based zero-splitting framework and also an arithmetic multiplication can be computed efficiently in the SPDZ framework if an auxiliary multiplication (Beaver) triple that is used to assist the multiplication is available, an efficient Beaver triple generator results in an efficient implementation of MPC trivially. An interesting research problem thus is – *how to construct highly scalable, high efficiency and secure multiplication (Beaver) triples for industry scale deployment?*

1.1 This Work

The Beaver triples deployed in the SPDZ framework are originally constructed from somewhat homomorphic encryptions [9]. Since computational cost of ciphertext multiplications is high if one leverages somewhat homomorphic encryptions, researchers are now studying alternative constructions such as oblivious-transfers [12], additively homomorphic encryptions [13,17], multiplicatively homomorphic encryption [24] and multiplicatively homomorphic key management protocol [25]. In this work, an efficient solution for generating Beaver triples leveraging asymmetric oblivious scalar product protocols is proposed and analyzed. The notation of asymmetric oblivious scalar protocol was first introduced and formalized by Zhu et al. in 2006 [21]. Suppose that Alice has an input vector $inp_A = (1, x_1, \cdots, x_l)$ $(x_i \in F, 1 \leq i \leq l, F$ is a finite field) while Bob has an input vector $inp_B = (y_0, y_1, \cdots, y_l)$ $(y_i \in F, 0 \leq i \leq l)$, asymmetric oblivious scalar protocol allows two parties Alice and Bob to collaboratively compute scalar-product obliviously so that at the end of the protocol execution, Alice learns $\sum_{i=0}^{l} x_i y_i$ while Bob learns nothing. As discussed above, we know that an asymmetric oblivious scalar protocol can be viewed as an abstraction of privacy-preserving scalar product in a convolution neural network.

It has already shown that the notion of oblivious scalar protocol in essence, is a shared scalar product protocol and an implementation of oblivious scalar protocol leveraging the Paillier's additively homomorphic encryption [16] has been proposed by Zhu et al. [22,23]. Instead of directly applying the state-of-the-art Paillier's encryption as a basis presented in our previous constructions, we investigate an alternative construction from the polynomial ring learning with error (ring-LWE, or RWLE for short). As demonstrated in Sect. 4, the proposed additive-only homomorphic encryption scheme is more efficient compared with the additive-only Paillier's encryption. It follows that we are able to propose an efficient implementation of Beaver triple generators.

Different from the state-of-the-art solutions proposed in [13,17], where a party (say, Alice) first splits her private input $\mathbf{x_A}$ to a share vector $\mathbf{x_A} = (x_{A,1}, \cdots, x_{A,l})$ and then invokes an AHE to compute shares s_A with Bob whose input is $\mathbf{x_B}$ and $\mathbf{x_B} = (x_{B,1}, \cdots, x_{B,l})$ and output is s_B such that $\mathbf{x_A} \cdot \mathbf{x_B} = s_A + s_B$, we formalize Beaver triple generators in the context of 2-party shared scalar product protocol to get $s_A + s_B = \mathbf{x_A} \cdot \mathbf{x_B}$ and then Alice (Bob resp.,) dispenses her shares $\mathbf{s_A} = (s_{A,1}, \cdots, s_{A,l})$ $(\mathbf{s_B} = (s_{B,1}, \cdots, s_{B,l})$ resp.,) to MPC servers. As such, the protocol presented in this paper can be viewed as a dual construction of the state-of-the-art AHE based solutions.

1.2 Secure CNN Computation Within SPDZ

We now provide a generic view to demonstrate why SPDZ is a promising framework for secure computation of CNN layered function $(b, w) \cdot (1, x)^T$. Let MPC_1, $\cdots$, MPC_m be m-party computational servers and $[x]$ be a share of x among the m-party. Assuming that MPC_1 (say, Alice) holds private data x (for simplicity, we view parameters of CNN as private data of Alice) and MPC_2 (say, Bob) holds private data y. Suppose m parties MPC_1, $\cdots$, MPC_m wish to compute $[xy]$ collaboratively. MPC_1 first selects an auxiliary value $x_A \in F$ uniformly at random and then invokes an additive-only homomorphic encryption (AHE) to perform a secure 2-party computation with MPC_2 whose auxiliary value is $x_B \in F$ (F is underlying finite field). Let $[s_A]$ ($[s_B]$ resp.,) be a share vector of s_A (s_B resp.,). Borrowing the notation from SPDZ, by ρ, we denote an opening of $[x] - [x_A]$ and by ϵ, an opening of $[y] - [x_B]$. Here an opening refers to a procedure where all participants send their shares to the initiator (either Alice or Bob in our case) via established secure channels. Since a secure channel between two parties can be easily implemented assuming the existence of public key infrastructure (PKI), we simply assume that there is a secure channel between each pair of MPC servers throughout the paper. Given ρ and ϵ, each party can compute secret

shares $[xy]$ of xy as follows: $[xy] = (\rho + [x_A]) (\epsilon + [x_B]) = [x_A x_B] + \epsilon[x_A] + \rho[x_B]$
$+ \rho\epsilon = [s_A] + [s_B] + \epsilon[x_A] + \rho[x_B] + \rho\epsilon$. As a result, if we are able to propose
an efficient yet secure computation of $x_A x_B = s_A + s_B$, the SPDZ provides a
promising framework to compute $[xy]$ indeed.

The rest of this paper is organized as follows: in Sect. 2, syntax, functionality and security definition of shared scalar product are proposed. An interesting
additive-only homomorphic encryption from polynomial ring learning with error
is constructed and analyzed in Sect. 3. We apply the developed additive-only
homomorphic encryption to generate Beaver triples and construct a secret sharing mechanism and dispensing protocol in Sect. 4. We conclude our work in
Sect. 5.

2 Syntax, Functionality and Security Definition of Shared Scalar Product

2.1 Syntax

A shared scalar product protocol consists of the following two probabilistic polynomial time (PPT) Turing machines:

- On input system parameter l, a PPT Turing machine A (say, Alice), chooses
  l elements $x_1, \cdots, x_l \in F$ uniformly at random (throughout the paper, we
  assume that $F = Z_m^*$, where m is a large prime number). The input vector of
  Alice is denoted by $inp_A = (x_1, \cdots, x_l)$;
- On input system parameter and l, a PPT Turing machine B (say, Bob),
  chooses l elements $y_1, \cdots, y_l \in Z_m$ uniformly at random. The input vector of
  Bob is denoted by $inp_B = (y_1, \cdots, y_l)$;
- On inputs inp_A and inp_B, Alice and Bob jointly compute the value
  $\sum_{i=1}^{l} x_i y_i \bmod m$;
- The output of Alice is s_A while Bob is s_B such that $\sum_{i=1}^{l} x_i y_i \bmod m = s_A + s_B$.

2.2 Functionality

The functionality $\mathcal{F}_{SSP}$ of shared scalar product protocol (SSP) can be
abstracted as follows:

- A player (say Alice) has her input vector $inp_A = (x_1, \cdots, x_l)$; Another player
  (say Bob) has his input vector $inp_B = (y_1, \cdots, y_l)$; Each participant sends the
  corresponding input set to $\mathcal{F}_{SSP}$ — an imaginary trusted third party in the
  ideal world via a secure and private channel.
- Upon receiving inp_A and inp_B, $\mathcal{F}_{SSP}$ checks whether $x_i \in Z_m$ and $y_i \in Z_m$
  $(1 \le i \le l)$.
  If the conditions are satisfied, then $\mathcal{F}_{SSP}$ computes $\sum_{i=1}^{l} x_i y_i \bmod m$;
  If there exists a subset $\tilde{s_A} \subset inp_A$ such that each $x_i \in \tilde{s_A}$ but $x_i \notin Z_m$, then
  $\mathcal{F}_{SSP}$ chooses an element $x_i' \in_r Z_m$ and substitutes x_i with x_i'. Similarly,

if there exists a subset $\tilde{s}_B \subset inp_B$ such that each $y_i \in \tilde{s}_B$ but $y_i \notin Z_m$, then $\mathcal{F}_{SSP}$ chooses an element $y_i' \in_r Z_m$ and substitutes y_i with y_i'. By $inp_A = (x_1, \cdots, x_l)$ (using the same notation of the input vector of Alice) we denote the valid input set of Alice which may be modified by $\mathcal{F}_{SSP}$ and by $inp_B = (y_1, \cdots, y_l)$ (again using the same notation of the input vector of Bob), we denote the valid input set of Bob which may be modified of the original input values by $\mathcal{F}_{SSP}$. Once given the valid input sets inp_A and inp_B, $\mathcal{F}_{SSP}$ computes $\sum_{i=}^{l} x_i y_i$ mod m.

Finally, $\mathcal{F}_{SSP}$ sends s_A to Alice via the secure and private channel while Bob learns s_B such that $\sum_{i=1}^{l} x_i y_i$ mod m.

- The output of Alice (Bob resp.) is s_A (s_B resp.) which is sent by $\mathcal{F}_{SSP}$ via the secure and private channel between them.

Remark 1. Notice that for semi-honest adversary, $\mathcal{F}_{SSP}$ does not check the input from Alice or Bob. That is, upon receiving inp_A and inp_B from a PPT distinguisher, $\mathcal{F}_{SSP}$ assumes that both Alice and Bob follow the protocol. $\mathcal{F}_{SSP}$ simply selects s_A and s_B uniformly at random such that $s_A + s_B = \sum_{i=1}^{l} x_i y_i$ mod m.

2.3   Security Definition

The security definition of shared scalar product protocols is defined in terms of the ideal-world vs. real-world framework. In this framework, we first consider an ideal model in which two dummy participants join in the imaginary trusted third party (TTP) ideal world. Note that the task of our construction is to remove such an imaginary TTP which does not exit in the real world. All performances are then computed via this trusted party. Next, we consider the real model in which a real two-party protocol is executed. A protocol in the real model is said to be secure with respect to certain adversarial behavior if the possible real execution with such an adversary can be simulated in the ideal model. That is, we want to show that there exists a polynomial time transform of adversarial behavior in the real conversation into corresponding adversarial behavior in the ideal model. We follow the security definitions of our previous work [21–23].

Definition 1. A shared scalar product protocol is secure against malicious (semi-honest resp.) Alice A, there exists a simulator sim_A that plays the role of A in the ideal world such that for any probabilistic polynomial time distinguisher D, the view of D when it interacts with A in real conversation is computationally indistinguishable from that when it interacts with sim_A in the ideal world.

Definition 2. A shared scalar product protocol is secure against malicious (semi-honest resp.) Bob B, if there exists a simulator sim_B that plays the role of B in the ideal world such that for any probabilistic polynomial time distinguisher D, the view of D when it interacts with B in real conversation is computationally indistinguishable from that when it interacts with sim_B in the ideal world.

Definition 3. A shared scalar product protocol is secure for any static probabilistic polynomial time (PPT) adversary if it is secure for any PPT Alice and any PPT Bob.

3 Additive-Only Homomorphic Encryption

Additively homomorphic encryption scheme can be inherited from somewhat fully homomorphic encryption scheme. The state-of-the-art somewhat fully homomorphic encryption is efficient if only additive property is deployed. With this observation in mind, we construct our AHE from the RLWE assumption.

3.1 Additive-Only Homomorphic Encryption Based on Ring-LWE

Let $f(x) = x^n + 1$ be a cyclotomic polynomial, i.e., the minimal polynomial of primitive roots of unity with $n = 2^d$. Let $R = \mathbb{Z}[x]/(f(x))$ and elements of the ring R will be denoted in lowercase bold, e.g. $\mathbf{a} \in R$. The coefficient of an element in R will be denoted by a_i such that $\mathbf{a} = \sum_{i=0}^{n-1} a_i x^i$. The infinite norm $||\mathbf{a}||$ is defined as $\max_i |a_i|$, and the ratio of expansion factor of R is defined as $\delta_R = \max\{||\mathbf{a} \cdot \mathbf{b}||/||\mathbf{a}|| \cdot ||\mathbf{b}|| : \mathbf{a}, \mathbf{b} \in R\}$.

Let q be an integer and by $\mathbb{Z}_q$ we denote the set of integers $(-q/2, q/2]$. Notice that $\mathbb{Z}_q$ is simply considered as a set and thus the notion of $\mathbb{Z}_q$ is different from that of $\mathbb{Z}/q\mathbb{Z}$. Let R_q be the set of polynomials in R with coefficients in $\mathbb{Z}_q$. For $a \in \mathbb{Z}$, we denote $[a]_q$ be the unique integer in $\mathbb{Z}_q$ with $[a]_q = a \mod q$. For $a, q \in \mathbb{Z}$, we define the remainder modulo q by $r_q(a) \in [0, q-1]$. For $\mathbf{a} \in R$, we denote $[\mathbf{a}]_q$ the element of R obtained by applying $[\cdot]_q$ to all its coefficients. Similarly, for $x \in R$, we use $[x]$ to denote rounding to the nearest integer, $\lceil x \rceil$ rounding up to the nearest integer and $\lfloor x \rfloor$ rounding down the nearest integer.

Definition 4. Decision $RLWE_{d,q,\chi}$ problem: for security parameter λ, let $f(x)$ a cyclotomic polynomial with degree $def(f) = \phi(m)$ depending on λ and $R = \mathbb{Z}[x]/(f(x))$ and $q = q(\lambda)$. For a random $\mathbf{s} \in R_q$ and a distribution χ over R, by $A_{\mathbf{s},\chi}^{(q)}$ we denoted a distribution obtained by choosing a uniformly random element $\mathbf{a} \leftarrow R_q$ and a noise term $\mathbf{e} \leftarrow \chi$ and outputting $[\mathbf{a}, \mathbf{a} \cdot \mathbf{s} + \mathbf{e}]$. The decision $RLWE_{d,q,\chi}$ problem aims to distinguish between distributions $A_{\mathbf{s},\chi}^{(q)}$ and the uniform distribution $U(R_q^2)$. The hardness of decision $RLWE_{d,q,\chi}$ problem assumes that there is no PPT distinguisher for the decision $RLWE_{d,q,\chi}$ problem.

Remark 2. As noted in [10], the distribution χ in general is not as simple as just sampling coefficients according to the Gaussian distribution $D_{\mathbb{Z},\sigma}^n$ ($\mu = 0$ and standard deviation σ). However, for polynomial $f(x) = x^n + 1$ and $n = 2^d$, we can indeed define χ as $D_{\mathbb{Z},\sigma}^n$. Also, notice that we can assume that $[\mathbf{a}, \mathbf{a} \cdot \mathbf{s} + \mathbf{e}] \approx U(R_q^2)$ for $\mathbf{s} \in \chi$ chosen uniformly at random.

3.2 A New Construction of Polynomial-Ring LWE Encryption

Motivated by the work of Fan and Vercauteren [10], an efficient additively homomorphic scheme is presented and analysed. Our polynomial ring LWE encryption consists of the following algorithms (KeyGen, Enc, Dec and Eval):

- the key generation algorithm KeyGen: on input a security parameter λ, KeyGen samples $\mathbf{a} \in R_q$ and $\mathbf{s}, \mathbf{e} \leftarrow \chi$; the public key $pk \stackrel{def}{=} ([-\mathbf{a} \cdot \mathbf{s} + t\mathbf{e}]_q, \mathbf{a})$ and secret key $sk \stackrel{def}{=} \mathbf{s}$. The output of KeyGen is (pk, sk).
- the encryption algorithm Enc: on input a message $\mathbf{m} \in R_t$, let $\mathbf{p_0} = pk[0]$ and $\mathbf{p_1} = pk[1]$, Enc samples $\mathbf{u}, \mathbf{e_0}, \mathbf{e_1} \leftarrow \chi$, and returns $ct = ([\mathbf{p_0}\mathbf{u} + t\mathbf{e_0} + \mathbf{m}]_q, [\mathbf{p_1}\mathbf{u} + t\mathbf{e_1}]_q)$.
- the decryption algorithm Dec: on input ct, let $\mathbf{c_0} = ct[0]$, $\mathbf{c_1} = ct[1]$ and the message $\mathbf{m} \in \mathbb{Z}_t$ is computed from $[[\mathbf{c_0} + \mathbf{c_1}\mathbf{s}]_q]_t$.
- the evaluation algorithm Eval: on input two ciphertexts ct_1 and ct_2, Eval outputs a ciphertext c such that $\text{Dec}(c) = \text{Dec}(ct_1) + \text{Dec}(ct_2)$.

The correctness is following from the lemma described below:

$$\begin{aligned}
\mathbf{c_0} + \mathbf{c_1}\mathbf{s} &= \mathbf{p_0}\mathbf{u} + t\mathbf{e_0} + \mathbf{m} + q\mathbf{k_0} + (\mathbf{p_1}\mathbf{u} + t\mathbf{e_1})\mathbf{s} + q\mathbf{k_1} \\
&= \mathbf{m} + (-\mathbf{a}\mathbf{s} + t\mathbf{e} + q\mathbf{k})\mathbf{u} + t\mathbf{e_0} + (\mathbf{a}\mathbf{u} + t\mathbf{e_1})\mathbf{s} + q(\mathbf{k_0} + \mathbf{k_1}\mathbf{s}) \qquad (1) \\
&= \mathbf{m} + t(\mathbf{e}\mathbf{u} + \mathbf{e_0} + \mathbf{e_1}\mathbf{s}) + q(\mathbf{k}\mathbf{u} + \mathbf{k_0} + \mathbf{k_1}\mathbf{s})
\end{aligned}$$

Since $[\mathbf{c_0} + \mathbf{c_1}\mathbf{s}]_q = \mathbf{m} + t(\mathbf{e}\mathbf{u} + \mathbf{e_0} + \mathbf{e_1}\mathbf{s})$, it follows that $[[\mathbf{c_0} + \mathbf{c_1}\mathbf{s}]_q]_t = \mathbf{m}$. The proof of security is extract the same as that presented in [10] and thus it is omitted.

3.3 Additive and Scalar Properties

Let $\mathbf{c} = (\mathbf{ct}[0], \mathbf{ct}[1])$ be an encryption of message $\mathbf{m} \in \mathbb{Z}_t$. Let $\mathbf{c'} = (\mathbf{ct'}[0], \mathbf{ct'}[1])$ be an encryption of message $\mathbf{m'} \in \mathbb{Z}_t$. It follows that $\mathbf{ct}[0] + \mathbf{ct'}[0] = \mathbf{p_0}\mathbf{u} + t\mathbf{e_0} + \mathbf{m} + q\mathbf{k_0} + \mathbf{p_0}\mathbf{u'} + t\mathbf{e_0'} + \mathbf{m'} + q\mathbf{k_0'} = \mathbf{p_0}(\mathbf{u} + \mathbf{u'}) + t(\mathbf{e_0} + \mathbf{e_0'}) + (\mathbf{m} + \mathbf{m'}) + q(\mathbf{k_0} + \mathbf{k_0'})$ and $\mathbf{ct}[1] + \mathbf{ct'}[1] = \mathbf{p_1}(\mathbf{u} + \mathbf{u'}) + t(\mathbf{e_1} + \mathbf{e_1'}) + q(\mathbf{k_1} + \mathbf{k_1'})$. Applying the decryption procedure above, one gets addition $(\mathbf{m} + \mathbf{m'})$ from the aggregated ciphertext with the help of the secret key $\mathbf{s}$. This means that the proposed encryption scheme is additively homomorphic.

4 The Construction and Proof of Security of Beaver Triples

In this section, we are able to propose a new Beaver triple generator and dispensing protocol leveraging the proposed additive-only homomorphic encryption.

4.1 Beaver Triple Functionality

We write $[x]$ to mean that each party P_i holds a random, additive sharing x_i of x such that $x = x_1 + \cdots + x_n$, where $i = 1, \cdots, n$. The values are stored in the dictionary Val defined in the functionality $\mathcal{F}_{\text{Triple}}$ [12]. Please refer to the Table 1 for more details

Table 1. Beaver triple functionality

The functionality of Beaver triple generator $\mathcal{F}_{\mathrm{BTG}}$
• The functionality maintains a dictionary, Val to keep track of assigned value, where entry of Val lies in a fixed field.   • On input (Triple, id_A,id_B, id_C) from all parties, sample two random values $a, b \leftarrow F$, and set [Val[id_A], Val[id_B], Val[id_C]] $\leftarrow (a, b, ab)$.

4.2   The Construction

We propose an efficient implementation of SSP based on additive-only homomorphic encryption described above. We assume that public and secret key pair of AHE are generated by Alice. We consider a semi-honest adversary and our protocol is described as follows:

1. Let $x_A \in F$ be a random value and c_A be a ciphertext $\mathrm{AHE}(x_A)$. Alice sends c_A to Bob;
2. Upon receiving c_A, Bob randomly selects a value $r_B \in F$ and then computes $c_B \leftarrow x_B \times c_A + \mathrm{AHE}(r_B)$;
3. Upon receiving c_B, Alice decrypts c_B to get $s_A = x_A x_B + r_B$.
4. Alice output s_A while Bob outputs $s_B := -r_B$.

The correctness of protocol can be easily verified and thus it is omitted. The rest of our work is to show that the proposed scheme is secure against static semi-honest adversary.

4.3   The Proof of Security

Theorem 1. *The proposed shared product protocol is secure against static and semi-honest adversary assuming that the underlying* AHE *is semantically secure.*

Proof. Suppose Alice gets corrupted. We construct a simulator as follows:

- The simulator sim_A invokes Alice to generate a pair of public and secret keys (pk_A, sk_A). sim_A is given pk_A and sk_A (in this paper, we are considering the static adversary).
- sim_A then corrupts the corresponding dummy Alice in the ideal world, and gets to know the input x_A and randomness r_A used for the real world protocol execution.
- sim_A sends x_A to and gets s_A from the functionality $\mathcal{F}_{\mathrm{BTG}}$ on behalf of the corrupted Alice.

sim_A then generates a random ciphertext c_B on as a simulation of honest Bob's transcript. Notice that in the real world, the plaintext of c_B is defined by $x_B x_A + r_B$ while in the simulation c_B is an encryption of a random value.

Since the underlying AHE is semantically secure, it follows that any probabilistic polynomial distinguisher cannot distinguish the simulated transcript from that of generated by the real world protocol.

Now, we assume that Bob gets corrupted and construct a simulator as follows:

- Whenever Bob gets corrupted (in this paper, we are considering the static adversary), the simulator sim_B corrupts the corresponding dummy Bob in the ideal world, and gets to know the input x_B and the randomness r_B assigned for the protocol execution.
- Upon receiving c_A, sim_B gets s_B from the functionality $\mathcal{F}_{BTG}$ on behalf of the corrupted Bob. sim_B then generates a ciphertext c_B by computing $x_B c_A +$ AHE($-s_B$).

A probabilistic polynomial time distinguisher's view on the simulated transcript is computationally indistinguishable from that generated in the real world. By combining sim_A and sim_B, we know that the proposed scheme is secure against static semi-honest adversary.

4.4   The Dispense Protocol

Very recently, a block-chain based solution called TaaS (triple as a service) for dispensing Beaver triples leveraging commodity-based cryptography (CBC) [4], has been proposed. In the TaaS framework [19], the concept of ledger is introduced to make sure that the service providers do not reshare twice the same Beaver triple and every transaction has to be logged on a ledger. In this paper, we allows the data owner to control dispensations of Beaver triples and provide two solutions to dispense shares among MPC servers: one is public key based solution and another based on hybrid encryption scheme.

- public-key solution: suppose that Alice gets s_A and Bob gets s_B such that $s_A + s_B = x_A x_B$. Alice (Bob resp.,) can split x_A and s_A by randomly selecting l-tuple $(x_{A,2}, s_{A,2}), \cdots, (x_{A,l}, s_{A,l})$. Alice then computes $x_{A,1}$ from the equation $x_{A,1} + \cdots + x_{A,l} = x_A$ and $s_{A,1}$ from the equation $s_{A,1} + \cdots + s_{A,l} = s_A$. Each of shares is sent to Alice by an encryption of $x_{A,j}$ and $s_{A,j}$, where Alice holds private key of the encryption.
- hybrid solution: alternative solution is to establish a secure channel between each pair of MPC servers and then encrypts shares using the shared session keys. Since these are standard crypto techniques, we omit the detail here.

We insist on the simple dispensing protocol above since a data owner control his/her randomness and hence his/her our data.

4.5   Experiment

In this section, we provide experiment result of our additive-only homomorphic encryption scheme. The test environment is depicted below: the Python 3.8.1 works in the Window 10 with processors: Intel(R) Core(TM) i7-8665U CPU

1.90 GHz 2.11 GHz; installed memory (RAM) 16.0 GB (15.8 usable); and system type: 64-bit operating system, x64-based processor. The test parameters are described below:

- polynomial modulus degree: $n = 2^4$;
- ciphertext modulus: $q = 140737488356903$ (48-bit safe prime number);
- plaintext modulus: $t = 32843$ (16-bit safe prime number);
- polynomial modulus: $poly\_mod = x^n + 1$

For generating 1 million ciphertexts, the encryption algorithm costs about 300 s. To the best of our knowledge, this could be at least 2 to 3 orders efficiency gained compared with that leveraging the Paillier encryption scheme. As such, our design meets with the industrial deployments within the framework of numpy datatype int64. However, more work should be continuously investigated for very larger integer vectors beyond the scope of numpy datatype int64. This interesting task is opened to the research communities.

5   Conclusion

We have proposed a new construction of multiplication triple generators leveraging asymmetric oblivious scalar product protocols and developed a new implementation of the scalar product protocol with the help of the proposed additive-only homomorphic encryption scheme leveraging the ring-learning with error assumption. Our experiment results have shown that the proposed scalar product protocol leveraging the proposed additive-only encryption scheme is more efficient than our previous work based on the Paillier's homomorphic encryption.

References

1. Demmler, D., et al.: ABY - a framework for efficient mixed-protocol secure two-party computation. In: NDSS 2015 (2015)
2. Araki, T., et al.: High-throughput semi-honest secure three-party computation with an honest majority. In: 23rd ACM SIGSAC Conference on Computer and Communications Security, Austria, Vienna, October 2013, pp. 805–817 (2016)
3. Beaver, D.: Efficient multiparty protocols using circuit randomization. In: Feigenbaum, J. (ed.) CRYPTO 1991. LNCS, vol. 576, pp. 420–432. Springer, Heidelberg (1992). https://doi.org/10.1007/3-540-46766-1_34
4. Beaver, D.: Commodity-based cryptography (extended abstract). In: 29th Annual ACM Symposium on Theory of Computing, pp. 446–455. ACM Press, TX, USA (1997)
5. Bogdanov, D., et al.: High-performance secure multi-party computation for data mining applications. Int. J. Inf. Sec. **11**(6), 403–418 (2012)
6. Bogdanov, D., et al.: Students and taxes: a privacy-preserving study using secure computation. Proc. Priv. Enhanc. Technol. **3**, 117–135 (2016)
7. Cramer, R., Damgård, I., Escudero, D., Scholl, P., Xing, C.: SPDZ$_{2^k}$: efficient MPC mod 2^k for dishonest majority. In: Shacham, H., Boldyreva, A. (eds.) CRYPTO 2018. LNCS, vol. 10992, pp. 769–798. Springer, Cham (2018). https://doi.org/10.1007/978-3-319-96881-0_26

8. Damgård, I., Keller, M., Larraia, E., Pastro, V., Scholl, P., Smart, N.P.: Practical covertly secure MPC for dishonest majority – or: breaking the SPDZ limits. In: Crampton, J., Jajodia, S., Mayes, K. (eds.) ESORICS 2013. LNCS, vol. 8134, pp. 1–18. Springer, Heidelberg (2013). https://doi.org/10.1007/978-3-642-40203-6_1

9. Damgård, I., Pastro, V., Smart, N., Zakarias, S.: Multiparty computation from somewhat homomorphic encryption. In: Safavi-Naini, R., Canetti, R. (eds.) CRYPTO 2012. LNCS, vol. 7417, pp. 643–662. Springer, Heidelberg (2012). https://doi.org/10.1007/978-3-642-32009-5_38

10. Fan, J., Vercauteren, F.: Somewhat practical fully homomorphic encryption. IACR Cryptol. ePrint Arch. **2012**, 144 (2012)

11. Goldreich, O.: The Foundations of Cryptography. II: Basic Applications. Cambridge University Press, UK (2004)

12. Keller, M., et al.: MASCOT: faster malicious arithmetic secure computation with oblivious transfer. In: 23rd ACM SIGSAC Conference on Computer and Communications Security, Austria, Vienna, pp. 830–842 (2016)

13. Keller, M., Pastro, V., Rotaru, D.: Overdrive: making SPDZ great again. In: Nielsen, J.B., Rijmen, V. (eds.) EUROCRYPT 2018. LNCS, vol. 10822, pp. 158–189. Springer, Cham (2018). https://doi.org/10.1007/978-3-319-78372-7_6

14. Lindell, Y.: How to simulate it - a tutorial on the simulation proof technique. IACR Cryptol. ePrint Arch. **216**, 46 (2016)

15. Orsini, E., et al.: Overdrive2k: efficient secure MPC over $\mathbb{Z}_{2^k}$ from somewhat homomorphic encryption. In: CT-RSA 2020, pp. 254–283 (2020)

16. Paillier, P.: Public-key cryptosystems based on composite degree residuosity classes. In: Stern, J. (ed.) EUROCRYPT 1999. LNCS, vol. 1592, pp. 223–238. Springer, Heidelberg (1999). https://doi.org/10.1007/3-540-48910-X_16

17. Rathee, D., Schneider, T., Shukla, K.K.: Improved multiplication triple generation over rings via RLWE-based AHE. In: Mu, Y., Deng, R.H., Huang, X. (eds.) CANS 2019. LNCS, vol. 11829, pp. 347–359. Springer, Cham (2019). https://doi.org/10.1007/978-3-030-31578-8_19

18. Smart, N.P., et al.: Commodity MPC via triples-as-a-service. In: CCSW@CCS, pp. 105–116 (2019)

19. Smart, N.P., et al.: TaaS: commodity MPC via triples-as-a-service. In: ACM SIGSAC Conference on Cloud Computing Security Workshop, London, UK, pp. 105–116 (2019)

20. Yao, A.C.-C.: How to generate and exchange secrets (extended abstract). In: FOCS 1986, pp. 162–167 (1986)

21. Zhu, H., Bao, F.: Oblivious scalar-product protocols. In: Batten, L.M., Safavi-Naini, R. (eds.) ACISP 2006. LNCS, vol. 4058, pp. 313–323. Springer, Heidelberg (2006). https://doi.org/10.1007/11780656_26

22. Zhu, H., Bao, F., Li, T., Qiu, Y.: More on shared-scalar-product protocols. In: Chen, K., Deng, R., Lai, X., Zhou, J. (eds.) ISPEC 2006. LNCS, vol. 3903, pp. 142–152. Springer, Heidelberg (2006). https://doi.org/10.1007/11689522_14

23. Zhu, H., Li, T., Bao, F.: Privacy-preserving shared-additive-inverse protocols and their applications. In: Fischer-Hübner, S., Rannenberg, K., Yngström, L., Lindskog, S. (eds.) SEC 2006. IIFIP, vol. 201, pp. 340–350. Springer, Boston, MA (2006). https://doi.org/10.1007/0-387-33406-8_29

24. Zhu, H., et al.: Privacy-preserving weighted federated learning within the secret sharing framework. IEEE Access **8**(2020), 198275–198284 (2020)

25. Zhu, H.: A lightweight, anonymous and confidential genomic computing for industrial scale deployment. CoRR abs/2110.01390 (2021)

The Impact of the Blockchain Technology on the Smart Grid Customer Domain: Toward the Achievement of the Sustainable Development Goals (SDGs) of the United Nations

Omid Ameri Sianaki[✉] and Sabeetha Peiris

Victoria University Business School, Melbourne, VIC, Australia
Omid.AmeriSianaki@vu.edu.au, sabeetha.peiris@live.vu.edu.au

Abstract. Our world needs access to affordable, reliable, and sustainable energy, along with climate adaptation and protection if we are to be sustainable, equitable, and inclusive. The purpose of this survey is to determine how blockchain technology can transform the smart grid in the customer domain of the maturity model, and how it can enhance achievement of the Sustainable Development Goals of the United Nations. We concluded from our study that the blockchain has the potential to contribute significantly to the achievement of goal number seven, Affordable and Clean Energy, and its interlinkages with goals number eleven, Industry, Innovation, and Infrastructure, and goal number eleven, Sustainable Cities and Communities, and goal number twelve, Responsible Consumption and Production, and finally goal thirteen, Climate Action. In this paper we provide a strong foundation upon which future research can be built and various elements can be combined for achieving the sustainable goals.

1 Introduction

To achieve a sustainable, equitable, and inclusive world with great economic opportunities, by creating jobs for women, providing great opportunities for children, and an efficient, reliable, and sustainable energy supply, there is a need to provide access to affordable, reliable, and sustainable energy alongside adaptation to and protection of the climate. It is nevertheless important to note that the United Nations General Assembly adopted the Sustainable Development Goals (SDGs) in 2015, which marked a major milestone for the 2030 Agenda for Sustainable Development as the blueprint for peace and prosperity for the planet. All countries, developed and developing, as part of a global partnership, must take action based on 17 SDGs.

By incorporating two-way communication, a smart grid provides energy to end-users and ensures efficient consumption. Various hardware devices are combined with ICT infrastructure, as well as software tools. It is crucial to build an infrastructure to coordinate and operate different parts of the smart grid for it to be sustainable, creative, and intelligent. Digital information technology with its controls and commands is being used to make the electric grid more reliable, secure, and efficient in a smart grid. Optimization of grid operations, as well as the pervasiveness of cyber security, have

L. Barolli et al. (Eds.): AINA 2022, LNNS 451, pp. 515–530, 2022.
https://doi.org/10.1007/978-3-030-99619-2_49

been utilized dynamically across the grid. The production of electricity is based on the combination of renewable and nonrenewable resources. A demand-side management program integrates demand response measures with other energy-saving measures and a range of technologies, including real-time, automated, interactive methods, are being used to optimize the physical operation of appliances, communication regarding grid operations, and distribution automation. By using the latest technologies, it will be possible to shift the demand for electric vehicles (EVs), air conditioners (AC) and modern storage systems from peak to off-peak (peak shaving) [1].

Although there are many components and complex connections with a decentralized smart grid, it could also pose security, privacy, and trust concerns that will require new and innovative technologies to solve. The blockchain, however, is an emerging technology that has great potential for creating decentralized systems. Due to the advent of blockchain technology, economic transaction systems have been completely replaced in various organizations. This innovation has the potential to reshape heterogeneous business models across a range of sectors. Even though it promises to facilitate the sharing, exchanging, and integration of information across users and third parties, decision-makers and planners would be wise to assess its suitability to various industries and business applications in depth. Blockchain technology offers numerous benefits, including decentralization, persistency, anonymity, and auditability. The array of blockchain applications ranges from cryptocurrency, financial services, risk management, the internet of things (IoT), and a variety of other services. A blockchain technology, due to these distinctive characteristics, is emerging as an important element for the next generation of internet interaction systems, such as smart contracts, public services, IoT, reputation systems, and security services [2].

In order for blockchain to function successfully, no central trusted authority is needed; instead, multiple entities in the network can share resources between themselves for the purpose of creating, maintaining, and storing a chain of blocks. The chain order is provided for everyone to ensure that it is not tampered with and that no data has been altered. A decentralized system increases resiliency and reliability by making any system redundant and resilient to system failures and cyberattacks while eliminating many of the problems of a centralized system. Due to its excellent properties, blockchains have been introduced and populated as digital currencies, however, they are gaining considerable attention in other applications that are non-monetary in nature as well, since the technology exhibits several excellent characteristics [3].

There has been an explosion of interest in blockchain technology applied to smart grids in recent years with new research areas coming about all the time. In addition, [3] conducted studies on the challenges that can be faced through the use of blockchain contributions to smart grids, such as blockchain for Smart Metering Infrastructure, blockchain in Decentralized Energy Trading and Market, blockchain for Monitoring, Measurement, and Control, blockchain for EVs and Charging Unit Management, and blockchain for Microgrids.

Our study demonstrates, blockchain technology will facilitate the smart grid to achieve four of the SDGs and we will provide further explanation of this contribution below. These goals include goals nine, eleven, twelve, and thirteen.

2   The Smart Grid and Sustainability

According to [4]'s study, smart technologies are considered intermediary elements that facilitate the translation of sustainability strategies identified by governments, industries, and universities into concrete pathways that encompass the domains of economy, environment, and society. The study examines how the smart grid contributes to sustainability in two major domains. With regards to the first domain, the increased supply of electric energy into the grid through the integration of renewable resources, ICT effectively plays an essential role in monitoring the state of the smart grid in real time and in diagnosing grid failures, as well as controlling the grid in order to redistribute the available energy within the local area. Another domain involves increasing resource efficiency through consumer participation in energy management. As outlined by the Natural Resources Defense Council (NRDC) [5], smart grid systems may result in significant benefits for the nation. As a result, consumers and utilities would both benefit from more energy efficiency, reduced carbon dioxide and other pollutants in the air, enabling greater use of plug-in vehicles, as well as lower utility prices. Through carefully crafted programs and efficient deployment, the NRDC encourages utilities and regulators to promote smart grid technology. According to the NRDC, utilities should be held accountable for adhering to the cost estimates and delivering the benefits of smart grid investments. It should enhance consumer protections where necessary. To access the benefits of meter data-driven services and products, smart grid consumers must be protected from unauthorized access to their meter information. The same conclusion has also been made by [6], whose study indicates that SG will enhance energy efficiency through optimization technologies and improve the power quality of the grid, in addition to designing incentive mechanisms to modify ordinary electricity consumption.

According to a UN report [7], action on sustainable energy can contribute to achieving all other Sustainable Development Goals. In this paper, we outline the connections between SDGS 9, 11, 12, and 13. Energy is the basis for all other SDGs, according to the UN report. With SDG 9 focusing on industry, innovation, and infrastructure, economic forces are directed to produce positive development outcomes. Industry contributes a large portion of GHG emissions. But if SDG 9 is aligned with related goals for clean or carbon-neutral development, including SDG 7, it can significantly contribute to attaining a carbon-neutral economy. SDG 7 is critical for the quality of life of future generations. Providing access to affordable and clean energy is inherently linked with the goals of eradicating poverty and hunger, expanding health care, ensuring gender equality, and ensuring the availability of clean water. In addition, enhancing inclusive and sustainable growth, creating decent work for all, tackling inequities, advancing sustainable production, fostering sustainable urban environments, fostering peace and justice, and strengthening institutions, enhancing biodiversity, and reducing greenhouse gas emissions are all inextricably connected to the availability and affordability of clean energy. City governments are able to play a role in achieving the SDG 7 and it is closely related to achieving SDG 11, which is focused on transforming cities and communities into sustainable and prosperous communities. Investing in renewable energy, as well as increasing energy and resource efficiency, are vital to ensuring that cities are sustainable. [7] discusses a variety of ways that city governments have leveraged the synergy between SDG 7 and SDG 11. Taking advantage of integrated urban planning and collaborating with utilities to achieve sustainable energy goals are two examples. Furthermore, the collaboration between the public and private sectors on carbon-neutralization of high-pollution urban systems and the procurement, installation, and/or contracting of clean energy capacity, and fostering citizen engagement in the energy transition are other initiatives. According to the report, SDG 7 and SDG 12 can make complementary contributions to climate change mitigation and environmental sustainability. Figures 1 and 2 provide an overview of the results of our study on the interrelatedness of goal number 7 with other four SDGs drawn from reference [7].

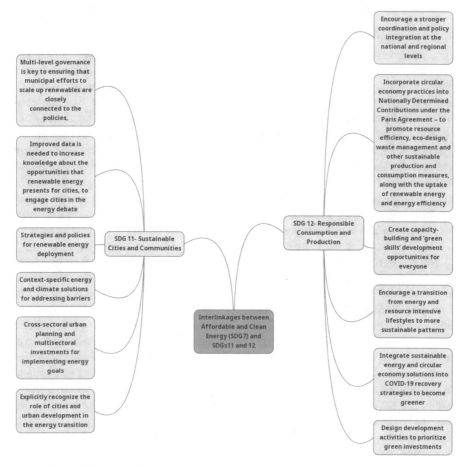

Fig.1. The Interlinkages between SDG7 and SDG 11 and SDG 12 based on [7]

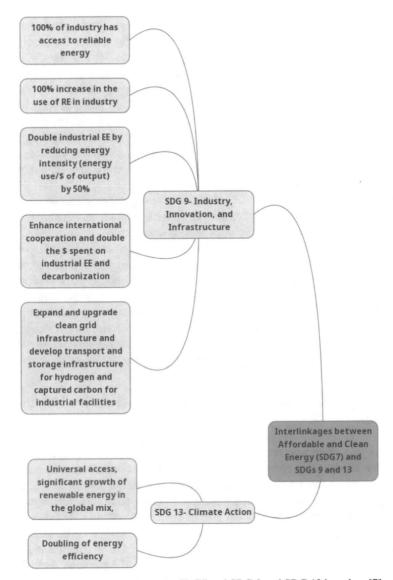

Fig.2. The Interlinkages between SDG7 and SDG 9 and SDG 13 based on [7]

3 The Blockchain and SDGS

An extensive literature review has been conducted by [8] to examine the relationship between blockchain and the Sustainable Development Goals. In light of the authors' conclusion, the relationship that has been debated the most is that between blockchains and "Affordable and Clean Energy (SDG 7)", Responsible Consumption and Production (SDG 12), Climate Change (SDG 13), and Sustainable Cities and Communities

(SDG 11). Ultimately, blockchains can be used to facilitate the development of alternative energy by enabling smart contracts that can exchange quotes for energy, including renewable energy, in order to recharge electric vehicles. Researchers suggest that blockchain technology has the potential to positively impact Climate Action (SDG 13), by, perhaps, providing a means of tracking and exchanging carbon emissions, or by using cryptocurrency for creating incentives to encourage pollution-defying behaviors. The authors discussed the potential and challenges associated with the application of blockchain technology to SDG 7's environmental-related goals. Potential applications include developing smart contracts for renewable energy producers and consumers. It discusses the challenges in developing blockchain systems that utilize alternative energy sources or reduce energy consumption.

SDG 11 has the potential to create more live-able cities through the implementation of platforms for tracking energy consumption, waste, and so on. The challenge is to define a blockchain model that combines various technologies. With the SDG 12 goal, tracking and tracing of supply chains and natural resource usage is made possible. A requirement for complementary assets is necessary for the goal to be achieved. Similarly, developing platforms for monitoring and exchange greenhouse gas emissions quotes are the potentials for climate Action (SDG 13) and developing blockchain systems based on alternative energy is the challenge. In a similar vein, blockchain platforms could be utilized to monitor and evaluate greenhouse gas emissions, which represents a significant opportunity for achieving SDG 13 (Climate Action), while developing blockchain applications based on alternative energy represents a particular challenge.

4 The Smart Grid Maturity Model and Blockchain: Literature Review

The smart grid came into reality due to the growing demands of electricity and the compulsion in producing energy using non-renewable resources, not only force energy suppliers to utilize these resources and meet the demands but also creates the necessity for more trustworthy and safe energy system. In other words, energy suppliers have transformed their normal grid to smart grid in order to optimize the resources and gain more profits as well as to support the sustainability of the grid [3, 9]. In the smart grid, it is mandatory to be aware of the environment and senses in order to comprehend the status of the grid and also to ensure safety, secured, effective and dependable functionality of the grid that facilitates by the automated data system [1].

This modernism simplifies and allows the characteristics of the smart grid functionality such as demand and supply administration, real-time price estimations, programmed meter and computerized reading. This is facilitated by the immense use of computer-based automation [10]. Therefore, the smart grid becomes a robust system and automatic responses to attackers ensures the solidity of the grid structure. Further smart grid can predict and identify potential issues at an initial phase and slight amendments could be easily resolved. The Smart grid technology mainly builds on two-way communication that enables the senses to work in a mechanism that involves great attention to detail. This facilitates effective progress and performance, versatile functionality of the system, and optimized operation. This also includes a supportive market-driven approach and incorporating substitute resources.

However, this research is focusing on the smart grid maturity model which is established by the Software Engineering Institute of the US. This smart grid maturity model [11] consists of eight domains such as grid operation, customer, technology, organization and structure, value chain integration, strategy, management and regulatory, societal and environmental and work asset management and then consists of six maturity levels form lower to highest level such as level 0-default, level 1- initiating, level 2-enabling, level 3-integrating, level 4-optimizing and level 5-pioneering. Every individual level has some features that aid a standard grid in revamping itself into a smart grid. These maturity levels and their respective features are required to be equipped to shift from low level to higher level of maturity.

All six levels of the smart grid maturity model are as follows. Level 0 – Default, is the very early phase of maturity and signifies the primary steps taken by the organization, hence the features are not defined well at this level.

Level 1 – Initiating, this level signifies the opening phase of the organization in the execution of the smart grid. At this level measures for performance are recognized and measures are placed at their suitable places to monitor the performance. Level 2 – Enabling, this level of maturity signifies the execution of the characteristics by the organization that facilitates the accomplishment of the grid transformation.

Level 3 – Integration, this level of maturity signifies the organization extensive execution of the grid transformation strategic proposals. Level 4 – Optimizing, at this level of maturity, the execution of the smart grid at a specific domain is modified and features such as contemplating issues and automatic amendments commence to occur in real-time. Level 5 – Pioneering, is the highest level of maturity and the organization is forward-moving its activities and is equipped with all the innovative characteristics of grid transformation and commencing to operate organization-wide.

Because the model is large-scale and expansive, this research is focusing only on one domain as the customer to narrow down the research. Another main part of the research is a blockchain technology-based smart contract which has cutting-edge features and is broadly used these days in the energy market due to the enormous benefits smart contract brings in Peer-to-Peer (P2P) energy trading. Therefore, the main purpose of this paper is to integrate the blockchain technology-based smart contract features with energy trading related expected features from the selected three domains of SGMM such as customer and examines how smart contract enhances the related features of levels and sub-levels. Also, we aim to depict a picture for readers that how blockchain in the smart grid can enhance SDGs achievement.

Blockchain is a combination of software applications, a database, and ledgers which is a series that is built from various blocks that are comprised of data [12]. While various programming languages could be used to develop blockchain, several developers have adopted the high-level object-oriented as the key programming language known as solidity. The main characteristics of this technology are that it records and monitor various blocks created, hence after the block or the information is entered in the blockchain, they cannot be altered or detached. This makes blockchain technology an immensely reliable and secured mechanism for property, currency, and agreements transferals and eliminates intermediaries such as financial institutions [13]. Generally, security mechanisms depend on a specific reliable domain to reserve and verify accumulated converted data. This allows attackers to carry out hazardous activities such as entering unsafe data, threatening the user and theft data. Hence these systems become unprotected. Whereas blockchains solve the existing security issues because it is decentralized and does not require authorization to rely on one single domain or network. Blockchain solves these trust issues by obtaining consent from most of the members when the new information is added to the chain of blocks and providing every member with the entire present and past data. Importantly, blockchains accomplish the consensus process in various ways particularly when attacks occur within the group or in some circumstances from the controller itself [14].

This leads to the fact that most members in the group having access to similar data are more protected than the group that trust a single controller for their data. Blockchain can defeat the security challenges with its primary features such as invariability, transparency, appropriateness and adaptability that helps to identify data meddling hence zero failure. On the other hand, obtaining consent from users helps to intercept fraudulent activities and the data is secured by utilizing digital signatures and asymmetric encryption algorithms. However, previous researchers have identified some risks involved in blockchain technology [15]. For example, when the user's private key is not verified throughout the signature proceeding and the attacker can recover it. Other situations are that the system is exposed to the possibility of risk when an individual hash capacity is higher than the total hash capacity of the whole blockchain, illegal smart contracts, the occurrence of double-spending when a blockchain network is disrupted and cryptocurrency is essentially stolen and data exposed of the deals [16]. Smart grid technology has examined various security methods for guaranteed data protection and safety. However, security aspects of smart grids have been evaluated that underlines primary security concerns based on the critical characteristics of information security such as confidentiality, integrity, availability, authenticity, authorization, and non-repudiation [17–19].

Further to the implementation of blockchain in cryptocurrency such as bitcoin in 2011, blockchain-based smart contracts were initiated in 2013. Smart contract developments made a remarkable impact on diverse industries such as health, and the internet of things (IoT) [9]. A smart contract is a software application deployed on Ethereum and formulated in the solidity programming language that can implement a predefined agreement utilizing blockchain, when and if particular preconditions, terms, rules and legal constraints are met. The primary objective of the smart contract is to facilitate two parties to perform trustworthy transactions without the involvement of a middleman. Another function of a smart contract is that it acts as a record keeper, with the relative reward method. Furthermore, smart contracts derive the features of blockchains, hence have no downtime, censorship or third-party interference [20].

Authors [17] have argued that significantly blockchain can offer unique trading platforms and enables automated execution of smart contracts in P2P networks [21], where prosumers and consumers could market or interchange their excess energy or workable demand on a P2P basis. User engagement could be secured and registered into unchangeable, transparent and tamper-resistant smart contracts. Facilitating such a systematized trading platform is an effective way of sending tariff indications and details on energy costs to users [22], at the same time offering them incentives for demand response and strategically managing their energy requirements. A smart contract could equip local energy and consumer marketplaces or microgrids, hence strengthening local energy production and utilization [23]. The smart contract enables a smart grid energy trading mechanism. This method facilitates automatic energy detection inputs via the controller issuing coins with a smart contract. These coins are the virtual tokens on Ethereum and distributed, and once they are reclaimed, they are returned to the controller's address, and then they communicate with the electric grid to transmit an equal amount of the energy. Further, the coin movements and addresses' balance are public, and any interested party could monitor the same. Whenever a user commits energy in the grid, the smart contract releases coins that match the energy produced and automatically sends them to the energy producer [20].

Ethereum facilitates user-generated smart contracts and intends to construct a technology platform for all uses, where transaction-based applications are established [24]. As per the recent survey by Eurelectric, the Union of the electricity industry, over 1000 schemes are utilizing Ethereum at the moment including various start-ups that are utilizing Ethereum based coins and cryptocurrencies for Initial Coin Offerings (ICOs). Smart contracts have been embedded with cryptographic keys and equip the contracts to encrypt the reports created from the stimulating actions. Cryptocurrencies intend to materialize patents and motivate the growth of energy storage solutions. A primary application of Ethereum is smart contracts and Decentralized Applications (DApps). DApps are open source, decentralized applications that function independently without human interference. DApps utilize cryptocurrencies or tokens and are implemented in a network of computers and reserve outputs in public ledgers [25].

Smart contracts and smart meters could enable automatic invoices for users and distribute generators enabling instantaneous invoicing. Utility providers could also gain an advantage for energy micropayments, pay-as-you-go or payment platforms for prepaid meters. The smart contract simplifies and accelerates the process of moving between energy suppliers. With the improved flexibility in the trade, competitiveness could grow hence possible lower energy rates can be offered. Furthermore, smart contracts track which generator and user-generated disparity in agreements. In addition to the above, after installing a smart meter in the location of a user, a smart contract sends the special meter ID of the user to the smart grid and it is accepted by the authenticator for validation. Once the user verification is done, the user obtains access to electricity [26]. Smart contracts and automated transaction implementation facilitate real-time agreement of payments for electricity as energy is being produced in real-time and also utilized to obtain geolocation of nodes that ensure precise forecast of power failure. The solution intends to resolve the other problems such as payment delays, debts population and identity fraud. Further, smart contracts are prompted and implemented depending on the malicious activity if detection on the meter or a relevant action is undertaken [26]. Smart contracts are used to enhance smart grid cyber resiliency and ensure the protection of energy transactions which is tremendously useful for energy trading mechanisms [17]. Smart contracts are self-enforced and tamper-resistant that provide enormous advantages such as removing the intermediaries and reducing the cost of transaction, administration and regulations. Also, it provides cost-efficient low-value transactions, with the support of blockchains ensuring integration among transaction systems [27]. Business operations and proceedings are administered via smart contracts and could potentially impact energy company operations.

5   Smart Contract of Blockchain Technology in Customer Domain

Customer domain is the vital domain in the smart grid maturity model that involves customers in energy management. This domain delineates the user engagement that maintains and protects their data in the smart grid [11, 28]. This domain portrays the company's abilities and features of customer engagement for gaining smart grid benefits. The key features of this domain are associated with the smart contract and how the smart contract enhances these features is detailed below.

5.1   Customer Domain Level 2 - Security and Privacy Requirements for Customer Protection Are Specified for Smart Grid

The smart contract is built-in blockchain technology that enables the customers to keep track of the energy usage and eliminates influences from other parties [26] and it is utilized to enhance the cyber resiliency of the smart grid and protects energy transactions. Therefore, a smart contract is immensely useful for trading energy applications. Because the electricity meter reading is done through the internet, there could be possibilities that the information is exposed to the attackers. [26] discusses the issues that customers face in their energy usage environment, for example, due to customers' lack of knowledge in meter readings, they are not aware of the reasons behind their huge utility bills and which devices they use the most in their homes. Whereas through the smart contract, transactions should be able to be performed based on real usage. Further, the implementation of a smart contract ensures trust and a visibility-based system among the users within the network for it to reduce the risks of these aforementioned problems [17].

Smart contracts ensure effective transactions and enable prosumers to sell excessive electricity to other members of the grid [29]. A model based on blockchain that utilizes smart contract as the middleman connecting the customers who consume energy and energy producers to lessen the cost, accelerate the transactions process and reinforce the reliability and protection of the transaction data. This model further emphasizes that when a transaction activity occurs, a blockchain-built meter notifies the blockchain that generates a unique timestamp in a distributed ledger for validation purposes. Thereafter, customers are invoiced by the system operators depending on the data stored in the smart contract. Firstly, the smart meter measures electricity usage data and encrypts the data utilizing the private key, i.e., private key A. Then, the smart contract decrypts the encrypted data utilizing the public key of the smart meter, i.e., public key A, and check the credibility. In terms of verifying the credibility, if public key A successfully decrypts the encrypted data, then the encrypted data have not been compromised since the private key of the smart meter cannot be acquired by adversaries, which protects the data integrity. Additionally, the utility company utilizes the public key for data encryption, i.e., public key B, therefore only the utility company with the matching private key, i.e., private key B, should be able to access the data to protect the privacy of consumers data [17].

5.2   Customer Domain Level 3 - Customer Products and Services Have Built-In Security and Privacy Controls that Follow Industry and Government Standards

This sub-level of the customer domain evaluates the current smart grid framework to understand if there are any standards or government regulations are required for utilizing blockchain-based smart contract consumer products and services in the smart grid. Energy user products and services must be built with security and privacy controls at an early stage that adheres to industry and government standards. According to the National Institute of Standards and Technology (NIST) of the smart grid framework 4.0 [30], the draft delineates the goals for the applicability of blockchain technology and smart contracts to the various smart grid sub-domains as well as standards are outlined for the growth of emerging smart grid technologies. This is to follow the directives and instructions provided by NIST. Sub-domain operations in the customer domain have been explained with pre-conditions, blockchain technology requirements and objectives from NIST below.

Smart contracts are used for reporting and authentication of energy loads, electricity usage pattern, cooperative usage of power sources in the sub-domain of smart meter application, where every appliance and methods are required to be registered with service providers to achieve decentralized energy load validation, views and opinions based on consensus and decentralized operation framework [31]. Smart contracts also used for decentralized electricity demands and incentive mechanisms for Peer-to-Peer (P2P) load administration in the sub-domain of managing loads demands for Demand Response Management System (DRMS), where every customer are required to be registered with service providers to achieve optimum P2P energy demand-side management [32]. A cryptographic algorithm is used to create hashes of accumulated metadata is required in the sub-domain of data management for load prediction where every storage equipment must be registered with service providers to achieve distributed decentralized verifiable data management. A consensus algorithm, DLT, and cryptographic algorithm for decentralized storage are required in the sub-domain of data management for electric vehicles (EVs)s, where every EV must be registered with service providers to achieve decentralized reporting of electricity usage and custom-made energy trading with EVs. A consensus algorithm is also required in the sub-domain of cybersecurity operations, where all actors and participants must be registered with service providers to achieve decentralized threat notifying and managing cybersecurity [30, 33, 34]. To impose regulations on data owners, the General Data Protection Regulation (GDPR) legislation [35] have been executed by the European Union (EU) since 2018. Evolving blockchain-based smart contract technology is following GDPR legislation and provides a decentralized solution to enable trusted communication and relationships among data owners and energy service providers for individual information sharing and data utilization assessment with visibility and traceability[36]. Apart from a fine-granular private data management system with GDPR compliance is expanded in [37] for data owners to detail user-centric access policies through programmable smart contracts.

5.3 Customer Domain Level 5- the Organization Can Assure Security and Privacy for All Customer Data Stored, Transmitted, or Processed on the Grid

This sub-level of the customer domain is to ensure the protection of customer data collected and transferred on the grid. The understanding of how smart contracts combined with blockchain technology could help an organization to assure security and privacy for collected data and transferred on the grid [11]. Smart contracts protect economic transactions against threats within the power storage systems in auction markets. A decentralized assessment approach is recommended following the Ethereum blockchain, to guarantee safe transactions and to eliminate cyber-attacks, where the rules are explained utilizing smart contracts [38]. Further, Ethereum smart contracts have been utilized for renewable energy transaction structures for blockchain-based smart homes where the energy transactions are enabled by the data flow, among smart homes [39]. Moreover, [10] discussed the unreliability in various loads, however, the renewable energy trading process and data flow are secured by smart contracts. Smart contracts decide the cost and the period of utilizing various devices from each entity, such as the decentralized storage applied for the data transaction during the decentralized energy management in SGs [40].

In addition to this, smart contracts are employed for prosumers to notify their price and deliver with a reliable format, while security problems are addressed [41]. Smart contracts ensure data protection, transmission and genuine financial transactions among various sectors in distributed energy management. Privacy issues including EV hijacking are resolved with the use of smart contracts cryptocurrencies in the blockchain [19]. A smart contract is also utilized for the charging format of EVs to mitigate attacks on the charging network[42]. The persistent and genuine pricing of electricity to charge EVs is addressed in [43]. Smart contract server nodes are in charge of executing and binding the contract script to the blocks in the blockchain. Smart contracts incorporate the optimal strategy controls by edge devices for energy resource allotments to customers by taking electricity usage, latency, and security into consideration [3]. However, trustworthiness becomes a matter of interest when the super-nodes (SNs) are compromised. A secured and trustworthy energy scheduling framework is known as Privacy-Preserving Energy Scheduling (PPES) for energy service companies (ESCOs) is implemented by [44]. In this framework, the smart contract points out expanding privacy issues of centralized ESCOs which may encompass financial and behavioral data that could cause privacy issues for the distributed energy market. Further, the consensus and smart contract will resolve the independent scheduling issues in the network and enables reducing energy costs and privacy protection.

6 The Developments in the Existing SG Work and the Importance of the Smart Grid Maturity Model (SGMM), the Blockchain Technology, and SDGs

Despite the initiatives and advantages that the smart grid offers, renewable energy sources present obstacles to the power generation and distribution system [45] as the intermittent nature of renewable energy sources and the facilitation of energy trading. The focal point of the smart grids is to enable the local energy production and consumption steadily for prosumers and consumers and enable them to trade electricity among themselves in a P2P manner that helps to lessen the transmission losses. However, facilitating these trading transactions within the smart grid in a centralized system would be high cost and demands complex communication infrastructure [46]. The aforementioned problems could be solved by developing the maturity model in smart grid systems along with the adoption of a decentralized method. SGMM is a strategic management tool established by the Software Engineering Institute of the US and provides utility companies with a standard framework to evaluate the current position and the problems and helps to face the challenges of the smart grid [11].

It will be easier to understand the importance of blockchain in Smart Grid when we examine Sect. 2 in which we discussed the interconnections of SDG 7 with the other four Sustainable Development Goals. As an example, we mentioned that city governments can contribute to achieving SDG 7, which is closely linked to SDG 11, which focuses on transforming cities into sustainable and prosperous communities. To ensure the sustainability of cities, it is vital to invest in renewable energy and increase resource efficiency. Additionally, we discovered that blockchain technology offers unique trading platforms and can be used to automate the execution of smart contracts in P2P networks. Consequently, the city government can gain a good understanding as to how blockchain can be utilized in this investment. As a result, if blockchain technology is capable of tracking emission trading, then a smart grid can provide this functionality. A second example is the complementary contribution that SDG 7 and SDG 12 may be able to make to climate change mitigation and environmental sustainability. It must be noted, however, that blockchain technology can positively impact climate action (SDG 13), perhaps by facilitating the tracking and exchanging of carbon emissions, or by using cryptocurrency to create incentives to encourage pollution-defying behaviours - something that can occur in electricity energy when the smart grid matures to a higher level of maturity, and which can be made possible through blockchain technology.

7 Conclusion

In order for the world to be a sustainable, equitable, and inclusive place, we must provide access to affordable, reliable, and sustainable energy alongside climate adaptation and protection. We conducted this survey in order to determine how blockchain technology can transform the smart grid in the customer domain of the maturity model and how it might enhance attainment of the United Nations Sustainable Development Goals by making them more feasible. A study conducted by us found that the blockchain will have a significant impact on the goals in the interlinking of SDG 7 and other four SDGs 9, 11, 12 and 13. This paper provided a strong foundation on top of which future research can build and combine more elements for achieving the sustainable goals in future studies. This study did not cover other smart grid maturity domains such as technology or value chain, but such a study is crucial and useful for governments, corporations, and policy makers who pursue achieving the Sustainable Development Goals.

References

1. Sianaki, O.A.: Intelligent decision support system for energy management in demand response programs and residential and industrial sectors of the smart grid. Curtin University (2015)
2. Daneshgar, F., Sianaki, O.A., Guruwacharya, P.: Blockchain: a research framework for data security and privacy. In: Workshops of the International Conference on Advanced Information Networking and Applications, pp. 966–974 (2019)
3. Mollah, M.B., Zhao, J., Niyato, D., Lam, K.-Y., Zhang, X., Ghias, A.M., et al.: Blockchain for future smart grid: a comprehensive survey. IEEE Internet Things J. **8**, 18–43 (2020)
4. Caputo, F., Buhnova, B., Walletzký, L.: Investigating the role of smartness for sustainability: insights from the smart grid domain. Sustain. Sci. **13**, 1299–1309 (2018)
5. Succar, S., Cavanagh, R.: The promise of the smart grid: goals, policies, and measurement must support sustainability benefits. NRDC Issue Brief (2012)
6. Hu, Z., Li, C., Cao, Y., Fang, B., He, L., Zhang, M.: How smart grid contributes to energy sustainability. Energy Procedia **61**, 858–861 (2014)
7. United Nations. Leveraging Energy Action For Advancing The Sustainable Development Goals. United Nations Department of Economic and Social Affairs (2021)
8. Parmentola, A., Petrillo, A., Tutore, I., De Felice, F.: Is blockchain able to enhance environmental sustainability? a systematic review and research agenda from the perspective of Sustainable Development Goals (SDGs). Bus. Strat. Environ. **31**, 194–217 (2021)
9. Musleh, A.S., Yao, G., Muyeen, S.: Blockchain applications in smart grid–review and frameworks. IEEE Access **7**, 86746–86757 (2019)
10. Hasankhani, A., Hakimi, S.M., Bisheh-Niasar, M., Shafie-khah, M., Asadolahi, H.: Blockchain technology in the future smart grids: a comprehensive review and frameworks. Int. J. Electric. Power Energy Syst. **129**, 106811 (2021)
11. University, C.M.: Smart Grid Maturity Model. Software Engineering Institute, Pittsburgh, Pennsylvania (2010)
12. Nakamoto, S.: Bitcoin: a peer-to-peer electronic cash system. Bitcoin, vol. 4 (2008). https://bitcoin.org/bitcoin.pdf
13. Tama, B.A., Kweka, B.J., Park, Y., Rhee, K.-H.: A critical review of blockchain and its current applications. In: 2017 International Conference on Electrical Engineering and Computer Science (ICECOS), pp. 109–113 (2017)
14. Taylor, P.J., Dargahi, T., Dehghantanha, A., Parizi, R.M., Choo, K.-K.R.: A systematic literature review of blockchain cyber security. Digital Commun. Netw. **6**, 147–156 (2020)
15. Li, X., Jiang, P., Chen, T., Luo, X., Wen, Q.: A survey on the security of blockchain systems. Futur. Gener. Comput. Syst. **107**, 841–853 (2020)
16. Möser, M., Soska, K., Heilman, E., Lee, K., Heffan, H., Srivastava, S., et al.: An empirical analysis of traceability in the monero blockchain. Proc. Priv. Enhancing Technol. **2018**, 143–163 (2018)
17. Mylrea, M., Gourisetti, S.N.G.: Blockchain for smart grid resilience: Exchanging distributed energy at speed, scale and security. In: 2017 Resilience Week (RWS), pp. 18–23 (2017)
18. Komninos, N., Philippou, E., Pitsillides, A.: Survey in smart grid and smart home security: issues, challenges and countermeasures. IEEE Commun. Surv. Tutor. **16**, 1933–1954 (2014)
19. Dorri, A., Kanhere, S.S., Jurdak, R., Gauravaram, P.: Blockchain for IoT security and privacy: the case study of a smart home. In: 2017 IEEE International Conference on Pervasive Computing and Communications Workshops (PerCom Workshops), pp. 618–623 (2017)
20. Kounelis, I., Steri, G., Giuliani, R., Geneiatakis, D., Neisse, R., Nai-Fovino, I.: Fostering consumers' energy market through smart contracts. In: 2017 International Conference in Energy and Sustainability in Small Developing Economies (ES2DE), pp. 1–6 (2017)
21. Swan, M.: Blockchain: Blueprint for a New Economy. O'Reilly Media, Inc., Newton (2015)

22. Mengelkamp, E., Gärttner, J., Rock, K., Kessler, S., Orsini, L., Weinhardt, C.: Designing microgrid energy markets: a case study: the Brooklyn microgrid. Appl. Energy **210**, 870–880 (2018)
23. Pinson, P., Baroche, T., Moret, F., Sousa, T., Sorin, E., You, S.: The emergence of consumer-centric electricity markets. Distrib. Utiliz. **34**, 27–31 (2017)
24. Wood, G.: Ethereum: a secure decentralised generalised transaction ledger. Ethereum Proj. Yellow Paper **151**, 1–32 (2014)
25. Andoni, M., Robu, V., Flynn, D., Abram, S., Geach, D., Jenkins, D., et al.: Blockchain technology in the energy sector: a systematic review of challenges and opportunities. Renew. Sustain. Energy Rev. **100**, 143–174 (2019)
26. Gao, J., Asamoah, K.O., Sifah, E.B., Smahi, A., Xia, Q., Xia, H., et al.: GridMonitoring: Secured sovereign blockchain based monitoring on smart grid. IEEE access **6**, 9917–9925 (2018)
27. Guo, Y., Liang, C.: Blockchain application and outlook in the banking industry. Finan. Innov. **2**(1), 1–12 (2016). https://doi.org/10.1186/s40854-016-0034-9
28. Sianaki, O.A., Masoum, M.A.: Versatile energy scheduler compatible with autonomous demand response for home energy management in smart grid: a system of systems approach. In: 2014 Australasian Universities Power Engineering Conference (AUPEC), pp. 1–6 (2014)
29. Wang, X., Yang, W., Noor, S., Chen, C., Guo, M., van Dam, K.H.: Blockchain-based smart contract for energy demand management. Energy Procedia **158**, 2719–2724 (2019)
30. Gopstein, A., Nguyen, C., O'Fallon, C., Hastings, N., Wollman, D.: NIST framework and roadmap for smart grid interoperability standards, release 4.0: Department of Commerce. National Institute of Standards and Technology (2021)
31. Greer, C., Wollman, D.A., Prochaska, D., Boynton, P.A., Mazer, J.A., Nguyen, C., et al.: Nist framework and roadmap for smart grid interoperability standards, release 3.0 (2014)
32. Aderibole, A., Aljarwan, A., Rehman, M.H.U., Zeineldin, H.H., Mezher, T., Salah, K., et al.: Blockchain technology for smart grids: decentralized NIST conceptual model. IEEE Access **8**, 43177–43190 (2020)
33. Nguyen, C.T., Gopstein, A.M., Byrnett, D.S., Worthington, K., Villarreal, C.: Framework and Roadmap for Smart Grid Interoperability Standards Regional Roundtables Summary Report (2020)
34. FitzPatrick, J.: NIST Framework and Roadmap for Smart Grid Interoperability Standards, Release 2.0 (2011)
35. G.D.P. Regulation. General data protection regulation (GDPR)," Intersoft Consulting, vol. 24 (2018). Accessed Oct 2018
36. Wang, Y., Su, Z., Zhang, N., Chen, J., Sun, X., Ye, Z., et al.: SPDS: a secure and auditable private data sharing scheme for smart grid based on blockchain. IEEE Trans. Ind. Inf. **17**, 7688–7699 (2020)
37. Truong, N.B., Sun, K., Lee, G.M., Guo, Y.: GDPR-compliant personal data management: a blockchain-based solution. IEEE Trans. Inf. Forensics Secur. **15**, 1746–1761 (2019)
38. Wang, K., Zhang, Z., Kim, H.S.: ReviewChain: smart contract based review system with multi-blockchain gateway. In: 2018 IEEE International Conference on Internet of Things (iThings) and IEEE Green Computing and Communications (GreenCom) and IEEE Cyber, Physical and Social Computing (CPSCom) and IEEE Smart Data (SmartData), pp. 1521–1526 (2018)
39. Kang, E.S., Pee, S.J., Song, J.G., Jang, J.W.: A blockchain-based energy trading platform for smart homes in a microgrid. In: 2018 3rd International Conference on Computer and Communication Systems (ICCCS), pp. 472–476 (2018)
40. Benisi, N.Z., Aminian, M., Javadi, B.: Blockchain-based decentralized storage networks: a survey. J. Netw. Comput. Appl. **162**, 102656 (2020)

41. Foti, M., Greasidis, D., Vavalis, M.: Viability analysis of a decentralized energy market based on blockchain. In: 2018 15th International Conference on the European Energy Market (EEM), pp. 1–5 (2018)

42. Kim, M., Park, K., Yu, S., Lee, J., Park, Y., Lee, S.-W., et al.: A secure charging system for electric vehicles based on blockchain. Sensors **19**, 3028 (2019)

43. Pajic, J., Rivera, J., Zhang, K., Jacobsen, H.-A.: Eva: fair and auditable electric vehicle charging service using blockchain. In: Proceedings of the 12th ACM International Conference on Distributed and Event-based Systems, pp. 262–265 (2018)

44. Tan, S., Wang, X., Jiang, C.: Privacy-preserving energy scheduling for ESCOs based on energy blockchain network. Energies **12**, 1530 (2019)

45. Yoldaş, Y., Önen, A., Muyeen, S., Vasilakos, A.V., Alan, I.: Enhancing smart grid with microgrids: challenges and opportunities. Renew. Sustain. Energy Rev. **72**, 205–214 (2017)

46. Strasser, T., et al.: A review of architectures and concepts for intelligence in future electric energy systems. IEEE Trans. Ind. Electron. **62**, 2424–2438 (2014)

Analysis of Variants of KNN for Disease Risk Prediction

Archita Negi and Farshid Hajati[✉]

College of Engineering and Science, Victoria University Sydney, Sydney, Australia
Archita.negi@live.vu.edu.au, farshid.hajati@vu.edu.au

Abstract. As we all know, Supervised Machine learning algorithms are gaining a lot of attention these days and are used in various fields. These algorithms play a great role in classifying, handling and predicting the data with the help of machine learning and makes use of a labelled dataset to predict and classify any given unlabeled data. KNN is one of the most popular supervised learning algorithms that delivers excellent results and is widely used these days for predicting the risk of certain diseases. Different variants of KNN are being introduced by different researchers due to distinct inadequacies of the standard KNN. The goal of this research is to evaluate the performance of such KNN variants in disease risk prediction by taking into account various factors such as performance, accuracy and so on. These variants which are already being proposed by different researchers will be reviewed and analyzed in this study. All these variants of K Nearest neighbor are implemented in Python Programming Language on different medical datasets taken from Kaggle. Extensive efforts are being made to analyze the different variants of KNN with respect to disease risk prediction and after implementation, it was noticed that some variants worked really well for some datasets. Out of all, Generalized KNN turned out to be one of the most efficient variants of KNN as it gave a good accuracy of 90.15% for mole cancer dataset. Also, it gave a precision of 90.13%, recall of 94.67% and AUC of 88.61% which overall is a very good performance. It was noticed that the Genetic KNN with GA Algorithm gave an accuracy of 99% for both Chronic Kidney Disease datasets. This study will be really beneficial for other researchers as well in the health industry and will also help in improvising their studies.

1 Introduction

With the evolution of technology, lives of humans has become much easier now. Diseases are increasing day by day as the environment is getting effected due to unwanted changes and the lifestyle of people has also changed. It is very important to have an early prediction of the diseases to avoid any unforeseen outcome. Machine learning is one of the hottest fields in computer science with a wide range of applications [1]. The goal of machine learning technologies is to provide algorithms the ability to learn and adapt [1]. Machine Learning has evolved in past few years and is now gathering a lot of attention as people are getting used to using machines to get things done. As the data is increasing day by day and the demand is also increasing, it is expected that smart data analysis will become even more prevalent as a crucial component of technological advancement. Machine

L. Barolli et al. (Eds.): AINA 2022, LNNS 451, pp. 531–545, 2022.
https://doi.org/10.1007/978-3-030-99619-2_50

learning algorithms are classified into a taxonomy based on the algorithm's expected outcome [1]. In order to apply Machine learning, the Data Scientist usually follows four steps. Firstly, the dataset which is to be used is chosen and then training data is prepared from the same. After the dataset is prepared, the second step is to determine the algorithm which is to be used on that particular dataset. Then, the dataset is fed to that particular algorithm and the model is trained. In the end, based on the training set, the model then gives out the predictions.

Data analytics, pattern recognition and machine learning all require classification. When the data is categorized based on prior information, it is known as supervised learning [2]. Each testing instance's class is determined by combining features and identifying patterns that are common to each class in the training data. There are two stages to classification. To quantify model performance and accuracy, a classification technique is applied to the training data set first, and then the extracted model is validated against a labelled test data set [2]. Document categorization, spam filtering, picture classification, fraud detection, chum analysis and risk analysis are all examples of classification applications.

Generally, Supervised machine learning and unsupervised machine learning are the two kinds of machines learning algorithms which are widely used these days [1]. Supervised machine learning algorithms are those type of models where the algorithm learns from a labelled dataset and the evaluation of its accuracy is based on its training data [1]. Unsupervised machine learning on the other hand is a kind of algorithm in which the model is not needed to be supervised [2]. When compared to supervised learning algorithms, unsupervised learning algorithms allow users to complete more difficult processing tasks [2].

The use of labelled datasets to train algorithms that need to classify given inputs and accurately predict the outcomes is what supervised machine learning is all about. In supervised learning, we use data that has been labelled to train the machine [1]. It typically learns from labelled training data and aids in the prediction of outcomes for any unexpected input. When data mining is done, supervised learning is further divided into two categories: classification and regression [2]. The test data is appropriately assigned to certain groups using a method called classification. It's used to organize the output of a class. Binary classification refers to an algorithm that attempts to classify input into two distinct classes whereas multiclass classification refers to an algorithm that attempts to categorize input into more than two classes. To explore the relationship between dependent and independent variables, regression is used. The following Fig. 1 represents Supervised Machine Learning.

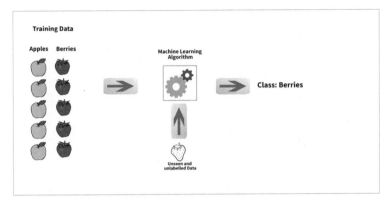

Fig. 1. Supervised Machine Learning

2   Research Background and Objectives

In this study, we aim to study the variants of K Nearest Neighbors in depth which are already being introduced by different researchers. The K-nearest neighbor (KNN) algorithm is a non-parametric classification approach that has been around for a long time. The KNN algorithm, given a query and a training set, finds the query's k-nearest neighbors in the training set and assigns a class label to the query using the majority voting rule. It is a very powerful algorithm which is robust and versatile in nature. There are various types of distance metrics that can be used with the KNN like Euclidean, Chebyshev, Manhattan, Hamming etc. This algorithm is majorly used because of its simple computation and performance [3]. The KNN technique has been widely used in various technologies, including pattern recognition, feature selection etc. because of its simplicity, efficacy, and intuitiveness. The KNN algorithm provides a lot of features and does not require any training because it is a nonparametric classification approach. It can classify the query directly based on the information provided by the training set and does not require prior knowledge of the statistical features of the training examples. Furthermore, it has been demonstrated that the classification accuracy of the KNN algorithm may converge to twice that of the ideal Bayesian classifier under the restriction where the total number of training instances are equal.

Let us take an example of two classes named as A and B as shown in the given Fig. 2. There is an unknown sample called X for which the classification needs to be done. For instance, we take the value of k as 5, we will check the five nearest neighbors of X and hence, based on the majority, the majority class is assigned to X.

Properties and Working of KNN Algorithm: The KNN Algorithm is one of the most fundamental and highly adaptable machine learning algorithms. It is mostly used to categorize anything which is of different types. It is simple to implement and can be done in simultaneously. When working with the KNN, there are several processes that must be followed. To begin, the number K of the neighbors is chosen. The Euclidean distance of K number of neighbors is calculated after selection. Then, based on the estimated Euclidean distance, K nearest neighbors are chosen [2]. The total amount of

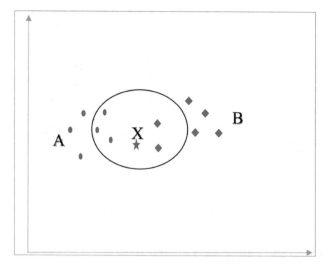

Fig. 2. KNN Algorithm

data points in each category among these k neighbors is determined, and the additional data points to that category for the maximum neighbor are calculated.

Some Related Studies

KNN Variants: In order to increase the accuracy and performance of KNN algorithm, a lot of variants of KNN have been introduced by different researchers. The performance of the traditional KNN can be improvised by applying additional techniques to the traditional KNN. All the variants mentioned below were found out using various Databases like IEEE, Google Scholar etc.

Fuzzy KNN: Keller, Gray [5] proposed a fuzzy version of KNN. In this paper, the idea of fuzzy sets is extended to the KNN algorithm in order to achieve better experimental results. The problem with the traditional KNN algorithm is that any labelled pattern, regardless of its desirable qualities, is given equal weight in the class membership decision. Three methods of assigning fuzzy memberships to labelled samples are proposed to solve the problem, including two new approaches: fuzzy K-nearest neighbor algorithm and fuzzy nearest prototype algorithm, with experimental results compared to crisp KNN algorithm. The proposed algorithm outperforms the current one [5].

Weighted KNN: It is also known as WKNN and is a modified version of the traditional KNN. One of the limitations of the traditional KNN is the selection of the hyperparameter K which has a direct impact on the efficiency of the algorithm. The algorithm becomes vulnerable to the outliers if the selected value of k is too low whereas if the same value of k is too large then there is a possibility that the neighborhood might include values from other classes as well. To overcome this particular problem of the traditional KNN, weighted KNN was introduced. In WKNN, the nearest k points are assigned with weight with the help of a function called kernel function. A kernel function is described as a function that decreases as the value of the distance increases. The main aim of this step if to assign more weights to the points that are closer and less weight to the points that are far from k. The distance function is the most commonly used function. All of the values in the traditional KNN have utmost importance but the attributes are not given equal importance. The main aim of this variant of K Nearest Neighbors is to first learn from its different attributes and then each attribute is assigned with the weight which affects the classification. Yigit [6] proposed the WKNN in which the motivation behind this variant of K nearest neighbor was to find the optimal weight of the attributes with the help of ABC Algorithm.

Generalized KNN: A generalized mean distance-based k-nearest neighbor algorithm was proposed to overcome the sensitivity of k and improve the KNN-based classification performance (GMDKNN) Gou, Ma [7]. The multi-local mean vectors of k nearest neighbors in each class can reflect the different local sample distributions of its own class, as argued in LMPNN Gou, Ma [7] and MLMKHNN (Pan et al. 2020) and prevent the single and unique value of k from which the sensitivity of k readily results. As a result, in the proposed GMDKNN, the query sample's k nearest neighbors is first found from each class, and then the associated k local mean vectors are calculated. The harmonic mean distance is then extended to its general form which is called generalized distance mean in the GMDKNN because of its ability to reflect the diverse weighted contributions from distinct multi-local mean vectors for classification (Pan et al. 2020).

Hybrid KNN: A novel KNN technique called as the hybrid KNN algorithm was proposed to address the drawbacks of the standard KNN algorithm such as noise, sparseness, and imbalance which are present in certain datasets (HBKNN) [8]. To overcome the imbalance problem, the proposed HBKNN algorithm uses the fuzzy relative transform method. In this method, local and global data are integrated. To manage datasets with noisy properties, the random space ensemble structure (RS-HBKNN) technique is utilized. The proposed variant outperform the standard KNN algorithm [8].

Locally Adaptive KNN: In standard KNN algorithm, global value of input parameter k is used. But according to this proposed algorithm, we can try using different values of k for different portions of input space. The value of K is calculated each time by using cross validation in its local neighborhood. This variant came into existence when it was realized that the KNN majority rule ignores the differences in classification ability of different nearest neighbor [9]. Also, the nearest neighbor selection is not reliable as it only uses the distance function to measure the similarity between the query and training instances and completely ignores the spatial distribution. Since the KNN algorithm is too sensitive to the value of k, if the value of k is too small then the classification accuracy is disregarded, and the noisy instances are amplified or if the value of k is too large, excessive noisy instances are selected as the nearest neighbors resulting in bad classification of data [9]. In order to address the two issues of the KNN algorithm i.e., the arbitrariness of directly taking the label of majority class as the classification decision result and the irrationality of using a fixed and single value of k for all queries, a new variant of KNN was introduced known as locally adaptive KNN [9]. This variant of KNN is based on discrimination class (DC-LAKNN). In this method, second majority class in classification is considered. There are two major advantages of using this variant of KNN. The DC-LAKNN algorithm focuses on the role of the second majority class used in classification of data by introducing the concept of the discrimination class based on the information fetched from not only majority class but also from second majority class.

The quantity and distribution information on both majority classes in the k neighborhood is considered for a given query to select the class which is most similar to the query from majority and second majority class as the discrimination class of the query at the current value of k. Hence, this DC-LAKNN algorithm considers the discrimination class which is selected by the majority class and the second majority class as the class most similar to the k-neighborhood [9].

K Tree algorithm: By incorporating a training stage into the classical KNN method, a new version of the KNN algorithm known as the k Tree algorithm was introduced, which is used to find the ideal K value for each test object [10]. Despite the fact that both algorithms have a similar running cost, experimental results reveal that the suggested method outperforms the standard KNN in terms of accuracy. Furthermore, the k Tree algorithm is proposed as an improved k Tree algorithm for speeding up the testing process by storing extra information about testing artefacts in the k Tree's leaf node.

Improved KNN: An improved version of the existing KNN technique was proposed for imbalance datasets that concentrated on the minority class sample rather than the majority class sample [11]. In usual, all samples are given the same weight, but in this method, the weight of the minority class sample is enhanced based on the minority class sample distribution's local features. The proposed approach is compared to the currently used weighted distance KNN algorithm (WDKNN). Experiments show that the suggested approach outperforms WDKNN when dealing with unbalanced datasets.

Stepwise KNN algorithm: Zang, Huang [12] proposed the Stepwise KNN algorithm, which is a modified version of the KNN algorithm with a feature reduction and kernel approach that can be used to handle duplicate attributes and uneven data distribution. The SWKNN uses a kernel based KNN method to gradually reduce extraneous features. The proposed approach outperforms the classic KNN algorithm in terms of decreasing high-dimensional features, according to the results of the experiments.

Genetic KNN: The performance of pattern or image recognition systems might be harmed by a large feature collection. In other words, having too many features can impair the identification system's classification accuracy because some of them are redundant and non-informative [13]. A variety of combinatorial sets of features should be obtained in order to maintain the ideal combination for optimal accuracy. Feature selection, also known as variable selection, attribute selection, or variable subset selection in machine learning and statistics, is the process of obtaining a subset of relevant characteristics (usually optimal) for use in machine model creation. Principal Component Analysis (PCA), Particle Swarm Optimization (PSO), and Genetic Algorithm (GA) are examples of these techniques [13]. In recent years, a growing number of academics have used the WEKA (Waikato Environment for Knowledge Analysis) program to reduce dimensionality. However, the feature selection approach of WEKA software is static, as users are unable to change the configuration of the feature selectors in question [13]. GA has a reputation for being an extremely adaptable and efficient feature selection method. A feature selection strategy based on GA was described [13]. The technique shown here involved selecting a combinatorial collection of features from an initial feature set using a unique fitness function. Features chosen by both the WEKA Feature Selectors and the GA were fed into a number of WEKA classifiers for benchmarking [13]. In more cases, GA-based features outperformed WEKA-based features. There are just two approaches.

Overall, the GA method and WEKA-CFS, which are wrapper-based feature selectors, performed better than the WEKA ranker (IG), which is a filter-based feature selector. The method's key advantage is controllability, as the GA can be fine-tuned to provide better outcomes all of the time by adjusting the fitness functions [13].

Reduced time complexity KNN: Singh [2] introduced an approach that is an upgraded version of the KNN algorithm that performs considerably better than the traditional KNN algorithm and reduces the time complexity. This method does a pre-classification before implementing the proposed algorithm. A threshold is used to divide the training set into various portions. Experiments revealed that the suggested KNN method with pre-classification outperforms traditional KNN while also having a lower time complexity.

Clustering based KNN: Zhou, Wang [14] presented a new clustering based KNN (CKLNN) for improving training samples. When there is an unequal distribution of training samples and a defined border, this method is applied. CLKNN pre-processes training data with clustering and adjusts the neighborhood number K dynamically in each iteration. In terms of execution time and accuracy rate, the suggested approach outperforms the traditional KNN algorithm.

SVM KNN: Sonar, Bhosle [15] suggested a hybrid SVM/KNN algorithm. This method works with both linear and nonlinear data. SVM is trained using k nearest neighbors in this approach. The goal was to use kernel to map feature points to kernel space and then determine the k closest neighbors for a given test data point within the training dataset. The search for support vectors is then reduced down. For benign and malignant classes, the suggested approach has a 100% accuracy for DDSM and a 94% accuracy for MIAS database.

Decomposed KNN: Xiao and Chaovalitwongse [16] suggested a deconstructed KNN technique that uses the distance function between the KNNs' centroids to classify an unknown pattern. In the Decomposed KNN technique, a learning method is employed to train the local-optima-free distance function by solving a convex optimisation problem. This technique facilitates in the collecting of critical features as well as the removal of unnecessary features by employing L1 regularisation. The strategy given can be applied to problems involving many classes. The authors have established the suggested method's theoretical relationship with the classical KNN algorithm, SVM algorithm, and LDA algorithm. When compared to other classification algorithms, experimental findings show that this variant provides accurate results.

KNN using different distance functions: The efficiency of the KNN method was explored by Hu, Huang [17], who presented an approach based on distance measures. Experiments are conducted on a variety of medical databases that contain three different forms of data: category, numerical, and mixed. The experiment's results are based on four different distance metrics: Euclidean, Cosine, Chi square, and Minkowsky. The Chi square distance metric performs the best when compared to three other distance metrics. On the other hand, the Euclidean distance metric works well with both numerical and categorical data.

Euclidean distance for normalized Data: In the current KNN algorithm, Pawlovsky and Nagahashi [18] provided a method for determining a successful parameter setting. The proposed method analysed a dataset of prognostic breast cancer using the KNN algorithm. A few settings are changed while using the KNN algorithm, such as the number of neighbours (K), the amount of data used, and the pre-processing method. Data can be normalized or used without being normalized or standardized. Using this strategy with Wisconsin's breast cancer prediction data, the KNN classification algorithm delivers greater accuracy according to the results provided in this research.

Method for fixing missing values: UshaRani and Sammulal [19] suggested a strategy for dealing with missing categorical and numerical attribute values in medical datasets. This is accomplished by mining medical datasets for secret information. The authors of this paper suggested a novel imputation method for repairing missing values by clustering medical records that are missing values-free. The suggested method for dealing with missing values also resulted in the records' dimensionality being reduced. This method may be applied to some other field of study.

3   Research Problems and Questions

KNN is an algorithm which is widely used and is often used for pattern classification for handling binary and multiclass problems in various domains. Although KNN is a simple algorithm used in various industries, it might give inaccurate results when the training datasets are not complete and are imbalanced. It becomes challenging to deal with the noisy data, uncertain and missing values in the datasets due to the limitations of traditional KNN [2]. In order to handle this discrepancy and inaccuracy, different variants of KNN algorithm are developed. These KNN variants have helped in improving the efficiency and accuracy in KNN and also with the help of these variants pre-processing can be done to handle imbalanced and incomplete datasets [2].

There are various issues in the traditional KNN algorithm. Firstly, the choice of the value of k. The selection of the value of k is very important and can be done by inspecting the data. If the chosen value of k is too small, then the result might get sensitive to noise points and if the value of the k is too large then the neighborhood may have many objects from other classes. If the value of k is large, it may reduce the overall noise [2]. The choice of distance metric is also one of the aspects as the most commonly used metric is Euclidean distance which gives equal importance to all the attributes which might not be good fit for the high dimensional datasets and results might not be that effective and accurate. For example, for categorical dataset hamming distance metric can be used. If in some cases some of the attributes values are missing in the dataset then the objects holding incomplete data may be discarded in the classification and the results might turn out to be inaccurate and biased. In some cases, datasets might have a class which has a very small number of objects whereas there can be another class which has major objects. Based on the number of objects, a class is termed as majority of minority class. There can also be cases when some of the attributes are not available in a data set. In some cases, the data is derived or modelled by some probability distribution functions which can make the data inaccurate due to such uncertain values. In order to resolve such issues of traditional KNN, various researchers have proposed different variants of KNN which helped in eliminating such issues and made the model even more efficient [2]. The main aim of this research is to perform the analysis of different types of KNN variants that helps in reducing the cons of the traditional KNN. These variants try to enhance the performance of traditional KNN and helps in giving more accurate results.

For each KNN variant, its performance and accuracy will be analyzed with the help of data used for the prediction of diseases. With the help of this research, we will have more clarity on which KNN variant is best suited for any disease prediction and will be of great help to other researchers as well who are doing their research in the similar domain and would help them in making their research better.

4 Research Methodology and Experiments

The research methodology used in this research is quantitative research. The goal is to examine the performance of several KNN variants which are already being introduced by different researchers using a dataset related to medical informatics. Machine learning focuses on those sorts of algorithms that do analysis on input data and produce predictions within an acceptable range in order to learn and optimize their performances. With the help of more data, these algorithms are able to make more accurate predictions. Similarly, significant research efforts are made to examine different forms of KNN variants which will be valuable in disease risk prediction. All KNN variations are implemented in Python, which is one of the most widely used machine learning languages. Python is a great choice for machine learning since it's so versatile, allowing you to use either OOPs or scripting. All the variants of KNN are implemented in python using PyCharm 2020.3.4 Professional Edition. Python is a high-level, general-purpose programming language that is interpreted.

4.1 Data Collection

Different datasets are used for this Research which are taken from Kaggle. In Python, Scikit-learn also known as Sklearn is the most usable and robust machine learning library. It uses a Python consistency interface to give a set of efficient tools for machine learning and statistical modelling, such as classification, regression, clustering, and dimensionality reduction. This library has been extremely helpful in implementing all these variants of KNN. The main aim of this research is to implement the variants of KNN on different types of medical datasets taken from Kaggle. Kaggle, a Google LLC subsidiary, is an online community of data scientists and machine learning experts. Users can use Kaggle to search and publish data sets, study and construct models in a web-based data-science environment, collaborate with other data scientists and machine learning experts, and compete in data science competitions. All these datasets that are used for this research are taken from Kaggle Repository only (Table 1).

Table 1. Datasets used for this research

Dataset	Source	Description
Heart Attack Possibilities	Kaggle	Used to predict the presence of Heart Disease
Heart Failure Outcomes	Kaggle	Used to predict survival of patients with heart failure from serum creatinine and ejection fraction alone
Diabetes	Kaggle	Used to predict diabetes in patients
Heart Disease Prediction	Kaggle	Used to predict which patients are most likely to suffer from a heart disease in the near future
CKD Preprocessed	Kaggle	Used to predict Chronic Diseases
Pima Indians Dataset	Kaggle	Used to predict whether or not each patient will have an onset of diabetes
CKD Prediction	Kaggle	Used to predict whether a patient is having chronic kidney disease or not, based on certain diagnostic measurements included in the dataset
Mole Cancer Dataset	Kaggle	Used to Predict the Cancer

5 Research Significance

People these days are facing numerous types of problems due to environmental changes and their living habits. It is extremely important to predict the diseases in the early stage so that the proper treatment can be given to the patients at the right time before it gets too late. It gets very difficult for the doctor to predict the symptoms accurately which is the most challenging task. In order to overcome this issue, data mining plays a vital role in predicting the diseases. There is a large amount of data which is growing every year in the medical field and with the help of that data a lot of predictions can be done which can be really beneficial to the medical industry. With the help of this data and various machine algorithms, the early prediction of diseases can be performed. In this research, one of the widely used machine learning supervised algorithms is used to do the same prediction of diseases which will be really beneficial for the medical industry. This research mainly focuses on KNN algorithm and its variants. This study will help other researchers to figure out the best variant of KNN for their research based on several parameters like accuracy, precision etc. The major goal of this study is to examine the performance of each and every KNN variant based on those parameters, and to do so with the help of those parameters and with the help of that it can be concluded at the end which variant of KNN works best for any which disease risk prediction.

6 Results

In this part of this research, results of each and every variant are discussed and analysed based on its performance and different performance metrics like accuracy, precision, recall and AUC. After carrying out the Performance Analysis of all the variants of KNN, the following results are achieved with respect to each variant of KNN as mentioned below:

Traditional KNN: Traditional KNN is the standard KNN and after measuring it against each of the four metrics, following results are achieved as mentioned in Table 2. It is clear from the results that the precision of 100% was achieved for the Heart Failure Outcomes and the best accuracy was achieved for 79.22%. The best Recall was for Mole Cancer Dataset which was 92.53% and best AUC was for Pima Indians Dataset which was 72.50%.

Table 2. Results of traditional KNN

Dataset	Accuracy	Precision	Recall	AUC
Heart Failure Outcomes	65.00%	100.00%	86.95%	54.34%
Diabetes	79.22%	70.27%	55.31%	72.51%
Heart Disease	65.57%	72.41%	61.76%	65.57%
Heart Disease Prediction	70.37%	68.18%	62.50%	69.58%
CKD Preprocessed	60.00%	75.00%	57.69%	60.98%
Pima Indians Dataset	79.22%	70.27%	55.31%	72.51%
CKD Prediction	62.50%	86.60%	50.00%	67.85%
KNN Dataset	71.05%	68.88%	92.53%	66.48%

Weighted KNN: Weighted KNN is one of the majorly used variant of KNN and after measuring it against each of the four metrics, following results are achieved as mentioned in Table 3. It is clear from the results that the precision of 81.81% was achieved for the CKD Preprocessed Dataset and the best accuracy of 80.70% was achieved for Mole Cancer Dataset. The best Recall was for Mole Cancer Dataset which was 95.52% and best AUC was for Mole Cancer Dataset which was 77.54%.

Fuzzy KNN: Fuzzy KNN is other majorly used variant of KNN and after measuring it against each of the four metrics, following results are achieved as mentioned in Table 4. It is clear from the results that the precision of 65% was achieved for the CKD Preprocessed and CKD Prediction Dataset and the best accuracy of 65% was achieved for CKD Preprocessed and CKD Prediction Dataset. The best Recall was for Mole Cancer Dataset which was 100% and best AUC was for Diabetes and Pima Indians Dataset which was 50.05%.

Table 3. Results of weighted KNN

Dataset	Accuracy	Precision	Recall score	AUC
Heart Failure Outcomes	43.83%	26.66%	17.39%	43.83%
Diabetes	74.67%	58.00%	61.70%	71.03%
Heart Disease	62.29%	67.74%	61.76%	62..36%
Heart Disease Prediction	68.51%	64.00%	66.66%	68.33%
CKD Preprocessed	70.00%	81.81%	69.23%	70.32%
Pima Indians Dataset	74.67%	58.00%	61.70%	71.03%
CKD Prediction	67.50%	77.08%	71.15%	65.93%
KNN Dataset	80.70%	77.10%	95.52%	77.54%

Table 4. Results of fuzzy KNN

Dataset	Accuracy	Precision	Recall score	AUC
Heart Failure Outcomes	61.60%	0.00%	0.00%	50%
Diabetes	42.20%	30.55%	70.21%	50.05%
Heart Disease	44.26%	0.00%	0.00%	50%
Heart Disease Prediction	42.59%	43.39%	95.83%	47.91%
CKD Preprocessed	65%	65%	100%	50%
Pima Indians Dataset	42.20%	30.55%	70.21%	50.05%
CKD Prediction	65%	65%	100%	50%
KNN Dataset	58.77%	58.77%	100%	50%

Generalized KNN: Generalized KNN is another variant of KNN and after measuring it against each of the four metrics, following results are achieved as mentioned in Table 5. It is clear from the results that the precision of 98.13% was achieved for the Mole Cancer Dataset and the best accuracy of 90.15% was achieved for Mole Cancer Dataset again. The best Recall was for Mole Cancer Dataset which was 94.67% and best AUC was for Diabetes and Pima Indians Dataset which was 88.61%. This variant of KNN performed really well for the Mole Cancer Dataset and gave some really good results for this dataset as it was a large dataset as well as compared to other datasets.

Genetic KNN: For Genetic KNN, only accuracy was calculated with and without Genetic Algorithm and this variant of KNN gave some really good results as it can be seen from the Table 6 that Genetic Algorithm has increased the accuracy quite good with the Genetic Algorithm and maximum of 99% accuracy was achieved for CKD Preprocessed and CKD Prediction Dataset.

Table 5. Results of generalized KNN

Dataset	Accuracy	Precision	Recall score	AUC
Heart Failure Outcomes	48.49%	16.27%	14.58%	39.55%
Diabetes	73.43%	62.50%	59.70%	70.25%
Heart Disease	62.70%	12.10%	11.51%	57.57%
Heart Disease Prediction	64.07%	59.66%	59.16%	63.58%
CKD Preprocessed	22%	36.97%	35.20%	17..6%
Pima Indians Dataset	73.40%	62.50%	59.70%	70.25%
CKD Prediction	38%	50.33%	60.08%	30.40%
Mole Cancer Dataset	90.15%	90.13%	94.67%	88.61%

Table 6. Results of genetic KNN

Dataset	Accuracy (with GA)	Accuracy (without GA)
Heart Failure Outcomes	82.75%	58.62%
Diabetes	76.31%	71.71%
Heart Disease	62.70%	12.10%
Heart Disease Prediction	87.03%	75.92%
CKD Preprocessed	99%	97.43%
Pima Indians Dataset	76.31%	71.71%
CKD Prediction	99%	97.46%

7   Conclusion

In this study, we aim to provide the overview of the performance of the variants of K Nearest Neighbor which are being proposed by different researchers and can be used for the prediction of diseases. In this research, we have performed analysis of different variants of K nearest Neighbors based on different performance metrics. KNN is one of the most powerful supervised machine learning Algorithms and has been really useful for disease risk prediction. A lot of researchers have used KNN for their research related to disease risk prediction. Since the variants of KNN performs much better than the traditional KNN, the main focus of this study are the variants of KNN. Also, after completion of this research, we are able to address our research problem and figure out which variant works best for which type of Disease. It is very important to have an early prediction of diseases to avoid the risk. After performing analysis for each variant, it can be concluded that Genetic KNN worked best for most of the datasets and the use of genetic algorithm with KNN has enhanced the performance of all the data sets really well. For heart related diseases, diabetes and mole cancer dataset Generalized KNN and Genetic KNN both performed really well. For Chronic Kidney Diseases both datasets, Genetic KNN worked the best. In future we can perform the same experiment on more variants of KNN and with more datasets. For this study only eight datasets and four variants are considered but for future work we can extend this research and collect more disease related datasets from other sources and other variants can also be implemented with the same. Also, this study will be really helpful for the fellow researchers to determine the most efficient variant for their respective studies.

References

1. Mitchell, T.M.: Machine learning (1997)
2. Singh, A., Thakur, N., Sharma, A.: A review of supervised machine learning algorithms. In: 2016 3rd International Conference on Computing for Sustainable Global Development (INDIACom). IEEE (2016)
3. Singh, A.P.: Analysis of variants of KNN algorithm based on preprocessing techniques. In: 2018 International Conference on Advances in Computing, Communication Control and Networking (ICACCCN). IEEE (2018)
4. Sinha, P., Sinha, P.: Comparative study of chronic kidney disease prediction using KNN and SVM. Int. J. Eng. Res. Technol. 4(12), 608–612 (2015)
5. Keller, J.M., Gray, M.R., Givens, J.A.: A fuzzy k-nearest neighbor algorithm. IEEE Trans. Syst. Man Cybern. 4, 580–585 (1985)
6. Yigit, H.: A weighting approach for KNN classifier. In: 2013 International Conference on Electronics, Computer and Computation (ICECCO). IEEE (2013)
7. Gou, J., et al.: A generalized mean distance-based k-nearest neighbor classifier. Expert Syst. Appl. 115, 356–372 (2019)
8. Kelly Jr., J.D., Davis, L.: A hybrid genetic algorithm for classification. In: IJCAI (1991)
9. Pan, Z., Wang, Y., Pan, Y.: A new locally adaptive k-nearest neighbor algorithm based on discrimination class. Knowl. Based Syst. 204, 106185 (2020)
10. Zhang, S., et al.: Efficient KNN classification with different numbers of nearest neighbors. IEEE Trans. Neural Netw. Learn. Syst. 29(5), 1774–1785 (2017)
11. Xueli, W., Zhiyong, J., Dahai, Y.: An improved KNN algorithm based on kernel methods and attribute reduction. In: 2015 Fifth International Conference on Instrumentation and Measurement, Computer, Communication and Control (IMCCC). IEEE (2015)
12. Zang, B., et al.: An improved KNN algorithm based on minority class distribution for imbalanced dataset. In: 2016 International Computer Symposium (ICS). IEEE (2016)
13. Babatunde, O.H., et al.: A genetic algorithm-based feature selection (2014)
14. Zhou, L., et al.: A clustering-Based KNN improved algorithm CLKNN for text classification. In: 2010 2nd International Asia Conference on Informatics in Control, Automation and Robotics (CAR 2010). IEEE (2010)
15. Sonar, P., Bhosle, U., Choudhury, C.: Mammography classification using modified hybrid SVM-KNN. In: 2017 International Conference on Signal Processing and Communication (ICSPC). IEEE (2017)
16. Xiao, C., Chaovalitwongse, W.A.: Optimization models for feature selection of decomposed nearest neighbor. IEEE Tran. Syst. Man Cybern. Syst. 46(2), 177–184 (2015)
17. Hu, L.-Y., Huang, M.-W., Ke, S.-W., Tsai, C.-F.: The distance function effect on k-nearest neighbor classification for medical datasets. Springerplus 5(1), 1–9 (2016). https://doi.org/10.1186/s40064-016-2941-7

18. Pawlovsky, A.P., Nagahashi, M.: A method to select a good setting for the KNN algorithm when using it for breast cancer prognosis. In: IEEE-EMBS International Conference on Biomedical and Health Informatics (BHI). IEEE (2014)
19. UshaRani, Y., Sammulal, P.: A novel approach for imputation of missing values for mining medical datasets. In: 2015 IEEE International Conference on Computational Intelligence and Computing Research (ICCIC). IEEE (2015)

Covert Timing Channels Detection Based on Image Processing Using Deep Learning

Shorouq Al-Eidi[1]([✉]), Omar Darwish[2], Yuanzhu Chen[3],
and Mahmoud Elkhodr[4]

[1] Computer Science Department, Memorial University of Newfoundland,
St. John's, Canada
shorouqa@mun.ca
[2] Information Security and Applied Computing, Eastern Michigan University,
Ypsilanti, USA
[3] School of Computing, Queen's University, Kingston, ON, Canada
[4] School of Engineering and Technology, Central Queensland University,
Rockhampton, Australia

Abstract. With the development of the Internet, covert timing channel attacks have increased exponentially and ranking as a critical threat to Internet security. Detecting such channels is essential for protection against security breaches, data theft, and other dangers. Current methods of CTC detection have shown low detection speeds and poor accuracy. This paper proposed a novel approach that used deep neural networks to improve the accuracy of CTC detection. The traffic inter-arrival times are converted into colored images; then, the images are classified using a CNN that automatically extracts the image's features. The experimental results demonstrated that the proposed CNN model achieved better performance than other detection models.

Keywords: Covert timing channels detection · Deep learning ·
Convolutional neural networks · Image processing

1 Introduction

Covert timing channels (CTCs) provide a mechanism to transmit unauthorized information across different processes. It utilizes the inter-arrival times between the transmitted packets to hide the communicated data. Due to the CTCs' ability to transfer data without being detected by traditional detection methods such as firewalls and intrusion detection systems, these channels have become a serious threat to the general community of internet users [1].

As a key part of security protection, detecting CTCs is particularly challenging. CTC detection methods are mostly focused on the statistical analysis of network traffic by observing network traffic behavior and extracting statistical features of covert and normal traffic, and comparing those properties to recognize anomalies and detect covert communication [3–7,11]. Unfortunately,

© The Author(s), under exclusive license to Springer Nature Switzerland AG 2022
L. Barolli et al. (Eds.): AINA 2022, LNNS 451, pp. 546–555, 2022.
https://doi.org/10.1007/978-3-030-99619-2_51

methods based on statistical feature analysis are often disrupted. The effectiveness of statistical feature analysis can be hampered by sensitivity to the high variation of network traffic.

Recently, rather than focusing on non-visible features for CTC detection, Al-Eidi et al. [2] proposed CTC visualization, a new approach based on image processing techniques. This work transformed the traffic inter-arrival time samples into two-dimensional colored images. Then, the image features were used for classification. More powerful detection methods based on various machine learning techniques also use the image features to uncover covert communication and its variations. However, the visualization method can handle covert communication obfuscation problems, but it suffers from the high time cost needed for image feature extraction. Moreover, these feature extraction methods demonstrate low efficiency when exposed to a large dataset because their quality substantially impacts classifiers' performance. The challenge for building CTC detection models is to find a means for extracting features effectively and automatically.

Unlike machine learning, deep learning avoids the manual step of extracting data features. For instance, the images can be input directly to the deep learning algorithms for predicting objects. In this way, the deep learning model is more powerful than traditional machine learning techniques. A convolutional neural network (CNN) is one of the most common deep learning algorithms that have been used in the computer vision research domain. CNN can automatically extract image features and avoid the drawbacks of manually extracting inappropriate features. For this reason, this work utilizes the CNN technique to classify colored images and detect CTC. Based on our best knowledge, the CNN technique has not been applied in the literature to detect such channel attacks.

The main contributions of the paper are summarized as follows:

- Propose a novel method for detecting CTC based on a convectional neural network (CNN).
- Determine the optimum model architecture that provides high-performance outcomes in CTC detection.
- Run extensive experiments to demonstrate the effective and efficient proposed approach for CTC detection.
- Compare the performance of the proposed approach with three machine learning approaches, namely, Support Vector Machine, Naive Bayes, and Decision Tree.

The rest of this paper is organized as follows. Section 2 illustrates the CTC detection model. Section 3 describes the experimental setup. Section 4 shows the results and analysis of this research. Finally, Sect. 5 concludes this work.

2   CTC Detection Based on CNN Model

This section presents our improved CTC detection method based on a CNN, which included: 1) collecting the dataset and transforming the traffic inter-arrival

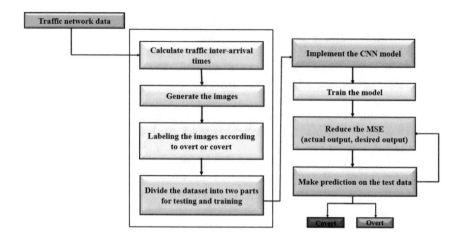

Fig. 1. Proposed model overview.

times into colored images; 2) designing and employing the CNN model to classify the image and detecting CTC. We describe the details of each module next.

2.1 Traffic Inter-Arrival Times to Colored Images

As shown in Fig. 1, we first collect a network traffic dataset that includes a legitimate packet exchange (between sender and receiver devices) and a malicious traffic process that uses a covert channel to transfer secret messages between the two devices. Then, we calculate the inter-arrival times (the time duration between two consecutive packets) of the traffic flow captured and save in our dataset. Finally, after calculating the inter-arrival times, we transform these values into colored images for utilizing the image processing technique in the deep learning domain.

To transform the inter-arrival times to colored images, we used a visualization module in [2]. The module generates a 2D matrix of size 28×28 to automatically place each inter-arrival time value in the matrix row by row. Then, each matrix is viewed as a colorful image by transforming each of its normalized values to a color pixel. At the end of this step, all the images will have binary labels, either overt or covert.

2.2 CNN Model Design and Image Classification

CNNs have developed quickly as a hot topic in image classification and recognition. Their local perception and weight-sharing network structure reduce the complexity of network models and the number of parameters. A CNN has even further advantages when the input is multidimensional. Compared with traditional recognition algorithms, this method is not hampered by complicated feature extraction and data reconstruction. Convolutional networks are multilayer perceptron designed for identifying 2D shapes, and they are robust concerning image deformation (e.g., translation, scale, and rotation).

Decrease the computation and parameters in CNN, allowing the depth model to be implemented. Many deep CNNs are being created in response to the fast growth in computer capacity and have shown satisfying results in computer vision tasks. On the other hand, these deep CNN algorithms contain complicated network architecture with many parameters, such as kernel size and layers. For example, VGG16 [12] has 16 layers and over a million parameters, requiring a significant amount of time and memory.

As a result, utilizing deep networks with millions of parameters does not always efficient, particularly in basic problems with less complicated data patterns. According to the literature, tackling complex pattern recognition issues seems feasible only with specialized hardware or optimization techniques such as GPU computations. Such options do not satisfy the network traffic, which is dynamic, quickly changing, and diverse due to the long training time of deep networks. It is critical to have a quick and light technique for working with large data to make decisions at an appropriate time.

This paper developed a shallow CNN architecture with fewer layers and smaller convolution kernels to classify CTC. By developing the shallow CNN detection model, specific image classification issues can be addressed, and CTC can quickly detect without the need for special hardware or complex computations.

The structure of the CNN model consists of several components, as shown in Fig. 2. The input image is a vector with 784-pixel values. It is input into the CNN model, where the convolution layers and filters generate the feature maps using the local receptive field. The two convolutional layers are designed for feature extraction tasks, including the ReLU function. The ReLU activation function is used instead of the non-linearity function because it is faster than tanh or sigmoid and helps in the vanishing gradient problem, which arises in lower network layers. The first filter in the CNN model is of size $3 \times 3 \times 32$ (32 features to learn in the first hidden layer), and the filter size of $3 \times 3 \times 64$ for the second convolution layer (64 features to learn from the second hidden layer). To decrease data size and computational complexity, max-pool layers with size 2×2 are followed the convolution layers. The stride is 1 for convolution layers and 2 for max-pool layers.

The batch normalization technique is used with CNN to normalize each feature map obtained after convolution layers. Batch normalization strategy is utilized in neural networks to improve the classification results by increasing the stability of the network and speeding up the training process. The presence of an internal covariate shift phenomena in training conventional deep neural networks [8] is the primary reason for adopting batch normalization. The data distribution in each intermediate layer would differ by updating the model training parameter before the changes. Because of this behavior, the network is continuously forced to adapt to changing data distributions, making the training very challenging. The batch normalization technique [9] is introduced to cope with

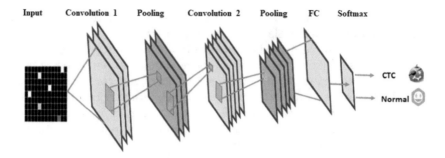

Fig. 2. Architecture structure for CNN model.

parameters that vary during CNN training and enhances network convergence and generalization capabilities [10].

After applying the batch normalization, the model features are then converted into a one-dimensional feature that meets the classifier requirements via the fully connected layer. ReLU is employed with the fully connected layer to improve the CNN model's nonlinear capabilities. Then, the dropout is used to minimize overfitting and enhance the model's generalization capabilities. Dropout is a regularization technique that is used to reduce overfitting. In dropout, the network inactive some of its nodes randomly based on a parameter. The probability parameter determines whether the node should remain in the network or not. In the final step of the CNN model, the Softmax classifier identifies images into two categories, covert and overt, based on their probability. Softmax needs minimal calculations and memory, which saves precious time resources.

The proposed model architecture contains various hyper-parameters that determine the network structure and how the network model is trained, such as the type of optimizer and learning rate. The model performance can vary considerably according to the selected set of hyper-parameters. For this reason, several hyper-parameters are tested to determine the best classification model parameters. For example, when the number of training epochs reaches 30, the model test results did not continually improve, and the training time was longer. The learning rate was another important parameter determining whether the objective function converged to a local minimum.

A suitable learning rate may cause the objective function to converge to a local minimum in a reasonable amount of time. As the number of iterations grew, the learning rate progressively dropped, speeding up model training in the early stages. The learning rate is adjusted for the proposed model at 0.001. Moreover, the batch size is 128 and 256 for model training and testing.

3   Experiment Setup

This section detailed the experimental work environment and dataset. Then, it shows the model results compared with other machine learning models.

3.1 Experimental Environment

We ran the training and testing experiments of the proposed model using a $x86-64$-ThinkStation-P920 server running Linux version $55-18.04.1$−Ubuntu, with 24 cores and 62.42 GB RAM. Moreover, we use Python libraries such as Tensor Flow and matplotlib for implementing the model and experiments.

3.2 Experimental Dataset

The dataset in this work build on the work of Al-Eidi et al. [2], which studied the effect of time-delays of covert traffic on the detection process. This work focuses on detecting one of the most sophisticated CTC attacks known as Cautious Covert Timing Channels (CCTC) [2]. CCTC is a more sophisticated cyberattack that attempts to mimic overt traffic behavior to transmit secret data between two communicating systems. To simulate the CCTC attack, the packet delay was set to be equal to half of the mean inter-arrival time of overt traffic, making it harder to detect by the security detection system.

In this work, we generated a dataset containing 5000 images: 2500 images for overt data and 2500 images for covert data. The image size is 28 × 28 pixels. For covert communication data, 64 bits of the cover message are injected into the traffic flow, where the mean inter-arrival times of overt traffic is adjusted as 0.0664 s.

3.3 Measures of CTC Detection

Different metrics have been calculated to evaluate the proposed model performance. Our evaluation aims at providing a comprehensive analysis of various accuracy measures to provide more details about the proposed model classification and detection performance. Therefore, the most popular accuracy measures in security domains have been used. The following list presents and explains each one of these measures.

$$Accuracy = \frac{(TP + TN)}{(TP + TN + FP + FN)} \tag{1}$$

$$Precision = \frac{TP}{(TP + FP)} \tag{2}$$

$$Recall = \frac{TP}{TP + FN} \tag{3}$$

$$F_1 score = \frac{2 \times (Precision \times Recall)}{(Precision + Recall)} \tag{4}$$

where True Positive (TP) is the number of images that are classified correctly as CTCs. True Negative (TN) is the number of images that are correctly identify as non-CTC. False Positive (FP) is the number of image that are incorrectly classified as CTCs. False Negative (FN) is the number of images that are incorrectly classified as non-CTC while, in fact, they are CTCs.

4   Experiment Results and Analysis

4.1   Model Accuracy and Loss

To validate the accuracy and efficiency of the proposed approach, several experiments are conducted. The dataset was divided into 70% for training and 30% for testing (validation). Then, the model is trained on the training dataset; While training the model, some fixed set of hyperparameters is selected. Finally, the testing dataset is used for prediction. The results obtained from the experiments show that the proposed model achieved an accuracy of 95.42% as shown in Fig. 3.

The learning behaviour of the proposed model is also analyzed. Figure 4 shows the loss of model training and testing against epochs. By observing the training loss values in the figure, it can be noticed that the model shows a fast decrease in loss and converges rapidly. In addition, it can be noticed from the curve of test loss that the model offers better learning behavior to avoid overfitting and gain generalization ability. These observations validate the optimal design configuration adopted for the CNN model and signify their CTC detection capabilities on the testing dataset.

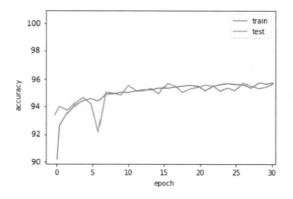

Fig. 3. Accuracy of CNN model.

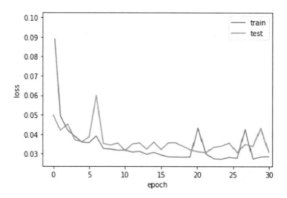

Fig. 4. Loss of CNN model.

4.2 A Comparison of CNN Model with Other Conventional Machine Learning Methods

To evaluate and compare the CNN model performance with other machine learning algorithms, we adopted three machine learning algorithms: Support Vector Machine (SVM), Decision Trees (DT), Naïve Bayes (NB), as shown in Fig. 5. As shown in the figure, the CNN classifier achieved the highest CTC detection accuracy, which achieved an accuracy of 95.42%. The SVM achieved the second-highest CTC detection, which reached 92.22%. The DT approach achieved third place with an accuracy of 91.30%, whereas the NV approach achieved fourth place with an accuracy of 90.4%.

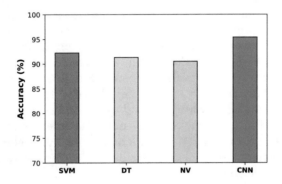

Fig. 5. Comparison accuracy results of CNN model machine learning approaches.

Moreover, Table 1 shows the classifiers performance in detecting CTC. The CNN classifier has achieved the highest score of correct predictions, where the F_1 score showed that it could predict 94.87% of the dataset correctly. Also, the SVM classifier has achieved the second level of recall, in which it can predict 90% of the CTC samples correctly. In contrast, the NV classifier has reached the lowest level of performance measures after DT. It can identify 89% of precision and 88.24% of recall for the whole dataset during the testing process.

Table 1. The performance of the tested algorithms in detecting CTC.

Classifier	Precision (%)	Recall (%)	$F_1 score$ (%)
SVM	91.50	90.00	90.74
DT	90.02	89.00	89.50
NV	89.00	88.24	88.62
CNN	**95.00**	**94.74**	**94.87**

5   Conclusion

This paper presented a novel shallow convolutional neural network with a batch normalization technique for image classification and CTC detection. The batch normalization strategy can accelerate the convergence speed and improve image classification accuracy. The CNN model has four layers as a simple structure with less time complexity on the benchmark image dataset. The proposed shallow network is characterized by small parameters, resistance to overfitting with a limited dataset, and stronger generalization ability. For specific CTC classification problems and based on the training and test sets that we generated and labeled, the proposed architecture performs better than other machine learning models, verifying the effectiveness and feasibility of this shallow neural network architecture. The proposed model presents an alternative solution for CTC detection tasks in different applications such as the smart visual Internet of Things and reduces the complexity of deep neural networks while maintaining their strength.

In the future, we will utilize the convolutional neural network model to localize the CTC in the network traffic. Discovering the segments of traffic flows containing the covert traffic is an essential objective that can drop only the part of traffic flows while allowing the rest of the traffic flow to pass through. This accurate identification substantially minimizes overt traffic interruptions caused by non-malicious applications. Moreover, we will evaluate using other deep learning methods in CTC classification and detection to achieve better industrial security defense.

References

1. Al-Eidi, S., Darwish, O., Chen, Y.: Covert timing channel analysis either as cyber attacks or confidential applications. Sensors **20**(8), 2417 (2020)
2. Al-Eidi, S., Darwish, O., Chen, Y., Husari, G.: SnapCatch: automatic detection of covert timing channels using image processing and machine learning. IEEE Access **9**, 177–191 (2020)
3. Archibald, R., Ghosal, D.: A comparative analysis of detection metrics for covert timing channels. Comput. Secur. **45**, 284–292 (2014)
4. Borders, K., Prakash, A.: Web tap: detecting covert web traffic. In: Proceedings of the 11th ACM Conference on Computer and Communications Security, pp. 110–120 (2004)
5. Cabuk, S.: Network covert channels: design, analysis, detection, and elimination. Ph.D. thesis, Purdue University (2006)
6. Cabuk, S., Brodley, C.E., Shields, C.: IP covert timing channels: design and detection. In: Proceedings of the 11th ACM Conference on Computer and Communications Security, pp. 178–187. ACM (2004)
7. Gianvecchio, S., Wang, H.: An entropy-based approach to detecting covert timing channels. IEEE Trans. Dependable Secure Comput. **8**(6), 785–797 (2010)
8. He, K., Zhang, X., Ren, S., Sun, J.: Deep residual learning for image recognition. In: Proceedings of the IEEE Conference on Computer Vision and Pattern Recognition, pp. 770–778 (2016)

9. Ioffe, S., Szegedy, C.: Batch normalization: accelerating deep network training by reducing internal covariate shift. In: International Conference on Machine Learning, pp. 448–456. PMLR (2015)
10. Lu, L., Yang, Y., Jiang, Y., Ai, H., Tu, W.: Shallow convolutional neural networks for acoustic scene classification. Wuhan Univ. J. Nat. Sci. **23**(2), 178–184 (2018)
11. Shrestha, P.L., Hempel, M., Rezaei, F., Sharif, H.: Leveraging statistical feature points for generalized detection of covert timing channels. In: 2014 IEEE Military Communications Conference, pp. 7–11. IEEE (2014)
12. Simonyan, K., Zisserman, A.: Very deep convolutional networks for large-scale image recognition. arXiv preprint arXiv:1409.1556 (2014)

Internet of Things and Microservices in Supply Chain: Cybersecurity Challenges, and Research Opportunities

Belal Alsinglawi[1]([⊠]), Lihong Zheng[1], Muhammad Ashad Kabir[1], Md Zahidul Islam[1], Dave Swain[2], and Will Swain[2]

[1] School of Computing, Mathematics and Engineering, Charles Sturt University, Bathurst, NSW 2795, Australia
{balsinglawi,lzheng,akabir,zislam}@csu.edu.au
[2] TerraCipher Pty. Ltd., Alton Downs, QLD 4702, Australia
{dave.swain,will.swain}@terracipher.com

Abstract. The livestock sector contributes to agricultural development, poverty reduction, and food security globally. The digitization of the supply chain entails cost savings and improved flexibility, both of which are essential ingredients for boosting resilience in the world food chains. Recently, microservice architectures are becoming popular alternatives to existing software development paradigms, particularly for developing complex and distributed applications. Microservices and supply chains are built upon the enabling technologies of the Fourth Industrial Revolution (Industry 4.0), including the Internet of Things (IoT). However, integrating supply chain infrastructure with IoT and microservices can create a new set of security threats that can defeat this objective without adequate awareness and countermeasures. This study presents an overview of current and potential threats to data-sharing systems in the smart dairy and livestock industries. This paper presents a multi-layered microservices architecture, which emphasizes the importance of communications across layers and multiple bridgeheads. Afterwards, it describes the security aspects of microservices architecture, the threats to data-sharing systems, and the link between attacks and security. Finally, this paper identifies areas that need further study to improve security requirements for IoT microservices in the supply chain and data sharing systems.

1 Introduction

The world is very different from its place six years ago when it committed to ending hunger, food insecurity, and all forms of malnutrition by 2030 [1]. The world has been not making any significant progress neither towards ensuring access to safe, nutritious and sufficient food for all people all year round nor towards eradicating all forms of malnutrition [1]. Conflicts, climate variability and extremes, and the economic slowdowns and downturns are the major factors slowing down progress, especially when inequality is high. COVID-19 made the path towards food sustainability even steeper, according to a report by the FAO organization [1]. The world bank report [2] indicated the livestock sector is an integral part of the global food system, contributing to poverty reduction, food security, and agricultural development. About 1.3 billion people depend on livestock for livelihoods, food and nutrition security. Another report by FAO stated that livestock contributes 40% of the global value of agricultural output [3].

In Australia, every two in three Australians have a positive experience with the red meat industry, according to Meat & Livestock Australia's (MLA) the latest consumer sentiment research [4]. Hence, mapping and optimizing the cattle supply chain operational process is integral for reducing logistics costs [5]. However, the operational factors of meat and livestock go beyond the logistics, transportation of livestock movements, buying or selling cattle and meeting the compliance procedures for the National Livestock Identification System (NLIS) [6]. The most integral requirement for this industry is to ensure transparency, traceability, and security of these digital transactions, which aims to facilitate the operational aspects of this industry and deliver food or share information among farmers or other herd and livestock authorities and third parties. In the digital transformation era, the Internet of Things (IoT) is becoming more and more prevalent in agriculture and food supply chain [7]. As a result, more governments and companies are reported to be implementing advanced IoT techniques in agriculture [8]. Typically, In a farming IoT-based industry, the IoT sensors can be attached to the farm animals to monitor their health and log performance. Livestock tracking and monitoring help collect data on stock health, well-being, and physical location. For example, such sensors can identify sick animals so that farmers can separate them from the herd and avoid contamination. Moreover, using drones for real-time cattle tracking also helps farmers reduce staffing expenses [9].

Fig. 1. IoT- meat and livestock supply chain.

Another example is smart herd sensors (collar tags) to deliver each cow's temperature, health, activity, and nutrition insights and collect information about the herd. Securing farming digital connectivity infrastructure is a critical operational task for the success of meat and livestock. Data sharing depends on functional, end-to-end supply chains Fig. 1. In all phases of the meat and livestock supply chain, from data producers/suppliers to the plate of consumers, it is of significant importance that data sharing, data flow, and services remain secure, trusted and controlled by data producers. The data producing requirements ensure that the data owners can turn on/off data sharing with a button click. Any risk breaking the meat and livestock supply chain, including data vulnerabilities and cyber-attacks through the journey of data sharing from the physical sensing layer to the end-users, will result in loss of data. Therefore, loss of millions of dollars if security requirements are not designed in secure, trustworthy and efficient mechanisms to protect microservices and data sharing systems for the supply chain and livestock industry.

Consequently, securing a supply chain in food and meat industries is a critical factor in sustaining food supplies and reaching its end journey efficiently to the plates of billions of people globally. This paper makes the following key contributions:

- We present cybersecurity for IoT microservices architecture for the supply chain: research trends and state-of-the-arts.
- We propose an overview of the smart herd (SH) focused multi-layered microservices architecture, highlighting the features of multiple bridgeheads and communications across layers.
- We analyze the microservices architecture security aspects, security threats in data-sharing systems and map between attacks and security aspects.
- To improve future research in IoT supply chain microservices for data-sharing systems, we propose a road map that highlights areas still needing investigation in related emerging areas.

The rest of the paper is organized as follows. Section 2 reviews the current state-of-the-art supply chain security and microservices. Then, Sect. 3 provides an overview of the microservices supply chain architecture in the smart herd and Livestock industry. Section 4 discusses the categorization and the itemization of supply chain security aspects then maps between cyberattacks and security aspects in data-sharing systems. Finally, Sect. 5 provides open research opportunities to improve the security of IoT supply chain microservices for the data-sharing systems.

2   Current Research

The IoT microservices architecture for the supply chain is still an unformed research area for smart herd despite recent research attempts. However, we take the opportunity to mention some of the most notable research projects in the literature in the IoT microservices architectures in the supply chain for data sharing regardless of the area of the work contribution. The itemization of research works we have analyzed and categorized is in the interest of Industrial 4.0 and the works toward the fifth industrial revolution.

Krämer et al. [10], Shi et al. [11], Maia et al. [12], Rahman [13], Taneja et al. [14], Zhang et al. [15], and Hu et al. [16] are notable works in IoT microservices, supply chain and data-sharing systems. The majority of these studies studied the scalability of the solution aspect. Most of the research interest in microservices is built upon IoT technologies. However, IoT security at physical, communication and network layers are not discussed. This justifies the research needs to secure IoT devices and data communication across different layers in heterogeneous IoT infrastructures. Blockchain technology is of interest to many researchers in the IoT microservices domain. However, fewer research attempts exploited the potential of Blockchain technology and smart contracts at the edge/computing layer. It is noted that the majority of blockchain-based studies focused on the application implementation of technology. At the same time, optimization of blockchain algorithms and techniques are not examined under complex security aspects. Undoubtedly, the security aspects of microservices and supply chain are neither discussed nor explored thoroughly. This leaves considerable space for improvement for the security requirements of microservices over-cloud and IoT/Edge computing.

3   IoT Microservices Architecture

The majority of research studies considered cloud/blockchain/or edge computing layers for the microservice paradigm of the proposed solution. Hence, Fig. 2 depicts the simplified IoT-based microservices architecture based on relevant research projects in the literature review [11, 13–15]. The cattle industry data-sharing infrastructure and supply chain comprise IoT technologies such as RFID sensors and tags [17], Drones [18] and farming/ agriculture robots [19], GPS, cattle behaviour motion and movement tracking devices [20], WiFi, LoRAWAN, NB-IoT [15], and many more. These technologies transfer smart herd data to the cloud and microservices via the emerging digital connectivity technologies such as 5G [21] and Fog nodes/Edge gateways etc. The following subsections discuss the smart herd microservices architecture elements for data sharing in the supply chain.

3.1   Physical Layer (Remote Sensing Layer)

The first layer of IoT Smart Herd Microservices Architecture is the physical layer (data plane layer). This layer consists of the physical IoT infrastructure, including various types of IoT devices, communication mechanisms and sensor networks to acquire data from the herd. For instance, Radio Frequency Identification (RFID) tags are attached and affixed to the cattle ear. Each tag is assigned with a unique ID to identify cows. The tags are connected wirelessly to IoT antennas and to the RFID reader is interfaced with a Fog node that is deployed across the farm and all connected to IoT LoRaWAN Gateway or NB-IoT. In addition, other sensory devices assist in collecting more information about the cows' locations (GPS sensor), behavioural activities using the Long Range Pedometer (LRP). The LRP is attached to the cow leg and used to count each step of the cow. Further, it collects data such as cow movements and biomarker input such as heat detection on the leg.

3.2   Edge Layer (Middleware Layer)

The Edge Layer is one of the integral data security components to secure different data inputs generated from heterogeneous devices from the physical layer. A variety of physical IoT devices at the physical layer are deployed at different farm locations, where it collects different data types in different dimensions. Further, IoT devices create heterogeneous communication through the gateways to the edge computing layer. This heterogeneity comes from the fact that each hardware manufacturer provides different data connectivity protocols and communication algorithms for IoT connected devices. Therefore, this poses a fundamental security problem due to the incompatibility of data formats generated from these devices. Opportunely, edge computing can be deemed a middleware layer to convert data from one format. Some research studies [15, 16] exploited the potential of the edge/fog computing layer by incorporating the blockchain security aspect to this layer to the security requirements for data transmitted from the physical layer to the cloud-microservices layer.

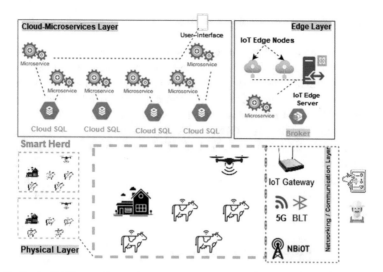

Fig. 2. Typical IoT smart herd microservices architecture for data sharing in supply chain

3.3 Cloud - Microservices Layer

Microservices break down large applications into small independent services, and each service has its own realm of responsibility [22]. The cloud - microservices layer is virtualized in high-end servers, serverless and containers (dockers). The microservices are deployed in this plane in the proposed non-monolithic architecture approach. This monolithic architecture supports data-sharing in supply chain infrastructure. The end-user layer (microservice supply chain) provides the end-users (e.g. other farmers, third party regulators, decision-makers, buyers) with the shared information by farmers who can control the previous layers how the data can be shared and to whom the data is shared or data sharing per request. Therefore, the access for the acquired data in this

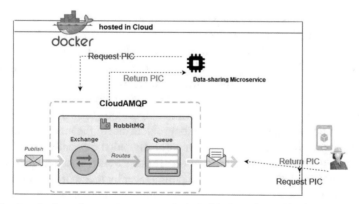

Fig. 3. The data sharing microservice communicated with the end-user (the farmer) via messaging broker (e.g. RabbitMQ)

layer via user-interface mechanisms such as mobile apps (freight and logistic users, customers ordering and purchasing), or web-interface and analytical dashboards (wholesalers, retailers, regulators) and microservice tech services via APIs (for a third party such as developers), or smart terminals. The data sharing is managed in microservices for livestock via a message broker (e.g. RabbitMQ) [23], as shown in Fig. 3.

4 Cybersecurity Threats in IoT Microservices Architecture

Although the IoT microservices architecture for the supply chain (Fig. 2) brings potential and benefits for the smart herd industry. However, the communication mechanisms and the interaction between different layers of components trigger cybersecurity threats and sources of cyberattacks in microservices-oriented data sharing systems. Therefore, we are motivated to discuss security aspects and classify cybersecurity attacks according to the security issues identified in the literature. The security aspects of IoT microservice and data sharing infrastructure are listed as follows:

1. **Privacy**: user protection from unauthorized access to other users' information. Examples of attacks that affect privacy security requirements in IoT microservices and supply chain and data-sharing systems include (The physical attack, The masquerade Attack, the Replay Attack) [24, 25].
2. **Authenticity**: It prevents users from spoofing other identities. Examples of these attacks such as (Cloud attacks, Ransomware, Cyber-terrorismnd Indirection Attacks) [25, 27].
3. **Confidentiality**: It protects data from unauthorized access. Tracing Attack and Brute Force Attack [28, 29] are types of attacks that impact the confidentiality of IoT microservice and data sharing systems.
4. **Integrity**: Ensures that information will not be altered during storage or transmission. Man-In-The-Middle Attack (MITM), Trojan Horse Attack, Biometric, Botnets, False Data Injection, and Data Fabrication [30, 31] are common forms for attacks that disturb IoT microservice and data sharing systems.
5. **Availability**: Ensures continuity of the services provided. The Denial of Service (DoS) attacks, Attacks on devices (networks and hardware), and Attacks on Software [25, 32] can impact the availability of IoT-microservices and Livestock data-sharing systems.
6. **Non-Repudiation:** It prevents users from repudiating their actions when using the system—for instance, Repudiation, Attacks, Malicious Code, and Attacks on computers and equipment [25, 33] are common types of Non-Repudiation.

We have investigated the pertinent categorization of cybersecurity threats in IoT data-sharing systems and microservices in the literature. Therefore, we group potential security threats based on relevance and cyber security resemblances between IoT microservices in data sharing systems and supply chain [34–36]. This motivated our research to narrow down, classify, and list various cyberattacks in IoT-based microservices architecture for the supply chain Fig. 2.

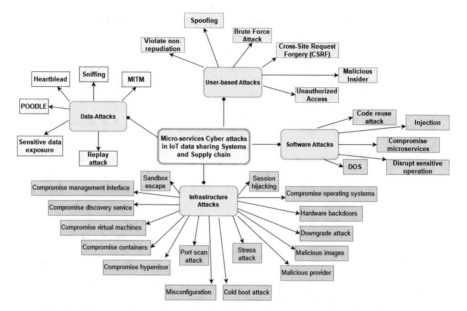

Fig. 4. Microservices cyberattacks in IoT data sharing systems and supply chain.

Figure 4 shows the cyber security threats per category [34] for microservices architecture in IoT data sharing systems and supply chain. The summary of the main four categories is discussed below:

1. **User-based attacks** - where users are directly attacked, such as malicious user actions, or indirectly (inadvertent insider actions).
2. **Data attacks** - impact data during storage, transmission, or processing in the system. Cyberattacks are aimed at stealing sensitive data that attackers can misuse.
3. **Infrastructure attacks** - attacks focused on microservices architecture components and platforms, including monitors, discovery services, message brokers, load balancers, and others.
4. **Software attacks** - attempts to transform or inject code for malicious purposes.

5   Open Research Challenges and Future Directions

Security threats and IoT security concerns for the microservices supply chain (e.g. meat and livestock) are common, unpredictable. They may occur at any point in time if systems security requirements are not established and implemented efficiently. The following are open research challenges [35] and future directions to be explored by computer scientists and cybersecurity researchers and propose solutions to secure IoT and microservices-based supply chain:

– **Access Control from a Security Perspective:** Access control is a traditional concern among data-sharing systems by data owners (e.g. livestock industry). However, they should adopt a security perspective for their property. To prevent unauthorized access to data sharing and supply chain microservices, authentication, authorization, and accounting should be incorporated. However, the state-of-the-art and cybersecurity market comes with some relevant solutions. However, there are still opportunities for more (access control, security privacy-preserving, data integrity, authentication, data confidentiality, and blockchain-based) solutions to address cybersecurity issues in data sharing microservices and supply-chain systems.

- **Data Protection:** A well-known characteristic of data sharing and microservices-supply chain solutions is that they acquire data and communicate enormous amounts of data thanks to smart devices and IoT, edge computing, and cloud computing. Therefore, there is a need for effective data protection mechanisms. There is still a practical challenge associated with data protection in this area. This brings a great potential to employ data provenance and blockchain technology using smart contracts for trusted supply chain microservices in the Internet of Things (IoT).

- **Network Infrastructure and Physical Layer Protection:** Implementing supply chain IoT microservices successfully depends on the physical layer and the network infrastructure. However, they are targeted by several attacks. Although there are various network infrastructure and physical hardware solutions in the market, it still needs to be developed and consider robust cyber infrastructure-physical hardware solutions, including Penetration testing and measures to control and mitigate the risk of cyber-attacks at infrastructure network, hardware levels.

- **Secure Communication Protocols:** Communication protocols are critically important because IoT microservices and supply chain data-sharing infrastructure, where sensors, actuators, drones, autonomous tractors, and other smart devices are spread out in various locations. Although several networks, Machine-to-Machine (M2M) and IoT protocols such as Bluetooth, WiFi, NFC, SSL, TLS, UDP, TCP and 6LoWPAN that have been adapted with other technologies, there is still room for improvement on secure protocols that are designed specifically for IoT microservices and data sharing.

- **Education as Risk Reduction:** Education and cyber enablers for meat-livestock markets have received relatively little attention in the literature when it comes to reducing cyber-risks. Also, cybersecurity education is of interest to data producers because of its threats to their businesses. Education remains an essential tool to reduce security risks such as Ransomware attacks. Introducing cyber security into IoT-enabled supply chains should provide opportunities to assess whether farmers are utilizing cyber-safe behaviors. It can be determined from this what educational formats, content, and dissemination would be most effective and appreciated by these stakeholders.

In technical and implementation settings, this research encourages researchers to consider different types of cyber-attacks that are apparent as per our taxonomy classification of their different types in Fig. 4 and their categorization according to The IoT Microservices Architecture in Fig. 2. Thereupon, attackers use sophisticated and advanced methods, in-mean, while cyberattacks have become popular in the latest years. Therefore, we encourage researchers and cybersecurity specialists to study and design develop security models to meet security needs in a more reliable, authentic, confidential, integral, trustworthy and safer IoT microservice in the supply chain and data-sharing systems. More proactive practical approaches and protective security frameworks are needed to establish and meet the security requirements in the AgeTech sector rather than being defensive in dynamic and rapid decision-making environments.

6 Conclusion

The adoption of IoT microservices architecture in the supply chain and data-sharing systems in smart herds and livestock has the potential to enhance global food security and sustain food programs globally. However, these technologies need to be protected from cyber-attacks to realize these potentials. Unfortunately, cyberattacks and security threats are present and rife in the IoT microservices for supply chain data-sharing systems. Moreover, these attacks may severely disrupt global supply chains, including livestock markets, especially countries that heavily rely on food and meat industries. This paper has examined the security of IoT microservices architecture for data-sharing in the supply chain, which is a critical challenge in the field of smart herd and livestock. Our paper has highlighted the important security aspects and security threats that need to be considered. As a result of our assessment, we have proposed a future security road map highlighting some related emerging areas that still need to be explored in IoT microservices and data-sharing systems in the supply chain.

References

1. FAO, IFAD, UNICEF, WFP and WHO. https://www.fao.org/documents/card/en/c/cb4474en
2. The World Bank. https://www.worldbank.org/en/topic/agriculture/brief/moving-towards-sus tainability-the-livestock-sector-and-the-world-bank
3. FAO. https://www.fao.org/animal-production/en/
4. MLA- Meat & Livestock Australia. https://www.mla.com.au/news-and-events/industry-news/positive-perceptions-of-red-meat-growing-among-consumers/
5. CISRO. https://www.csiro.au/en/research/animals/livestock/livestock-logistics
6. Department of Industry, NSW, Australia. https://www.dpi.nsw.gov.au/__data/assets/pdf_file/0007/723625/NLIS-Compliance-Procedures-for-Property-Identification-Codes.pdf
7. Mateo-Fornés, J., Plà-Aragonés, L.M., Castells-Gasia, J.P., Babot-Gaspa, D.: An internet of things platform based on microservices and cloud paradigms for livestock. Sensors 21, 5949 (2021)
8. Ruan, J., et al.: A life cycle framework of green IoT-based agriculture and its finance, operation, and management issues. IEEE Commun. Mag. 57, 90–96 (2019)
9. Singh, C., Khilari, S.H., Nair, A.N.: Farming-as-a-service (FAAS) for a sustainable agricultural ecosystem in India: design of an innovative farm management system 4.0. In: Digital Transformation and Internationalization Strategies in Organizations, pp. 85–123. IGI Global (2022)
10. Krämer, M., Frese, S., Kuijper, A.: Implementing secure applications in smart city clouds using microservices. Future Gener. Comput. Syst. 99, 308–320 (2019)
11. Shi, P., Wang, H., Yang, S., Chen, C., Yang, W.: Blockchain-based trusted data sharing among trusted stakeholders in IoT. Softw. Pract. Exp. 51, 2051–2064 (2021)
12. Filev Maia, R., Ballester Lurbe, C., Agrahari Baniya, A., Hornbuckle, J.: IRRISENS: an IoT platform based on microservices applied in commercial-scale crops working in a multi-cloud environment. Sensors 20, 7163 (2020)

13. Rahman, M.A., Abuludin, M.S., Yuan, L.X., Islam, M.S., Asyhari, A.T.: EduChain: CIA-compliant blockchain for intelligent cyber defense of microservices in education industry 4.0. IEEE Trans. Ind. Inform. **18**, 1930–1938 (2021)
14. Taneja, M., Jalodia, N., Byabazaire, J., Davy, A., Olariu, C.: SmartHerd management: a microservices-based fog computing–assisted IoT platform towards data-driven smart dairy farming. Softw. Pract. Exp. **49**, 1055–1078 (2019)
15. Zhang, J., et al.: A blockchain-based trusted edge platform in edge computing environment. Sensors **21**, 2126 (2021)
16. Hu, S., Huang, S., Huang, J., Su, J.: Blockchain and edge computing technology enabling organic agricultural supply chain: a framework solution to trust crisis. Comput. Ind. Eng. **153**, 107079 (2021)
17. Alsinglawi, B., Elkhodr, M., Nguyen, Q.V., Gunawardana, U., Maeder, A., Simoff, S.: RFID localisation for internet of things smart homes : a survey. Int. J. Comput. Netw. Communi. **9**, 81 (2017)
18. Behjati, M., Mohd Noh, A.B., Alobaidy, H.A., Zulkifley, M.A., Nordin, R., Abdullah, N.F.: LoRa communications as an enabler for internet of drones towards large-scale livestock monitoring in rural farms. Sensors **21**, 5044 (2021)
19. Brown, J., Sukkarieh, S.: Design and evaluation of a modular robotic plum harvesting system utilizing soft components. J. Field Robot. **38**, 289–306 (2021)
20. Taneja, M., Byabazaire, J., Jalodia, N., Davy, A., Olariu, C., Malone, P.: Machine learning based fog computing assisted data-driven approach for early lameness detection in dairy cattle. Comput. Electron. Agric. **171**, 105286 (2020)
21. Tang, Y., Dananjayan, S., Hou, C., Guo, Q., Luo, S., He, Y.: A survey on the 5G network and its impact on agriculture: Challenges and opportunities. Comput. Electron. Agric. **180**, 105895 (2021)
22. Developers & Practitioners. https://cloud.google.com/blog/topics/developers-practitioners/microservices-architecture-google-cloud
23. Akbulut, A., Perros, H.G.: Performance analysis of microservice design patterns. IEEE Internet Comput. **23**, 19–27 (2019)
24. Idoje, G., Dagiuklas, T., Iqbal, M.: Survey for smart farming technologies: challenges and issues. Comput. Electr. Eng. **92**, 107104 (2021)
25. Ferrag, M.A., Shu, L., Yang, X., Derhab, A., Maglaras, L.: Security and privacy for green IoT-based agriculture: review, blockchain solutions, and challenges. IEEE Access **8**, 32031–32053 (2020)
26. Bothe, A., Bauer, J., Aschenbruck, N.: RFID-assisted continuous user authentication for IoT-based smart farming. In: 2019 IEEE International Conference on RFID Technology and Applications (RFID-TA), pp. 505–510. IEEE (2019)
27. Nesarani, A., Ramar, R., Pandian, S.: An efficient approach for rice prediction from authenticated Block chain node using machine learning technique. Environ. Technol. Innov. **20**, 101064 (2020)
28. Ametepe, A.F.-X., Ahouandjinou, S.A.R., Ezin, E.C.: Secure encryption by combining asymmetric and symmetric cryptographic method for data collection WSN in smart agriculture. In: 2019 IEEE International Smart Cities Conference (ISC2), pp. 93–99. IEEE (2019)
29. Yu, S., Lou, W., Ren, K.: Data Security in Cloud Computing. Morgan Kaufmann/Elsevier, Book section 15, pp. 389–410 (2012)
30. Chamarajnagar, R., Ashok, A.: Integrity threat identification for distributed IoT in precision agriculture. In: 2019 16th Annual IEEE International Conference on Sensing, Communication, and Networking (SECON), pp. 1–9. IEEE (2019)
31. Davcev, D., Mitreski, K., Trajkovic, S., Nikolovski, V., Koteli, N.: IoT agriculture system based on LoRaWAN. In: 2018 14th IEEE International Workshop on Factory Communication Systems (WFCS), pp. 1–4. IEEE (2018)

32. Bisogni, F., Cavallini, S., Di Trocchio, S.: Cybersecurity at European level: the role of information availability. Commun. Strateg., 105–124 (2011)
33. Holkar, A.M., Holkar, N.S., Nitnawwre, D.: Investigative analysis of repudiation attack on MANET with different routing protocols. Int. J. Emerg. Trends Technol. Comput. Sci. (IJETTCS) **2** (2013)
34. Hannousse, A., Yahiouche, S.: Securing microservices and microservice architectures: a systematic mapping study. Comput. Sci. Rev. **41**, 100415 (2021)
35. Yazdinejad, A., et al.: A review on security of smart farming and precision agriculture: security aspects, attacks, threats and countermeasures. Appl. Sci. **11**, 7518 (2021)
36. Demestichas, K., Peppes, N., Alexakis, T.: Survey on security threats in agricultural IoT and smart farming. Sensors **20**, 6458 (2020)

An Architecture for Autonomous Proactive and Polymorphic Optimization of Cloud Applications

Marta Różańska[1]([⊠]), Paweł Skrzypek[2], Katarzyna Materka[2], and Geir Horn[1]

[1] Department of Informatics, University of Oslo, P.O. Box 1080, Blindern, 0316 Oslo, Norway
martaroz@ifi.uio.no, Geir.Horn@mn.uio.no
[2] 7Bulls.com and AI Investments, Al. Szucha 8, 00-582 Warsaw, Poland
pawel.skrzypek@aiinvestments.pl, kmaterka@7bulls.com

Abstract. Existing autonomous Cloud application management platforms continuously monitor the load and the environment of the application react when optimization is needed. This paper introduces the concepts of polymorphic architecture optimization and proactive adaptation of Cloud applications, which is a significant improvement of the standard reactive optimization. Polymorphic architecture optimization considers the change of the technical form of the component while proactive adaptation uses the predicted future workload and context. Based on this, we propose an architecture for a Cross-Cloud application management platform that supports complex optimization of Cloud applications in terms of architecture, resources, and available offers.

1 Introduction

Optimization of Cloud applications is a complex problem: there are many offers and possible application configurations, and therefore there is a need for a management tool that can work as middleware and support the optimization process of Cross-Cloud applications. Reactive optimization is well known and there are middleware platforms like Multi-cloud Execution ware for Large scale Optimised Data Intensive Computing[1] (MELODIC) available [5]. However, reactive optimization may be insufficient as reconfiguring the application and starting new Virtual Machines (VMs) do take some time. The goal of the proactive adaptation feature is to provide the ability to optimize the application's deployment in a given time horizon in the future. Proactive adaptation is based on forecasting future resource needs This may solve the timeliness of the application management, but it is not sufficient to allow the application to exploit various hardware accelerators. The goal of the polymorphic adaptation is to provide the ability to optimize the application's deployment by changing the application architecture at runtime.

This paper proposes a high-level architecture for a platform supporting reactive, proactive, and polymorphic autonomic Cloud application management. This is currently being implemented as the Modelling and Orchestrating heterogeneous Resources

---

[1] https://melodic.cloud/.

L. Barolli et al. (Eds.): AINA 2022, LNNS 451, pp. 567–577, 2022.
https://doi.org/10.1007/978-3-030-99619-2_53

and Polymorphic applications for Holistic Execution and adaptation of Models In the Cloud (MORPHEMIC) software platform.[2] We present the concept of the Cloud application optimization and its desired features as well as some similar platforms in Sect. 2. Section 3 presents the architecture of the MORPHEMIC platform while in Sect. 4 the proactive and polymorphic adaptation of Cloud applications is described. In Sect. 5, we present an example of simple application to demonstrate what polymorphic and proactive adaptation means in practice.

2 Cloud Application Optimization

2.1 Utility Based Reactive Optimization

Cloud Computing revolutionizes the usage of ICT infrastructure because if offers unlimited capabilities and flexibility. The new resources can be added on-demand in a matter of minutes and removed when not needed. These capabilities create new challenges for an application owner. It is necessary to watch the application's performance and the value it offers to the company's clients to scale the application when the demand increases or reduce the deployment size when it decreases. As the application is watched in real-time and decisions are based on the current performance and execution context, this type of adaptation is called reactive optimization.

Optimization of the application configuration is a complex task that is difficult, or even impossible, to be performed and handled by a human. The only solution is to let the application be managed by an autonomic computing deployment platform building on the ideas from the vision of *autonomic computing* [8]. The core building block in this architecture is the Monitor, Analyse, Plan, Execute—with Knowledge (MAPE-K) [7] feedback loop. This concept has been used to build management frameworks for mobile applications [12], for intrusion detection systems [15], and recently for autonomic management of applications deployed simultaneously across multiple Cloud providers in the MELODIC optimization platform [5]. An autonomic application management platform will, by definition, make decisions on behalf of the organization owning the deployed application. These decisions are valid only if they concur with the decisions that otherwise would have been made by the human DevOps engineers in these organizations. The challenge for all autonomic computing platforms is therefore to capture the application owners' utility and maximise this utility given the current use of the application and its execution context. Kephart and Das advocated that the best way to represent utility in autonomic systems was as a utility *function* [9].

In order to alleviate these challenges, the MELODIC project proposed an advanced autonomous middeleware that can act as a DevOps engineer and make the necessary adaptations to the deployment configuration as the demand changes [5]. The application does not need to be changed to be deployed and managed using MELODIC because the DevOps engineer just describes the application components, the data sets, the utility function, and the monitoring sensors in a Domain Specific Language (DSL) [2] Cloud Application Modelling and Execution Language (CAMEL) [1]. Based on this description, the goals for the deployment, and the deployment constraints, the MELODIC

---

[2] https://morphemic.cloud.

platform is able to find a good initial deployment and then continuously optimise and adapt this deployment as the application's execution context changes. Experiments performed on real applications confirmed that Cloud applications can benefit from such autonomous Cloud optimization [6].

2.2   Proactive Adaptation

The goal of the proactive adaptation is to optimize the application's deployment for a future time point. This goal is particularly important from a practical point of view. The reconfiguration of an application takes time, usually some minutes. During the reconfiguration period the runtime execution context may have changed, which can lead to another immediate reconfiguration. There are some critical applications, for instance from the medical or networking fields, that must run with minimal downtime. If proactive adaptation is supported, then the optimization platform can adapt application resources ahead of the time they will be needed to prevent situations that are critical, and to ensure the expected performance of the application.

A platform that supports proactive adaptation must have the ability to predict future conditions and behaviour of an application as a key enabler for proactive adaptation. Furthermore, the platform has to be able to select the best deployment configuration for a time point at the needed time horizon. The platform must provide the ability to verify the future performance of the selected deployment in conjunction with the traditional approach based on the actual values of monitored metrics.

2.3   Polymorphic Adaptation

The goal of the polymorphic adaptation is to optimize the application's deployment by changing the application architecture at runtime so scaling the application's components not only in the number of instances but also in the form they are deployed. It is possible only when a component can run in different technical forms. In particular, as a runtime decision, the optimization platform should decide which form or configuration of a component should be selected for deployment: VM, container (deployment as a container within a VM), serverless as a deployed function, to hardware accelerators, *etc.*. Depending on the application's requirements and its current workload, its components could be deployed in various forms in different environments to maximize the utility of the application deployment. Choosing the right form to deploy for a component can be done *a priori* at the application modelling stage, or as a run-time decision.

The modelling language must support polymorphic modelling of application components that have the corresponding binary implemented artefacts available for the hardware variants to consider. There should be some kind of recommender system to inform the application owner about potential benefits from implementing a component for different hardware. Comparing the application form with similar applications may reveal that other applications of the same class normally provides an accelerator in a particular form for a given component type. One example can be that encryption components are normally delegated to dedicated hardware, and the recommendation could be to implement the encryption component in the current architecture for the same type of accelerator. Finally, there should be a pre-selection of component forms to use for the initial deployment, and a regular reconsideration of the component forms based on the monitored application execution context to detect bottleneck components and potentially change their deployed form to better performing variants, if that change maximizes the application's utility.

2.4   Autonomous Cross-Cloud Application Optimization Platforms

Apart from MELODIC, there are other platforms with similar goals and possibilities [11]. For instance, Rainbow framework for creating self-adapting Information Technology (IT) systems is a complete and popular solution [4]. However, it requires a tight integration of the application to enable self adaptation capabilities. Furthermore, the adaptation in the Rainbow framework is based on predefined rules instead of a utility function, and the proactive and polymorphic adaptation is not covered. Another popular project, CYCLONE, introduces a set of tools, which need to be deployed and integrated to support Cross-Cloud application deployments [17]. The CYCLONE project is focused on several aspects of Cross-Cloud inter-operability like networking and security, but does not cover application deployment optimization. The PrEstoCloud project introduces a complete solution to optimize Big Data applications including edge devices [19]. It contains components for predicting a workload and optimizing resources based on these predictions, so it supports proactive adaptation. The work in MORPHEMIC can thus be considered as further extensions of the results of the PrEstoCloud project. The polymorphic adaptation is not covered in all aforementioned platforms.

3   Architecture for Autonomous Optimization Platform

The MORPHEMIC platform follows the microservice architecture principles [14], as it can be seen in Fig. 1. Components are deployed using Docker[3] containers [13]. The platform is built using the baseline frameworks MELODIC[4] and the ProActive Scheduler.[5] The MORPHEMIC platform is mainly responsible for the polymorphic and proactive adaptation, both of which are essentially done by performing the architecture selection and predicting future metrics values. As a next step, this information is provided to MELODIC to find the optimal deployment solution, which is then orchestrated. The deployment itself is performed by the ProActive Scheduler.

The MORPHEMIC platform communicates with baseline frameworks through their public Application Programming Interface (API). For synchronous communication, the Representational State Transfer (REST) [3] API with the JavaScript Object Notation (JSON) [16] scheme is used. An integration solution based on an Enterprise Service Bus (ESB) with Business Process Management (BPM) orchestration has been chosen due to the architecture and the characteristics of the MORPHEMIC platform [10]. The

---

[3] https://www.docker.com/.

[4] https://melodic.cloud.

[5] https://proactive.activeeon.com/.

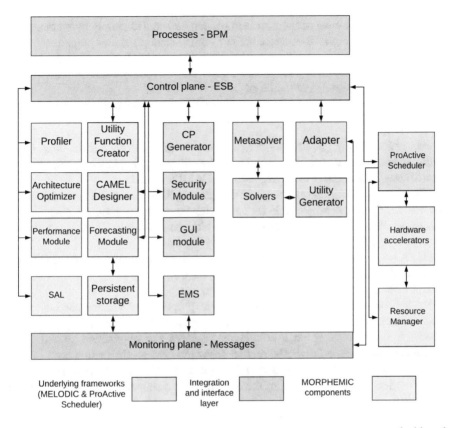

Fig. 1. Integration view of the MORPHEMIC components. New components are marked in yellow colour, and components of underlying frameworks are marked in blue colour.

chosen implementation of the ESB with the BPM is MuleESB[6] with Camunda.[7] It includes an Active Message Queue[8] (AMQ) broker, a Messages Oriented Middleware (MOM) integration solution, which is re-used for the Monitoring Plane. The Integration layer allows for reliable and monitorable invocation of methods. Also, it supports re-usability of the methods exposed by underlying components and avoid any point-to-point communication.

The orchestration of the data and the action flows within the system is modelled as processes. The dedicated BPM processes used in MORPHEMIC include the deployment process, which is a process responsible for orchestrating the deployment of a new application, from the upload of the user's CAMEL model, until the final deployment in the Cloud. Another process is the reconfiguration of the application. This is based on a new optimal application configuration found by a solver, and handles all events generated by the system components to address properly the application reconfigura-

---

[6] https://www.mulesoft.com/platform/soa/mule-esb-open-source-esb.
[7] https://camunda.com/.
[8] https://activemq.apache.org/.

tion. Thanks to this, most of the new features can be handled simply by reconfiguring or creating a new process flow, instead of performing changes in the components' code. This integration-oriented architecture introduces an abstraction layer between the business flow and the domain systems, and thanks to that, the components could be used or re-used by any process or any platform.

Figure 1 shows all the top-level components of the MORPHEMIC platform. The platform includes:

Graphical User Interface (GUI) module used to configure and operate the platform,
Utility Function Creator used for creation of the utility function in a user-friendly way,
CAMEL Designer used for the creation of the CAMEL model,
Profiler responsible for constructing and maintaining the profile of an application,
Architecture Optimizer responsible for optimizing the application architecture.
Constraint Programming (CP) Generator which converts the CAMEL application model to a CP problem,
Metasolver which triggers the reconfiguration when any the constraints of the CP problem is violated, and selects the appropriate Solver,
Solvers which are capable of solving the CP problem,
Utility Generator that calculates the utility of the proposed CP solution candidate,
Adapter responsible for orchestrating the deployment of the application.
ProActive Scheduler that deploys and reconfigures the application to the Cloud providers using *Scheduling Abstraction Layer (SAL)* as an abstraction layer.
Resource Manager which is a sub-component of ProActive Scheduler responsible for deployment and configuration of the infrastructure resources exploited by the applications,
Hardware Accelerator that deploys to hardware accelerated Cloud computing resources.
The Event Management Services (EMS) is used for distributed event processing and collection of metric values,
Forecasting module forecasts metric values into the EMS infrastructure,
Persistent storage is responsible for storing real and forecasted metric values, and
Performance module is responsible for maintaining a performance model for an application and Cloud resources.

4 Optimization Flow

This section provides a description of the optimization flow of the application managed by the MORPHEMIC platform with emphasis on the proactive and polymorphic optimization. Figure 2 presents the MORPHEMIC optimization flow using MELODIC as the real-time application manager. The result of executing this flow is a deployed and continuously optimized application configuration. The flow contains four main parts that can be mapped to the autonomic computing MAPE-K loop:

Profiling is the pre-processing step as it handles the preparation of the CAMEL model, the utility function creation based on the information in the model, and the application profiling used for the polymorphic adaptation.

Reasoning is the core process of the platform. It includes both Analysis and Planning from the MAPE-K loop finding the best deployment configuration within the given operational constraints and considering the application's execution context for the desired reconfiguration time point. The history of execution contexts represents the *knowledge* required for the reasoning.

Executing deals with the adaptation and orchestration of the deployment model to be deployed across the chosen Cloud providers.

Monitoring is responsible for gathering data used in the autonomous optimization.

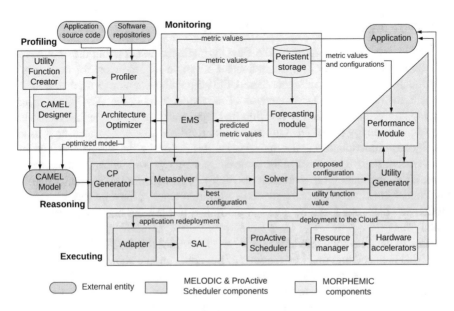

Fig. 2. Workflow view on the MORPHEMIC high-level components. Colour boxes indicate mapping of MAPE-K loop to components.

4.1 Proactive Adaptation Implementation

The proactive adaptation uses the MELODIC reasoning process [5], but running on a prediction of the application's execution context for a future time point. After the successful initial deployment of the application, the measurements are collected. Then, the Persistent Storage stores the measurements as time-series and prepares the datasets for the Forecasting Module. Once a minimum volume of data is available, the Forecasting Module starts forecasting. It is possible to use many forecasting methods in the Forecasting Module. The representative prediction value for one metric value for the future time point of interest ensembles the multiple predictions coming from the forecasters available for that metric value. All the metric values are finally combined into one vector representing the application's predicted execution context for the future time point of interest. This predicted execution context is used by a sub-module of the Forecasting Module, the *Severity-based Service-Level Objective (SLO) violation Detector* to predict the likelihood of a violation of the CP constraints [18]. If an SLO violation is likely for the future time point of interest, the proactive reconfiguration is triggered: the predicted metric values are passed to a solver starting the reasoning for the future time point of interest. The reasoning process involves the Performance Module to provide an estimate of the impact from a candidate application configuration on the utility metric values for the future time point of interest. When the best deployment configuration for this time point is found, the re-deployment process is proceeded as in MELODIC.

4.2   Polymorphic Adaptation Implementation

The deployment process starts by the DevOps engineer first creating the application's CAMEL topology model [20], including the resource requirements for each component, using the CAMEL Designer. The Utility Function Creator is used in a GUI to simulate the deployment and adaptation of the model helping the DevOps engineer to develop and validate the utility function that represents the trade-offs that would have been done manually by the DevOps engineer at run-time. The utility function is used as the objective function during the optimization that is done at the component type level, which means that all instances of the same component *type* will be deployed in the same *form*. If the application can deal with polymorphic components concurrently deployed in different *forms* they must be provided as different *types* in the application's CAMEL model, and constraints can be given in the model to limit the number of instances of each form.

   The application code provided by the DevOps engineer is first analysed by the Profiler. This means extracting key *features* of the application's source code enabling the classification of the application against a corpus of open source software. The purpose is to suggest alternative *forms* of the components used in similar applications, for instance recommend that cryptographic components are also implemented for Graphics Processing Units (GPUs) or Field-Programmable Gate Arrays (FPGAs) as this is common for cryptographic components, or that the components can be implemented using a library that is already compiled for the various polymorphic forms. The information from the Profiler and the database of binary component forms available will then be used by the Architecture Optimizer using information from the EMS about the application's current execution context if the application is already deployed. The result is a CAMEL model containing only the polymorphic component forms that can actually be deployed, and with appropriate preferences for the forms that will give the highest utility for the application's current or predicted execution context.

   During run-time, the optimization of the application's deployment configuration will either be reactive when the application's context changes or proactive as the predicted context for a future time point likely creates an SLO violation, *i.e.*, violations of the optimization problem constraints. Polymorphic components with only one type defined in the CAMEL model will have all instances of the component deployed with this type. However, if the model contains different *types* of the component for each of its polymorphic *forms*, then the instances of the component can be of different forms depending on what maximizes the utility function.

5   Example Application

Consider a simple application aimed at secure storage of documents in a company's database. In addition to the database it consists of one server responsible for decoding documents fetched from the database with a company secret key, and then recoding the document to be delivered to a user with a user specific key. When the document is committed back to the server, the inverse decoding and recoding process happens. The application's load depends on the requests for documents generated by the application's users. The deployment goal is to provide the best possible response to the users ensuring that each user should have the decrypted document opened in a reasonable time, and at the same time the deployment should be done at minimal price for the company owning the application. The server can be implemented as four component *types*: a front-end user manager taking care of the incoming requests, and two cryptographic component types one for encoding documents, one for decoding documents, and a database. In a minimal installation, all components are deployed as containers in a VM.

Reactive adaptation can be performed by monitoring the number of requests and the average response time to the user. When the average response time is above an acceptable threshold, the reconfiguration is triggered. Then, the goal of the reactive autonomous optimization platform is to keep the average response time close to the nominal value. As the number of users grow, one may easily scale multiple instances of the interface component as separate VMs and add a VM with a load balancer for the user requests to the interface components. However, the encoder and decoder are likely to be the performance bottleneck of the application. Even though they can be scaled horizontally with more VMs serving each type, it may be more cost-efficient to scale them vertically getting more performance out of fewer instances.

Proactive adaptation can be performed by triggering the reconfiguration before the average response time reaches the maximum acceptable value. What is more, during the reasoning, the predicted number of coming requests can be used so the application will be optimized not for the current situation, but for the future moment in time. This will ensure that the new needed application components are available when they are needed, and prevent the application from having any requests delayed, which is the goal of the application owner and which is expected by the end users.

Polymorphic adaptation is to consider scaling not only in the number of instances, but also in the form of deployment. The *polymorphic* interface component may be morphed from its container *form* to a native VM *form*. Cryptographic algorithms can be implemented on Central Processing Units (CPUs), GPUs, and FPGAs, and instead of implementing many instances, one may rather implement few more powerful instances by deploying the same component in different *forms*. The *polymorphic* cryptographic components may have even more *forms* to select from when making the decision on which configuration will be best for the application's current execution context. Choosing the right form to deploy for a component can be done *a priori* at the application modelling stage finding the set of component forms that maximizes the expected number of user requests, or as a run-time decision reacting to the current measurement of the application's execution context or proactive reacting to the forecast execution context for a future time point. Hence, the concept of polymorphic adaptation is orthogonal to the concept of proactive adaptation.

6   Conclusion and Future Work

This paper has discussed the architecture of autonomic optimisation platform for proactive and polymorphic adaptation of Cloud application resources that is implemented in the MORPHEMIC platform. We have described the design of the platform that can proactively and reactively adapt Cloud resources according to user utility. According to our knowledge, this is the first complete architecture of such platform with all desired features. We have indicated how an application deployed in the Cloud can benefit from reactive, proactive, and polymorphic adaptation. Furthermore, it is discussed how these adaptations can be combined together. The implementation of the MORPHEMIC platform is still in progress so the full evaluation of the platform will be conducted with MORPHEMIC use case providers in forthcoming publications.

Acknowledgements. This work has received funding from the European Union's Horizon 2020 research and innovation programme under grant agreement No 871643 MORPHEMIC *Modelling and Orchestrating heterogeneous Resources and Polymorphic applications for Holistic Execution and adaptation of Models In the Cloud.*

References

1. Rossini, A., et al.: The cloud application modelling and execution language (CAMEL). OPen Access Repositorium der Universität Ulm (2017). https://doi.org/10.18725/OPARU-4339
2. Bergmayr, A., et al.: The evolution of CloudML and its manifestations (2015). https://doi.org/10.18725/OPARU-4339
3. Fielding, R.T.: REST: architectural styles and the design of network-based software architectures. Doctoral dissertation, University of California (2000). https://doi.org/10.18725/OPARU-4339
4. Garlan, D., Cheng, S.-W., Huang, A.-C., Schmerl, B., Steenkiste, P.: Rainbow: architecture based self-adaptation with reusable infrastructure. Computer **37**(10), 46–54 (2004). https://doi.org/10.1109/MC.2004.175
5. Horn, G., Skrzypek, P: MELODIC: utility based cross cloud deployment optimisation. In: Proceedings of the 32nd International Conference on Advanced Information Networking and Applications Workshops (WAINA), pp. 360–367. IEEE Computer Society (2018). https://doi.org/10.1109/WAINA.2018.00112
6. Horn, G., Skrzypek, P., Materka, K., Przeździęk, T.: Cost benefits of multi-cloud deployment of dynamic computational intelligence applications. In: Barolli, L., Takizawa, M., Xhafa, F., Enokido, T. (eds.) WAINA 2019. AISC, vol. 927, pp. 1041–1054. Springer, Cham (2019). https://doi.org/10.1007/978-3-030-15035-8_102
7. IBM: An architectural blueprint for autonomic computing. White Paper, 3rd edn., p. 34. IBM (2005). https://doi.org/10.18725/OPARU-4339
8. Kephart, J.O., Chess, D.M.: The vision of autonomic computing. Computer **36**(1), 41–50 (2003). https://doi.org/10.1109/MC.2003.1160055
9. Kephart, J.O., Das, R.: Achieving self-management via utility functions. IEEE Internet Comput. **11**(1), 40–48 (2007). https://doi.org/10.1109/MIC.2007.2
10. Kritikos, K., Skrzypek, P., Różńska, M.: Towards an integration methodology for multi-cloud application management platforms. In: Proceedings of the 12th IEEE/ACM International Conference on Utility and Cloud Computing Companion - UCC 2019 Companion, pp. 21–28. ACM Press (2019). https://doi.org/10.1145/3368235.3368833

11. Kritikos, K., Skrzypek, P., Zahid, F.: Are cloud platforms ready for multi-cloud? In: Brogi, A., Zimmermann, W., Kritikos, K. (eds.) ESOCC 2020. LNCS, vol. 12054, pp. 56–73. Springer, Cham (2020). https://doi.org/10.1007/978-3-030-44769-4_5
12. Geihs, K., et al.: A comprehensive solution for application-level adaptation. Softw. Pract. Exp. **39**(4), 385–422 (2009). https://doi.org/10.1002/spe.900
13. Merkel, D.: Docker: lightweight Linux containers for consistent development and deployment. Softw. Pract. Exp. **39**(4), 5 (2009). https://doi.org/10.1002/spe.900
14. Nadareishvili, I., Mitra, R., McLarty, M., Amundsen, M.: Microservice Architecture: Aligning Principles, Practices, and Culture. O'Reilly Media, Inc., Newton (2016). https://doi.org/10.1002/spe.900
15. Papamartzivanos, D., Gomez Marmol, F., Kambourakis, G.: Introducing deep learning self-adaptive misuse network intrusion detection systems. IEEE Access **7**(4), 13546–13560 (2019). https://doi.org/10.1109/ACCESS.2019.2893871
16. Pezoa, F., Reutter, J.L., Suarez, F., Ugarte, M., Vrgoč, D.: Foundations of JSON schema. In: Proceedings of the 25th International Conference on World Wide Web, WWW 2016, pp. 263–273. International World Wide Web Conferences Steering Committee (2016). https://doi.org/10.1145/2872427.2883029
17. Slawik, M., et al.: CYCLONE unified deployment and management of federated, multi-cloud applications. In: 2015 IEEE/ACM 8th International Conference on Utility and Cloud Computing (UCC), pp. 453–457 (2015). https://doi.org/10.1002/spe.900
18. Tsagkaropoulos, A., et al.: Severity: a QoS-aware approach to cloud application elasticity. J. Cloud Comput. **10**(1), 45 (2021). https://doi.org/10.1186/s13677-021-00255-5
19. Verginadis, Y., Alshabani, I., Mentzas, G., Stojanovic, N.: PrEstoCloud: proactive cloud resources management at the edge for efficient real-time big data processing. In: Proceedings of the 7th International Conference on Cloud Computing and Services Science, pp. 611–617. SCITEPRESS - Science and Technology Publications (2017). https://doi.org/10.5220/0006359106110617
20. Verginadis, Y., Kritikos, K., Patiniotakis, I.: Data and cloud polymorphic application modelling in multi-clouds and fog environments. In: La Rosa, M., Sadiq, S., Teniente, E. (eds.) CAiSE 2021. LNCS, vol. 12751, pp. 449–464. Springer, Cham (2021). https://doi.org/10.1007/978-3-030-79382-1_27

Fault Tolerance in Cloud: A Brief Survey

Kamal K. Agarwal$^{(\boxtimes)}$ and Haribabu Kotakula

Birla Institute of Technology and Science, Pilani 333031, Rajasthan, India
{h20200286,khari}@pilani.bits-pilani.ac.in

Abstract. The Cloud computing environment has seen tremendous growth in terms of enterprises migrating to the cloud. The Cloud computing environment has been characterized by its support to multi-tenancy as well as rapid elasticity. A cloud service provider needs to adhere to service level agreements agreed with its tenants. As cloud infrastructure is composed of heterogeneous, it ought to have faults when different components interact with each other. Fault tolerance is one critical task in order to overcome faults in a timely manner and smooth operations. Through this paper, we present a brief survey of fault tolerance approaches available in the literature for the cloud computing environment. The fault tolerance techniques in literature have been broadly classified into three categories viz. Reactive approaches, proactive approaches, and resilient approaches. Reactive methods try to recover the system after faults. Proactive approaches try to prevent the system from entering in failure state by predicting faults beforehand. Resilient approaches are relatively new in the area of cloud fault tolerance predict faults for dynamically changing environments to prevent failures. Resilient methods are have used machine learning and artificial intelligence as a key factors for predicting faults.

Keywords: Cloud computing · Fault-tolerance · Resilience

1 Introduction

The Cloud computing environment has seen rapid growth in use cases [1]. Enterprises are making more and more services available on the cloud. The Cloud computing environment has been characterized by a multi-tenancy-based model, which allows tenants to share pooled resources in a rapid and on-demand elastic manner. Resource provisioning is supplemented by standard protocols. These shared resources are of different types, such as hardware, software application, etc. And all satisfy different requirements. Cloud services provided are basically divided into three layers viz. Infrastructure as a Service (IaaS), Platform as a Service (PaaS), and Software as a service (SaaS). As cloud computing infrastructure is comprised of many heterogeneous component interactions, among these gives rise to faults in the cloud. A fault in a system is described as a defect or inability of a system to perform normally. A fault in a system causes errors in

© The Author(s), under exclusive license to Springer Nature Switzerland AG 2022
L. Barolli et al. (Eds.): AINA 2022, LNNS 451, pp. 578–589, 2022.
https://doi.org/10.1007/978-3-030-99619-2_54

execution and leads to failures causing degraded performance and service delivery [2,3]. These failures can have large effects on complete infrastructure, which can lead to an outage, which causes a client's service delivery to be longer than expected. The primary reasons for these faults include software failures such as incorrect upgrades, uneven workload, less strict security, etc., or hardware failures such as network failures, power failures, etc. The major types of faults in the cloud environment are Network faults, Physical Faults, Process Faults, Service Expiry Faults, Media Faults, Processor Faults, Ordering faults, restrictions faults, parametric faults, contributor faults, time restriction faults, and resource contention faults [4].

As the number of cloud users and criticality of services offered increases, the need for fault- tolerance also increases [5,6]. The computing components, storage systems and networks all bring their own challenges but in cloud all of these are virtualized also bringing in additional challenges of communication and security.

This complex cloud environment may experience faults in any sub-part of the whole cloud system [7]. A fault tolerance has been defined as the ability of a system, that enables systems to continue their anticipated operations regardless of faults [4].

Regarding fault tolerance as a phenomenon related to reliability and successful operation, a fault-tolerant system should be able to recognize the faults in a particular component or other unexpected adversities in order to fulfil its specification. There are many fault tolerance techniques each working on a subset of failures and differs in their approach such as Reactive FT, Proactive FT and Resilient FT.

Reactive approaches of fault tolerance are used to recover system after failure has happened on the other hand proactive approaches try to avoid failure by proactively identifying fault prone components and replacing them in timely manner. Resilient fault tolerance approaches are comparatively new and uses emerging technologies of machine learning and artificial intelligence to predict faults in system based on historical operational data of system. This new approach i.e. resilient is quite similar to proactive approach's objective of minimizing faults in system by predicting but there exist a subtle difference between the two as in resilient methods continuously updates their model depending on the dynamics of the environment. We would like to inform the reader that we have focused on resilient methods, for more work on reactive and proactive approaches reader may refer [4,7,21].

2   Faults in Cloud

A fault, in general, is a basic impediment to regular system operation, and it causes errors. Errors, in turn, result in system failure. System faults can be classified into four categories: Transient faults, Intermittent faults, Permanent faults and Byzantine Faults [4].

– Transient faults occur when a momentary circumstance affects the system. This encompasses problems such as network connectivity challenges such as partitions in network and service unavailability in cloud computing. Transient faults can be corrected easily.

- Intermittent failures occur at random i.e. irregular periods, and they typically resemble a system, hardware device, or component malfunctioning. These issues are typically difficult to diagnose and resolve permanently. As an example, consider a hard drive that fails to function as a result of temperature variations but eventually resumes its normal operation.
- Permanent defects persist until the underlying cause is addressed. They usually arise as a result of a complete failure of a system component, and they are usually easy to identify.
- Byzantine faults are exceedingly difficult to identify because they cause a system component to behave irrationally and deliver inaccurate results to the client. This can occur as a result of a system's internal state being corrupted, data corruption, or erroneous network paths. They are incredibly difficult and costly to manage. In other cases, replicas of system components are used, and a voting approach is used to identify the suitable response.

3   Challenges

As we have already pointed out that a distributed system such as cloud comprising of heterogeneous components brings in many challenges when these components communicate with each other. Here we will discuss some of the challenges in cloud:

- Multi-Dimensional QoS requirement: As we know cloud supports multiple tenants to use pooled resources but each tenant has different QoS requirement for different components of the system. For example some would prefer low latency, or availability or consistency [8].
- Priorities for fault tolerance: When developing fault tolerance goal is to make as much as possible part of the whole system fault tolerant because of its trade-off with costs. And system architects need to consider components based on their priorities.
- Monitoring: Virtualization in cloud makes it possible for pooling resources (both Hardware and Software) from different vendors. When these resources are connected together detecting faults in them requires proper monitoring which is a difficult tasks as it may need to devise many methods to monitor these resources.
- Power Awareness: As the cloud market is growing, number and sizes of these data warehouses are also growing day by day. These large data warehouses have large amount of costs attributed to power consumption. In recent years some of the works in fault tolerance has also contributed to power awareness to reduce overall costs.
- Cloud Native vs. Cloud Enabled: Cloud enabled also known as "born in cloud". Cloud native application are developed using multiple services in cloud. Cloud native workload in these application consist of workload generated by cloud native services while cloud enabled workload is something

when a part of the application has not been migrated to cloud and computing workload differs. So fault tolerance mechanism needs to monitor these in single view [4].

- Automation: The number of virtual machines (VMs) hosted in cloud systems is rapidly increasing. Manual control of such systems by humans will become nearly impossible. As a result, it is vital to consider automation to manage fault tolerance solutions for large systems. The lack of common frameworks (APIs) that can be easily applied to any cloud system for FT is the main cause of worry for automation (need for plug and play FT).

4 Fault Tolerance Approaches

We seek to provide a brief survey of fault tolerance approaches in cloud. In our study we group different methods proposed for fault tolerance in three broad categories reactive approaches, proactive approaches and resilient approaches.

- **Reactive Approaches:** Reactive approach deals with recovering the system from aftereffects of failures. It works like a restoration process on components. Mostly it uses redundancy i.e. data replication for restoration from failures. Some of the reactive approaches available in literature are:
  - Checkpointing\Restarting: The checkpoint/restart technique periodically saves state of a component and in the event of failure component is restarted by last saved state. These kinds of technique are mostly suitable for long running batch jobs. Different techniques under this approach differ based on objective such as minimizing storage overhead or reducing restart time. Some of the papers reviewed under this category are summarized in this report. Cheng wang et al. [9] proposes a storage replication protocol GANNET which incorporates a lightweight checkpointing algorithm to save primary VM's state. Protocol reduces IO contention by buffering disk states from both primary and secondary VM in secondary's VM. During the checkpoint secondary will flush only the primary's updated state in its disk. Also, to overcome problem of ever growing size of memory checkpoint algorithm compares disk contents of primary and secondary periodically, and apply identical modification to secondary's storage. This allows to free up the buffers. Zhigang et al. [10] proposes a checkpoint based solution for asynchronous systems FR-WORB and FR-WAC. The proposed solution tries to leverage surviving data for recovery without rolling back entirely. FR-WORB builds a restarting point which provides a way to interrupted computations to go forward. It accelerates recovery by using surviving data. This FR-WORB does not checkpoint states of nodes. A variation of FR-WAC which recomputes lost data from last checkpoint instead of scratch. Authors have proposed some optimization to lower overhead, message pruning to reduce number of messages exchanged, Non-blocking recovery if standby node are not available and new data reassignment policy if recovery workload has been load balanced between multiple standby nodes.

- Replication The replication approach employs redundancy by deploying duplicated system component on different resources. In the event of failure duplicated components are activated for handling workload. With this approach main challenges lie in optimal resource usage i.e. controlling redundancy, bandwidth conservation, selecting component to replicate etc. Yanchun Wang et al. (2018) [11] purposed a criticality-based fault tolerance for multi-tenant service based systems (CFT4MTS) which determines criticality of components in system and then formulates FT. For identifying important component services in a cloud environment based on service criticality, which is computed using the three indicators listed below: a) quality-based criticality, which is determined using sensitivity analysis techniques and is based on the impacts of service anomalies on the SBS's multidimensional quality, b) tenant-based criticality, which is determined using sensitivity analysis techniques and is based on the impacts of service anomalies on tenants, and c) tenants' differentiated multidimensional quality preferences.

  Futian wang et al. (2021) [8] purposes a probability based fault tolerance strategy (PFT4MTS) for multi-tenant based systems. PFT4MTS first evaluates each component's criticality by considering that quality of each component over time changes and in a multi-tenant based system each tenant has different quality preferences. An optimal fault tolerant strategy is built using criticalities of different components obtained as result of PFT4MTS. Criticality of a component has been modeled by considering response time and throughput of components. Authors have suggested incorporating more QoS parameters using one of the mathematical models used for response time or throughput.

- Retry: The requests that failed are simply retried. The retry technique is based on the system's indepotence. Indepotence is defined as a condition that ensures that the results of an application obtained in the presence of duplicate requests and failures are same to the results obtained in the absence of duplicate requests and failures. Before it achieves its goal, a job is executed in such manner on a constant basis. The failure/incompetent mission is retried with the same tool. Retry was employed to manage failures in those systems. The retry method determine failed queries in order to execute again.

- Task re-submission: When a failed task is recognized, it is resubmitted to the same or a different resource for execution in task re-submission. Such techniques are commonly employed to achieve fault tolerance in cloud-based workflow systems. The core concepts of re-submission and replication are commonly utilized in distributed systems for fault tolerance. Re-submission works by resubmitting a single job to a failed resource after correcting or a new one, which in turn lengthen the task completion time.

- Rescue Workflow: The term "rescue workflow" refers to a strategy for addressing fault tolerance in workflow-based systems. Even if a task fails, the process can continue until it is impossible to proceed without attending to the failed job.

- Load Balancing: Load balancing is essential for cloud systems' failure tolerance and functionality. After being inundated by client requests, many servers might fail due to expended computational resources (CPU or RAM). Load balancing must be implemented as the first line of load prevention in cloud systems to reduce such failures. Traditional centralized load balancing solutions are ineffective in cloud environments. Cloud systems are very scalable and are constructed on distributed servers that may be located across numerous data centers (depending on the architecture of the cloud system). The centralized assignment of computing demands to servers is deemed infeasible due to the magnitude and complexity of such systems.

- N-version and Recovery Block: The N-version programming paradigm is a multi-version programming methodology. Separate teams working on the same set of requirements generate functionally similar programs. The teams work in isolation and do not interact with one another. The N-version concept is built upon the assumption that separately built programs have less likelihood of comparable defects in two or more versions, significantly. Tolerance for design defects is provided via the Recovery Block (RB), which uses various representations of input data. Recovery block is a reactive and high-level task technique for reducing the impact of an unpleasant occurrence after it occurs (as opposed to safety, which involves pre-deployed superfluous capacity). Lei wang (2019) [12] purposes ARCMeas, Architecture Reliability-sensitive critical measure. Author has suggested ARCMeas which considers criticality of each of the component of composite system as described in SLAs. This approach has some improvement over CFT4MTS as it also considers the underlying architecture of the composite system. Author first determines Reliability-sensitive measure i.e. RMeas for each component based on the perturbation sensitive QoS parameters analyzed using sensitivity analysis. Second part includes determining Architecture-based Criticality Measure i.e. AMeas which analyses dependence among components to determine criticality and then combine RMeas and AMeas to obtain ARCMeas. ARCMeas helps to rank components in order to formulate a cost effective FT strategy. In order to demonstrate effective of ARCMeas author have used RB as FT strategy but proposes to uses any other strategy to be used based on system.

- **Proactive Approaches:** Proactive approaches aids in failure avoidance by predicting failures. They do so by monitoring the system and executing failure prediction algorithm periodically to determine health of the system. Some of the well-known proactive approaches are described here for completeness.

  - Software Rejuvenation Software rejuvenation is a process of gracefully terminating a system and restarting it. It can be used with N-version programming and error counting. Error counting helps in escalation of errors for faster recovery during failure.

- Self-Healing Self-healing is a characteristic of a component or system which allows it to automatically find errors, diagnose them and correct in timely manner. It can be used for both hardware and software fault.
- Preemptive Migration Preemptive migration approach avoids system failure through migration of workload from fault prone resources by predicting faults beforehand. In the current scenario of virtualized environment preemptive migration is most suitable approach. As it monitors the system as a whole in order to predict components which are about to crash. In today's scenario, VM/Container migration techniques having minimum possible overhead of migration aids this process of workload migration.
- Prediction Prediction forms the foundation of proactive fault tolerance approach. It allows system architects to consider wide number of parameters to monitor. These quantified resource parameter aids in formulating an efficient fault tolerance framework.
- Monitoring The primary objective of monitoring is to aid other proactive methods in order to take appropriate decisions. It is used to collect and control system's run time state which will include resource usage statistics, indicating system performance in real time. As this is one crucial part of any proactive method system architect needs to design system by keeping obeservability in mind.
- SGuard Rollback and recovery based method most prominently used in real time video streaming systems. It is relatively new approach used for fault tolerance in distributive stream processing engines. It uses combination of reactive approaches for fault mitigation.

– **Resilient Approaches:** Resilient approaches are relatively new in the area of fault tolerance. This approach empowers systems to be smart and learn from environment to adapt dynamically. Resilient approaches available in literature provided here for a brief overview. In essence resilient methods are composed of reactive and proactive methods only but have some additional advantage of dynamic adaptability in changing environment.

  - Machine Learning As in recent year ML/AI has found many promising applications, this is also one of them. This is to be considered as emerging area for machine learning/artificial intelligence based method. Reinforcement learning is most common technique in fault tolerance area, as it uses reward and penalty based learning. Some other approaches from ML has not been found directly linked to FT of Cloud computing environment but they can also be extended to learn about systems. Hussaini Adamu et al. [13] proposed a method to predict faults in cloud using machine-learning based models. Authors have used failure data of The National Energy Research Scientific Computing Center (NERSC) available on Computer Failure Data Repository (CDFR). Formulation of two models have been completed to find correlation between hardware failure and time. Two machine learning models Linear Regression Model (linear model) and Support Vector Machine (Non Linear relation) of hardware failure (response variable) and time as predictor variable. The prepossessing step for training above model involved categorising components in infrastructure and then finding correlation, which can used as a good predictor of hardware failure. Zhixin Li et al. [14] proposed a method of predicting security state of VM. Hidden Markov Model used for realizing security state of Vm and then observation state prediction by using AdaBoost. The state data of VM as observed by cloud platform was collected to feed in AdaBoost ensemble in order to feed in HMM. Appropriate parameter for both models was obtained using training on historical data. HMM model predicts hidden state of VM which can further be used for appropriate action as fault tolerance strategy. Jiechao Gao et al. [15] proposed an algorithm to for prediction of job failure and task failure in cloud environment using deep learning methods. The proposed method Bi-LSTM based on LSTM (a artificial recurrent neural network architecture). The proposed algorithm uses multiple layer model to layer to learn from data. Authors proposed to use task priority, task re-submission and scheduling delay in addition to CPU, Memory usage data. The architecture of Bi-LSTM model contains one input layer, two Bi-LSTM layers, one output layer and the Logistic Regression (LR) layer to classify whether the tasks and jobs are failed or finished. Shujie Han et al. [16] purposes STREAMDFP a stream mining framework for prediction of disk failure and fault tolerance. STREAMDFP removes the need for having logs available apriori and formulates the problem as stream mining problem. The problem modelled here predicts failure either as classification problem or regression problem for likelihood of failure. Authors have make an improvement in terms of identification of concept drift (a change in statistical pattern over time). STREAMDFP has three key techniques online labelling, concept-drift aware training, prediction with support for various machine learning algorithms. STREAMDFP incorporates incremental algorithms to detect and change according concept drift. Bhargavi K et al. [17] purposes a reinforcement learning based algorithm to load balance dynamically. The algorithm requires two preliminary work such as formulation of task subsets for each task set received and form PMs by grouping subset of VMs. RROP-LBA(Raven Roosting Optimization Policy based Load Balancing Agent) placed on each PM distributes subset of tasks formed in preliminary step. It uses reinforcement learning based approach with the raven roosting optimization policy. The simulation results shows successful task completion rate to be high while blocking probability and response time to be low. Pengpeng Zhou et al. [18] purposes a log based anomaly detection model LogSayer. LogSayer represents system using suitable statistical features which are not sensitive to exact log sequence. It uses LSTM neural network to identify historic correlation patterns and then applies BP neural network for dynamic anomaly decisions. LogSayer models for each of the component based on their individual characteristics. The architecture of LogSayer consist of two phases: offline and online. The offline stage parses the log into structured form with different groups and then train a model for each group using this stored data. In online phase newly arrived logs are divided into groups similar to what was done in offline phase to extract feature and identify abnormality. LogSayer reports these component anomaly to admin. The updation of model built is done in accordance with the feedback from admins in order to reduce false positives and false negatives. To improve the efficiency and availability of diagnosis model during update phase, online stage incorporates a controller module which will keep a backup model (replication of last trained model). Jing Yue et al. [19] purposes method for determining micro-service aging and rejuvenation policy based on CVA (container vertical autoscaling) architecture. The purposed work aids in container vertical expansion as well as contraction to improve utilization of resources and availability of microservice based system. The purposed method uses CNN and LSTM based neural network to determine QoS violation probability. A microservice is considered to the aging if QOS violations probabilities are higher and resource (CPU, memory, disk) usage exceeds a threshold. The CVA method uses CPU, memory and disk usage to determine resource requirement every 60 s and scale up or down. Avinab Marahatta et al. [20] purposes an Artificial Intelligence based "Prediction based Energy-aware Fault-tolerant Scheduling scheme" (PEFS). PEFS has two stages, in first stage task failure is predicted and then in next stage formulates scheduling strategy. Task failure prediction stage uses deep neural network trained on historical data to predict failure of each arriving task and classify each task as either failure-prone or non-failure-prone task. The scheduling of the two tasks is done differently, The failure-prone tasks are transformed into super tasks using elegant vector reconstruction method to schedule replicated failure-prone tasks on different host.

- Fault Induction Fault induction is an approach where faults are introduced purposefully in order to detect errors and learn from these errors. This can be considered as synonym of mock drill exercise, where not only the system but also the persons involved in business processes goes through a simulated disaster incident in order to identify flaws in system, both in software as well as hardware. This exercise also helps personnel involved to be better prepared for any major outages. Through this exercise all the system as well as personnel learns about deficiency of the system.

5   Brief Discussion and Future Directions

In the study, we note how different approaches emphasize different faults from three broad categories of fault tolerance viz. Reactive, Proactive, and Resilient Methods, Resilient methods are a relatively new field for fault tolerance strategy in a cloud computing environment. We can also note that reactive approaches are dominant; one reason for its dominance can be attributed to lower overhead compared to proactive approaches. Also, the proactive approaches require more control over resources to aid in proper monitoring. As Reactive approaches have been studied in the literature for a long time, they have improved a lot. In our survey, we can see how CFT4MTS uses the criticality of components by having a sensitivity analysis of different parameters such as response time, throughput, etc.; this includes the mean QoS value of a parameter. PFT4MTS is an improvement over CFT4MTS as it considers the fact that the quality of components deteriorates over time. ARCMeas further improves the criticality analysis by incorporating analysis of architecture to include dependency between components to find the overall criticality of components. These types of criticality analysis help us to overcome major challenges of controlling redundancy or other cost associated with FT. Also, the authors purpose to use any fault tolerance method after determining criticality. Proactive methods owing to greater planning and monitoring overhead, are less preferred. Also, the requirement of greater control and difficulty in monitoring resources blocks the way for proactive methods. As of now, approaches such as preemptive migration have been applied to a cloud environment by using VM/Container migration features. Resilient methods try to overcome shortcomings of traditional proactive methods to have flexibility in learning and adapting to the environment. This aids in reduced reconfiguration overhead and study of the system for determining parameter and their effects. In the future, we speculate resilient methods to dominate FT as advances in ML/AI for a low-cost model building will help in easy and cost-effective deployment in the cloud environment. As the cloud computing environment has seen growth similarly, the field of machine learning has also seen tremendous growth due to tremendous growth in technology. Similar to reactive and proactive approaches, resilient methods could also focus on a particular type of fault. The work presented in [13] can be taken as a good starting point as it demonstrates a simple scenario of predicting the correlation between hardware fault and time. It provides a good insight into how the failure prediction can be modeled as a machine-learning problem when proper data is available. Similarly, work in [14] presents a method to predict the hidden state of VM in order to take corrective action beforehand. The works presented in [15] and [20] use an artificial neural network and deep neural network to isolate failure-prone tasks and schedule them separately in order to minimize task/job failures. The work presented in [16] and [18] uses logs to identify anomalous behavior, where [16] removes the need for having historical logs by modeling the problem as a stream analysis problem along with the incorporation of concept-drift; similarly, [18] uses a feedback loop to minimize the false positive and false negative. The work presented in [19] has taken up a new area of research to determine the aging of complex microservice architecture and apply rejuvenation as a fault-tolerance policy. The proposed method uses neural network-based architecture to predict the aging of a microservice. And [17] presents load balancing as a way of fault-tolerance where load balancing is done in accordance with the prediction from reinforcement learning. As we can see, all the proposed resilient methods use one of the methods of reactive or proactive approaches to provide fault-tolerance but differ in terms of fault prediction and applicability in a dynamic environment.

6 Conclusion

Fault tolerance is one of most important factor to consider today's growing cloud market. With increase in enterprise migrating to cloud service provider need to adhere with SLA to provide high availability. We have reviewed fault tolerance approaches available in literature. Three broad categories of methods are reactive, proactive and resilient methods. Reactive approaches are most commonly used approach owing to less configuration and uncertainty issues. Greater overhead and complicated design make proactive approaches less desirable. Replication and Checkpoint/Restart are the two most preferred way of fault tolerance in FT for reactive approach, similarly preemptive migration is one preferred choice for proactive fault tolerance. Resilient methods being relatively new area has not seen many works directly attributed to fault tolerance in cloud but existing works in distributed environment based on AI/ML can be extended to cloud to make systems resilient against fault. On future direction we anticipate more works on resilient methods for fault tolerance.

References

1. Armbrust, M., et al.: A view of cloud computing. Commun. ACM **53**(4), 50–58 (2010)
2. Amin, Z., Singh, H., Sethi, N.: Article: review on fault tolerance techniques in cloud computing. Int. J. Comput. Appl. **116**(18), 11–17 (2015)
3. Singh, G., Kinger, S.: A survey on fault tolerance techniques and methods in cloud computing. Int. J. Eng. Res. Technol. **2**(6), 1215–1217 (2013)
4. Shahid, M.A., Islam, N., Alam, M.M., Mazliham, M.S., Musa, S.: Towards resilient method: an exhaustive survey of fault tolerance methods in the cloud computing environment. Comput. Sci. Rev. **40**, 100398 (2021). ISSN 1574-0137. https://doi.org/10.1016/j.cosrev.2021.100398
5. Gokhroo, M.K., Govil, M.C., Pilli, E.S.: Detecting and mitigating faults in cloud computing environment. In: 2017 3rd International Conference on Computational Intelligence & Communication Technology (CICT), pp. 1–9 (2017). https://doi.org/10.1109/CIACT.2017.7977362
6. Nazari Cheraghlou, M., Khadem-Zadeh, A., Haghparast, M.: A survey of fault tolerance architecture in cloud computing. J. Netw. Comput. Appl. **61**, 81–92 (2016)
7. Kumari, P., Kaur, P.: A survey of fault tolerance in cloud computing. J. King Saud Univ. Comput. Inf. Sci. **33**(10), 1159–1176 (2021). ISSN 1319-1578. https://doi.org/10.1016/j.jksuci.2018.09.021
8. Wang, F., Hong, T., Wang, D., Zhang, C.: A probability-based fault tolerance strategy for service-based systems. In: 2021 3rd International Conference on Advances in Computer Technology, Information Science and Communication (CTISC), pp. 92–97 (2021). https://doi.org/10.1109/CTISC52352.2021.00025
9. Wang, C., Chen, X., Wang, Z., Zhu, Y., Cui, H.: A fast, general storage replication protocol for active-active virtual machine fault tolerance. In: 2017 IEEE 23rd International Conference on Parallel and Distributed Systems (ICPADS), pp. 151–160 (2017). https://doi.org/10.1109/ICPADS.2017.00031

10. Wang, Z., Gao, L., Gu, Y., Bao, Y., Yu, G.: A fault-tolerant framework for asynchronous iterative computations in cloud environments. IEEE Trans. Parallel Distrib. Syst. **29**(8), 1678–1692 (2018). https://doi.org/10.1109/TPDS.2018.2808519

11. Wang, Y., He, Q., Ye, D., Yang, Y.: Formulating criticality-based cost-effective fault tolerance strategies for multi-tenant service-based systems. IEEE Trans. Softw. Eng. **44**(3), 291–307 (2018). https://doi.org/10.1109/TSE.2017.2681667

12. Wang, L.: Architecture-based reliability-sensitive criticality measure for fault-tolerance cloud applications. IEEE Trans. Parallel Distrib. Syst. **30**(11), 2408–2421 (2019). https://doi.org/10.1109/TPDS.2019.2917900

13. Adamu, H., Mohammed, B., Maina, A.B., Cullen, A., Ugail, H., Awan, I.: An approach to failure prediction in a cloud based environment. In: 2017 IEEE 5th International Conference on Future Internet of Things and Cloud (FiCloud), pp. 191–197 (2017). https://doi.org/10.1109/FiCloud.2017.56

14. Li, Z., Liu, L., Kong, D.: Virtual machine failure prediction method based on AdaBoost-Hidden Markov model. In: 2019 International Conference on Intelligent Transportation, Big Data & Smart City (ICITBS), pp. 700–703 (2019). https://doi.org/10.1109/ICITBS.2019.00173

15. Gao, J., Wang, H., Shen, H.: Task failure prediction in cloud data centers using deep learning. In: 2019 IEEE International Conference on Big Data (Big Data), pp. 1111–1116 (2019). https://doi.org/10.1109/BigData47090.2019.9006011

16. Han, S., Lee, P.P.C., Shen, Z., He, C., Liu, Y., Huang, T.: Toward adaptive disk failure prediction via stream mining. In: 2020 IEEE 40th International Conference on Distributed Computing Systems (ICDCS), pp. 628–638 (2020). https://doi.org/10.1109/ICDCS47774.2020.00044

17. Bhargavi, K., Babu, B.S.: Load balancing scheme for the public cloud using reinforcement learning with raven roosting optimization policy (RROP). In: 2019 4th International Conference on Computational Systems and Information Technology for Sustainable Solution (CSITSS), pp. 1–6 (2019). https://doi.org/10.1109/CSITSS47250.2019.9031053

18. Zhou, P., Wang, Y., Li, Z., Wang, X., Tyson, G., Xie, G.: LogSayer: log pattern-driven cloud component anomaly diagnosis with machine learning. In: 2020 IEEE/ACM 28th International Symposium on Quality of Service (IWQoS), pp. 1–10 (2020). https://doi.org/10.1109/IWQoS49365.2020.9212954

19. Yue, J., Wu, X., Xue, Y.: Microservice aging and rejuvenation. In: 2020 World Conference on Computing and Communication Technologies (WCCCT), pp. 1–5 (2020). https://doi.org/10.1109/WCCCT49810.2020.9170005

20. Marahatta, A., Xin, Q., Chi, C., Zhang, F., Liu, Z.: PEFS: AI-driven prediction based energy-aware fault-tolerant scheduling scheme for cloud data center. IEEE Trans. Sustain. Comput. **6**(4), 655–666 (2021). https://doi.org/10.1109/TSUSC.2020.3015559

21. Mukwevho, M.A., Celik, T.: Toward a smart cloud: a review of fault-tolerance methods in cloud systems. IEEE Trans. Serv. Comput. **14**(2), 589–605 (2021). https://doi.org/10.1109/TSC.2018.2816644

Load Distribution for Mobile Edge Computing with Reliable Server Pooling

Thomas Dreibholz[1]([✉]) [iD] and Somnath Mazumdar[2] [iD]

[1] Simula Metropolitan Centre for Digital Engineering,
c/o OsloMet – Storbyuniversitetet, Pilestredet 52, 0167 Oslo, Norway
dreibh@simula.no
[2] Department of Digitalization, Copenhagen Business School,
Howitzvej 60, 2000 Frederiksberg, Denmark
sma.digi@cbs.dk

Abstract. The energy-efficient computing model is a popular choice for both, high-performance and throughput-oriented computing ecosystems. Mobile (computing) devices are becoming increasingly ubiquitous to our computing domain, but with limited resources (true both for computation as well as for energy). Hence, workload offloading from resource-constrained mobile devices to the edge and maybe later to the cloud become necessary as well as useful. Thanks to the persistent technical breakthroughs in global wireless standards (or in mobile networks), together with the almost *limitless* amount of resources in public cloud platforms, workload offloading is possible and cheaper. In such scenarios, Mobile Edge Computing (MEC) resources could be provisioned in proximity to the users for supporting latency-sensitive applications. Here, two relevant problems could be: *i*) How to distribute workload to the resource pools of MEC as well as public (multi-)clouds? *ii*) How to manage such resource pools effectively? To answer these problems in this paper, we examine the performance of our proposed approach using the Reliable Server Pooling (RSerPool) framework in more detail. We also have outlined the resource pool management policies to effectively use RSerPool for workload offloading from mobile devices into the cloud/MEC ecosystem.

1 Introduction

Mobile devices are becoming an indispensable part of our daily life. Such devices are used for a large list of Internet-based activities (such as financial transactions, online conferencing, route navigation, Internet browsing, text/video messaging, and online gaming). These modern devices (including smart phones) are denoted as User Equipment (UE). Recent UEs have decent computational as well as

This work has partly been supported by the 5G-VINNI project (grant no. 815279 within the H2020-ICT-17-2017 research and innovation program) and also partly by the Research Council of Norway (under project number 208798/F50).
The authors would like to thank Ann Edith Wulff Armitstead for her comments.

L. Barolli et al. (Eds.): AINA 2022, LNNS 451, pp. 590–601, 2022.
https://doi.org/10.1007/978-3-030-99619-2_55

storage capabilities. In some scenarios, they are limited by the steadily increasing amount of user data and computation demand. In such events, cloud computing become a viable solution to offload tasks that require large computation power and storage. Apart from that, the pay-as-you-go pricing model also offers cloud services at cheaper costs than overall on-premise clusters management. However, cloud platforms do not offer low latency between UEs, which may impact latency-sensitive applications performance. To counter such performance-related issues, Mobile Edge Computing (MEC) is becoming popular, thanks to the progress in the mobile network ecosystem. MEC can be placed between the cloud and the user, while resourcefully supported by the cloud.

MEC could be seen as a small subset of the cloud. Managing such geographically distributed resource pools without adding a large amount of management overhead is not easy. To date, MEC is not mature enough, thus offloading applications to MEC efficiently is not a trivial matter. In this work, we did *not* aim for a task offload solution where we add multiple abstraction layers and try to find the *almost* perfect mapping of tasks to resources. Instead, our proposed solution was inspired by the "Keep It Simple, Stupid" (KISS) approach, where we reuse the existing, lightweight Reliable Server Pooling (RSerPool) framework [1,2]. In addition to that, we have adapted server selection policies by incorporating the structure of the MEC.

In our previous work [3], we already presented how RSerPool manages resource pools and also how it handles the application sessions effectively. Now, as an extension of the work, in this paper, we examine the use of different server selection policies in more detail (over a large parameter range). We conducted the performance evaluation of resource selection policies, using simulations of a MEC and (multi-)cloud setup with the RSerPool Simulation (RSPSIM) [2] model. We provided insights into how the proper choice and configuration of pool member selection policies could result in good performance. We also believe such a setup realistically resembles the serverless computing setups.

The rest of the paper is structured as follows: we first introduce RSerPool in Sect. 2. Then, we describe our approach in Sect. 3, followed by the description of our simulation setup and result in Sect. 4. Finally, we conclude our paper in Sect. 5.

2   Reliable Server Pooling (RSerPool)

The management of server pools, as well as sessions (between clients and server pools) is a traditional problem in computer networking. To avoid "reinventing the wheel" for each application, the Internet Engineering Task Force (IETF) founded the Reliable Server Pooling (RSerPool) [1] working group to develop a generic standard. Therefore, RSerPool is an application-independent and open-source framework, with the goals of being simple and lightweight. RSerPool is also suitable for devices with very limited resources, which is ideal for MEC environments.

RSPLIB[1] [2, Chapter 5] is the most widespread open-source implementation of RSerPool. Apart from that, a simulation model RSPSIM[2] [2, Chapter 6] has also been developed.

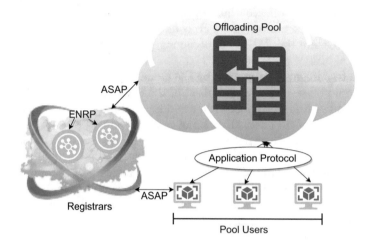

Fig. 1. Illustration of the Reliable Server Pooling (RSerPool) Architecture

Figure 1 illustrates the RSerPool architecture. A resource pool constitutes a set of servers, providing a certain service. Servers in a pool are denoted as Pool Elements (PE). A pool is identified by its unique Pool Handle (PH, e.g. a string like "Offloading Pool") within its operation scope. The handlespace is the set of all pools of an operation scope. It is managed by Pool Registrars (PR, also denoted as Registrars). To avoid a single point of failure, RSerPool setups should consist of at least two registrars. The PRs synchronise the handlespace by using the Endpoint haNdlespace Redundancy Protocol (ENRP) [2, Section 3.10]. An operation scope is limited to an organisation or company. This means that it does not scale to the whole Internet, in contrast to the Domain Name System (DNS). This is a significant simplification, which keeps RSerPool very lightweight concerning the administrative overheads. Pools can be distributed over large geographic areas to achieve a high resilience of services.

PEs use the Aggregate Server Access Protocol (ASAP) [2, Section 3.9] to dynamically register to, or deregister from, a pool at one of the PRs of their operation scope. The PR chosen for registration becomes the Home-PR (PR-H) of the PE. It also monitors the availability of the PE by a keep-alive mechanism [4]. Clients, which are denoted as Pool Users (PU) in the context of RSerPool, use ASAP to access the resources of a pool. A PU can query a PR of the operation scope to select PE(s). This selection is performed by using a pool-specific pool

---

[1] RSPLIB: https://www.uni-due.de/~be0001/rserpool/#Download.
[2] RSPSIM: https://www.uni-due.de/~be0001/rserpool/#Simulation.

member selection policy [4], which is usually just denoted as pool policy (see more details in Subsect. 3.2). ASAP may also be used between PU and PE, then realising a *Session Layer* functionality between a PU and a pool. In this case, ASAP can also be used to assist the actual *Application Layer* protocol to handle failovers and help with state synchronisation [4].

3 Our Proposed Offloading Approach

In this section, we are explaining our offloading approach in two parts. The first part is primarily focused on setting up the RSerPool (refer to Subsect. 3.1). Next, we will discuss the server pool member selection policies for our MEC setup (refer to Subsect. 3.2).

3.1 Setting Up Basic RSerPool

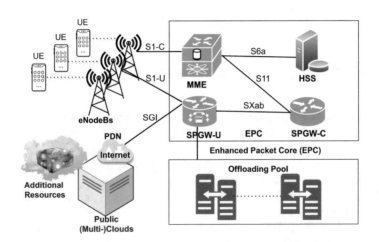

Fig. 2. Application scenario with MEC and Public (Multi)-Clouds (PMCs)

Figure 2 shows the basic MEC scenario [3,5]. Here, UEs are running applications, which aim at offloading tasks to the cloud instead of using the scarce and battery-powered UE resources. Particularly, this could be a serverless computing setup, where the application pushes a certain task for processing into the cloud. The cloud can instantiate an application-specific container that runs the assigned task and frees up the container after execution. That is, the compute node is stateless, which allows for high flexibility of workload offloading for different application types. Cloud resources are provided by the local MEC, as well as by cheaper public (multi)-clouds (PMC). Particularly, it is from a cost-perspective an advantage to use PMCs, which are in vicinity to the user. Distant cloud resources offer a high round trip time (RTT), as distance adds network latency to the service request handling. Therefore, the local MEC resource is (if available) preferred. So, now the questions could arise: *how to manage such server pools, consisting of MEC as well as PMC resources?*

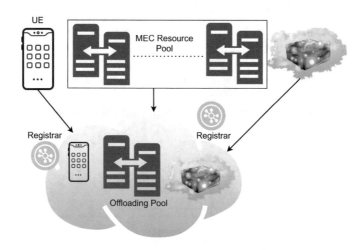

Fig. 3. RSerPool with mobile edge computing and (multi-)cloud resources

In Fig. 3, we describe our approach [3,5] to applying RSerPool. Resources are added into a pool, identified by its PH (here: "Offloading Pool"), which means the pool consists of PEs in the MEC as well as in a PMC setup. Since the original purpose of RSerPool is high availability, it may be straightforward to also run a PE instance on the UE itself. RSerPool is also lightweight, which means it has a very small management overhead. With RSerPool, it is also possible to add the UE resources to the pool. If everything fails, then the application on the UE is still able to "offload" a task to itself (i.e. the PE on the UE). This would be the case when there is no network coverage, which is not so unlikely for a mobile usage scenario. In such events, the application may still provide a useful service for its user with reduced performance and also with increased energy consumption. However, MEC or PMC resources are always preferable in these scenarios.

Overall, the pool consists of three different types of resources: MEC, PMCs, as well as UE resources. In addition, PRs are needed for managing the handlespace. To avoid a single point of failure, it needs to be at least two, e.g. one for MEC and another for PMCs. For standalone operations in case of loss of network coverage, also a PR instance needs to run on the UE itself. This is possible, since PRs are lightweight, meaning they have only low memory and CPU requirements. Now, in this case, the relevant question is: *how can the PRs finally handle the different resource types in the pool?*

3.2 Pool Member Selection Policies for MEC

The selection of a suitable PE is performed by the pool policy (as described in Subsect. 3.3). In case of our UE/MEC/PMC setup, it has to achieve the following four goals [3]:

- Goal 1: Only use UE resources if there is no other possibility (such as no network coverage or lack of MEC resources).
- Goal 2: Use of PMCs resources only when they are a suitable choice (such as when the MEC resources are highly utilised).
- Goal 3: Otherwise, use the MEC resources.
- Goal 4: Apply load balancing.

3.3 Policies

Two simple resource allocation algorithms, Random (RAND) and Round Robin (RR), neither have information about load state nor about resource type [2, Subsection 3.11.2]. They are used in this work for comparison purposes only. Apart from these two, we also have used different resource allocation policies with load states:

- Least Used (LU) [2, Subsection 3.11.3] selects the PE p where its load L_p is lowest. In case of multiple PEs with the same lowest load (e.g. three PEs with load of 0%), round robin or random selection is applied among these least-loaded PEs. Therefore, it will not differentiate between MEC, PMC and UE resources. In other words, it is not able to satisfy the first three goals mentioned above.
- Priority Least Used (PLU) [2, Subsection 8.12.2] adds a PE-specific load increment constant I_p to LU. This means, PEs are chosen based on the lowest sum $L_p + I_p$. Setting $I_{p_{\mathrm{MEC}}} < I_{p_{\mathrm{PMC}}}$ for all MEC PEs p_{MEC} and PMCs PEs p_{PMC}, as well as $I_{p_{\mathrm{UE}}} = 100\%$ for the UE PE. Then, PLU should achieve our four goals.
- Least Used with Distance Penalty Factor (LU-DPF) [2, Subsection 8.10.2] adds a PE-specific Distance Penalty Factor (DPF) constant D_p to LU. Then, the PEs are chosen based on the lowest sum

$$L_p + \mathrm{RTT}_p * D_p$$

where RTT_p is the approximated[3] Round-Trip Time (RTT) to the PE. For simplicity, D_p can be the same for all PEs, then making it a pool-specific constant. Assuming

$$\mathrm{RTT}_{p_{\mathrm{MEC}}} \ll \mathrm{RTT}_{p_{\mathrm{PMC}}}$$

for all MEC PEs p_{MEC} and PMCs PEs p_{PMC}, LU-DPF should achieve the goals 2 to 4. However, goal 1 is likely to be violated in this case, since $\mathrm{RTT}_{p_{\mathrm{UE}}}$ may (mostly) be minimal. LU-DPF can therefore be extended to a new policy Priority Least Used with Distance Penalty Factor (PLU-DPF), by adding a PE-specific load increment constant I_p (as for PLU).

---

[3] On PR-H: $\mathrm{RTT}_{\mathrm{PR-H}\leftrightarrow\mathrm{PE}}$; on other PR: $\mathrm{RTT}_{\mathrm{PR}\leftrightarrow\mathrm{PR-H}} + \mathrm{RTT}_{\mathrm{PR-H}\leftrightarrow\mathrm{PE}}$.

- LU and its variants are adaptive policies, i.e. they require up-to-date load information from the PEs in the handlespace, which is managed by the PRs. A PR-H has to be updated with the load states of its PEs (by re-registration via ASAP). Then, it distributes the updates to the other PRs (via ENRP). So, propagating load updates takes time, leading to temporary inaccuracy. The Least Used with Degradation (LUD) [6] policy adds a load degradation variable X_p for each PE p. Each time a PE is selected by a PR, it increases the load degradation by the load increment constant I_p. On update from the PE, X_p is reset to zero. Then, PEs are chosen based on the lowest sum $L_p + X_p$. This can be combined with the idea of PLU to a new policy Priority Least Used with Degradation (PLUD) choosing a PE by lowest sum of $L_p + I_p + X_p$, and with distance penalty factor to a new policy Priority Least Used with Degradation and DPF (PLUD-DPF) selecting a PE by lowest sum of

$$L_p + I_p + X_p + \mathrm{RTT}_p * D_p.$$

Now, one question arising from these different policies is about their performances: *Which policies are useful for our use case, and which policy should be used for a UE/MEC/PMC setup?*

4   Simulation and Results

For our simulation, we use the RSPSIM [2, Chapter 6] model for RSerPool to create a setup as depicted in Fig. 3. RSPSIM is based on OMNeT++ 6.0pre15 and has been extended to support our new polices (PLUD, PLU-DPF and PLUD-DPF, see Subsect. 3.3).

4.1   Application Description

As application, we use the CALCAPPPROTOCOL model from [2, Section 8.3]. This model is part of RSPSIM, as well as of the RSerPool implementation RSPLIB, which was used for our initial proof-of-concept real-world measurements in [3]. In CALCAPPPROTOCOL, a PE has a request handling *capacity* given in the abstract unit of calculations per second (calculations/s). An arbitrary application-specific metric for capacity may be mapped to this definition (such as CPU operations, processing steps, disk space usage). Each request has a request size, which is the number of calculations consumed by the processing of the request. Following the multi-tasking principle, a PE can process multiple requests simultaneously. The user-side performance metric is the handling speed. The total time for handling a request d_{Handling} is defined as the sum of queuing time, start-up time (dequeuing until reception of acceptance acknowledgement) and processing time (acceptance until finish). The *handling speed* (in calculations/s) is defined as:

$$\mathrm{HandlingSpeed} = \frac{\mathrm{RequestSize}}{d_{\mathrm{Handling}}}.$$

4.2   Simulation Setup

For our setup as depicted in Fig. 3, we use below parameters, unless otherwise stated. For each simulation scenario, 64 runs are performed.

- n PU instances may run on the UE. Each PE generates requests with an average size of 1,000,000 calculations, at an average interval of 10 s (negative exponential distribution for both).
- There is one PE for *each* PU on the UE side, with a capacity of only 200,000 calculations/s (i.e. n PU instances mean n UE PEs).
- Four PEs, each with a capacity of 1,000,000 calculations/s, are deployed as MEC resources, having a one-way delay between UE and PE within 5 ms and 15 ms (uniform distribution). From testbed experiments [3,5], these delays are realistic in local 4G setups.
- In total, ten PEs are deployed as PMC resources, each with a capacity of 1,000,000 calculations/s, with one-way delay between UE and PE between 30 ms and 300 ms (uniform distribution), with delays based on Internet measurements in [7].
- Minimum processing speed per PE is 200,000 calculations/s, i.e. PEs in MEC and PMC accept up to five requests in parallel, while the UE PE just can run at most a single one. When fully loaded, further requests get rejected, and a new PE has to be selected.
- For policies with load increment:

$$I_{p\text{MEC}} = 10\%, I_{p\text{PMC}} = 20\%, I_{p\text{UE}} = 100\%.$$

- For DPF policies: DPF $D_p = 0.0001$ for all PEs.
- One PR in the UE network, one in the MEC cloud, and one in the PMC.

4.3   Pool Member Selection Policy Performance

Our first simulation is examining the general properties of the different pool policies in our MEC setup (defined in Subsect. 4.2). Figure 4 presents the utilisation of different PEs on UE in MEC as well as in PMC (vertical direction) for the different pool policies (horizontal direction). On the x-axis, the number of PUs is varied. Note, that n PUs also means n low-performance PEs at the UE side ("the more UEs, the more resources on UEs in total"), while the number of PEs in MEC (4) and PMC (10) remain fixed. There is a line for the average utilisation of each PE, which mostly overlap (explained later). Thick error bars mark the 10%- and 90%-quantiles, together with grey ribbons for better visibility. Thin error bars show the absolute minima and maxima. The corresponding average handling speed (as defined in Subsect. 4.2) is shown in Fig. 5, again with 10%- and 90%-quantiles (thick error bars and ribbons) as well as absolute minima and maxima (thin error bars).

From the utilisation of Random (RAND) and Round Robin (RR) (refer to Fig. 4), it can be observed that all three types of resources, such as UE, MEC and PMC, are used. Furthermore, the utilisation of UE PEs is significantly larger

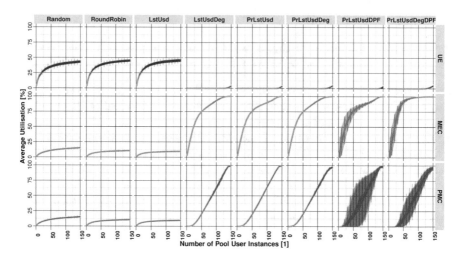

Fig. 4. Average utilisation of the three pool element types

than for MEC and PMC. This is due to the fact that a UE PE only performs one request at a time (due to lower UE performance), while MEC and PMC PEs can run up to five requests in parallel. This clearly violates goal 1 of Subsect. 3.2, which states that UE resources should not be used unless there is no other choice. This is confirmed by a low handling speed (refer to Fig. 5). Nevertheless, up to around 20 PUs, there is still a better performance ($\geq$200,000 calculations/s) than running on the UE itself, in addition to reduced battery consumption on the UE. Least Used (LU) is not much better, as expected (see Subsect. 3.2), so these three policies (RAND, RR and LU) are not a good choice.

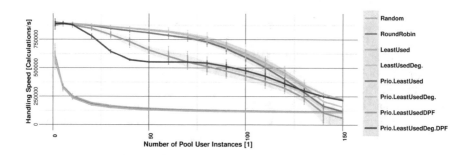

Fig. 5. Average request handling speed

Comparing the utilisation results of LU to the Least Used with Degradation (LUD), Priority Least Used (PLU) and Priority Least Used with Degradation (PLUD) policies (refer to Fig. 4), there is a significant difference: at low loads, only MEC resources are used (as intended). With the MEC resources utilisation increasing, there is an increase in PMC utilisation as well. However, MEC is the preferred choice, with higher utilisation than for PMC. Apart from that, there is no usage of slow, expensive, battery-powered UE resources, as long as sufficient MEC/PMC resources are available. The usefulness is confirmed by the handling speed (Fig. 5), with superior values over the whole x-axis range. The two variants of policies with degradation (LUD and PLUD) perform slightly better than plain PLU. Note, that the difference is subtle: multiple degradations for a PE p may occur before a load update resets the degradation variable X_p to zero, while plain PLU always adds *one* fixed load increment. As shown here, the compensation of handlespace load inaccuracy due to network delay performs slightly better. And, there is an advantage for PLUD compared to LUD at high loads (here: ≥ 130 PUs). On the utilisation side, the degradation policies show an increased utilisation variation for higher loads: inaccuracy occurs due to network delay, so there is a difference between the PEs having low and high PU↔PE RTT. Nevertheless, it can easily be seen that LUD, PLU, and particularly PLUD fulfil all four goals (for goals refer to Subsect. 3.2).

Our setup defined in Subsect. 4.2 is based on our multi-country Internet testbed scenario already mentioned in our previous work [3]. Therefore, particularly the RTT between PU and PMC PEs significantly varies, from ca. 60 ms to ca. 600 ms. With the DPF policy variants PLU-DPF and PLUD-DPF, the PE choice takes the RTT into account ($X_p = 0.0001$). In this case, the utilisation lines (Fig. 4) for the different PEs become distinct, leading to a significant variation of the 10%- and 90%-quantiles, as intended: nearer PEs are preferred over far-away PEs with respect to RTT. However, in this scenario, the effect on the handling speed (Fig. 5) is not large, even leading to a lower performance than PLU, LUD and PLUD for a larger number of PUs. Such findings could lead to the question *whether the DPF policies are useful at all for a MEC/PMC scenario?*

4.4   Performance with Distance Penalty Factor

We performed simulations for LUD, PLU, PLUD, PLU-DPF and PLUD-DPF to further examine the usefulness of the DPF policies. Figure 6 presents the utilisation results for MEC and PMC PEs (vertical direction; UE PEs omitted, as they are unused, i.e. utilisation at 0%), while Fig. 7 presents the corresponding handling speed results. We varied the number of PUs (horizontal direction). On the x-axis, the average request size in calculations is varied. Note, that we use particularly small requests here, from 50,000 (5e4) to 500,000 (5e5) calculations, which is small compared to a PE capacity of 1,000,000 (1e6) calculations/s.

A benefit of using PLUD-DPF in comparison to LUD or PLU is visible with small requests, where the network latency significantly affects the handling speed performance (Fig. 7). This could be the case for real-time cloud processing of interactive applications (e.g. handling audio/video data) on the UE. For request sizes of up to 150,000 (1e5) calculations, PLUD-DPF achieves a better handling speed, even with 100 PUs. On the other hand, PLU-DPF (without degradation) performs worse than PLU, LUD and PLUD, even with only 25 PUs. Because of

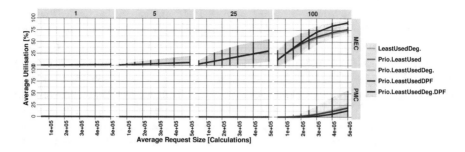

Fig. 6. Average utilisation for varying request sizes

the short requests, load information in the handlespace becomes inaccurate, e.g. a request is already finished when its increased load state gets propagated to a PR selecting a new PE. So, policies with degradation are essential for handling inaccurate load information here.

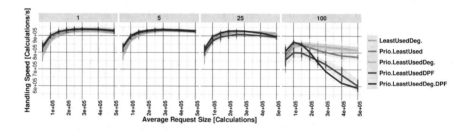

Fig. 7. Average request handling speed for varying request sizes

As expected from the handling speed results above, the utilisation of the MEC PEs (Fig. 6) is highest for PLUD-DPF with a large number of PUs. That is, PLUD-DPF combines the information from DPF and degradation variable X_p to prefer MEC PEs. In summary, PLUD-DPF can provide better performance than for instance PLUD in scenarios with short requests. However, the setup has to be configured carefully. For longer-running requests, LUD, PLU or particularly PLUD will be better and less complicated choices.

5   Conclusion and Future Work

Mobile devices are resource-constrained. Serverless computing offers the possibility to offload tasks into a public cloud platform. But cloud computing does not offer low latency. To counter the latency issue, another resource level abstraction has been created known as Mobile Edge Computing (MEC) that sits in proximity to its users. However, load distribution and the management of computing resource pools with different resource types is a challenging task. Our proposed approach is to reuse the lightweight, simple, yet powerful Reliable Server Pooling (RSerPool) framework to distribute the tasks onto the resource pool by satisfying four basic goals. It also supports multiple resource allocation policies, and we added three new ones: PLUD, PLU-DPF and PLUD-DPF. We have presented our simulated results and showed that three policies – LUD, PLU and particularly PLUD – are better for longer-running requests, while PLUD-DPF achieves a better handling speed for short requests. As future work, we plan to extend the analysis with real applications/benchmarks set up in a *true* geographically distributed scenario including OPENAIRINTERFACE-based EPC. We also intend to contribute our results into the IETF standardisation process of RSerPool, as well as the development of orchestration frameworks, particularly OPEN SOURCE MANO.

References

1. Lei, P., Ong, L., Tüxen, M., Dreibholz, T.: An overview of reliable server pooling protocols. Informational RFC 5351. IETF, September 2008
2. Dreibholz, T.: Reliable server pooling – evaluation, optimization and extension of a novel IETF architecture. Ph.D. thesis, University of Duisburg-Essen, Faculty of Economics, Institute for Computer Science and Business Information Systems, March 2007
3. Dreibholz, T., Mazumdar, S.: Reliable server pooling based workload offloading with mobile edge computing: a proof-of-concept. In: Proceedings of the 3rd International Workshop on Recent Advances for Multi-Clouds and Mobile Edge Computing (M2EC) in conjunction with the 35th International Conference on Advanced Information Networking and Applications (AINA), Toronto, Ontario, Canada, vol. 3, May 2021
4. Dreibholz, T., Rathgeb, E.P.: Overview and evaluation of the server redundancy and session failover mechanisms in the reliable server pooling framework. Int. J. Adv. Internet Technol. (IJAIT) **2**(1), 1–14 (2009)
5. Dreibholz, T., Mazumdar, S.: A demo of workload offloading in mobile edge computing using the reliable server pooling framework. In: Proceedings of the 46th IEEE Conference on Local Computer Networks (LCN), Edmonton, Alberta, Canada, October 2021
6. Zhou, X., Dreibholz, T., Rathgeb, E.P.: A new server selection strategy for reliable server pooling in widely distributed environments. In: Proceedings of the 2nd IEEE International Conference on Digital Society (ICDS), Sainte Luce, Martinique, February 2008
7. Dreibholz, T.: HiPerConTracer - a versatile tool for IP connectivity tracing in multipath setups. In: Proceedings of the 28th IEEE International Conference on Software, Telecommunications and Computer Networks (SoftCOM), Hvar, Dalmacija, Croatia, September 2020

A Survey on Advances in Vehicular Networks: Problems and Challenges of Architectures, Radio Technologies, Use Cases, Data Dissemination and Security

Ermioni Qafzezi[1]([✉]), Kevin Bylykbashi[2], Phudit Ampririt[1], Makoto Ikeda[2], Keita Matsuo[2], and Leonard Barolli[2]

[1] Graduate School of Engineering, Fukuoka Institute of Technology (FIT), 3-30-1 Wajiro-Higashi, Higashi-Ku, Fukuoka 811–0295, Japan
{bd20101,bd21201}@bene.fit.ac.jp
[2] Department of Information and Communication Engineering, Fukuoka Institute of Technology (FIT), 3-30-1 Wajiro-Higashi, Higashi-Ku, Fukuoka 811-0295, Japan
kevin@bene.fit.ac.jp, makoto.ikd@acm.org, {kt-matsuo,barolli}@fit.ac.jp

Abstract. Vehicular Ad hoc Networks (VANETs) have had a massive development over the last few years. The aim of this paper is to discuss different aspects of VANETs, their architecture, emerging applications, radio access technologies, information dissemination and security issues that they face. We give details of traditional and recent architectures and present an overview of use cases and applications brought to attention by various standardization bodies. In addition, we discuss the challenges of existing wireless technologies that support V2X communications and provide a comparison between short-range radio technologies and cellular networking technologies. Lastly, we present several surveys regarding data dissemination and security problems in VANETs.

1 Introduction

According to World Health Organization, around 1.3 million people die every year because of road traffic crashes [36]. Vehicular Ad hoc Networks (VANETs) promise to serve the needs of this fast-paced world by preventing road traffic deaths, improving traffic management and efficiency of driving. However, vehicular networks face unique challenges that come from the unique characteristics of vehicular environments. Large and dynamic topologies, variable capacity wireless links, bandwidth and hard delay constraints, and short contact durations are some of the characteristics of these networks. These challenges are caused by the high mobility and high speed of vehicles, and the frequent changes in density happening even in the same area.

In this regard, the research community is actively developing new technologies and methods to address all identified issues while trying to predict challenges that may come with the deployment of these technologies. This paper focuses

L. Barolli et al. (Eds.): AINA 2022, LNNS 451, pp. 602–613, 2022.
https://doi.org/10.1007/978-3-030-99619-2_56

on these technologies and discusses different concepts of vehicular networks, details the technical aspects of some existing and emerging network architectures, describes various applications enabled through the network communications applied in vehicles, and outlines aspects related to the security and privacy of such networks.

The remainder of this paper is as follows. Section 2 focuses on the distinction between several architectures of vehicular networks widely used by many researchers. Section 3 and Sect. 4 are dedicated to applications and radio access technologies, respectively. Information dissemination in these networks and users' security and privacy are covered in Sect. 5 and Sect. 6. The last section (Sect. 7) concludes the paper and gives some directions of future work.

2   Network Architectures

In the following, we describe the evolution of vehicular network architectures and some popular terms found in literature.

2.1   Vehicular Ad Hoc Networks

VANETs can be defined as networks of vehicles spontaneously created, able to connect vehicles with other vehicles in the network and with the infrastructure via Vehicle-to-Vehicle (V2V) and Vehicle-to-Infrastructure (V2I) communication links [18]. They are a subclass of Mobile Ad hoc Networks (MANETs), and as such, are based on inter-vehicle communications and do not rely on central coordination [19]. The vehicles behave as sensor nodes and relay the messages via one-hop or multi-hop communications. The infrastructure includes Road Side Units (RSUs), road signs, Electronic Toll Collection (ETC), and so on. These types of communications support many applications spanning from road safety and traffic optimization to rural and post-disaster scenarios connectivity.

2.2   Internet of Vehicles

The advances in vehicle manufacturing and communication technology, together with the full integration of Artificial Intelligence (AI) and cloud platforms in VANETs, are expected to enhance road safety and driving experience. Vehicles will be able to exchange information with many more entities, such as pedestrians, infrastructure, and networks, via Vehicle-to-Everything (V2X) communications [21]. With all these entities connected through vehicles and with many others being designed for the Internet of Things (IoT), the term ad hoc was considered obsolete by many researchers as it does not comprehensively cover the wide range of technologies involved within/connecting these entities [22–25]. The concept of the Internet of Vehicles (IoV) has emerged as a broader concept to better represent the new era of vehicular networks [26]. A typical VANET communication scenario is presented in Fig. 1.

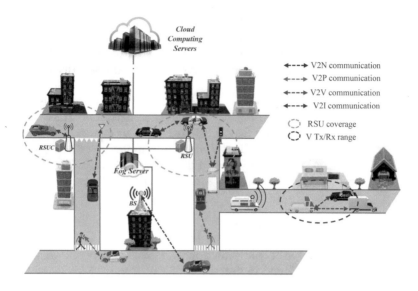

Fig. 1. Illustration of a typical IoV scenario.

2.3   Cellular V2X

The evolution of the 5th Generation of cellular networks (5G) marks a huge leap in the advance of VANETs. 5G base stations (5G-NR gNodeB) can serve as a gateway to the Internet and therefore enable big data storage, processing, and analyzing in the cloud infrastructure [20]. The leverage of the infrastructure of cellular networks enables the integrated operation between V2V, V2I, V2P (Vehicle-to-Pedestrian), and V2N (Vehicle-to-Network) with the network of vehicles [2,4,6]. Different from other technologies, cellular networks provide ubiquitous coverage for V2I and V2N services, support high mobility of vehicles as well as support a high number of vehicles within the same cell. Moreover, 5G supports high data rate applications. From the financial perspective, cellular based V2X Communications (C-V2X) reduce the cost by utilizing the already placed infrastructure of cellular networks. C-V2X communications provide support for enhanced safety and traffic management applications of transportation systems. However, the main challenge of C-V2X is latency. Due to their centralized nature, the data need to go through a base station, which can add latency in communication. The time is critical, especially for safety applications. Another challenge is that of data dissemination to a particular group of vehicles in need for that information out of all the vehicles inside the cell [19].

3   Applications

In this section, we survey the work of different standardization bodies, including 3rd Generation Partnership Project (3GPP) [2,4,5], ITU [24], European Telecommunications Standards Institute (ETSI) [13], and other launched projects of The 5G Infrastructure Public-Private Partnership (5GPPP) such as Fifth Generation Communication Automotive Research and innovation (5GCAR) [17] and 5G for Connected and Automated Road Mobility in the European UnioN (5GCARMEN) [8] to make an up-to-date list of the applications and use cases that have emerged over the years. The applications are classified into three main use case groups: 1) Road safety and advanced vehicle control, 2) Advanced traffic management, and 3) Comfort and infotainment services. Each of these groups is presented in Table 1, Table 2 and Table 3, respectively.

Road safety and advanced vehicle control applications purpose is to reduce the risk of driving and save the lives of road users either by directly/indirectly preventing collisions or by mitigating the severity of the accident, if for any reason that accident cannot be avoided.

The main objective of advanced traffic management applications is to optimize traffic flow and provide more efficacy for the road systems, for the vehicles, pedestrians, and commuters using public transportation. However, safety and traffic management are correlated to each other in many ways. For instance, traffic congestion leads to higher risk for car crashes, as drivers are prone to drive faster in order to compensate the delay caused by traffic.

The purpose of comfort and infotainment services is to enrich the travel experience by providing on-demand information to the drivers and passengers. Drivers can be more comfortable when their requested information is provided in real-time.

4   Radio Access Technologies

Every application has its own QoS requirements to be satisfied in order to run smoothly and serve its purpose. The exponential growth of the proposed applications has proportionally increased the complexity of meeting their QoS demands due to the limited allocated radio spectrum. Various radio access technologies have been proposed over the past years to exploit every single slot of the allocated radio spectrum, and more are under development. The radio access technologies implemented in vehicular networks include WiFi, DSRC and cellular networks.

4.1   Short-Range Radio Technologies

Short-Radio Radio Technologies (DSRC/WAVE) are defined in IEEE 802.11p [23] as wireless communication technologies used within a small diameter region. Traditionally, they have been used for low-mobility scenarios, e.g., Wifi, Bluetooth and ZigBee. However, since DSRC was designed by the U.S. Department of Transportation for vehicular networks, they support data communication between high-speed vehicles and between the vehicles and the roadside infrastructure [31]. Such applications included toll collections, commerce transactions via cars and vehicle safety services. From technical perspective, DSRC has many limitations, such as limited range of communication, congestion due to broadcasted messages, unreliable delivery of messages and intermittent connectivity

Table 1. Road safety and advanced vehicle control use cases.

No.	Use case	Source
1	Vehicle Platooning	3GPP/ETSI
2	Automated cooperative driving for short distance grouping	3GPP
3	Cooperative collision avoidance of connected automated vehicles	3GPP
	Longitudinal collision avoidance	ITU-R/ETSI
4	Changing driving-mode	3GPP
5	Emergency trajectory alignment	3GPP
6	Cooperative lane change of automated vehicles/Lane merge	3GPP/5GCAR
7	Road safety services/V2X Road safety service via infrastructure	3GPP
8	Curve Speed Warning	3GPP
9	Enhancing Positional Precision for traffic participants	3GPP
10	Teleoperated support	3GPP
11	eV2X Remote driving	3GPP
	Remote driving for automated parking	5GCAR
12	V2V Emergency Stop/V2I Emergency Stop	3GPP
	Stationary vehicle - vehicle problem	ETSI
13	Emergency notification and personal security/SOS service	ITU-R/ETSI
14	Forward Collision Warning	3GPP
15	Intersection safety information provisioning for urban driving	3GPP
	Intersection collision avoidance	ITU-R/ETSI
16	Collective perception of environment	3GPP
17	Cooperative Adaptive Cruise Control	3GPP/ETSI
18	Video data sharing for assisted and improved automated driving	3GPP
	See-through	5GCAR
	Vision Enhancement Systems	ITU-R
19	V2V Emergency vehicle warning	3GPP
	Back-situation awareness of an emergency vehicle arrival	5GCARMEN
	Emergency vehicle management	ITU-R/ETSI
20	Wrong way driving warning	3GPP/ETSI
21	Pre-crash Sensing Warning/Pre-crash restraint deployment	3GPP/ITU-R
22	Control Loss Warning	3GPP
23	Warning to Pedestrian against Pedestrian Collision	3GPP
24	Vulnerable Road User Safety	3GPP
25	Pedestrian Road Safety via V2P awareness messages	3GPP
	Network assisted vulnerable pedestrian protection	5GCAR
	Vehicle-pedestrian accident avoidance	ITU-R
26	Motorbike awareness	5GCARMEN
	Motorcycle approaching indication	ETSI
27	Event horizon	5GCARMEN
28	Lateral collision avoidance	ITU-R
29	Mixed Use Traffic Management	3GPP
	Vehicle sensors and state sharing	5GCARMEN
	Safety readiness	ITU-R
	Decentralized floating car data	ETSI
30	Public travel security	ITU-R
31	Slow vehicle indication	ETSI
32	Signal violation warning	ETSI
33	Roadwork warning	ETSI

[9,20]. As a result, this technology saw its end in the United States as FCC decided to split the 75 MHz of DSRC spectrum (5.850–5.925 GHz), allocating the lower 45 MHz of the band for Wi-Fi and other unlicensed uses and the upper 30 MHz for ITS that must use C-V2X technology [15]. Europe, on the other hand,

Table 2. Advanced traffic management use cases.

No.	Use case	Source
1	Sensor and state map sharing/Map download and update definition local map acquisition/Route guidance	3GPP/ETSI 5GCAR/ITU-R
2	Queue Warning	3GPP
3	V2N Traffic Flow Optimization Traffic network monitoring and control	3GPP ITU-R
4	3D video composition for V2X scenario	3GPP
5	Green Driving	5GCARMEN
6	Travel demand management	ITU-R
7	Incident detection and management	ITU-R
8	Emissions testing and mitigation/Environment analysis	ITU-R/5GCARMEN
9	Electric zones	5GCARMEN
10	Dynamic Speed Limit	5GCARMEN
11	Pre-trip travel information	ITU-R
12	En-route transit information	ITU-R
13	Dynamic ride sharing/Ride matching and reservation Car rental/sharing assignment/reporting	3GPP/ITU-R ETSI
14	Public transportation management	ITU-R
15	Pedestrians route guidance	ITU-R
16	Regulatory/contextual speed limits notification	ETSI
17	Traffic light optimal speed advisory	ETSI
18	Limited access warning, detour notification	ETSI
19	In-vehicle signage	ETSI
20	Vehicle and RSU data calibration	ETSI

did not abolish this technology. Instead, they adopted a technology-neutral approach that supports both Cooperative ITS (C-ITS) and C-V2X [11].

4.2 Cellular Networks

Although the use of cellular networks infrastructure for vehicular networks was proposed since when 3G was becoming a reality, it is just recently that cellular networks have come into play in enabling present and future vehicular networking applications. The shift started with LTE, and now with 5G is closer than ever. This technology is named C-V2X and is specified in the 3GPP Release 14 [18] for LTE, whereas the support of 5G and the specification of the service requirements are specified in the 3GPP Release 15 [3].

Due to many advantages it offers, C-V2X meets the requirements of use cases of today and tomorrow. Based on 3GPP Release 14 specification [18] this technology has been introduced as a substitute of 802.11p (DSRC) as it has shown a superior performance in terms of coverage, delay, mobility support, scalability and reliability [7]. While both technologies can co-exist, C-V2X covers all services in an end-to-end manner offered by 802.11p. So, instead of using hybrid "solution" that incorporates IEEE 802.11p-based technology, it is better

Table 3. Comfort and infotainment services use cases.

No.	Use case	Source
1	Automated Parking System	3GPP
	Parking management	ITU-R/ETSI
2	Video streaming	5GCARMEN
3	En-route driver information	ITU-R
4	Personalized public transportation	ITU-R
5	Vehicle administration	ITU-R
6	Fleet management	ITU-R/ETSI
7	Vehicle preclearance	ITU-R
8	Automated roadside safety inspections	ITU-R
9	Electronic payment services/Electronic toll collect	ITU-R/ETSI
10	Point of Interest notification	ETSI
11	Insurance and financial services	ETSI
12	Media downloading	ETSI
13	Remote diagnosis and just in time repair notification	3GPP/ETSI
14	Vehicle software/data provisioning and update	ETSI
15	Stolen vehicle alert	ETSI
16	Tethering via Vehicle	3GPP
17	Personal data synchronization	ETSI

to use the one that offers better performance and supports adaptive evolution. C-V2X short-range communications is known as the direct mode or Sidelink and uses the PC5 interface, whereas C-V2X long-range communications is known as network mode or UP/Downlink and is implemented over the Uu interface for LTE and 5G NR Uu URLLC. Both satisfy stringent latency and reliability requirements [5].

5   Information Dissemination

Information dissemination has crucial role in effective communication in vehicular networks. It is important to develop an efficient and reliable data distribution scheme, that will successfully deliver the messages to the appropriate vehicles. Data dissemination is intended to transmit data over dispersed networks and serve the needs of all vehicles through various methods of dissemination, depending on the application requirements. For example, for safety messages it is important to deliver them in real-time to all vehicles in the vicinity of the possible danger, whereas for traffic management messages is more important to send them to a group of vehicles, instead of one specific vehicle.

Many communication methods and protocols are proposed over the years, grouped mainly according to the strategy of transmitting the information to the desired destination. They can be categorized as follows.

- The unicast strategy refers to point-to-point communications, which means that data is transmitted from a single source to a single destination node [16].

This strategy is achieved through a hop-by-hop greedy forwarding mechanism that relays the information immediately or a carry-and-forward mechanism that stores the data in case of a lack of continuous connectivity and forwards it when a decision is made.

- The multicast/geocast strategy is used to deliver data from a single node to a set of nodes that lie within a specific geographical region, also known as zone of relevance [10, 25, 29]. Geocast is considered among the most feasible approaches for safety-related applications in vehicular networks since it can inform all nodes traveling close to the event location. Beacon messages with Cooperative Awareness Message (CAM) [14] and Decentralized Environmental Notification Message (DENM) [1] are typical examples. The former provides periodical information of a vehicle to its neighbors about its presence, position, speed, etc. whereas the latter is an event-triggered message delivered to alert road users of a hazardous event.

- Broadcast strategy disseminates information to all vehicles in the network without exception [12, 35]. This approach can be used for data sharing, weather information, road condition, traffic, entertainment, and advertisement announcements.

6 Security, Privacy and Trust

The unique characteristics of vehicular networks make the development and implementation of security, privacy, and trust management solutions a challenging issue. Secure communication between all involved entities in vehicular networks is of the utmost importance since any successful attack from a malicious hacker could result in a traffic accident [21, 34]. Therefore, a good security framework must consider the basic security requirements in terms of authentication, availability, confidentiality, integrity, non-repudiation and protect these services from different threats and attacks. Some of the most hazardous attacks in vehicular networks are Denial of Service, Distributed Denial of Service, Sybil attack, Black and Gray Hole, Wormhole, Injection of erroneous messages, Replay, and Eavesdropping.

Only dedicated individuals should have the right to access and control the vehicle information and one of the most common methods to protect the privacy of individuals is anonymity [30, 32]. Anonymity is defined as the state of being not identifiable within a set of subjects, which can be provided by employing an authentication scheme. Anonymous authentication schemes can be identified as: public-key infrastructure, symmetric cryptography, identity-based, group, and certificateless signatures.

Trust management deals with the trustworthiness of the received information and the trustworthiness of the vehicles that have sent that information [22, 28]. There are three popular trust management models: entity-centric, data-centric, and combined trust models. In order to establish an effective trust management model, some properties need to be considered, such as decentralization, real-time constraints, information sparsity, scalability, and robustness. Looking at the growth and feasibility of blockchain technology, perhaps vehicular networks are not that far from establishing the long-awaited trust model that addresses those existing challenges [37]. In Table 4, we present the work of different authors on this topic.

Table 4. Recent surveys related to security, privacy, and trust.

Survey	Aspect(s)	Contribution	Year
[32]	Privacy	Comprehensive overview of existing privacy-preserving schemes	2015
[26]	Trust	An adversary-oriented survey of different trust models and their evaluation against cryptography	2016
[21]	Security	Overview of security challenges and requirements, and a novel classification of different attacks along with their corresponding solutions	2017
[33]	Security	A review of services, challenges and security threats evolving from the software-defined approach	2018
[28]	Security, Privacy, Trust	A Survey covering well-studied topics of security, authentication schemes, location privacy protection mechanisms, and existing trust managements models	2019
[34]	Security	Deep analysis of various security aspects such as requirements, challenges, and attacks together with the evaluation of their respective solution	2019
[30]	Security, Privacy	A comprehensive review of various research works that address privacy, authentication and secure data dissemination	2020
[27]	Security, Privacy	Introduces security and privacy aspects of C-V2X, and discusses open security challenges and issues	2020
[22]	Trust	A review, analysis, and comparison of current trust establishment and management solutions, and a discussion of future opportunities	2021

7 Conclusions

In this paper, we presented a review of vehicular networks. We introduced a brief overview of the evolution of vehicular network architectures from VANETs to IoV. We gave a detailed overview of use cases supported by V2X communications and the advancement of LTE and 5G technology. The uses cases include work of different standardization bodies, such as 3GPP, ITU, ETSI and 5GPPP. We compared and outlined the key features of different radio access technologies implemented in VANETs. We reviewed various methods of data dissemination and their appropriate application. We also explained challenges that vehicular network have in terms of security, privacy and trust, and review the work of different authors in this field. Despite the latest advancements, there is still a lot of work to be done regarding resource management, resource allocation, handover issues, service migration, QoS provisioning and priority application selection.

References

1. European Telecommunications Standards Institute (ETSI): Intelligent Transport Systems (ITS); Vehicular Communications; Basic Set of Applications; Part 3: Specifications of Decentralized Environmental Notification Basic Service. Technical report, vol. 102, pp. 633–637 (2010)
2. 3rd Generation Partnership Project (3GPP): Technical specification group services and system aspects; study on LTE support of vehicle to everything (V2X) Services (Release 14). Technical report 22.885 (2015). V14.0.0
3. 3rd Generation Partnership Project (3GPP): Technical specification group services and system aspects; Enhancement of 3GPP Support for V2X Scenarios; Stage 1 (Release 15). Technical report 22.186 (2018). V15.4.0
4. 3rd Generation Partnership Project (3GPP): Technical specification group services and system aspects; Study on Enhancement of 3GPP Support for 5G V2X Services (Release 15). Technical Report 22.886 (2018). V15.3.0
5. 3rd Generation Partnership Project (3GPP): Technical specification group services and system aspects; Architecture Enhancements for 5G System (5GS) to Support Vehicle-to-Everything (V2X) Services (Release 16). Technical Report 23.287 (2020). V16.5.0
6. 5G Automotive Association (5GAA): Explore the technology: C-V2X. https://5gaa.org/5g-technology/c-v2x/. Accessed 15 May 2021
7. 5G automotive association (5GAA): C-V2X use cases methodology, examples and service level requiremets. White Paper FHWA-HRT-18-027 (2019)
8. 5G for connected and automated road mobility in the European UnioN (5GCARMEN): 5G CARMEN use cases and requirements. Deliverable d2.1, The 5G infrastructure public private partnership (5GPPP) (2019)
9. Abboud, K., Omar, H.A., Zhuang, W.: Interworking of DSRC and cellular network technologies for v2x communications: a survey. IEEE Trans. Veh. Technol. **65**(12), 9457–9470 (2016). https://doi.org/10.1109/TVT.2016.2591558
10. Bachir, A., Benslimane, A.: A multicast protocol in ad hoc networks inter-vehicle geocast. In: The 57th IEEE Semiannual Vehicular Technology Conference, 2003. VTC 2003-Spring., vol. 4, pp. 2456–2460 (2003). https://doi.org/10.1109/VETECS.2003.1208832
11. C-ITS Deployment Group: C-ITS deployment takes off, increasing road safety and decreasing congestion. https://c-its-deployment-group.eu/mission/statements/december-2019-constitutive-act-of-c-its-deployment-group/. Accessed 15 May 2021
12. Durresi, M., Durresi, A., Barolli, L.: Emergency broadcast protocol for inter-vehicle communications. In: 11th International Conference on Parallel and Distributed Systems (ICPADS 2005), vol. 2, pp. 402–406 (2005). https://doi.org/10.1109/ICPADS.2005.147
13. European telecommunications standards institute (ETSI): Intelligent Transport Systems (ITS); Vehicular Communications; Basic Set of Applications; Definitions. Technical report, vol. 102, p. 638 (2009). V1.1.1
14. European telecommunications standards institute (ETSI): Intelligent Transport Systems (ITS); Vehicular communications; basic set of applications; Part 2: Specification of cooperative awareness basic service. Technical report 102 637-2 (2011)
15. Federal Communications Commission (FCC): FCC 20–164. FCC Report and Order on Use of the 5.850-5.925 GHz Band. Report and order (2020)

16. Ferreiro-Lage, J.A., Gestoso, C.P., Rubiños, O., Agelet, F.A.: Analysis of unicast routing protocols for vanets. In: 2009 Fifth International Conference on Networking and Services, pp. 518–521 (2009). https://doi.org/10.1109/ICNS.2009.96
17. Fifth Generation Communication Automotive Research & innovation (5GCAR): 5GCAR Scenarios, Use Cases, Requirements and KPIs. Deliverable D2.1 V2.0, The 5G Infrastructure Public Private Partnership (5GPPP) (2019)
18. 3rd Generation Partnership Project (3GPP): Technical specification group radio access network; Study on LTE-Based V2X Services (Release 14). Technical report 36.885 (2016). V14.0.0
19. Gyawali, S., Xu, S., Qian, Y., Hu, R.Q.: Challenges and solutions for cellular based v2x communications. IEEE Commun. Surv. Tutorials **23**(1), 222–255 (2021). https://doi.org/10.1109/COMST.2020.3029723
20. Hameed Mir, Z., Filali, F.: LTE and IEEE 802.11p for vehicular networking: a performance evaluation. EURASIP J. Wirel. Commun. Netw. **2014**(1), 1–15 (2014). https://doi.org/10.1186/1687-1499-2014-89
21. Hasrouny, H., Samhat, A.E., Bassil, C., Laouiti, A.: VANET security challenges and solutions: a survey. Veh. Commun. **7**, 7–20 (2017). https://doi.org/10.1016/j.vehcom.2017.01.002
22. Hussain, R., Lee, J., Zeadally, S.: Trust in VANET: a survey of current solutions and future research opportunities. IEEE Trans. Intell. Transp. Syst. **22**(5), 2553–2571 (2021). https://doi.org/10.1109/TITS.2020.2973715
23. IEEE Computer Society: 802.11p-2010 - IEEE standard for information technology- local and metropolitan area networks- specific requirements- Part 11: Wireless LAN Medium Access Control (MAC) and Physical Layer (PHY) Specifications Amendment 6: Wireless Access in Vehicular Environments (2010). https://doi.org/10.1109/IEEESTD.2010.5514475
24. International Telecommunication Union. Rec. ITU-R M.1890-1: Operational radio-communication objectives and requirements for advanced Intelligent Transport Systems (2019). Mobile, radiodetermination, amateur and related satellite services (M Series)
25. Kaiwartya, O., Kumar, S., Kasana, R.: Traffic light based time stable geocast (T-TSG) routing for urban vanets. In: 2013 Sixth International Conference on Contemporary Computing (IC3), pp. 113–117 (2013). https://doi.org/10.1109/IC3.2013.6612173
26. Kerrache, C.A., Calafate, C.T., Cano, J.C., Lagraa, N., Manzoni, P.: Trust management for vehicular networks: an adversary-oriented overview. IEEE Access **4**, 9293–9307 (2016). https://doi.org/10.1109/ACCESS.2016.2645452
27. Lai, C., Lu, R., Zheng, D., Shen, X.: Security and privacy challenges in 5G-enabled vehicular networks. IEEE Netw. **34**(2), 37–45 (2020). https://doi.org/10.1109/MNET.001.1900220
28. Lu, Z., Qu, G., Liu, Z.: A survey on recent advances in vehicular network security, trust, and privacy. IEEE Trans. Intell. Transp. Syst. **20**(2), 760–776 (2019). https://doi.org/10.1109/TITS.2018.2818888
29. Maihofer, C., Eberhardt, R.: Geocast in vehicular environments: caching and transmission range control for improved efficiency. In: IEEE Intelligent Vehicles Symposium, 2004, pp. 951–956 (2004). https://doi.org/10.1109/IVS.2004.1336514
30. Manivannan, D., Moni, S.S., Zeadally, S.: Secure authentication and privacy-preserving techniques in vehicular Ad-hoc NETworks (VANETs). Veh. Commun. **25**, 100–247 (2020). https://doi.org/10.1016/j.vehcom.2020.100247

31. Naik, G., Choudhury, B., Park, J.M.: IEEE 802.11bd amp; 5g NR v2x: evolution of radio access technologies for v2x communications. IEEE Access **7**, 70169–70184 (2019). https://doi.org/10.1109/ACCESS.2019.2919489

32. Petit, J., Schaub, F., Feiri, M., Kargl, F.: Pseudonym schemes in vehicular networks: a survey. IEEE Commun. Surv. Tutorials **17**(1), 228–255 (2015). https://doi.org/10.1109/COMST.2014.2345420

33. Shafiq, H., Rehman, R.A., Kim, B.S.: Services and security threats in SDN based VANETs: a survey. Wireless Commun. Mobile Comput. 2018(8631), 851 (2018). https://doi.org/10.1155/2018/8631851

34. Sharma, S., Kaushik, B.: A survey on internet of vehicles: applications, security issues & solutions. Veh. Commun. **20**, 100–182 (2019). https://doi.org/10.1016/j.vehcom.2019.100182

35. Tonguz, O., Wisitpongphan, N., Bai, F., Mudalige, P., Sadekar, V.: Broadcasting in VANET. In: 2007 Mobile Networking for Vehicular Environments, pp. 7–12 (2007). https://doi.org/10.1109/MOVE.2007.4300825

36. World Health Organization (WHO): Global status report on road safety 2018: summary. Geneva, Switzerland (2018). (WHO/NMH/NVI/18.20). Licence: CC BY-NC-SA 3.0 IGO)

37. Yang, Z., Yang, K., Lei, L., Zheng, K., Leung, V.C.M.: Blockchain-based decentralized trust management in vehicular networks. IEEE Internet Things J. **6**(2), 1495–1505 (2019). https://doi.org/10.1109/JIOT.2018.2836144

Intelligent Blockchain-Enabled Applications for Sharing Economy

Alkhansaa A. Abuhashim[1,2(✉)]

[1] Computer and Information Science, Temple University, Philadelphia, USA
`a.a.alkhansaa@gmail.com`
[2] Information Systems, Al-Imam Muhammad Bin Saud University,
Riyadh, Kingdom of Saudi Arabia

Abstract. Sharing economy is one of the fastest-growing businesses in the last century. Therefore, it is a major benefit to utilize blockchain technology to enhance sharing economy applications' transparency, security, and decentralization. We propose a framework for data management to support blockchain-based sharing economy applications. It supports indexing and querying blockchain data to facilitate the retrieval process instead of the conventional sequential scanning of the blockchain ledger. Furthermore, we implement an on-demand service for indexing future requests.

Keywords: Blockchain · Sharing economy · Smart contract · Ledger · Indexing · Querying · Cache

1 Introduction

Sharing economy is one of the significant business models nowadays for individuals and investors. It connects users, businesses, and companies worldwide in innovative ways to provide/receive goods and services. By the sharing economy, individuals exchange goods/services by renting or buying items such as homes, cars, or services. Airbnb and Uber, the most popular platforms, allow service providers and customers to connect to the advantage of each other in a peer-to-peer manner [1].

Nowadays, blockchain has significantly impacted different sectors such as the healthcare sector [2], Internet of Things (IoT) [3,4], tracking and authenticity fields (e.g., supply chain applications [5,6], tracking and tracing systems [7–9], certifications [10,11]), and many other sectors. The decentralized feature of the blockchain has solved the limitations of the third-party authentication processes by reducing the time and speed of transactions through peer-to-peer communications [9]. Furthermore, blockchain features ensure transparency, enhance security, and provide traceability of goods/services for sharing economy applications.

Based on what has been said, blockchain technology is an effective technology to support security and privacy concerns for millions of connected Internet of

Things (IoT) devices in a sharing economy environment [12,13]. It has decent features of immutability, security, and privacy that have been used in many sectors, including finance, healthcare, computer science, law, and many others. Besides, the technology has no central authority controlling the chain; however, the distributed virtual machine nodes participate in accessing and updating the current state of the blockchain network.

IoT devices have numerous constraints regarding their storage and communication capabilities, while blockchain needs high resources for mining and storing processes. Based on the previous centralized environment scenarios, blockchain would not have the retrieving abilities as the central system because the blockchain does not provide the concept of relations/tables. Therefore, querying the ledger to retrieve a transaction record is never an efficient operation on the blockchain database since the process requires scanning the blocks in the blockchain database to find the needed data. Additionally, different types of queries, such as complex and time-based queries, would add a layer of inefficiency and complexity. On the other hand, the limitations of IoT devices in sharing economy environments add another layer of complexity to the blockchain platforms and vice versa. However, the existing solutions to support the sharing economy are not mature yet. Our proposed solution will fill the gap in this field by providing a secure, transparent, and efficient data management model that utilizes blockchain technology.

2   BCQuery Framework

BCQuery is a blockchain-based framework for sharing economy applications to facilitate the operations of querying and retrieving historical and future data from the blockchain. The following subsections depict the framework architecture with three tiers for indexing, storing, and retrieving the data.

2.1   Architecture Overview

The framework answers historical queries in which a user enquires blockchain data that happened previously and has been mined with a past timestamp. A graph and relational databases are utilized to index blocks and transactions data in an application-specific way to retrieve and answer historical queries efficiently. Figure 1 shows the framework's tiers to prepare and index sharing economy related data to be accessed whenever needed.

Any mined block has multiple transactions with different purposes. For sharing economy applications, the transactions are for registering a new agreement between a customer and an owner (e.g., rental session) or updating user data such as IoT data. The framework's tiers index these transactions and their related data as the following: (1) Blockchain and Smart Contracts tier, (2) Indexes and database tier, and (3) Data management Tier.

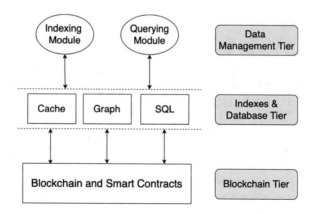

Fig. 1. The proposed BCQuery architecture

2.2 Blockchain and Smart Contracts Tier

Generating smart contracts and updating user data are the ***main operations*** in this tier, as the following:

- *Generating smart contracts* are initiated by a user to either (1) start a new agreement with another party or (2) to register a new asset for the owner. For instance, initiating a smart contract for a rental session between an owner and a customer for a unit rental and, for the other case, initializing a smart contract to register a new IoT device in the owner's rental home. To initiate an agreement, i.e., a rental session, the owner generates an instance of her/his *Rental Session Contract* by passing agreement information to the smart contract. To register a new device, the owner sends device specifications as attributes to *Registration Contract* to record the device in the blockchain.
- *Updating user data* is practically attained by triggering a function of the user's agreement smart contract to update, for example, assets usage data. Specifically, calling a function in the *rental session contract* to update IoT data in the owner rental home. The generated smart contract for the rental session is a controller to maintain house assets usage and the payment required. It stays active until the end of the session.

2.3 Indexes and Databases Tier

The framework has different indexer storage units where blockchain data is stored for indexing. The tier has *Graph indexes* which primarily index relationships of blockchain data and *SQL Relations* to organize blocks and transactions data to be easily retrieved.

- *Graph Nodes* index the relationships between the following: (1) Block ID, (2) Tx ID, (3) Log ID, and addresses (for miner, sender, receiver, smart contract, and IoT devices).

– *SQL Database* has relations for the following application and blockchain data. **Note.** the relations scheme matches the sharing economy application needs and design: (1) block data (block#, timestamp, ..), (2) Contract data (created tx_{hash}, created time, ..), (3) External Owner Account (EOA) data, (4) transactional data (normal transactions, ERC20 transactions, ERC721 transactions), (5) Log/Event data, and (6) Iot devices information.

Graph Indexing. We adopt a graph database to index the addresses and hashes for users' accounts (owners and customers), smart contracts, blocks, and transactions for the application retrieval. When transactions and smart contracts are generated, they are connected to their generators (users/smart contracts addresses). Therefore, indexing these addresses as relationships between users, smart contracts and/or IoT devices facilitate retrieving what user would like to access from the blockchain, especially if the query relates to querying a pack of transactions for a particular user or an IoT device. The increasing number of relationships increases the number and the complexity of join operations which degrades the performance of inserting and retrieving. In this case, relational databases do not have the property of representing relationships between blockchain hashes and addresses. Therefore, indexing these data by the graph database would assist the retrieving process for related information of blockchain addresses and hashes.

SQL Relations. On the other hand, some queries require access to a particular address or a hash category (e.g., account address) for specific conditions. For example, a time-based query q_x to retrieve all renters who rent a home h owned by owner O in the last two months. In this scenario, retrieving the query response will be more efficient if the addresses are categorized into tables (owner, customer, and IoT device). These types of queries will not be served efficiently by the graph indexing that maintains the relationship between these data without categorizing them to perform range or time queries. SQL relations let the query middleware decide where to retrieve based on query conditions instead of retrieving all the transactions connected to home h without filtering the transactions based on the query condition. Therefore, the processing time for executing a query will consume unnecessary retrieving and comparison operations without categorizing the addresses.

2.4   Data Management Tier

This tier manages data after it's mined as a block in addition to the historical blockchain data. It has two modules for data indexing and querying the blockchain.

    a. Indexing Module. In this module, blockchain data (blocks info, transactions, logs, and user data) are indexed in a graph and a relational database to

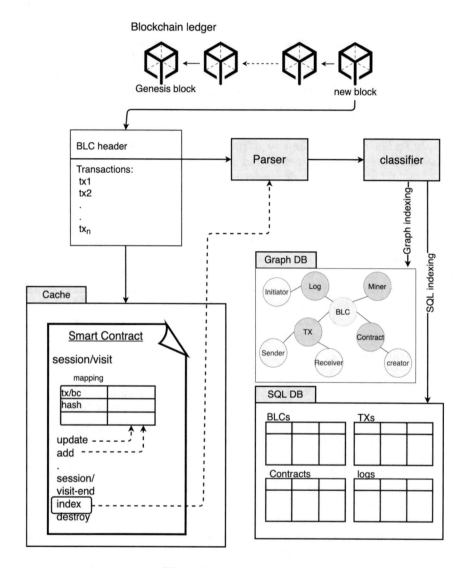

Fig. 2. Indexing framework

facilitate querying the blockchain, as described in Sect. 2.3. The indexing process is handled through a parser and classifier units of the module to process the incoming mined block data, shown in Fig. 2.

The parser unit is to parse the coming mined block to extract its components. It extracts the block header and recognizes addresses and transactions. Then, the classifier unit classifies the data of the bulk of transactions in the coming block. It categorizes every transaction and related data if it is a regular transaction or a deployed contract. After parsing and classifying the mined block, the index unit indexes the addresses and relationships in the graph database. In contrast, transactions, contracts, logs, and their related fields are indexed into the SQL database.

Different smart contract designs can be adapted to support blockchain applications in organizing applications' data and transactions [9]. However, it does not provide the operations for retrieving blockchain-related data and some transactional information from the ledger. Therefore, we design our framework to index mined blocks and transactions to be retrieved for different application purposes.

b. Querying Module. The indexing module stores blockchain data to be efficiently retrieved. After the data is inserted into the blockchain and then blockchain blocks are indexed, the user can retrieve the blockchain inquiries through Query Algorithm 1. Querying the data that has been indexed will be efficiently retrieved by the querying module since the time consumed will be direct access to the indexes that match the user's conditions. In contrast, the querying performance of the conventional sequential search of the blockchain ledger is exponentially increased when the length of the required scanned blocks increases [9].

The query algorithm (Algorithm 1) is applied to the indexes and database tier to extract the previously indexed blockchain data. For example, if the received query is about the user's current session/visit, the cache smart contract for on-demand query services answers the query, see Sect. 3. However, for complex queries, the algorithm applies the retrieving processes for each part of the complex condition ($C1$, $C2$, ..) on the corresponding indexer that answers each condition C_i.

3   On-Demand Query Services

Indexing and querying processes allow users to query the historical data from the blockchain; however, users sometimes need to query future information with certain conditions.

3.1   Cache Smart Contract

The previous indexing and querying framework retrieves users' past information stored in the blockchain. This service allows the user to query future data that might happen in the future based on some criteria. The On-Demand query is popular mainly in economic environments, especially if the condition relates to goods/services prices.

For example, a customer may want to get notifications for particular houses if their rental prices are decreased below a specific price. A temporary storage indexer, named Query On-Demand Service, is adopted to maintain users' data for future demand by caching smart contracts for the inquired assets in the temporary indexer. Specific-purpose smart contracts are used to cache the required services for users' future queries. Algorithm 1 shows the technical steps of querying blockchain data by integrating the indexing framework.

Algorithm 1: Query Algorithm

Data: Cache - Smart Contracts for current sessions (*sessionIndexer*, Fig. 3), G - Graph Indexing Database, SQL - SQL Indexing Database, Q - User Query

Result: R - Query Result

if Q *is about current session(s)* **then**
$\quad$ Apply Q on Cache;
$\quad$ $R \leftarrow Cache - Result$;
else if Q *is complex query* **then**
$\quad$ $C_1, C_2, .., C_n \leftarrow Q - ComplexCondition$;
$\quad$ **else**
$\quad\quad$ $C_1 \leftarrow Q - SimpleCondition$;
$\quad\quad$ $n \leftarrow 1$;

$\quad$ **for** $i \leftarrow 1$ *to* n **do**
$\quad\quad$ **if** C_i *is about current session(s)* **then**
$\quad\quad\quad$ Apply Q_i on Cache;
$\quad\quad\quad$ $R_i \leftarrow Cache - Result$;
$\quad\quad$ **else if** C_i *is relationship condition* **then**
$\quad\quad\quad$ Apply Q_i on G;
$\quad\quad\quad$ $R_i \leftarrow G - Result$;
$\quad\quad$ **else**
$\quad\quad\quad$ Apply Q_i on SQL;
$\quad\quad\quad$ $R_i \leftarrow SQL - Result$;
$\quad\quad$ $R+ \leftarrow R_i$;

return R;

3.2 Cache Implementation

Cache smart contract is implemented using Solidity language on the Rinkeby test network 3. It is deployed using Metamask [14] the gateway to Ethereum nodes by Infura remote provider. It includes an indexer smart contract which indexes a list of sessions inner smart contracts. A session smart contract is initialized when a customer, for example, asks an owner to be notified when a unit reaches a specific price. The required information would be registered through the session indexer smart contract to keep a record for all deployed sessions smart contracts by the owner. The owner can deploy different indexer smart contracts to match other asset criteria such as renting price, asset size, or renting time.

4 Related Work

Since blockchain ledger is growing fast and there is a lack of data management technologies for the ledger with the current state of the art, for this reason, querying blockchain ledger is inefficient, expensive, and time-consuming. The

```
contract sessionSC{
  uint public session_id;
  address public user_address;
  uint public requested_price;
  uint public current_price;

  constructor (uint s_id, address u_add, uint req_price, ..)
  {
    session_id = s_id;
    user_address = u_add;
    requested_price = req_price;
    current_price = 0;
    ...
  }

  function update_price(uint memory price) public
  {
    current_price = price;
    if (current_price <= requested_price) {
      notify(user_address);
      selfdestruct(msg.sender);
    }
    ...
  }
  function notify(address u_add) { ...}
}

contract sessionsIndexer{
  // session_id => sessionSC
  mapping (bytes32 => sessionSC) public session_list;

  function addSession (uint s_id, address u_add, uint req_price, ..)
          public  {
      session[country_c] = new sessionSC(s_id, u_add, req_price, ..);
      ...
  }

  function getSessionInfo (uint session_id) external view returns
          (address u_add, uint req_price, uint current_price) { ... }
  ...
}
```

Fig. 3. sessionIndexer Smart Contract is an index of a list of sessionSC smart contracts.

current state-of-the-art for querying the blockchain is not yet mature. For example, blockchain explorers such as Etherscan [15], the most popular blockchain explorers for Ethereum network, provide a web service with basic search functionalities to search blockchain data through REST API. By Etherscan, users can only explore and search for transactions, addresses, and blocks but not for more advanced queries such as range and join queries.

Other studies provide data tools or APIs to allow users to obtain blockchain data. For instance, Google BigQuery [16] imports Bitcoin and Ethereum data for online analysis to enable researchers to apply SQL queries on massive datasets. Moreover, EtherQL [17] proposes a query layer for retrieving blockchain data that includes range and Top-k queries for data analysis and research. However, most studies are designed for specific applications, which do not apply to the sharing economy applications, and others fail to provide well-processed up-to-date blockchain data. Our proposed framework presents a tiered architecture to support data management, indexing, and querying blockchain transactions for sharing economy applications.

5 Future Work

Our ongoing work evaluates our framework through Ethereum blockchain and smart contracts to measure the framework's indexing time consumption, storage usage, and query response time. xblock [18] datasets will be used that contain six datasets for blocks, normal and internal transactions, contracts, and two types of tokens. We will utilize the data from October 6, 2020, to March 8, 2021, to evaluate our framework's indexing and querying operations. In addition, we are evaluating "On-Demand Query Services" by implementing "cache smart contract" to show the performance (retrieving time) of "On-Demand" queries compared with the baseline scenario, which retrieves the required data by scanning the ledger without using any indexing technique.

6 Conclusion

It is a significant benefit to utilize blockchain technology to enhance the security and decentralization of sharing economy applications, which demands extensive data technology to support its operations. Therefore, we implement a blockchain-based sharing economy application framework to support data management for indexing and querying blockchain data.

References

1. Farmakia, A., Kaniadakisb, A.: Power dynamics in peer-to-peer accommodation: insights from airbnb hosts. Int. J. Hospital. Manage. **89**, 102571 (2020)
2. Hasselgren, A., Kralevska, K., Gligoroski, D., Pedersen, S.A., Faxvaag, A.: Blockchain in healthcare and health sciences-a scoping review. Int. J. Med. Inf. **134**, 104040 (2020)
3. Singh, S.K., Rathore, S., Park, J.H.: Blockiotintelligence: a blockchain-enabled intelligent IoT architecture with artificial intelligence. Futur. Gener. Comput. Syst. **110**, 721–743 (2020)
4. Abuhashim, A.A., Tan, C.C.: Blockchain-based internet-of-vehicle. In: Magaia, N., Mastorakis, G., Mavromoustakis, C., Pallis, E., Markakis, E.K. (eds.) Internet of Things: Technology, Communications and Computing. Springer, Cham, pp. 313–340 (2021). https://doi.org/10.1007/978-3-030-76493-7_10
5. Saberi, S., Kouhizadeh, M., Sarkis, J., Shen, L.: Blockchain technology and its relationships to sustainable supply chain management. Int. J. Prod. Res. **57**(7), 2117–2135 (2019)
6. Groenfeldt, T.: IBM and Maersk apply blockchain to container shipping, March 2017. https://www.forbes.com/sites/tomgroenfeldt/2017/03/05/ibm-and-maersk-apply-blockchain-to-container-shipping/?sh=572eb5843f05
7. Neisse, R., Steri, G., Nai-Fovino, I.: A blockchain-based approach for data accountability and provenance tracking. In: Proceedings of the 12th International Conference on Availability, Reliability and Security (2017)

8. Wang, Z., Tian, Y., Zhu, J.: Data sharing and tracing scheme based on blockchain. In: International Conference on Logistics, Informatics and Service Sciences (LISS) (2018)
9. Abuhashim, A., Tan, C.C.: Smart contract designs on blockchain applications. In: IEEE Symposium on Computers and Communications (ISCC) (2020)
10. Liu, L., Han, M., Zhou, Y., Parizi, R.M., Korayem, M.: Blockchain-based certification for education, employment, and skill with incentive mechanism. In: Choo, K.-K.R., Dehghantanha, A., Parizi, R.M. (eds.) Blockchain Cybersecurity, Trust and Privacy. AIS, vol. 79, pp. 269–290. Springer, Cham (2020). https://doi.org/10.1007/978-3-030-38181-3_14
11. Abuhashim, A.A., Shafei, H.A., Tan, C.C.: Blockvc: a blockchain-based global vaccination certification. In: The 4th IEEE International Conference on Blockchain (2021)
12. Rahman, A., Rashid, M., Hossain, S., Hassanain, E., Alhamid, M.F., Guizani, M.: Blockchain and IoT-based cognitive edge framework for sharing economy services in a smart city. IEEE Access **7**, 18611–18621 (2019)
13. Huckle, S., Bhattacharya, R., White, M., Beloff, N.: Internet of Things, blockchain and shared economy applications. Proc. Comput. Sci. **98** (2016)
14. Metamask. https://metamask.io/
15. Etherscan. https://etherscan.io/
16. Tigani, J., Naidu, S.: Google Bigquery Analytics. Wiley, Hoboken (2014)
17. Li, Y., Yan, Y., Zheng, K., Liu, Q., Zhou, X.: Etherql: a query layer for blockchain system. DASFAA, 556–567 (2017)
18. Zheng, P., Zheng, Z., Wu, J., Dai, H.-N.: Xblock-eth: extracting and exploring blockchain data from Ethereum. IEEE Open J. Comput. Soc. **1**, 95–106 (2020). http://dx.doi.org/10.1109/OJCS.2020.2990458

Lessons Learned from Demonstrating Smart and Green Charging in an Urban Living Lab

Shanshan Jiang[1]([✉]), Marit Natvig[1], Svein Hallsteinsen[1], and Karen Byskov Lindberg[2]

[1] SINTEF Digital, Trondheim, Norway
{Shanshan.Jiang,Marit.K.Natvig,Svein.Hallsteinsen}@sintef.no
[2] SINTEF Community, Oslo, Norway
karen.lindberg@sintef.no

Abstract. Smart and green electric vehicle charging needs digital support which integrates systems from the energy, transport and building sectors. The Green-Charge project has proposed, demonstrated, and evaluated such support in an urban living lab setting. The proposed solutions are documented in a Reference Architecture meant to act as a blueprint both facilitating the extension and integration of the involved systems in the prototype implementation and supporting replication. However, successful uptake also depends heavily on motivating and engaging relevant stakeholders. In this paper, we share our experience and lessons learned from the design, implementation, and deployment of the proposed solutions in an urban housing cooperative. Barriers and drivers regarding this innovation process are identified and recommendations to overcome the barriers are suggested. The findings are intended to help stakeholders and policy makers to develop successful strategies for sustainable electric mobility and electric energy supply.

1 Introduction

Electric mobility (eMobility) partially powered by local renewable energy sources (RES) is a powerful measure to decarbonise the transport sector and offer demand flexibility to the electric energy supply system [1]. However, several barriers need to be overcome to reach massive electric vehicles (EV) adoption. Prospective EV owners are concerned about the availability of the charging infrastructure where they can charge their EVs when needed. Charging of many EVs at the same time produces peak loads and pressure on existing electricity infrastructure for electricity providers and building owners. The charging demands need to be coordinated with other energy activities in the neighbourhood. It is also challenging to use local RES optimally due to a mismatch between the availability of locally produced renewable energy and the energy consumption patterns.

The GreenCharge project (https://www.greencharge2020.eu/) aims to address the above challenges and demonstrate how technological solutions and associated business models can be integrated and deployed to overcome barriers to wide scale adoption of EVs. The GreenCharge concept for *green and smart charging* is built upon cross sectorial collaboration involving business actors and supporting technical systems from the transport, electric energy supply and building sectors [2]. The idea is based on energy smart neighbourhoods (ESN), where EV charging is managed together with other local

© The Author(s), under exclusive license to Springer Nature Switzerland AG 2022
L. Barolli et al. (Eds.): AINA 2022, LNNS 451, pp. 624–636, 2022.
https://doi.org/10.1007/978-3-030-99619-2_58

energy demand and local RES in the neighbourhood, exploiting demand flexibility and local storage capacity to adapt local demand to local production in order to facilitate smart EV charging with renewable energy. The digital support for such an ESN is a system of systems (SoS) with complex functionality and interactions. To facilitate the integration into such SoS in a well-defined manner, GreenCharge has defined a Reference Architecture (RA) for a full-fledged specification of the ecosystem [3]. Different systems can be implemented or extended based on a subset of the specifications in the RA and collaborate to facilitate smart and green charging.

Urban living labs (ULL) in Oslo, Bremen and Barcelona have been selected to evaluate the solutions and facilitate learning and future replication. The implementation and demonstration of smart and green charging in an urban living lab can be seen as a realisation of the RA with selected functionality adapted to local context and needs. The adoption of the GreenCharge solutions in ULLs has to address a number of technical, social, and organisational challenges. In this paper, we present the design, implementation and deployment of the proposed solutions for smart charging with optimal energy management in a housing cooperative in Oslo, and lessons learned from this process. The focus is to learn from this process. The related research questions are:

- RQ1: Which enablers and barriers affected the work?
- RQ2: What are the lessons learned that can support re-implementations of smart and green charging?

In the following, Sect. 2 discusses related work and Sect. 3 describes the approach for the overall innovation process and the evaluation and learning. Section 4 presents the context of the Oslo ULL. Section 5 describes the RA and the prototype deployed in the ULL. Lessons learned from the process evaluation are presented in Sect. 6, before we conclude the paper in Sect. 7.

2   Background and Related Work

ENTSO-E has provided a deep analysis of eMobility and its impact on the power system [1], focusing on interfaces between transport and energy sectors. The GreenCharge concept and the Reference Architecture are consistent with the framework and recommendations proposed in [1]. In addition, the interfaces with the building sector are considered, and all types of energy use in neighbourhoods are included, charging included.

Smart charging refers to charging supervised by an exterior control system, as opposed to the state-of-the-art (SotA) charging [1]. The current SotA technology and services that facilitate charging of a fleet of EVs involve a *reactive* management system, which ensures that the aggregated electricity load for charging several EVs does not surpass the available capacity in the parking place. In this case, all EVs immediately start charging after plugging in. If the capacity limit is reached, all EVs will have their charging power reduced, either by reducing the charging speed or charging the cars randomly in sequence. Hence, the system provides a safe solution for charging of multiple EVs, but the EV users might become dissatisfied if the desired charging demand is not met.

Another solution is a *predictive* smart energy management system with optimal scheduling, which provides an optimal plan for scheduling the charging of individual EVs. The optimal plan is calculated based on the departure time of each EV, and its charging need (the difference between the actual State-of-Charge (SoC) at plug in, and the desired SoC when plugging out). In such a system, the charging of EVs that will be unplugged the next day can be postponed (temporarily stopped), and EVs that plan to leave within the next few hours can be prioritised. GreenCharge follows this approach.

In the literature, optimisation models are used in a simulation environment to investigate predictive smart energy management [4, 5]. However, only a few demos have been tested in real life, and the experiences are not always successful. For instance, in the INVADE project (https://h2020invade.eu/the-project/), the aim was to design and develop a flexibility management system based on optimisation. However, the demos in the project experienced large challenges in the transfer of real-time data between the different stakeholders, and the implementation of the optimal scheduler [6].

3   Approach

Urban living labs (ULL) have been a popular approach to address the challenges related to sustainable urban interventions. According to Steen and van Bueren [7], the characteristics of urban living labs include four dimensions: a) *aimed* at innovation, formal learning for replication and increasing urban sustainability; b) covering *activities* of development (all phases of the product and service development process), co-creation and iteration (feedback, evaluation and improvement); c) *participants* from public and private sectors, users and knowledge institutes, all with decision-making power; d) innovations taking place in the *real-life use context*. They further identified five overarching phases in the innovation process: research, development, testing, implementation and commercialisation. The GreenCharge innovation process adopts the urban living labs approach and covers the first four phases (except commercialisation), where demonstration and evaluation are performed in real-life settings with collaboration of a multidisciplinary development team and the active involvement of users in the whole process.

The design science method [8] is used and adapted to support the overall iterative research and innovation process. The activities are considered an integral part of the ULL concept, as illustrated in Fig. 1. The main principles of the approach are:

1. **Environment – context:** The current environment provides input on stakeholders, needs, requirements, barriers, existing models, etc.
2. **Design and Build:** The artefacts are established.
3. **Demonstrate and Evaluate:** The artefacts are tested and evaluated.
4. **Knowledge base:** The validated results and the knowledge gained enhance the knowledge base and are shared for further exploration and exploitation.

The artefacts designed and built are a Reference Architecture, business models and prototypes integrating hardware and software. The artefacts are developed through iterative and interlinked processes illustrated by the **relevance, design, and evaluate cycles** in the figure. The three cycles have been enriched with characteristics of ULL described above (indicated as a, b, c, d). The relevance is verified and validated through the environment to ensure that the requirements are met and that the artefacts are relevant/correct. A multidisciplinary project team, residents, and the housing cooperative administration have been actively participated in the co-creation of the innovative solutions. The business models and prototypes realise selected parts of the architecture, and they are designed, built, and tested in ULLs to demonstrate their feasibility and to facilitate evaluations. Evaluations are done based on data collected from the demonstrations and by means of simulations to investigate the possible impacts and to learn about the innovation process. This paper focuses on the latter, and covers the design, implementation and deployment stages for the prototype artefacts and the associated parts of the Reference Architecture.

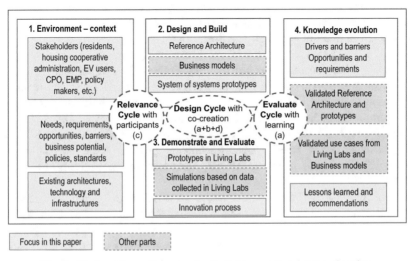

Fig. 1. Design science framework adapted to smart and green charging.

The process evaluation builds on the CIVITAS evaluation framework [9] and is based on qualitative analysis through collation and analysis of activities throughout the whole process to understand more clearly why new solutions succeed or fail. The evaluation focuses on identifying drivers, barriers and risks, as well as the required supporting activities. The analysis is based on the following input: 1) *pre-analysis* of stakeholders with a significant role in the process and their specific roles, risks, possible barriers and drivers; 2) *monitoring and assessment* of relevant actions and events to understand what has happened and why. The monitoring identifies supporting activities that have a significant influence, such as communication, introduction of a new design method, planning or decision-making methods, stakeholder involvement and engagement activities. The "noise" from irrelevant activities must be taken into consideration. The implementation of the automatic research data collection needed in the impact evaluation was for example a complex and comprehensive task that caused many discussions and actions that were not relevant to the process evaluation. 3) *Involvement of the stakeholders* to collect input on what they have experienced and learned. The input is collected through minutes from meetings and logs documenting challenges, events and decisions during this process. 4) *Focus groups*. A focus group was arranged with all actors involved in the ULL to investigate the barriers and drivers encountered and the effect of supporting activities. A neutral facilitator asked open questions, and the participants discussed. The input was analysed. The results are summarised in Sect. 6.

4   Relevance Cycle – the Context

The Oslo ULL addresses a housing cooperative, Røverkollen, where each apartment has its own, private parking space in a common garage. Røverkollen housing cooperative is located in a suburb in the eastern part of Oslo and has 246 apartments divided on 5 housing blocks. To provide convenient green charging to the residents, 60 new charge points (CP) were installed in the garage, 70 kWp of photovoltaic solar panels on the garage roof, and a stationary battery on the ground floor.

The stakeholders involved in the ULL are residents (users and owners of the private CPs), Røverkollen housing cooperative administration (facilitating that the residents can charge at their own parking place), the utility company/DSO and electricity provider (representing the public grid and the grid infrastructure), and the multidisciplinary project team - Fortum (technology provider acting the role of Charge Point Operator (CPO)), ZET (technology provider acting for eMobility provider (EMP) and part of the Local Energy Manager (LEM) roles), eSmart (technology provider acting the role of LEM), Oslo municipality (urban planning authority and policy maker) and SINTEF (research institute).

Before GreenCharge, the few EV owners in Røverkollen used 4 CPs outside the garage for charging. As more residents have bought or plan to buy EVs, the housing cooperative decided to allow EV owners to install their own private CPs in their parking garage. Røverkollen wants solutions for smart energy management to facilitate the charging of all EVs without high investment for upgrading the existing electricity infrastructure, and at the same time, ensure that the total charging demand is within the electricity capacity limits. Further, as there is a high peak load tariff for large consumers in Norway, Røverkollen wants to reduce the energy peaks by utilising local RES which at the same time increases the greenness of the charged energy.

The initially installed load management system (LMS) from the CP providers had a *reactive* control feature as described in Sect. 2. There were no incentives for "green" charging behaviour, i.e., no economic benefit and no technical support to shift charging to periods with high production from local solar panels and/or low demand on the grid.

The GreenCharge innovations offer a *predictive*, optimal and coordinated energy management with flexible and priority charging that can satisfy the above needs. Predictive smart energy management is supported, i.e., optimal distribution of the EV charging over time (accounting for other use of energy in the garage), adapted to PV availability, and to individual charging demand both regarding the amount of and timing of the energy requested. The residents and the Røverkollen administration were motivated and actively involved in the process of co-creation and testing the innovative solutions.

5   Design Cycle Activities

5.1   Reference Architecture (RA)

The GreenCharge innovations require cross sectorial collaboration involving business actors and supporting technical systems of the energy supply, transport and building sectors. The electric energy supply and the building sectors are mature and highly regulated sectors with well-established business structures and supporting technical systems. In the younger eMobility sector, a business structure with business actor roles, supporting appliances as well as technical and business systems has already emerged as well. Therefore, our approach to realising the GreenCharge concept is to extend the functionality of and the collaboration between these already existing systems.

In line with this approach, the RA aims to specify the participation of relevant existing systems in the realisation of the GreenCharge solution in terms of modified and/or added responsibilities and collaboration patterns necessary to support the GreenCharge concept. This is described in terms of UML stakeholder-, use case-, decomposition-, and collaboration models in the RA document [3]. An initial version was developed in the beginning of the project in co-design between the research and commercial partners to facilitate a common understanding of the GreenCharge concept and serve as a blueprint for the implementation of the digital support for prototypes used in ULLs. Then it has been refined towards the end of the project to reflect lessons learned during the implementation and use of the prototypes in ULLs.

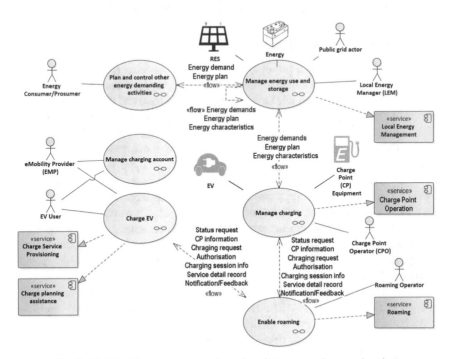

Fig. 2. Stakeholders, use cases and services for smart and green charging.

Figure 2 gives a high-level view of the RA with stakeholder roles (stickmen) and important technical equipment (pictograms), main use cases (ovals) and software *services* supporting them (rectangles). The key elements and the extension needed to support the GreenCharge innovative features include: *Charge Service Provisioning* supports EMPs in the selling of charging services to EV drivers. This involves authorising the use of selected CPs and taking care of billing and payment. In addition, it provides the user interface to the *Charge Planning Assistance* service assisting EV drivers in planning the charging of their EVs and booking charging at suitable CPs. This is an innovative service not commonly provided in SotA charging stations. It is to be provided by EMPs as an extension of their offer to their customers (EV drivers) and provides important information about the flexibility of the planned charging session, such as planned arrival and departure times, required amount of energy and whether they allow Vehicle-to-Grid (V2G). *Charge Point Operation* supports CPOs to operate their CPs to fulfil charging requests relayed through EMPs. It needs extensions to support advance booking of CPs. *Roaming* enables seamless access to CPs by connecting EMPs and CPOs. It needs to be extended with support for advance booking and for relaying the flexibility information. *Local Energy Management* automates shifting and shaping of flexible loads and use of local storage resources in a neighbourhood with the aim to optimise utilisation of local RES and reduction of peak power. Demand flexibility information is solicited from the building inhabitants. Establishing such ESNs is a shared responsibility of the building owners in the neighbourhood.

Each ULL implements a subset of services defined in the RA according to the local context and business needs. Altogether they cover all the features in RA.

5.2   System of System Prototype in the Oslo Living Lab

The garage is managed by a prototype of system of systems (SoS) implementing a selected subset of services defined in the RA adapted to the local context as illustrated in Fig. 3. Since the Fortum Charge & Drive management system and the eSmart Connected Prosumer system are commercial systems, special design and workaround has been done for the integration of this SoS prototype in order to reduce the changes of the operational systems. This prototype manages the CPs in the garage and offers flexible charging (default option) and priority charging (has priority when there is not enough energy to fulfil all demand). In addition, it supports *predictive, optimal and coordinated use of energy.* Information on energy demand from charging requests and heating cables, energy availability from the public grid, local RES and stationary battery, as well as historical data are used to dynamically calculate optimal energy distribution among energy demanding activities in the garage. The charging of individual EVs, use or storage of energy from local RES, and the use of energy from stationary batteries are then scheduled for optimal load balancing and optimal use of energy from RES.

As shown in Fig. 3, the prototype consists of four system components implementing four services defined in the RA. The *Fortum Charge & Drive management* implements *Charge Point Operation* service for the steering of the CPs. The *ZET App & Charge Management backend* is a new development and implements *Charge Planning Assistance* and *Charge Service Provisioning* services. It is used by the EV users to start the charging and to provide input on user profile information (such as information about the EV and default values to simplify the charging requests) and charging requests with charging constraints. The EV user can monitor the charging process and check the estimated SoC in the App. The *Local Service Management* service is provided by *eSmart Connected Prosumer* platform and the *ZET individual charge planning*: The *eSmart Connected Prosumer* monitors issues that may affect the energy availability and use (weather, RES production, stationary battery, heating cables, charging demands with varying flexibility), and calculates a dynamic *overall capacity plan* for optimal energy use at an aggregated level (i.e., predicted total capacity in the garage that can be used for the next 48 h with 15 min interval). When receiving the capacity plan, the *ZET Individual Charge Planning* generates *individual charge plan* for each CP, upon which the *Fortum system* controls the start/stop and the energy transferred at individual CPs. The eSmart system also controls the charging/discharging of the stationary battery.

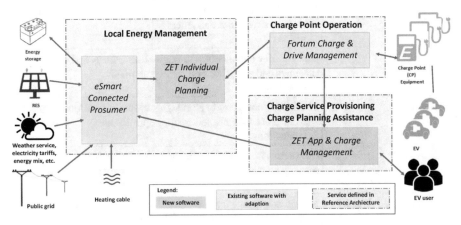

Fig. 3. The system components in the prototype and their interactions.

Initially the ambition was to include also a number of apartments in the ESN, but this was dropped due to regulatory constraints. Since each apartment and the garage has separate grid contracts and meters, the coordinated smart energy management of the apartment and the garage in such case would lead to unfair distribution of energy cost.

6 Evaluation Cycle with Lessons Learned

This section presents the findings from the process evaluation. Key drivers and barriers identified in the process are discussed here. A complete and detailed description of the drivers and barriers can be found in [10].

6.1 Key Drivers and Barriers

Table 1 provides an overview of the key drivers identified and their effects and Table 2 provides an overview of the key barriers identified and how they were handled.

Table 1. Key drivers and effects.

Driver type	Key drivers	Effects
Behavioural	**Positive attitude** from the housing cooperative administration to the new solutions	Input on needs and possibilities
	Subsidises of CP investments	Participants recruited
	Promotion activities like launch event, explanatory videos, information meetings and letters to residents	Increased awareness and knowledge among residents
	Stakeholder engagement through meetings and workshops on business models	Roles, focus and user needs clarified
Economic	**eMobility incentives** in Norway, such as tax reductions, toll road fee reductions	eMobility acceptance and positiveness
	Subsidises from municipality and housing cooperative association	Investments in CPs, stationary battery, and PV panels
Implementation capacity	**Multidisciplinary team** with technology providers (CPO, EMP, LEM), business model experts, municipality, researchers, housing cooperative administration, and residents	Cross-sectoral solutions, in line with the ULL characteristics
	Flexibility of partner with respect to additional tasks and responsibilities	Problems could be solved
Technical/economic	**Business models and technology alignments** in workshops with business model designers and technology providers	Technology support for the implementation of business models

The biggest barriers we experienced are associated with the lack of support in commonly used protocols for the novel features of GreenCharge solutions and the difficulties with making experimental modifications and extensions to 24/7 systems. For example, regarding the integration with CP equipment, the built-in CP load management system (LMS - with simple load balancing) blocked the scheduling done by the smart energy

Table 2. Barriers and how they were handled.

Barrier types	Barriers	How handled
Complexity	**Complex and novel solution.** Lack of off-the-shelf components and integration of systems from several partners	Requirement updated to match new insight. Unforeseen problems addressed when they occurred
Technical	**Integration with CP equipment.** Interface did not work as expected and blocked the smart energy management	Testing, investigations, and delays. Software workaround solved problem and safety risk, but reduced the quality
	Integration of outdoor heating cables. Difficult to control correctly. Cannot predict the consumption pattern due to unknown dependencies on temperature, delay, humidity, etc.	Direct control not implemented for safety reasons – cannot risk icing. The load integrated as a non-flexible residual load instead
	Integration between management systems. No standardised interfaces for integration of charge management and local energy management	Custom interfaces to the systems and their capabilities (as planned)
	Access to SoC is a problem since current protocols do not support access to current SoC from the EV's on-board systems	Collection of SoC via user interface in App. The user provides SoC values manually
Regulatory	**Regulatory barrier in local grid infrastructure** blocked the creation of a virtual smart meter and blocked billing of the planned ESN as one customer	Data from the garage and from some apartments are used as input to simulation of more comprehensive ESNs in order to investigate the possible effects on key performance indicators
Collaboration	**Coordination and communication problems** due to the multidisciplinary work and the many actors involved	Weekly telcos to follow up blockers and coordination between activities

management. At the end, the built-in LMS had to be disabled. However, this leads to a safety risk in case of technical problems (e.g., software errors, loss of Internet connection) where fuses may blow, equipment may be damaged, and charging may be blocked. To mitigate this, a software workaround was designed regarding smart energy management: a low speed redefault charging ensured some charging in any case, and capacity was reserved for low-speed charging, to prohibit overloading. The disadvantage of this approach is that the reservation leaves less capacity for use in optimisations, and the value of the solution is reduced.

Another serious barrier is that the electric energy sector is a highly regulated one with a regulatory regime lagging behind the rapid growth of decentralised RES and the increasing interest in local energy communities like ESNs. In Røverkollen, although it was technically feasible to create a virtual meter and bill the initially planned ESN as one entity, it was refused by the local DSO because it was against the regulation. We also suspect that DSOs are reluctant to accept such solutions because they would lose payment for the use of the grid to transfer locally produced energy between the members of the local community. Hopefully, this will change in the future as local RES and smart house technology and becomes more widespread.

6.2 Lessons Learned

Key lessons learned important for future sustainable eMobility strategies include:

Flexible Charging Can be Implemented, provided that the charge management system and the charge point equipment can be integrated and controlled in a detailed and flexible way. Charge point equipment control must support individual control of each charge point. It must be possible to start and stop the individual charge points at any time and to charge with different power from different charge points.

The Implementation of an ESN is a Challenge. Today, this is not done by easy plug and play. Off the shelf components from different providers cannot easily be combined due to the lack of standards and standardised interfaces. It may also be difficult to control the systems and equipment involved (charge points included).

Business Models Should Address More than Just the Money Flows. Price models may for example be used to encourage the desired charging behaviour. Flexibility should be rewarded, which is not common today.

6.3 Recommendations

This section provides recommendations derived from the process evaluation.

Stakeholder Involvement: Several types of actions must be considered to get input from and to involve the stakeholders, e.g., workshops and meetings, information letters to EV users, launch events creating publicity, interviews and questionnaires. Affected stakeholders must be involved whenever relevant, e.g., regarding the purchase of hardware, the design of the functionality supported by the technology (e.g., App), business models, and price models. Users must know how they can find information and how they can get support.

Business Models and Price Models: The business and price models must be designed in collaboration with all partners involved. The traditional approach to business models is not sufficient. The value proposition is also about sustainability with respect to environmental and societal aspects, e.g., to reduce energy peaks. The right combination of technical solutions, business models, and price models has the potential to motivate to a desired behaviour and to handle business related problems.

Design and Implementation: The implementation must be followed up at a weekly basis. All partners involved must participate. Blockers, problems, and potential problems must be identified at a detailed level, actions must be decided, and responsibilities must be assigned. Blockers and actions must be followed up.

Hardware and Equipment for ESNs: *The needs must be specified in detail. Statements from the providers regarding the ability of the devices cannot be trusted* unless they are based on a detailed specification of the needs. *The integration with the energy management system and the ability for equipment control* must be emphasized and verified. Many charge points are today provided with a built-in solution for simple load balancing that may cause problems in an ESN. The local energy management in the ESN may not be able to control the charge points as required. Thus, the details must be discussed with the provider of the software controlling the charge points to facilitate integration with the energy management system. *The involvement of experts is crucial.* Considering the problems described above, most building/property owners should use external expertise on the design and development of the total ESN solution.

Policy, Standardisation, and Harmonisation Issues: Charging protocols must provide the current SoC to facilitate optimal charge planning in ESNs. Navigation systems must facilitate the provision of desired SoCs, e.g., based on planned trips or artificial intelligence using input on the EV user's habits. Providers of charge point equipment must arrange for integration with local energy management in ESNs to facilitate an extended load balancing that takes predictions and the needs of the whole ESNs into account. Providers of devices such as stationary batteries must recognise the needs in ESNs and support the control mechanisms required. The software integration between local energy management and charge management must be standardised.

7 Conclusion

The GreenCharge Reference Architecture provides a full-fledged specification for implementing solutions for smart and green charging. When implementing and demonstrating the solutions in selected urban living labs, adaptions and workaround have to be done due to the constraints of the local context and the available technology. Moreover, a successful adoption of innovative solutions needs not only well-functioning technical systems, but also support in other aspects, such as user engagement and economic and policy incentives. This paper presents the lessons learned from demonstrating smart energy management and smart charging in a housing cooperative in Oslo. We experienced barriers related to technical, regulatory and collaboration issues, in particular, the integration with the energy management system, the ability for equipment control, the lack of support in commonly used protocols for the proposed innovative features and the difficulties with making experimental modifications and extensions to 24/7 systems. Recommendations to overcome such barriers are suggested. For instance, standards are crucial for facilitating the integration of local energy management and charge management.

Acknowledgments. Authors of this paper, on behalf of GreenCharge consortium, acknowledge the European Union and the Horizon 2020 Research and Innovation Framework Programme for funding the project (grant agreement n° 769016). The authors would like to thank all project partners for the collaboration and project management, in particular, ZET, eSmart and Fortum for the implementation of the prototype for Oslo ULL, as well as Røverkollen for the support and involvement.

References

1. ENTSO-E.: Position paper on electric vehicle integration into power grids (2021)
2. Natvig, M., Jiang, S., Hallsteinsen, S.: Stakeholder motivation analysis for smart and green charging for electric mobility. In: Barolli, L., Amato, F., Moscato, F., Enokido, T., Takizawa, M. (eds.) WAINA 2020. AISC, vol. 1150, pp. 1394–1407. Springer, Cham (2020). https://doi.org/10.1007/978-3-030-44038-1_127
3. Natvig, M., Jiang, S., Hallsteinsen, S.: Initial architecture design and interoperability specification. GreenCharge Project Deliverable D4.1 (2019)
4. Salpakari, J., Rasku, T., Lindgren, J., Lund, P.D.: Flexibility of electric vehicles and space heating in net zero energy houses: an optimal control model with thermal dynamics and battery degradation. Appl. Energy **190**, 800–812 (2017)
5. Iversen, E.B., Morales, J.M., Madsen, H.: Optimal charging of an electric vehicle using a Markov decision process. Appl. Energy **123**(C), 1–12 (2014)
6. Ottesen, S.Ø.: The need for data and algorithms. Solenergi webinar, 22 June 2021
7. Steen, K., van Bueren, E.: the defining characteristics of urban living labs. Technol. Innov. Manag. Rev. **7**(7), 21–33 (2017)
8. Hevner, A.R.: A three cycle view of design science research. Scand. J. Inf. Syst. **19**(2), 87–92 (2007)
9. Engels, D.: Refined CIVITAS process and impact evaluation framework (2015)
10. Natvig, M., Sard, R., Jiang, S., Hallsteinsen, S., Venticinque, S.: Intermediate result for innovation effects evaluation/intermediate evaluation result for stakeholder acceptance analysis. GreenCharge Project Deliverable D5.4/6.3 (2021)

Assessment of Rail Service Capacity Under the Current Regulations Aimed at Ensuring Social Distancing Conditions Against the COVID-19 Pandemic

Marilisa Botte[1](✉), Antonio Santonastaso[2], and Luca D'Acierno[2]

[1] Department of Architecture, Federico II University of Naples, 80134 Naples, Italy
marilisa.botte@unina.it
[2] Department of Civil, Architectural and Environmental Engineering,
Federico II University of Naples, 80125 Naples, Italy
antoniosantonastaso95@gmail.com, luca.dacierno@unina.it

Abstract. Due to the global pandemic we are experiencing, our lives and social behaviours in urban contexts are deeply changed. Many regulations and rules have been and still are enacted by the world's governments to minimise the level of risk to which the population is exposed. The majority of them aims at ensuring people social distancing and, in this context, the public transport sector plays a crucial role. Indeed, guaranteeing social distancing on buses or trains, while maintaining good service quality at the same time, represents a very challenging goal. Within this framework, the paper presents a methodology for optimising rail service operations thus satisfying mobility demand while complying with the maximum occupation coefficients allowed by current regulations. In order to show the goodness of the proposed method, it has been applied in the case of a real regional rail line in the south of Italy.

1 Introduction

The importance of promoting a smart and sustainable mobility system in our cities is universally acknowledged, thus assuring a high-quality service while reducing transport externalities, such as air and noise pollution [1–4].

In this context, promoting the use of public transport, with a rail system as backbone and bus feeder services as an add-on, turns out to be crucial. However, the global pandemic we are experiencing has deeply modified user perception and traveler choices, thus altering the ordinary demand and supply patterns within mobility frameworks. Specifically, on one side, demand patterns result to be dramatically changed, due to work from home, distance learning, e-commerce and e-grocery practices [5–9]. On the other side, public transport systems have experienced a severe capacity drop, due to several regulations which are enacted for ensuring social distancing [10–13].

This made public transport modes less attractive and generated disaffection among travelers, which feel uncomfortable in a closed space with strangers [14–16], although literary studies on such an aspect are not consolidated and came to conflicting conclusions [17–20]. Indeed, as widely shown in the literature [21–23], public transport has experienced a dramatic decrease in passenger volume, at the expense of sustainable levels of the mobility system.

Therefore, guaranteeing social distancing, while maintaining good service quality at the same time, represents a very challenging goal. Clearly, both demand and supply models, as well as evaluation methods, are involved in such an analysis [24–30]; however, the optimisation of transit carrying capacity according to the current regulations turns out to be the most crucial issue. The matter becomes even more challenging in the case of metro and rail systems; indeed, as widely shown in the literature, the definition of rail system capacity is a quite complex concept and its estimation needs to be into account several factors, i.e. the number of vehicles on the line, the average speed value, the stability and heterogeneity of the service [31,32]. Specifically, according to the existing works, three main methods may be identified for estimating railway capacity, such as analytical methods [33–37]; optimisation methods [32,38]; simulation methods [31,39].

Within this framework, this research proposes a Decision Support System (DSS) for helping transit service operators to optimize carrying capacity under the current legislation, thus maintaining good service quality and remaining attractive for users.

The remainder of the work is thus structured: Sect. 2 illustrates the proposed simulation-based methodology; Sect. 3 describes an application to a real rail context; finally, Sect. 4 shows the main findings.

2   The Proposed Methodology

The proposed approach relies on a simulation-based methodology for estimating rail service carrying capacity under different train occupation coefficients, thus being able to model limitations dictated by the social distancing legislation. Obviously, the compliance with such regulations depends on user common sense; however, it involves also rail operators which are forced to steadily re-schedule service for maintaining acceptable service quality levels. Therefore, the goal is to provide dispatchers with useful decision support systems for properly re-arranging service operations under different social distancing conditions imposed.

The maximum number of passengers that can be carried in an hour by a rail line l (i.e., Cap_l), can be computed as the product between the capacity of the single train t (i.e., Cap_t) and the hourly service frequency scheduled on the line (i.e., φ_l), as shown below:

$$Cap_l = Cap_t \cdot \varphi_l \tag{1}$$

However, under social distancing conditions, the number of users who can be on-board, at once, depends on the vehicle occupancy coefficient allowed by the current regulations, i.e.:

$$Cap_l = Cap_t \cdot \alpha_{SD} \cdot \varphi_l \qquad \text{with } \alpha_{SD} \in [0; 1] \qquad (2)$$

where α_{SD} is the maximum occupancy rate allowed by social distancing rules.
    By analysing variables involved in Eq. 2, it can be stated that:

- Cap_t is a physical constraint depending on vehicle types operating on the line;
- α_{SD} is an exogenous factor;
- φ_l is the only factor that can be handled by rail operators, by complying with infrastructural, technological and budget constraints.

In this context, we propose a what-if methodology providing the best design solution under specific occupancy rate values. Specifically, firstly, it is necessary to model travel demand conditions (by reproducing mobility patterns) and the analysed rail network (by reproducing infrastructure, signaling system, timetable and operating rolling stock). Once having a robust model which properly reproduces ordinary conditions, it is possible to simulate different scenarios and find the best strategy in correspondence of each specific occupancy rate. In particular, the proposed approach aims to evaluate, for each scenario, both user generalised cost and operating cost, thus modeling a two-fold perspective, i.e. passenger satisfaction and organisational and monetary effort of rail operators.

3   Application to a Real Regional Rail Line

In order to show the effectiveness of the proposed methodology, it has been applied to a real regional rail context in the South of Italy, i.e. Circumvesuviana network. It comprises 6 lines that share the same initial terminus, i.e. Porta Nolana station. It is worth noting that, although the entire network has been modeled, our evaluations focus on the Naples-Sorrento line since it presents an infrastructural criticality, i.e. the single-track section between Moregine and Sorrento stations. A detailed description of the line framework can be found in [40]. In addition, it is worth specifying that also features of the Porta Nolana terminus, where lines merge together, represent a key factor since the capacity of the node could strongly affect the threshold value of capacity on the lines.
    Therefore, according to the approach described in Sect. 2, travel demand patterns and features of the Circumvesuviana network (i.e., infrastructure, signalling system, timetable and operating rolling stock) have been properly reproduced, thus obtaining the simulation model at the basis of our DSS. Specifically, rail service has been modelled using the commercial software OPENTRACK® [41]; while the interaction with travel demand has been simulated with an ad-hoc assignment tool properly developed [42].
    In particular, according to Eq. 2, different values of service frequency (φ_l) and train occupancy rate allowed by social distancing lows (α_{SD}) have been tested. The simulated service configurations are the following:

- *Service configuration* 0, indicated as *Scenario* 0, consists in assuming the pre-COVID ordinary timetable (referred to December 2019);
- *Service configuration* 1, indicated as *Scenario* 1, consists of incorporating additional runs within the Naples-Sorrento line pre-COVID timetable, thus increasing line carrying capacity, by assuming the schedule of other overlapping lines as invariant;
- *Service configuration* 2, indicated as *Scenario* 2, consists in considering, for the Naples-Sorrento line, a build-up from the scratch timetable, thus maximising line carrying capacity, by modifying consequently the timetable of other overlapping lines.

Table 1. Operational parameters of the analysed service configurations.

Service configuration (Scenario)	*Total daily run on the Circumvesuviana network [#]*	*Total daily run on the Naples-Sorrento line [#]*	*Average daily service headway[a] on the Naples-Sorrento line [min]*
0	248	62	30
1	288	102	21
2	318	132	15

[a]Service headway is equal to $1/\varphi_l$

Table 1 shows service parameters of the above-mentioned configurations for the daily service; while, for the train occupancy rate, values of 100% (i.e., 1350 pax/train) and 50% (i.e., 675 pax/train) have been simulated.

By combining the 3 analysed service configurations with the 2 vehicle occupancy rate values tested, a total amount of 6 scenarios is derived.

Preliminary, it is necessary to highlight that, although between service configuration 0 and 1 there is an increase in service frequency, the graphs show the same numbers of runs and this is due to the analysed period. In other words, although as average daily value frequency increases, in the morning peak period no changes occur since there is no gap for adding further runs. Instead, in the case of service configuration 2, where the schedule has been created ex-novo, the doubling of service frequency is also reflected in the number of runs depicted. The analysed period also determines the fact that no issues occur in the outward trip, since the metropolitan area of Naples represents the socio-economic attracting pole. This implies that the Sorrento-Naples direction is the most loaded in the morning peak period; while, the opposite direction is the most loaded in the afternoon peak period, when commuters come back home. However, since the afternoon peak period relates to a wider time interval, the load diagrams are more smoothed. Details of the load diagrams during the morning peak hour are shown in Figs. 1, 2, 3, 4, 5 and 6.

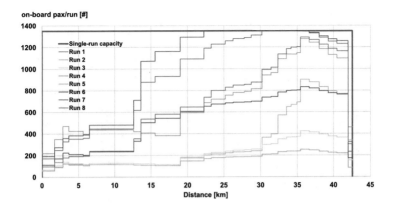

Fig. 1. Scenario 0 with an occupancy rate equal to 100% (Sorrento-Naples direction)

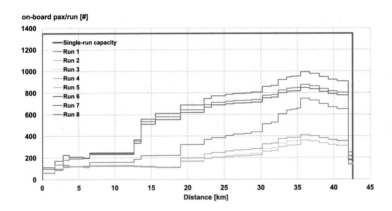

Fig. 2. Scenario 1 with an occupancy rate equal to 100% (Sorrento-Naples direction)

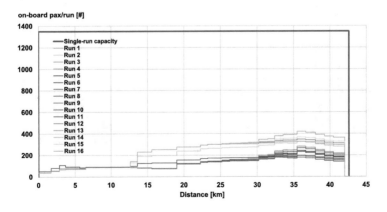

Fig. 3. Scenario 2 with an occupancy rate equal to 100% (Sorrento-Naples direction)

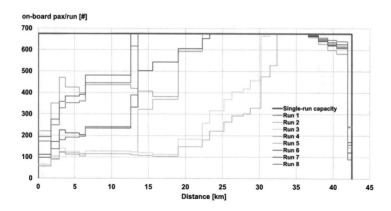

Fig. 4. Scenario 0 with an occupancy rate equal to 50% (Sorrento-Naples direction)

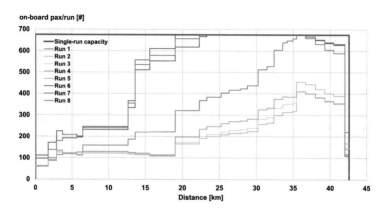

Fig. 5. Scenario 1 with an occupancy rate equal to 50% (Sorrento-Naples direction)

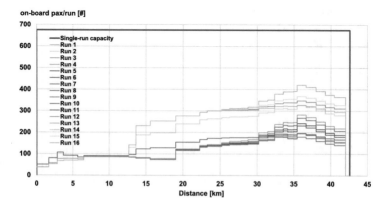

Fig. 6. Scenario 2 with an occupancy rate equal to 50% (Sorrento-Naples direction)

Table 2. KPIs of simulated scenarios.

| | Service configuration (Scenario) | | | | | |
	0		1		2	
Vehicle occupancy rate	100%	50%	100%	50%	100%	50%
Average on-board time [min]	31.1	31.1	31.2	31.2	29.0	29.0
Average waiting time [min]	18.0	48.4	10.6	12.7	7.5	7.5
Average user generalised cost [€]	18.0	48.4	10.6	12.7	7.5	7.5
Average system cost [€]	9.42	15.88	9.89	10.33	11.11	11.11

Therefore, as can be seen, even with an occupancy rate of 100%, the service configuration 0 results to be insufficient for satisfying the demand flow involved in the return trip (i.e. from Sorrento to Naples). Dropping down to an occupancy rate equal to 50%, instead, service configuration 2 needs to be implemented.

Moreover, for each scenario, the average value of user travel time and waiting time has been computed, as well as the average value of the user generalised cost. Finally, by adding the operational cost to the user generalised cost, an estimation of the average system cost has been carried out. Details on the computation of such variables can be found in [43].

In particular, Table 2 shows the daily average values of the above mentioned KPIs for each analysed scenario and, as can be seen, the most affected variable by social distancing regulations is the user waiting time. Indeed, a reduction in vehicle occupation rates, if not compensated by an increase in service frequency, could prevent users to board the first approaching train, with an evident impact on passengers' satisfaction and a reduction in public transport attractiveness.

4   Conclusions

In conclusion, the paper proposes a decision support methodology for optimising rail service operations thus satisfying mobility demand while complying with the maximum occupation coefficients allowed by current regulations. This results to be very useful in the COVID-19 pandemical era, where rail service companies find they have to deal with fast-changing regulations and reschedule service operations very quickly for going on to ensure service quality levels such that avoid users to choosing alternative transport modes.

Obviously, some issues still remain as the own perception of each user about onboard comfort levels, the need of re-arranging urban activities thus handling the demand peaks at terminals, and onboard convoys, or the necessity of public subsidies for transport operators; however, the provided application on a real rail regional line has provided a satisfactory outcome which confirms the effectiveness of the proposed approach.

As research prospects, we propose to apply the presented methodology to different network contexts, thus ulteriorly testing its goodness, and to model also platform occupation rates, thus performing a more exhaustive analysis. Moreover, it would be desirable to investigate the potentiality of adopting innovative mobility solutions for first/last mile services, thus making public transport still attractive despite the issues discussed above.

References

1. Cartenì, A.: Urban sustainable mobility. Part 1: Rationality in transport planning. Transp. Probl. 9(4), 39–48 (2014)
2. Shiller, P.L., Kenworthy, J.: An Introduction to Sustainable Transportation: Policy, Planning and Implementation. Routledge, London (2017). https://doi.org/10.4324/9781315644486
3. Botte, M., Pariota, L., D'Acierno, L., Bifulco, G.N.: An overview of cooperative driving in the European union: policies and practices. Electronics 8(6), 1–25 (2019). https://doi.org/10.3390/electronics8060616
4. Zhao, X., Ke, Y., Zhuo, J., Xiong, W., Wu, P.: Evaluation of sustainable transport research in 2000–2019. J. Clean. Prod. 256, 1–16 (2020). https://doi.org/10.1016/j.jclepro.2020.120404
5. Abou-Zeid, G.: Adoption and use of e-grocery shopping in the context of the COVID-19 pandemic: implications for transport systems and beyond. M.Sc. Thesis. Portland State University, Portland (OR), USA (2021). https://doi.org/10.15760/etd.7658
6. Mouratidis, K., Papagiannakis, A.: COVID-19, internet, and mobility: the rise of telework, telehealth, e-learning, and e-shopping. Sustain. Cities Soc. 74, 1–11 (2021). https://doi.org/10.1016/j.scs.2021.103182
7. Nguyen, M.H., Armoogum, J., Thi, B.N.: Factors affecting the growth of e-shopping over the COVID-19 era in Hanoi, Vietnam. Sustainability 13(16), 1–21 (2021). https://doi.org/10.3390/su13169205
8. Jain, T., Currie, G., Aston, L.: COVID and working from home: long-term impacts and psycho-social determinants. Transp. Res. Part A 156, 52–68 (2022). https://doi.org/10.1016/j.tra.2021.12.007
9. Music, J., Charlebois, S., Toole, V., Large, C.: Telecommuting and food e-commerce: socially sustainable practices during the COVID-19 pandemic in Canada. Transp. Res. Interdiscip. Perspect. 13, 1–7 (2022). https://doi.org/10.1016/j.trip.2021.100513
10. Mazzuccato, M., Kattel, R.: COVID-19 and public-sector capacity. Oxford Rev. Econ. Policy 36(S1), S256–S269 (2020). https://doi.org/10.1093/oxrep/graa031
11. Tirachini, A., Cats, O.: COVID-19 and public transportation: current assessment, prospects, and research needs. J. Public Trans. 22(1), 1–21 (2020). https://doi.org/10.5038/2375-0901.22.1.1
12. Gkiotsalitis, K., Cats, O.: Public transport planning adaption under the COVID-19 pandemic crisis: literature review of research needs and directions. Transp. Rev. 41(3), 374–392 (2021). https://doi.org/10.1080/01441647.2020.1857886
13. Hörcher, D., Singh, R., Graham, D.J.: Social distancing in public transport: mobilising new technologies for demand management under the Covid-19 crisis. Transportation, 1–30 (2021). https://doi.org/10.1007/s11116-021-10192-6

14. Campisi, T., Basbas, S., Al-Rashid, M.A., Tesoriere, G., Georgiadis, G.: A region-wide survey on emotional and psychological impacts of COVID-19 on public transport choices in Sicily. Italy. Trans. Transp. Sci. **2**, 1–10 (2021). https://doi.org/10.5507/tots.2021.010

15. Javid, M.A., Abdullah, M., Ali, N., Dias, C.: Structural equation modeling of public transport use with COVID-19 precautions: an extension of the norm activation model. Transp. Res. Interdiscip. Perspect. **12**, 1–10 (2021). https://doi.org/10.1016/j.trip.2021.100474

16. Przybylowski, A., Stelmak, S., Suchanek, M.: Mobility behaviour in view of the impact of the COVID-19 pandemic - public transport users in Gdansk case study. Sustainability **13**(1), 1–12 (2021). https://doi.org/10.3390/su13010364

17. Levy, A.: That mit study about the subway causing Covid spread is crap. Streets Blog NYC (online article) (2020). https://nyc.streetsblog.org/2020/04/17/that-mit-study-about-the-subway-causing-covid-spread-is-crap

18. Sadik-Khan, J., Solomonow, S.: Fear of public transit got ahead of the evidence - Many have blamed subways and buses for coronavirus outbreaks, but a growing body of research suggests otherwise. The Atlantic (online article) (2020). https://www.theatlantic.com/ideas/archive/2020/06/fear-transit-bad-cities/612979/

19. Harris, J.E.: The subways seeded the massive coronavirus epidemic in New York City. National Bureau of Economic Research, Working Paper (27021), pp. 1–22 (2021). https://doi.org/10.3386/w27021

20. Hu, M., et al.: The risk of COVID-19 transmission in train passengers: an epidemiological and modelling study. Clin. Infect. Dis. **72**(4), 604–610 (2021). https://doi.org/10.1093/cid/ciaa1057

21. Liu, L., Miller, H.J., Scheff, J.: The impacts of COVID-19 pandemic on public transit demand in the united states. PLoS ONE **15**(11), 1–22 (2020). https://doi.org/10.1371/journal.pone.0242476

22. Grechi, D., Ceron, M.: COVID-19 lightening the load factor in railway transport: Performance analysis in the north-west area of Milan. Res. Transp. Bus. Manag. 1–11 (2021, in press). https://doi.org/10.1016/j.rtbm.2021.100739

23. Vichiensan, V., Hayashi, Y., Kamnerdsap, S.: COVID-19 countermeasures and passengers' confidence of urban rail travel in Bangkok. Sustainability **13**(16), 1–21 (2021). https://doi.org/10.3390/su13169377

24. Bifulco, G.N., Cartenì, A., Papola, A.: An activity-based approach for complex travel behaviour modelling. Eur. Transp. Res. Rev. **2**(4), 209–221 (2010). https://doi.org/10.1007/s12544-010-0040-3

25. Gallo, M., D'Acierno, L., Montella, B.: A multimodal approach to bus frequency design. WIT Trans. Built Environ. **116**, 193–204 (2011). https://doi.org/10.2495/UT110171

26. De Martinis, V., Gallo, M., D'Acierno, L.: Estimating the benefits of energy-efficient train driving strategies: a model calibration with real data. WIT Trans. Built Environ **130**, 201–211 (2013). https://doi.org/10.2495/UT130161

27. Cartenì, A., Henke, I.: External costs estimation in a cost-benefit analysis: the new Formia-Gaeta tourist railway line in Italy. In: Proceedings of the 17th IEEE International Conference on Environment and Electrical Engineering (IEEE EEEIC 2017) and 1st Industrial and Commercial Power Systems Europe (I&CPS 2017). Milan, Italy (2017). https://doi.org/10.1109/EEEIC.2017.7977614

28. Gallo, M., De Luca, G., D'Acierno, L., Botte, M.: Artificial neural networks for forecasting passenger flows on metro lines. Sensors **19**(15), 1–14 (2019). https://doi.org/10.3390/s19153424

29. Ortega, J., Thòt, J., Palaguachi, J., Sabbani, I.: Optimization model for school transportation based on supply-demand analyses. J. Softw. Eng. Appl. **12**(6), 215–225 (2019). https://doi.org/10.4236/jsea.2019.126013

30. Henke, I., Cartenì, A., Molitierno, C., Errico, A.: Decision-making in the transport sector: a sustainable evaluation method for road infrastructure. Sustainability **12**(3), 1–19 (2020). https://doi.org/10.3390/su12030764

31. International Union of Railways (UIC): UIC Code 406: Capacity (2013). https://tamannaei.iut.ac.ir/sites/tamannaei.iut.ac.ir/files/files_course/uic406_2013.pdf

32. Prencipe, F.P., Petrelli, M.: Analytical methods and simulation approaches for determining the capacity of the Rome-Florence "Direttissima" line. Ingegneria Ferroviaria **73**(7–8), 599–633 (2018). https://www.ingegneriaferroviaria.it/web/en/content/analytical-methods-and-simulation-approaches-determining-capacity-rome-florence-

33. Schwanhäusser, W.: Die bemessung der pufferzeiten im fahrplangefüge der eisenbahn (in German) (1974). http://www.via.rwth-aachen.de/downloads/Dissertation_Schwanhaeusser_2te_Auflage_Anlagen.pdf

34. Bonora, G., Giuliani, L.: I criteri di calcolo di potenzialità delle linee ferroviarie (in Italian). Ingegneria Ferroviaria **37**(7), 1 (1982)

35. International Union of Railways (UIC): UIC Leaflet 405-1: Method to be used for the determination of the capacity of lines (1983)

36. Rete Ferroviaria Italiana - RFI (Italian National Railway Infrastructure Manager): Metodi di calcolo della capacità delle linee ferroviarie (in Italian). Technical report, pp. 1–49 (2011). http://www.dic.unipi.it/massimo.losa/TIV/Ferrovie/Metodi_di_calcolo_della_capacita.pdf

37. Schultze, K., Gast, I., Schwanhäusser, W.: Streckenleistungsfähigkeit simulation fahrplankonstruktion (in German). Technical report, pp. 1–38 (2015). https://docplayer.org/46924828-E-i-n-f-ue-h-r-u-n-g-sls-plus-streckenleistungsfaehigkeit-simulation-fahrplankonstruktion-stand-14-dezember-2015.html

38. Gonzalez, J., Rodriguez, C., Blanquer, J., Mera, J.M., Castellote, E., Santos, R.: Increase of metro line capacity by optimisation of track circuit length and location: in a distance to go system. J. Adv. Transp. **44**(2), 53–71 (2010). https://doi.org/10.1002/atr.109

39. Lindfeldt, A.: Railway capacity analysis: methods for simulation and evaluation of timetables, delays and infrastructure (2015). http://kth.diva-portal.org/smash/record.jsf?pid=diva2%3A850511&dswid=893

40. D'Acierno, L., Botte, M.: Passengers' satisfaction in the case of energy-saving strategies: a rail system application. In: Proceedings of the 18th IEEE International Conference on Environment and Electrical Engineering (IEEE EEEIC 2018) and 2nd Industrial and Commercial Power Systems Europe (I&CPS 2018). Palermo, Italy (2018). https://doi.org/10.1109/EEEIC.2018.8494575

41. Nash, A., Huerlimann, D.: Railroad simulation using open-track. WIT Trans. Built Environ. **74**, 45–54 (2004). https://doi.org/10.2495/CR040051

42. D'Acierno, L., Botte, M., Montella, B.: Assumptions and simulation of passenger behaviour on rail platforms. Int. J. Transp. Dev. Integrat. **2**(2), 45–54 (2018). https://doi.org/10.2495/CR040051

43. Botte, M., D'Acierno, L., Montella, B., Placido, A.: A stochastic approach for assessing intervention strategies in the case of metro system failures. In: Proceedings of the 2015 AEIT International Annual Conference. Naples, Italy (2015). https://doi.org/10.1109/AEIT.2015.7415258

A Floating Car Data Application to Estimate the Origin-Destination Car Trips Before and During the COVID-19 Pandemic

Armando Cartenì[1], Ilaria Henke[2(✉)], Assunta Errico[3], Luigi Di Francesco[1], Antonella Falanga[1], Mario Bellotti[4], Fabiola Filardo[4], and Giuseppe Cutrupi[4]

[1] Department of Engineering, University of Campania "Luigi Vanvitelli", Via Roma 29, 81031 Aversa, Caserta, Italy
luigi.difrancesco@unicampania.it, antonellafalanga95@outlook.it
[2] Department of Civil, Architectural and Environmental Engineering, University of Naples "Federico II", 80125 Napoli, Italy
Ilaria.henke@unina.it
[3] Department of Agriculture, University of Naples, 80055 Naples, Italy
assunta.errico@unina.it
[4] VEM Solutions S.p.A., Venaria Reale, Italy
mario.bellotti@viasatgroup.com, {fabiola.filardo, giuseppe.cutrupi}@vemsolutions.it

Abstract. In March 2020, the World Health Organization declared a global pandemic due to an unprecedented health crisis by COVID-19. In the first stage, all the Countries applied strict policies to limit the spreading of the virus, significantly reducing the mobility trips (reduction over than the 90% for the public transport modes), and possibly by structurally modifying mobility habits of citizens. With respect to the study of how (and how much) mobility habits are changing during the pandemic, the new technologies and the real-time traffic data for monitoring the travel demand can play a significant role, and this research tries to contribute in this sense. Within this issue, the aim of this research was twice: *i)* verify the applicability of the Floating Car Data (FCD) for the origin-destination (OD) car trips (OD matrices) estimation, proposing an ad-hoc methodology for the scope; *ii)* estimating and comparing the OD car trips before and during COVID-19 pandemic within the Campania Region in south of Italy, investigating both seasonal and yearly impacts of the pandemic on the mobility habits (e.g. lock-down periods vs. recovery periods; summer vs. winter periods). Estimated results confirm the ability of FCDs to reproduce OD car trips. With respect to the impacts produced by the pandemic on car mobility, the estimation results underline that during the periods of the main mobility restrictions, the structure of the regional demand has significantly changed with respect to a pre-pandemic period: extra-provincial car trips have decreased (between 23% and 42%) than the intra-provincial ones, which have even increased (up to +5%); the distance travelled was reduced up to the 24%.

Keywords: SARS-CoV-2 · Pandemic · New normal · Probe vehicle · Transportation · Mobility habits · Planning · Demand estimation

L. Barolli et al. (Eds.): AINA 2022, LNNS 451, pp. 647–656, 2022.
https://doi.org/10.1007/978-3-030-99619-2_60

1 Introduction

In March 2020, The World Health Organization (WHO) declared a global pandemic due to an unprecedented health crisis by COVID-19. In the first stage, all the Countries applied strict policies to limit the spreading of the virus, reaching a minimum value comparable for all Countries in the summer period. Once the first wave seemed to be concluded, first a second and then a third one hit even harder the world. Despite the administration of the vaccine, the spread of variants has made some restrictions indispensable. Indeed, after the peak of each wave, most of the countries removed restrictive measures (e.g. travel restrictions, lockdowns, smart working and education), re-allowing the spread of the virus (e.g. [1]).

The central role of physical distancing policies in reducing the spread of the virus has been clear since the start of the pandemic [2, 3], but it is very challenging in group situations such as those of transport public, schools and workplaces. Within this topic, many studies have observed a direct influence (correlation) between mobility habits and the spread of the COVID-19, especially in the second and third waves (e.g. [4–7]). Other transportation studies deal with the direct diffusion of the virus in group situations (where the minimum social distance is almost never guaranteed), pointing out how the mobility level (e.g. trips/day) is an indirect measure of social interactions (e.g. activities to be carried out), and therefore of infections (e.g. [6, 8–10]).

COVID-19 outbreak has dramatically impacted on people's lives, significantly changing daily habits, and this has had very significant repercussions on the transport and mobility sectors (e.g. [4, 6]). In addition, it has been observed that the average distance traveled each day decreased during the lockdown periods, but above all extra-regional and international trips significantly decreased. (e.g. [11]). Other studies show how users have changed their habits and mobility choices during the pandemic (e.g. [12, 13]). Furthermore, after periods of lockdown, a total trip reduction and an increase in car trips was also observed.

With respect to the study of how (and how much) mobility habits are changing during the pandemic, the new technologies and the real-time traffic data for monitoring the travel demand can play a significant role, and this research tries to contribute in this sense. For example, two of the biggest data players worldwide, Google LLC and Apple Inc., have made their data available to help researchers and policy-maker to contain the spread of the virus. Overall, these databases are based on a large amount of useful and trustworthy data (big data) able to study the trend of the travel demand over different territorial scales (e.g. city, region, nation); by contrast, these data are too aggregate and not allow a disaggregate spatial disaggregation useful for an origin-destination (OD) trips estimation commonly performed within the transportation planning applications (e.g. [14–20]). Within this topic hhigh quality data on traffic conditions are a precondition to provide relevant information about traffic conditions and mobility habits. Floating Car Data (FCD) are a valuable source of information based on vehicles equipped with a GPS and that moving in the traffic provide information on both the vehicle characteristics (e.g. speed and acceleration) and on initial origin and the final destination of the trip (e.g. [21–23]). The FCDs has become useful traffic data sources due to their lower cost (e.g. no costly infrastructure is needed) and higher spatial coverage (e.g. entire cities or regions covered with relatively few probe vehicles). Unfortunately, these data sources are still used only for car applications and some time for the freight one (e.g. [24]). The main FCDs applications focused on travel time estimation (e.g. [25–28]), traffic congestion evaluation and prediction (e.g. [22–30]) and, some time, OD estimation (e.g. [31, 32]).

Starting from these considerations, the aim of this research was twice: *i)* verify the applicability of the Floating Car Data (FCD) for the origin-destination (OD) car trips (OD matrices) estimation, proposing an ad-hoc methodology for the scope; *ii)* estimating and comparing the car trips before and during COVID-19 pandemic within the Campania Region in south of Italy, investigating both seasonal and yearly impacts of the pandemic on the mobility habits in Italy (e.g. lock-down periods vs. recovery periods; summer vs. winter periods). The paper is organized as follows. Section 2 reports the FCD estimation methodology proposed; Sect. 3 describes and argues the main results and discussion. Finally, conclusions are reported in Sect. 4.

2 Materials and Methods

2.1 A Methodology for the Origin-Destination Car Trips Estimation

Demand estimation is one of the core activities in transport planning as a decision support system for transport policies and/or interventions evaluation [16, 35–39]. Recently, easier-to-retrieve and low-cost data has opened up new opportunities for demand estimation (e.g. [40]). Among the new technologies, the use of FCDs, obtained from GPS equipped vehicles, can be an added value to better study mobility habits (e.g. [41, 42]). Figure 1 summarized the proposed FCD methodology for the origin-destination (OD) car trips estimation. Overall, the methodology consists in two phases: 1) **car demand level estimation**, that is the total number of car trips within a reference time period (e.g. business day); 2) **OD percentage distribution estimation**, that is an OD matrix quantification, where the generic element represents the percentage of car trips from the origin O towards the final destination D.

The car demand level estimation can be performed using different methods. Among these, many are based on a mobility survey that can be used for a "direct" or a "model" estimation (for details see, for example, [43]). In the proposed application for the demand level estimation the following formulation was applied:

$$N_{C,a,t} = (Pop_a \cdot r_{a,t} \cdot n_{trips,a,t} \cdot \%_{C,a,t})/lf_{C,a,t} \qquad (1)$$

where:

$N_{C,a,t}$ is the total number of car trips relative to the study area a, referring to the time period t;

Pop_a is the total population of the study area a;

$r_{a,t}$ is the fraction of population that makes at least one trips within the time period t (mobility rate), relative to the study area a;

$n_{trips,a,t}$ is the average number of trips per citizen within the time period t and the study area a;

$\%_{C,a,t}$ is the percentage of car trips (C) within the study area a and relative to the time period t;

$lf_{C,a,t}$ is the car's average load factor relative to the study area a, referring to the time period t.

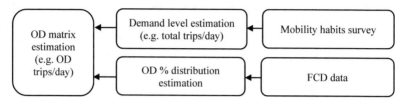

Fig. 1. The proposed Floating Car Data (FCD) methodology for the origin-destination (OD) car trips estimation.

The sources used for the car demand level estimation are:

the national socio-economic database [44] for the Pop_a quantification; the Official National Monitoring Observatory "Audimob" of Isfort [11] for the estimation of $r_{a,t}$, $n_{trips,a,t}$ and $\%_{C,a,t}$; precisely, this Observatory periodically carries out continuous sample surveys on the Italians mobility through telephone and computer interviews. Through this observatory was possible to analyze the mobility habits before and during the COVID-19 in Italy; an ad-hoc survey performed within this study in Campania region for $lf_{C,a,t}$ estimation.

Once the car demand level was estimated, its spatial representation (OD percentage distribution estimation in Fig. 1) was quantified in term of percentage distribution through the FDC data and described in the following section. Finally, the overall OD car trip demand matrix is obtained multiplied by the corresponding "car demand level estimation" and the "OD percentage distribution estimation" (as reported in Fig. 1). Results for the proposed case study are described in Sect. 3.

2.2 The Floating Car Data Considered

FCDs for traffic estimation have different advantages, for example it is no longer necessary to identify (limited) counting sections, as the vehicle fleet moves like probes on the territory, recording every detail of the single trip. Furthermore, there are no spatial or temporal limits in the data collection. Thanks to the GPS locator the precise movement of the vehicle (probe) is recorded from the starting of the trip up to the final destination. Therefore, thanks to the data transmitted via GPS, it is possible to know the entire route of the vehicle, since its position is known at any moment. The scientific community has long debated on the ability of FCDs to properly estimate demand for mobility (e.g. [41, 42]). For example, Nigro et al. [31], based on a database relating to the city of Rome, concluded that FCDs are not suitable to estimate directly the OD matrix due to the weakness of penetration rate. By contrast, the authors underline how travel times and route choice probabilities obtained by FCD are more reliable, "*since they represent the traffic conditions and behaviors that vehicles experiment along the path*". All researchers agree that only with an adequate FCD penetration rate is possible to obtain an acceptable percentage distribution of the OD car demand within the study area within a reference time period. Often, a preliminary cleanness and/or stratification (e.g. by origin-destination pair) of the dataset is required (as the one performed in this study). This consists in a cleanup activity aimed in obtain a single (clean) record relating to a trip from the origin up to the final destination, with the cleaning of possible intermediate stops along the trip. Often, it is necessary to establish a time interval such as to aggregate two consecutive trips if the intermediate stop between them is less than the fixed interval (5 min was the value used in this research). This aggregation/cleaning of the database prevents, for example, the switching of the vehicle's engine at traffic lights or during refueling, due to "start and stop", is mistakenly considered as the endpoint of the trip. Another clean/rebalance activity is the possible necessary stratification (e.g. weigh) of the sample by single OD (or macro-area) to take into account, for example, of small and/or unbalanced datasets not representative of the real travel demand distribution (e.g. through an a priori OD matrix). Knowing the clean and/or rebalanced routes of each vehicle in the fleet, it is possible to estimate the OD percentage distribution estimation (Fig. 1) for car trips within the study area and the time period considered. The dataset used in the proposed application was provided by the VEM Solutions S.p.A., an Italian Viasat Group company that has been operating for over 45 years in Design, Industrialization and Production of high-quality and mass-market electronic products. The capacity to design and develop hardware, firmware, and software for innovative technological solutions, overlooking the entire supply chain in the fields of electronics, IT, telecommunications and satellite telematics, has allowed it a strategic positioning on the Internet of Things (IoT), Big Data Analytics and Business Intelligence markets. The Company has one of the most complete datasets in Europe, providing a high FCD penetration rate for the case study considered useful pr the aim of the research. Precisely, the database considered contains records with specific information, including: Identification codes; starting and stopping data and time; origin and destination of the travel; total distance traveled by car.

2.3 The Application Case Study

As said, one of the aims of the paper was to estimate and compare the car trips before and during COVID-19 pandemic within the Campania Region in south of Italy, investigating both seasonal and yearly impacts of the pandemic on the mobility habits in Italy (e.g. lock-down periods vs. recovery periods; summer vs. winter periods). In Italy, from 2020, in order to contain the spread of the Covid-19 epidemic, the Government have enacted many directives also directly or indirectly impacting on mobility habits and on freedom of travelling. Overall, main measures that deeply affected the lifestyles of Italians may be identified, including: i) the "**National lockdown**" (8 March 2020); ii) the end of the high emergency phase and the beginning of so called "**Phase 2**" of the spread of the virus (4 May 2020), in which many retail businesses reopened and some restrictions are lifted, such as social isolation and regional trips; iii) the division of Italy into three areas (zones) according to the infection index (Rt) reached by the single Regions (3 November 2020):

Yellow zones, the range at moderate risk level characterized by closure of museums, exhibitions, cinemas; public transport vehicles allowed to run no more than half full, except for school buses. There is no limit to travel within municipalities, regions or between regions;

Orange zones, with high risk and severity, in which the following additional measures were undertaken with respect to yellow zones: strict travel bans to regions or municipalities other than those of residence, except for essential purposes; restaurants and bars closed to the public except for takeaway services and home deliveries before 10 p.m.;

Red zones, with the highest risk and severity, in which lockdown is declared. This corresponding by closure of all retail businesses except for grocers, chemists and stationers; all schools are closed; all team sports activities are suspended, as well as activities relating to personal care services, except for laundries, dry cleaners, barbers, hairdressers and funeral services; restricted access to public administration staff, and exclusively for necessary activities, promotion of the use of smart working for employees. Therefore, to travel in one's own region or municipality is not allowed. In this study, mobility choices were observed with respect to different times periods under different evolutionary periods of the pandemic and different government-imposed mobility restrictions.

The study area is the Campania Region (South Italy), a territorial area is divided into 500 municipalities, grouped into five administrative Provinces: Caserta, Naples and Salerno along the coast, while Benevento and Avellino in internal areas. Campania is the second Italian region in terms of population density, with a population (in 2019) of 5,740,291 inhabitant.

For the aim of the research, the investigated time period were:

- a Tuesday in March, June and November 2019 (pre-pandemic);
- a Tuesday in March 2020, during the "National lockdown" as defined before;
- a Tuesday in June 2020, during the "Phase 2" of the spread of the virus;
- a Tuesday in November 2020, in which Campania was in the "Red zone";
- a Tuesday in March 2021, again characterized by the "Red zone" restrictions;
- a Tuesday in June 2021, characterized by the "Yellow zone" restrictions.

3   Results and Discussions

The car demand level estimation referred to time periods defined before was estimated through the Eq. (1). Results reported in Fig. 2 underline how during the "National lockdown" of the 2020, the car mobility has significantly reduced of more than the 70% with respect to the same pre-pandemic period. By contrast, during the "Phase 2", in which many retail businesses reopen and some restrictions are lifted, the car trips increase up to -30% with respect to the pre-COVID-19.

In November 2020, the Campania Region experienced "Red zone" restrictions and a new, but less severe, lockdown was declared. In this period the car trips reduced up to about −50% compared to the ones estimated for the 2019. In March 2021, Campania has once again entered the "Red zone" but this time mobility habits and government restrictions have reduced car travel by only 35% compared to the pre-pandemic. Finally, in June 2021 Campania experienced the less-severe "Yellow zone" restrictions and a 25% (5%) reduction of car trips was observed with respect to June 2019 (2020). In Fig. 3 results of OD matrix estimation are reported, from which it is possible to observe a significant decrease in extra-provincial car trips during the pandemic. Table 1 reports the percentage variation of the OD daily car trips during (lock-down of March 2020) and before (March 2019) the COVID-19 pandemic, from which emerge how the extra-provincial car trips (often for non-essential purposes) have decreased (between 23% and 42%) than the intra-provincial ones, which have even increased (up to +5%), because of they are mainly characterized by study/work purposes and purchases of basic necessities.

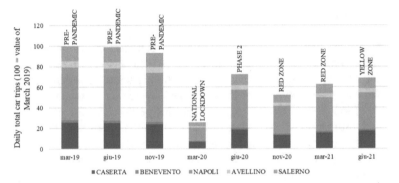

Fig. 2. Daily total car trips per provinces of origin and time period.

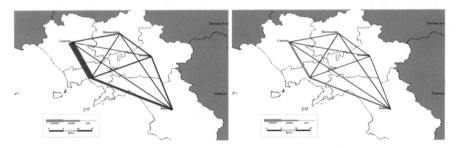

Fig. 3. Extra-provincial daily car trips in March 2019 (left) and March 2020 (right).

Finally, a reduction in the average car distance travelled was also found, which has decreased from 6.54 km in March 2019 up to 4.95 km (−24%) in March 2019 during the first main lock-down phase.

Table 1. Percentage variation of the OD daily car trips during (lock-down of March 2020) and before (March 2019) the COVID-19 pandemic.

	Caserta	Benevento	Naples	Avellino	Salerno	Total
Caserta	5.4%	−48.7%	−21.7%	−19.8%	−30.1%	3.1%
Benevento	−48.9%	0.8%	7.5%	−22.9%	−40.5%	−2.4%
Napoli	−21.8%	4.8%	0.8%	−45.0%	−27.6%	−0.5%
Avellino	−33.9%	−18.2%	−42.3%	−5.9%	−53.4%	−8.4%
Salerno	−17.8%	−69.0%	−26.8%	−53.3%	1.6%	0.1%
Total	**3.1%**	**−2.3%**	**−0.5%**	**−8.5%**	**0.1%**	**0.0%**
Total excluding intra-province car trips	*−23.2%*	*−31.0%*	*−24.0%*	*−42.3%*	*−32.9%*	–

4 Conclusion

With respect to the study of how mobility habits are changing during the pandemic, the real-time traffic data for monitoring the travel demand can play a significant role, and this research has tried to contribute in this sense. The study performed underline how FCDs are a useful traffic data sources for travel demand estimation also for their increasing availability at low cost. With respect to the impacts produced by the pandemic on car mobility, the estimation results underline that during the periods of the main mobility restrictions, the structure of the regional demand has significantly changed with respect to a pre-pandemic period. The research conducted should be considered as preliminary; some limitations have been highlighted and future evaluations/studies will make possible to refine the proposed methodology.

Acknowledgments. Research carried out within the funding program VALERE: VAnviteLli pEr la RicErca; SEND research project, University of Campania "Luigi Vanvitelli", Italy and VEM Solutions S.p.A. within the company Viasat Group, Italy.

References

1. Bontempi, E.: The Europe second wave of COVID-19 infection and the Italy "strange" situation. Environ. Res. **193**, 110476 (2021)
2. Islam, N., et al.: Physical distancing interventions and incidence of coronavirus disease 2019: natural experiment in 149 countries. BMJ **370**, m2743 (2020)
3. McGrail, D.J., Dai, J., McAndrews, K.M., Kalluri, R.: Enacting national social distancing policies corresponds with dramatic reduction in COVID19 infection rates. PLoS ONE **15**(7), e0236619 (2020)
4. Fang, H., Wang, L., Yang, Y.: Human mobility restrictions and the spread of the novel coronavirus (2019-ncov) in China. National Bureau of Economic Research (2020)
5. Muller, S.A., Balmer, M., Neumann, A., Nagel, K.: Mobility traces and spreading of COVID-19. MedRxiv (2020)

6. Cartenì, A., Di Francesco, L., Martino, M.: How mobility habits influenced the spread of the COVID-19 pandemic: results from the Italian case study. Sci. Total Environ. **741**, 140489 (2020)
7. Cartenì, A., Di Francesco, L., Martino, M.: The role of transport accessibility within the spread of the coronavirus pandemic in Italy. Saf. Sci. **133**, 104999 (2021B)
8. Cartenì, A., Di Francesco, L., Henke, I., Marino, T.V., Falanga, A.: The role of public transport during the second covid-19 wave in Italy. Sustainability **13**(21), 11905 (2021)
9. Lee, H., et al.: The relationship between trends in COVID-19 prevalence and traffic levels in South Korea. Int. J. Infect. Dis. **96**, 399–407 (2020)
10. Kraemer, M.U., et al.: The effect of human mobility and control measures on the COVID-19 epidemic in China. Science **368**(6490), 493–497 (2020)
11. ISFORT, Istituto Superiore di Formazione e Ricerca per i Trasporti. 17° Rapporto sulla mobilità degli italiani (2020)
12. Abu-Rayash, A., Dincer, I.: Analysis of mobility trends during the COVID-19 coronavirus pandemic: exploring the impacts on global aviation and travel in selected cities. Energy Res. Soc. Sci. **68**, 101693 (2020)
13. Long, J.A., Ren, C.: Associations between mobility and socio-economic indicators vary across the timeline of the Covid-19 pandemic. Comput. Environ. Urban Syst. **91**, 101710 (2022)
14. Cartenì, A.: Updating demand vectors using traffic counts on congested networks: a real case application. WIT Trans. Built Environ. **96**, 211–221 (2007)
15. Carteni', A.: A cost-benefit analysis based on the carbon footprint derived from plug-in hybrid electric buses for urban public transport services. WSEAS Trans. Environ. Develop. **14**, 125–135 (2018)
16. Cartenì, A.: Urban sustainable mobility. Part 1: rationality in transport planning. Transp. Probl. **9**(4), 39–48 (2014)
17. Cartenì, A.: Accessibility indicators for freight transport terminals. Arab. J. Sci. Eng. **39**(11), 7647–7660 (2014B)
18. Cascetta, E., Cartenì, A., Henke, I.: Stations quality, aesthetics and attractiveness of rail transport: empirical evidence and mathematical models [Qualità delle stazioni, estetica e attrattività del trasporto ferroviario: evidenze empiriche e modelli matematici]. Ingegneria Ferroviaria **69**(4), 307–324 (2014)
19. D'Acierno, L., Botte, M.: A passenger-oriented optimization model for implementing energy-saving strategies in railway contexts. Energies **11**(11), 2946 (2018)
20. D'Acierno, L., Gallo, M., Montella, B., Placido, A.: Analysis of the interaction between travel demand and rail capacity constraints. WIT Trans. Built Environ. **128**, 197–207 (2012)
21. Altintasi, O., Tuydes-Yaman, H., Tuncay, K.: Detection of urban traffic patterns from Floating Car Data (FCD). Transp. Res. Procedia **22**, 382–391 (2017)
22. Kong, X., Xu, Z., Shen, G., Wang, J., Yang, Q., Zhang, B.: Urban traffic congestion estimation and prediction based on floating car trajectory data. Fut. Gener. Comput. Syst. **61**, 97–107 (2016)
23. Nuzzolo, A., Comi, A., Polimeni, A.: Urban freight vehicle flows: an analysis of freight delivery patterns through floating car data. Transp. Res. Procedia **47**, 409–416 (2020)
24. Ruppe, S., Junghans, M., Haberjahn, M., Troppenz, C.: Augmenting the floating car data approach by dynamic indirect traffic detection. Procedia Soc. Behav. Sci. **48**, 1525–1534 (2012)
25. Rahmani, M., Jenelius, E., Koutsopoulos, H.N.: Non-parametric estimation of route travel time distributions from low-frequency floating car data. Transp. Res. Part C Emerg. Technol. **58**, 343–362 (2015)
26. Shi, C., Chen, B.Y., Li, Q.: Estimation of travel time distributions in urban road networks using low-frequency floating car data. ISPRS Int. J. Geo Inf. **6**(8), 253 (2017)

27. Rahmani, M., Koutsopoulos, H.N., Jenelius, E.: Travel time estimation from sparse floating car data with consistent path inference: a fixed point approach. Transp. Res. Part C Emerg. Technol. **85**, 628–643 (2017)

28. Kong, X., et al.: Mobility dataset generation for vehicular social networks based on floating car data. IEEE Trans. Veh. Technol. **67**(5), 3874–3886 (2018)

29. Chen, Y., Chen, C., Wu, Q., Ma, J., Zhang, G., Milton, J.: Spatial-temporal traffic congestion identification and correlation extraction using floating car data. J. Intel. Transp. Syst. **25**(3), 263–280 (2021)

30. Erdelić, T., Carić, T., Erdelić, M., Tišljarić, L., Turković, A., Jelušić, N.: Estimating congestion zones and travel time indexes based on the floating car data. Comput. Environ. Urban Syst. **87**, 101604 (2021)

31. Nigro, M., Cipriani, E., del Giudice, A.: Exploiting floating car data for time-dependent origin–destination matrices estimation. J. Intel. Transp. Syst. **22**(2), 159–174 (2018)

32. Jahnke, M., Ding, L., Karja, K., Wang, S.: Identifying origin/destination hotspots in floating car data for visual analysis of traveling behavior. In: Gartner, G., Huang, H. (eds.) Progress in Location-Based Services 2016. Lecture Notes in Geoinformation and Cartography. Springer, Cham (2017). https://doi.org/10.1007/978-3-319-47289-8_13

33. Botte, M., Pariota, L., D'Acierno, L., Bifulco, G.N.: An overview of cooperative driving in the European Union: policies and practices. Electronics **8**(6), 1–25 (2019)

34. Cartenì, A.: The acceptability value of autonomous vehicles: a quantitative analysis of the willingness to pay for shared autonomous vehicles (SAVs) mobility services. Transp. Res. Interdisc. Perspect. **8**, 100224 (2020)

35. Cartenì, A.: A new look in designing sustainable city logistics road pricing schemes. WIT Trans. Ecol. Environ. **223**, 171–181 (2017)

36. Cartenì, A.: Urban sustainable mobility. Part 2: Simulation models and impacts estimation. Transp. Probl. **10**(1), 5–16 (2015)

37. Carteni, A., Henke, I.: External costs estimation in a cost-benefit analysis: the new Formia-Gaeta tourist railway line in Italy. In: 2017 17th IEEE International Conference on Environment and Electrical Engineering and 2017 1st IEEE Industrial and Commercial Power Systems Europe, EEEIC/I and CPS Europe 2017 (2017). Art. no. 7977614

38. Henke, I., Cartenì, A., Molitierno, C., Errico, A.: Decision-making in the transport sector: a sustainable evaluation method for road infrastructure. Sustainability **12**(3), 764 (2020)

39. D'Acierno, L., Gallo, M., Montella, B., Placido A.: The definition of a model framework for managing rail systems in the case of breakdowns. In: Proceedings of 16th International IEEE Annual Conference on Intelligent Transportation Systems, IEEE ITSC 2013, The Hague, The Netherlands, 2013 October, pp. 1059–1064 (2013). Art. no. 6728372

40. Cantelmo, G., Viti, F.: A big data demand estimation model for urban congested networks. Transp. Telecommun. **21**(4), 245–254 (2020)

41. Croce, A.I., Musolino, G., Rindone, C., Vitetta, A.: Estimation of travel demand models with limited information: floating car data for parameters' calibration. Sustain. (Switz.) **13**(16), 8838 (2021)

42. Mitra, A., Attanasi, A., Meschini, L., Gentile, G.: Methodology for O-D matrix estimation using the revealed paths of floating car data on large-scale networks. IET Intell. Transp. Syst. **14**(12), 1704–1711 (2020)

43. Cascetta, E.: Transportation Systems Analysis. Springer, Boston (2009). https://doi.org/10.1007/978-0-387-75857-2

44. ISTAT (2021). [WWW Document]. https://www.istat.it/it/archivio/222527. Accessed 2 Feb 2020

Simulation and Evaluation of Charging Electric Vehicles in Smart Energy Neighborhoods

Rocco Aversa[1,2,3], Dario Branco[3(✉)], Beniamino Di Martino[1,2,3],
Luigi Iaiunese[3], and Salvatore Venticinque[3]

[1] Department Engineering, University of Campania "Luigi Vanvitelli", Caserta, Italy
rocco.aversa@unicampania.it
[2] Department Computer Science and Information Engineering, Asia University,
Taichung City, Taiwan
[3] Department of Engineering, University of Campania, via Roma 29,
81031 Aversa, Italy
{dario.branco,beniamino.dimartino,salvatore.venticinque}@unicampania.it
luigi.iaiunese@studenti.unicampania.it

Abstract. The GreenCharge simulator reproduces in a virtual environment the events that occur in a real smart micro-grid using a collection of real measured data. It allows to extend the evaluation capability in real trials, which are usually limited in the heterogeneity and number of devices. It supports the integration of optimization modules for evaluating different energy management strategies. The simulator reproduce the scenario as a discrete sequence of events in time. In this paper we present the usage and the capability of the simulation tool using data collected from trials, which are operated in the Bremen pilot of the Greencharge project. We quantify the room for improvement, which can be introduced exploiting the users's flexibility for charging their electric vehicles, evaluating a set of KPIs.

1 Introduction

Smart energy neighborhoods are promising innovative solutions to improve energy management and to leverage the spread of electric vehicles. In this context, the European project Greencharge has experimented and evaluated innovative solutions into three real pilots (in Oslo, Bremen and Barcelona) to investigate the effectiveness, scalability and replicability of smart energy neighborhoods. The evaluation methodology is based on the CIVITAS evaluation framework [4], which has been used and specialized according to the requirements of the GreenCharge project [3,6]. The framework defines Key Performance Indicators (KPIs), which are computed from data collected in project pilots, and which provide information on different categories, such as energy, transport, service availability and business [5]. Among the other tools, a simulator has been designed and developed to extend the evaluation possibility beyond the results

L. Barolli et al. (Eds.): AINA 2022, LNNS 451, pp. 657–665, 2022.
https://doi.org/10.1007/978-3-030-99619-2_61

obtained from real data [1,2]. Simulation scenarios can be defined to scale and to increase heterogeneity of devices in the real pilot, or to evaluate different energy management optimization strategies.

In this paper we present an example of evaluation process that starts from a preliminary analysis of the Bremen pilot from the available data. Some relevant KPIs and their computed value will be discussed and used to identify the strengths and weaknesses of operated measures. The energy management strategies applied in a simulated scenario, which replicates the real pilot as a sequence of discrete events, will generate new data to be evaluated as in the real case. The obtained result can be used as a valid decision support tool.

2   The Bremen Pilot

Bremen is one of the three pilot cities within the Greencharge project. It includes two demonstrators with both public charge stations used by residents, and private charge stations which are part of an industrial cluster and are used by employees. The demonstrator that we are going to analyze is composed of three geographically dislocated locations belonging to three different sections of the energy grid. Three main innovative measures are operated in this pilot. A booking and reservation mechanism (for public charge stations), based on priority levels, is used for minimizing waiting times and to make aware the energy management system about user's plan; The use of renewable energy sources for the supply of energy to charge sessions, especially photo-voltaic (PV) plants; The introduction of batteries in order to reduce the power peak and to store PV energy when it cannot be consumed.

For evaluation purpose many kinds of data are automatically connected from pilots, such as PC production, battery usage, charge booking information, EVs information, charge sessions and others. In Table 1 the subset of data used in the following sections to analyze the real pilot and to model the simulation scenarios is listed.

Table 1. Table data

Type	Name	Description
Solar plants	Power Peak	The power peak of solar plants installed on the site
	Timeseries	The timeseries of energy production for each solar panel installed in the period April-December 2021
Booking information	EST	The estimated start Time
	LFT	The latest finish time
	SOCStart	The estimated percentage of battery owned by the electric vehicle at plugin Time
	SOCend	The estimated percentage of battery owned by the electric vehicle at Plugout Time
	Priority	It is 0 if it is a not priority charge, 1 otherwise
	V2G	It is 1 if the user expresses his will to use the vehicle to grid technology, 0 otherwise

(*continued*)

Table 1. (*continued*)

Type	Name	Description
EV information	Capacity	The battery capacity
	CEfficiency	The battery efficiency in charging
	DisCEfficiency	The battery efficiency in discharging
	MaxChPower	The maximum power with which the battery can be charged
	MaxDisChPower	The maximum power with which the battery can be discharged
Charge session	Plugin Time	The actual start Time
	Plugout Time	The actual finish time
	SOCStart	The actual percentage of battery owned by the electric vehicle at the plugin Time
	SOCEnd	The actual percentage of battery owned by the electric vehicle at Plugout Time
	Time Series	The timeseries containing the charge profile, collected in the period April-December 2021

3 Data Analysis

Here we focus one location of the first demonstrator of the Bremen pilot. The connection to the grid is characterized by a power constraint of 30 kW and is monitored by a smart meter with a power constraint of 70 kW. The parking spots are covered by a PV plant. The power peak of PV production observed by the collected measures has been 11.76 kW. There are three charge points available for the electric vehicles. Two sockets can charge at 22 kW and third one at maximum 11 kW in AC. Besides, there is just one battery to eventually stock the surplus of PV energy production. Its capacity is 100 kWh, with a max power of 10 kW. Unfortunately, because of technical the measurement about the battery usages are affected by errors and are useless. At the time of writing, collected data include 206 charge sessions in 169 days by 10 electric cars. As it is shown in Fig. 1(a) we observed a not uniform distribution charges per car.

Even the distribution of charges per charge point (PC) is not equally distributed, as it is shown in Fig. 1(b). In Fig. 2(a) the distribution of connection time is shown. We can observe that most of charge sessions last from 9 to 10 h. Such a result depends on the users' behaviours, who are employers charging at work. They park their EV for a time that is longer than what is strictly necessary for a full charge. This a regular behaviour could allow to plan in advance an optimization strategy that can modulate the charge during the overall connection time.

The *charge at work* behaviour is confirmed by the distribution of plug-in and plug-out times shown in Fig. 3(a) and Fig. 3(b). We can observe how the statistical distribution shows that connections occur early in the morning and disconnections occur in the early afternoon. Moreover, in Fig. 2(b) we highlight

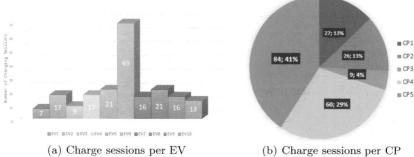

(a) Charge sessions per EV

(b) Charge sessions per CP

Fig. 1. Distribution of charge sessions

that no charges last during the night. From the data, we observed that the users does not book the charge in advance, but at plug-in time they provide the estimated departure time. However, we measured an average delay of 4.8 h between the planned departure time and the actual plug-out time.

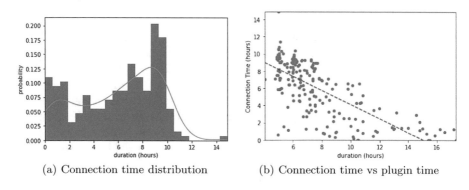

(a) Connection time distribution

(b) Connection time vs plugin time

Fig. 2. Connection time

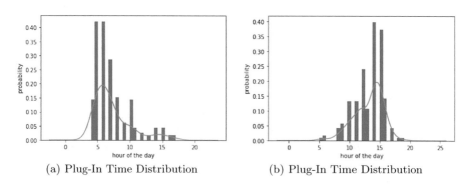

(a) Plug-In Time Distribution

(b) Plug-In Time Distribution

Fig. 3. Plug-in time and Plug-out distribution

4   KPIs Evaluation

We compare here the KPIs values computed in three different simulated scenarios with the KPIs values computed from real data. The subset of KPIs discussed in the following section are listed in Table 2. In Fig. 4, we can see the power production and the power consumption during the first 17 days of August. The red series in the chart represent the power consumed by the 11 charging sessions. It corresponds to the power consumed by the grid (blue) a part the amount of power self-consumed by the PV plant production (green). Among the notable KPIs in this time interval there is certainly the KPI GC5.5.5 whose value is about 73%, which means there is a high percentage of cars that finish their charge after the departure time planned in the reservation, with an average delay of 190 minutes. From the results we also know that, in the case that the Vehicle to Grid (V2G) technology was enabled, we would have been able to nominally accumulate and export 240 kWh by the batteries of the cars in order to reuse the energy produced by the solar panels to supply charges with energy when PV production is not enough. Nevertheless, considering the self-consumption (whose value is about 12%), there is not enough green energy in the EV batteries to exploit the V2G, and it would be more effective to increase self-consumption and self-sufficiency independently from the V2G mechanism. Finally, we measured a power peak of 16 kW on August 16th with an average power consumption of 0.43 kW. This is due to the fact that only EVs consume energy and they do not charge during the night.

Table 2. Subset of Greencharge KPIs

Code	Container name	Description
GC5.5.5	Charging Availability	Share of sessions that are not finished (plug out) in time
GC5.5.6	Charging Availability	Average delay in plug-out time
GC5.4	Share of battery capacity for V2G	The amount of energy in or share of capacity in EV batteries that can be used to accumulate energy-surplus, and to return it when needed
GC 5.10.1	Peak to average ratio	Maximum peak power demand per period (kW)
GC 5.10.2	Peak to average ratio	Average power demand per period (kW)
GC 5.14.1	Self-consumption	The share of energy produced locally that is consumed locally,
GC 5.14.2	Self-sufficiency	The share of the total energy consumption that is locally produced

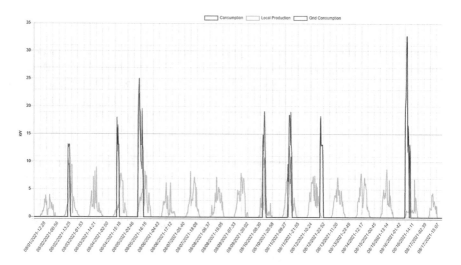

Fig. 4. First 17 days of August

5 The Simulated Scenarios

As a simple example of evaluation by simulation, we focus on only one day. In Fig. 5, we observe the total power consumed by three charge sessions (red) and the power produced by the PV plant (green) on August 16th, 2021. Detailed information about the charge sessions are listed in Table 3, where the second column include the energy demand, $MaxChPow$ is the maximum charging power, AAT is the actual arrival time, PDT is the predicted disconnection time and ADT is the actual disconnection time. In the last column we have the end of charge, that occurs before the disconnection time if the battery is full. Note that the charge sessions with id 5 and 6 share the same reservation. We can assume

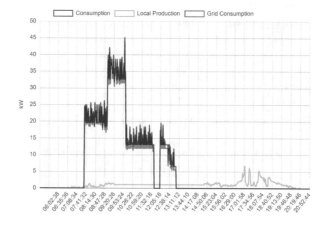

Fig. 5. August 16th in Bremen

Table 3. Charging sessions overview.

CSID	Energy	MaxChPow	AAT	PDT	ADT	Charge end
10	37.5 kWh	17.0 kW	07:34:24	12:00:00	16:00:31	10:01:24
5	12.27 kWh	7.2 kW	08:55:36	17:00:00	11:41:05	11:40:57
6	10.34 kWh	7.2 kW	12:02:31	17:00:00	17:20:55	12:58:36

that charge session was interrupted by the user during lunch time for 20 min and then resumed afterwards. In this scenario, analyzing the KPIs, it appears that approximately 67% of the cars finish their charging session beyond the expected time with an average delay time of 87 min. The measured power peak is 16 kW around 10AM and the average consumed power is 2.14 kW. Self-consumption is about 3% while self-sufficiency around 5%.

In order to evaluate the room for improvement of self-consumption, self-sufficiency and power peak on the same day, we defined three simulation scenarios. In the *First scenario* the user at plug-in time estimates correctly the departure time. Planned departure time correspond to the actual plug-out time recorded in the collected data. In the *Second scenario* all EVs are plugged-out at the estimated planned departure time that is recorded in the booking information. We assume the original booking information is correct and the users' behaviour complies with it. In the *Third scenario* all the EVs are charged as soon as possible at the nominal charging power. Booking times match the minimum required time to charge the energy demand. The details of the inputs to the simulator regarding the charge sessions are provided in Table 4. In Figs. 6(a), 6(b) and 6(c) we can see how the different configuration of simulated scenarios impact on the consumption profiles. In the first scenario, the charge power is calculated to ensure that the car reaches the required charge level at the same time the car is disconnected. It allows to lower the peak power, as we will see by the correspondent KPI value. It also increases the self-consumption and self-sufficiency since the recharges takes place during the day and because of the small size of the PV plant. In the second scenario, the average power measured into the collected data is set as charging power and the plug-out time is calculated as the plug-in time plus the necessary time to charge the energy demand at this power value. This behavior decreases the peak, but this happens only because there are not power spikes, due to measurement errors or to power fluctuations, in simulation. No particular improvement is expected in the KPIs that indicate the green behavior of this scenario. Finally, in the third simulation scenario, the cars charge at the maximum available power through an *as soon as possible* policy. We will see that this configuration involves the worsening of most of the KPIs since no smart energy management is applied. In Fig. 7 we can compare the KPIs value in the different scenarios. A zero value for KPIs GC5.5.6 and GC5.5.5 indicates that in simulation we assumed no plug-out delay respect to the planned time in reservation. In the first scenario, all evaluated KPIs have been improved, reaching a self-sufficiency of 24% and a self-consumption of 84%, decreasing the power peak and a peak-to-average power ratio 51% respect the

Table 4. Parameters of simulated charge sessions.

Real data				Scenario 1	Scenario 2	Scenario 3
EV	*Energy Demand*	*Max Ch. Pow*	*AAT*	*ADT/PDT*	*ADT/PDT*	*ADT/PDT*
10	37.5 kWh	17.0 kW	07:34:24	16:00:31	12:00:00	09:46:45
5	12.27 kWh	7.2 kW	08:55:36	11:41:05	17:00:00	10:37:51
6	10.34 kWh	7.2 kW	12:02:31	17:20:55	17:00:00	13:28:36

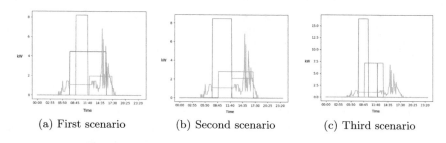

(a) First scenario   (b) Second scenario   (c) Third scenario

Fig. 6. Power series in simulated scenarios

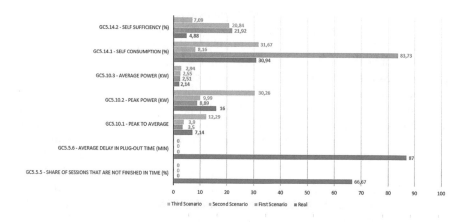

Fig. 7. KPIs values

baseline. In the second scenario, we observe and intermediate improvement with a worse self-consumption that is 8%. Finally, in the third scenario, there are no significant improvements and we can just observe that the power peak overcomes the power limit.

6   Conclusions

In this work we analyzed the real data collected in a pilot of the Greencharge project which aim at evaluating innovative solution for smart energy neighborhood. We presented a methodology for evaluate energy optimization strategies in simulation scenarios that are built by the collected data in real trials, based on a quantitative analysis of KPIs. We evaluated, by a simple use case, the room for KPIs improvement that a smart management system could obtaining exploiting the charge flexibility if the user's plan is notified in advance.

Acknowledgements. Authors of this paper, on behalf of GreenCharge consortium, acknowledge the European Union and the Horizon 2020 Research and Innovation Framework Programme for funding the project (grant agreement no. 769016).

References

1. Aversa, R., Branco, D., Di Martino, B., Venticinque, S.: Greencharge simulation tool. In: Advances in Intelligent Systems and Computing, vol. 1150 AISC, pp. 1343–1351 (2020)
2. Aversa, R., Branco, D., Di Martino, B., Venticinque, S.: Container based simulation of electric vehicles charge optimization. In: Barolli, L., Woungang, I., Enokido, T. (eds.) Advanced Information Networking and Applications. AINA 2021. Lecture Notes in Networks and Systems, vol. 227. Springer, Cham. https://doi.org/10.1007/978-3-030-75078-7_1
3. Venticinque, S., Aversa, R., Di Martino, B., Natvig, M., Jiang, S., Sard, R.E.: Evaluating technology innovation for e-mobility. In: Proceedings - 2019 IEEE 28th International Conference on Enabling Technologies: Infrastructure for Collaborative Enterprises, WETICE 2019, pp. 76–81. Institute of Electrical and Electronics Engineers Inc., (2019)
4. Marijuán, G.A., Etminan, G., Möller, S.: Smart cities information system key performance indicator guide version:2.0. Technical report, EU Smart Cities Information System (SCIS) (2017). ENERC2/2013-463/S12.691121
5. Natvig, M.K., Jiang, S., Hallsteinsen, S., Venticinque, S., Enrich Sard, R.: Evaluation approach for smart charging ecosystem - with focus on automated data collection and indicator calculations. In: Barolli, L., Woungang, I., Enokido, T. (eds.) Advanced Information Networking and Applications. AINA 2021. Lecture Notes in Networks and Systems, vol. 227. Springer, Cham. https://doi.org/10.1007/978-3-030-75078-7_65
6. Venticinque, S., Di Martino, B., Aversa, R., Natvig, M.K., Jiang, S., Sard, R.E.: Evaluation of innovative solutions for e-mobility. Int. J. Grid Util. Comput. **12**(2), 159–172 (2021)

Author Index

Printed in the United States
by Baker & Taylor Publisher Services